Welfare Risk and Exposure Assessment for Ozone

Second External Review Draft

Appendices

EPA-452/P-14-003b
February 2014

Welfare Risk and Exposure Assessment for Ozone
Second External Review Draft
Appendices

U.S. Environmental Protection Agency
Office of Air and Radiation
Office of Air Quality Planning and Standards
Health and Environmental Impacts Division
Risk and Benefits Group
Research Triangle Park, North Carolina 27711

DISCLAIMER

This draft document has been prepared by staff from the Risk and Benefits Group, Health and Environmental Impacts Division, Office of Air Quality Planning and Standards, U.S. Environmental Protection Agency. Any findings and conclusions are those of the authors and do not necessarily reflect the views of the Agency. This draft document is being circulated to facilitate discussion with the Clean Air Scientific Advisory Committee to inform the EPA's consideration of the ozone National Ambient Air Quality Standards.

This information is distributed for the purposes of pre-dissemination peer review under applicable information quality guidelines. It has not been formally disseminated by EPA. It does not represent and should not be construed to represent any Agency determination or policy.

Questions related to this preliminary draft document should be addressed to Travis Smith, U.S. Environmental Protection Agency, Office of Air Quality Planning and Standards, C539-07, Research Triangle Park, North Carolina 27711 (email: smith.jtravis@epa.gov).

APPENDIX 4A:

SPATIAL FIELDS FOR THE W126 METRIC

Appendix 4A
Spatial Fields for the W126 Metric

Table of Contents

LIST OF FIGURES

LIST OF TABLES

4A.1 OVERVIEW

EPA focused the analyses in the welfare risk and exposure assessments on the W126 O_3 exposure metric. The W126 metric is a seasonal aggregate of hourly O_3 concentrations, designed to measure the cumulative effects of O_3 exposure on vulnerable plant and tree species, with units in parts per million-hours (ppm-hrs). The metric uses a logistic weighting function to place less emphasis on exposure to low hourly O_3 concentrations and more emphasis on exposure to high hourly O_3 concentrations (Lefohn et al, 1988).

The first step in calculating W126 concentrations was to sum the weighted hourly O_3 concentrations within each month, resulting in monthly index values. Since most plant and tree species are not photochemically active during nighttime hours, only O_3 concentrations observed during daytime hours (defined as 8:00 AM to 8:00 PM local time) were included in the summations. The monthly W126 index values were calculated as follows:

$$\textit{Monthly W126} = \sum_{d=1}^{N} \sum_{h=8}^{19} \frac{C_{dh}}{1+4403*\exp{(-126*C_{dh})}} \qquad \textbf{(Equation 1)}$$

where N is the number of days in the month,

d is the day of the month (d = 1, 2, ..., N),

h is the hour of the day (h = 0, 1, ..., 23), and

C_{dh} is the O_3 concentration observed on day d, hour h, in parts per million.

Next, the monthly W126 index values were adjusted for missing data. If N_m is defined as the number of daytime O_3 concentrations observed during month m (i.e. the number of terms in the monthly index summation), then the monthly data completeness rate is $V_m = N_m / 12 * N$. The monthly index values were adjusted by dividing them by their respective V_m. Monthly index values were not computed if the monthly data completeness rate was less than 75% ($V_m < 0.75$).

Finally, the annual W126 index values were computed as the maximum sum of their respective adjusted monthly index values occurring in three consecutive months (i.e., January–March, February–April, etc.). Three-month periods spanning across two years (i.e., November–January, December–February) were not considered, because the seasonal nature of O_3 makes it unlikely for the maximum values to occur at that time of year. The annual W126 concentrations were considered valid if the data met the annual data completeness requirements for the existing standard.

The various assessments in the welfare REA have a need for complete spatial coverage of W126 index values. For example, the Forest and Agricultural Sector Optimization Model (FASOM) estimates economic changes in national agricultural and timber markets due to relative changes in spatially varying air pollution fields. Direct measurement of concentrations is the preferred method for generating such data, but prohibitive logistics and costs limit the possible spatial coverage and temporal resolution of such a database. Numerical methods that extend the spatial coverage of existing air pollution networks with a high degree of confidence have thus been a long-standing topic of investigation by researchers.

Appendix 4c of the 2nd draft O$_3$ Health REA describes the methodology of four different techniques for predicting air quality concentrations across space, and presents the results of an evaluation to determine which is the most appropriate for generating national-scale air quality spatial fields as inputs to those assessments. The four methods are: 1) Voronoi Neighbor Averaging (VNA; interpolating the monitoring data), 2) the Community Multi-scale Air Quality model (CMAQ; using modeled air quality concentrations), 3) enhanced Voronoi Neighbor Averaging (eVNA), and 4) Downscaler (DS). These last two methods combine, or "fuse" the air quality monitoring data with the modeled concentrations from CMAQ. In this appendix, we extend the evaluations of these four methods in the Health REA to the W126 metric, in order to determine which method is most appropriate for generating national air quality spatial fields for W126 under recent air quality conditions, air quality adjusted to just meet the existing O$_3$ standard, and air quality adjusted to meet three potential alternative secondary O$_3$ standards with forms of W126 and levels of 15 ppm-hrs, 11 ppm-hrs, and 7 ppm-hrs.

4A.2 AIR QUALITY SPATIAL FIELD TECHNIQUES

This section briefly describes the methodology of the four techniques considered for generating air quality spatial fields for W126, which are used as inputs to the biomass loss analyses presented in Chapter 6, and the foliar injury analyses presented in Chapter 7.

4A.2.1 VORONOI NEIGHBOR AVERAGING (VNA)

The Voronoi Neighbor Averaging (VNA; Gold, 1997; Chen et al, 2004) interpolation technique uses inverse distance squared weighted averages of the concentrations from a set of nearest neighboring monitors to estimate the concentration at a specified location (in this case a gridded field with 12km resolution covering the contiguous U.S.). VNA identifies the nearest neighboring monitors for each grid cell using a Delaunay triangulation algorithm, then takes the inverse distance squared weighted average of the concentrations from each neighboring monitor to estimate a concentration value for the grid cell. The following paragraphs provide a numerical example of the VNA technique applied to a model grid domain.

The first step in VNA is to identify the set of nearest monitors for each grid cell in the domain. The left-hand panel of Figure 4A-1 below presents a numerical example with nine model grid cells and seven monitoring sites, with the focus on identifying the set of nearest neighboring sites to grid cell "E", the center cell. The Delaunay triangulation algorithm identifies the set of nearest neighboring monitors by drawing a set of polygons called the

"Voronoi diagram" around the center of grid cell "E" and each of the monitoring sites. Voronoi diagrams have the special property that the each edge of the polygons are the same distance from the two closest points, as shown in the right-hand panel below.

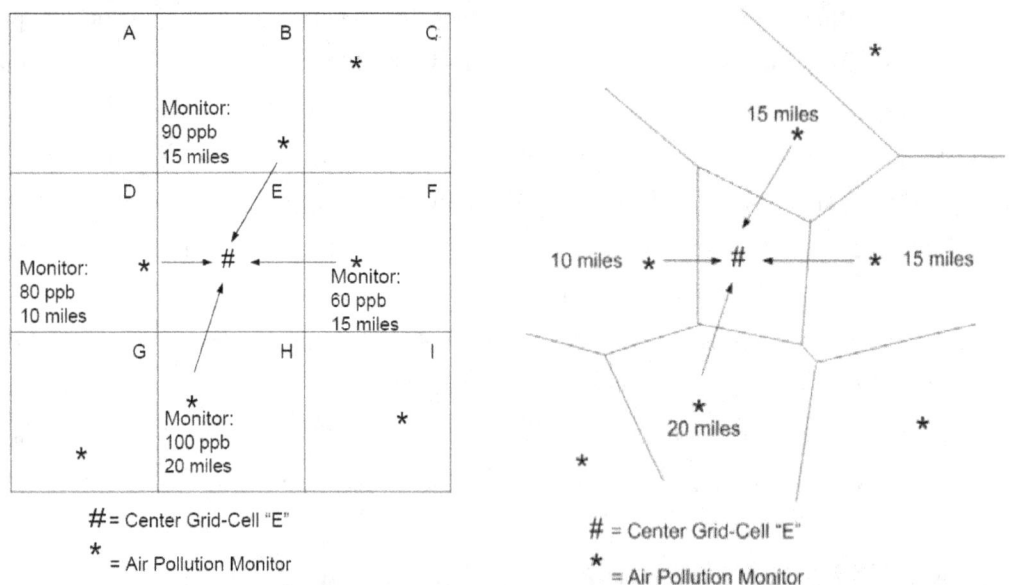

Figure 4A-1   **Numerical example of the Voronoi Neighbor Averaging (VNA) technique applied to a model grid domain**

VNA then chooses the monitoring sites that share a boundary with the center of grid cell "E". These are the nearest neighboring sites, which are used to estimate the concentration value for grid cell "E". The VNA estimate of the concentration value in grid cell "E" is the inverse distance squared weighted average of the four monitored concentrations. The further the monitor is from grid cell "E", the smaller the weight.

For example, the weight for the monitor in grid cell "D" 10 miles from the center of grid cell "E" is calculated as follows:

$$\frac{1/10^2}{1/10^2+1/15^2+1/15^2+1/20^2} = 0.4675 \quad \text{(Equation 2)}$$

The weights for the other monitors are calculated in a similar fashion. The final VNA estimate for grid cell "E" is calculated as follows:

$$VNA(E) = 0.4675 * 80 + 0.2078 * 90 + 0.2078 * 60 + 0.1169 * 100 = 80.3\ ppb$$

(Equation 3)

4A.2.2 COMMUNITY MULTI-SCALE AIR QUALITY (CMAQ) MODEL

For more than a decade, the Community Multi-scale Air Quality (CMAQ) model has been a powerful computational tool used by EPA and states for air quality management. The CMAQ system simultaneously models multiple air pollutants, including ozone, particulate matter, and a variety of air toxics to help regulators determine the best air quality management scenarios for their communities, states, and countries. CMAQ is also used by states to assess implementation actions needed to attain National Ambient Air Quality Standards.

The CMAQ system includes emissions, meteorology, and photochemical modeling components. Research continues in all of these areas to reduce biases and uncertainties in model simulations. CMAQ is a multi-scale system that has been applied over hemispheric, national, regional, and urban modeling domains with progressively finer resolution in a series of nested grids. The CMAQ modeling community includes researchers, regulators, and forecasters in academia, government, and the private sector with thousands of users worldwide.

Modeled air quality concentrations from CMAQ simulations have a twofold purpose in the generation and analysis of air quality spatial fields. First, the modeled concentrations are "fused" with the ambient measurement data using the eVNA and DS techniques. Second, the original modeled concentrations are evaluated against the resulting concentration estimates from the other spatial field techniques, to ensure that those techniques successfully reduce biases in the modeled air quality fields.

4A.2.3 ENHANCED VORONOI NEIGHBOR AVERAGING (EVNA)

Enhanced Voronoi Neighbor Averaging (eVNA; Timin et al, 2010) is a direct extension of VNA used to combine monitored and modeled air quality concentration data. Continuing from the previous numerical example for VNA, suppose the model grid cells containing monitors are associated with modeled concentrations as shown in Figure 4A-2 below. The modeled concentrations are used to weight the VNA estimates relative to the modeled concentration gradient:

$$eVNA(E) = \sum_{i=1}^{n_i} Weight_i * Monitor_i * \frac{Model_E}{Model_i} \quad \textbf{(Equation 4)}$$

where $Monitor_i$ represents the monitored concentration for a nearest neighboring monitor,

$Weight_i$ represents the inverse distance squared weight for $Monitor_i$,

$Model_E$ represents the modeled concentration for grid cell "E", and

$Model_i$ represents the modeled concentration in the grid cell containing $Monitor_i$.

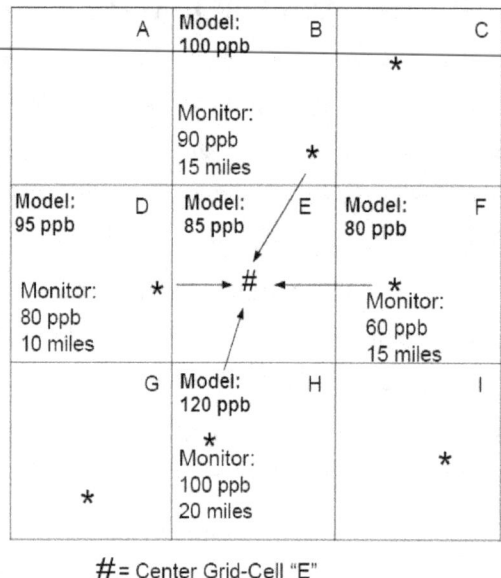

#= Center Grid-Cell "E"

* = Air Pollution Monitor

Figure 4A-2   Numerical example of the Enhanced Voronoi Neighbor Averaging (eVNA) technique applied to a model grid domain

Based on the values shown in Figure 4A-2, the eVNA estimate for grid cell "E" is calculated as follows:

$$eVNA(E) = \left(0.4675 * 80 * \frac{85}{95}\right) + \left(0.2078 * 90 * \frac{85}{100}\right) + \left(0.2078 * 60 * \frac{85}{80}\right) + \left(0.1169 * 100 * \frac{85}{120}\right) = 70.9\ ppb$$

(Equation 5)

In this example, eVNA adjusts the modeled concentration in grid cell "E" downward to reflect the tendency for the model to over-predict the monitored concentrations. In general, the eVNA method attempts to use the monitored concentrations to adjust for model biases, while preserving local gradients in the modeled concentration fields. The computations for VNA and eVNA were executed using the R statistical computing program (R, 2012), with the Delaunay triangulation algorithm implemented in the "deldir" package (Turner, 2012).

4A.2.4 DOWNSCALER (DS)

The Downscaler (DS) model is EPA's most recently developed method for spatially predicting air pollution concentrations. DS essentially operates by calibrating CMAQ data to the observational data, and then uses the resulting relationship to predict "observed" concentrations at new spatial points in the domain. Although similar in principle to a linear regression, spatial modeling aspects have been incorporated for improving the model fit, and a Bayesian[1] approaching to fitting is used to generate an uncertainty value associated with each concentration prediction. The uncertainties that DS produces are a major distinguishing feature from earlier fusion methods previously used by EPA such as the "Hierarchical Bayesian" (HB) model (McMillan et al, 2009). The term "downscaler" refers to the fact that DS takes grid-averaged data (CMAQ) for input and produces point-based estimates, thus "scaling down" the area of data representation. Although this allows air pollution concentration estimates to be made at points where no observations exist, caution is needed when interpreting any within-grid cell spatial gradients generated by DS since they may not exist in the input datasets. The theory, development, and initial evaluation of DS can be found in the earlier papers of Berrocal, Gelfand, and Holland (2009, 2010, and 2011).

DS develops a relationship between observed and modeled concentrations, and then uses that relationship to spatially predict what measurements would be at new locations in the spatial domain based on the input data. This process is separately applied for each time step (daily in this work) of data, and for each of the pollutants under study (ozone and PM2.5). In its most general form, the model can be expressed in an equation similar to that of linear regression:

$$Y(s,t) = \sim\beta_0(s,t) + \beta_1(t) * \sim x(s,t) + \varepsilon(s,t) \quad \textbf{(Equation 6)}$$

where:

$Y(s,t)$ is the observed concentration at point s and time t.

$\sim x(s,t)$ is the CMAQ concentration at time t. This value is a weighted average of both the grid cell containing the monitor and neighboring grid cells.

$\sim\beta_0(s,t)$ is the intercept, and is composed of both a global and a local component.

$\beta_1(t)$ is the global slope; local components of the slope are contained in the $\sim x(s,t)$ term.

$\varepsilon(s,t)$ is the model error.

DS has additional properties that differentiate it from linear regression:

1) Rather than just finding a single optimal solution to Equation 1, DS uses a Bayesian approach so that uncertainties can be generated along with each concentration prediction. This involves drawing random samples of model parameters from built-in "prior" distributions and assessing their fit on the data on the order of thousands of times. After each iteration, properties of the prior distributions are adjusted to try to improve the fit of the next iteration. The resulting collection of $\sim\beta_0$ and β_1 values at each space-time point are the "posterior" distributions, and the means and standard distributions of these are used to predict concentrations and associated uncertainties at new spatial points.

_____

[1] Bayesian statistical modeling refers to methods that are based on Bayes' theorem, and model the world in terms of probabilities based on previously acquired knowledge.

2) The model is "heirarchical" in structure, meaning that the top level parameters in Equation 1 (ie ~$\beta_0(s,t)$, $\beta_1(t)$, ~$x(s,t)$) are actually defined in terms of further parameters and sub-parameters in the DS code. For example, the overall slope and intercept is defined to be the sum of a global (one value for the entire spatial domain) and local (values specific to each spatial point) component. This gives more flexibility in fitting a model to the data to optimize the fit (i.e. minimize $\varepsilon(s,t)$).

4A.3 EVALUATION OF SPATIAL FIELD TECHNIQUES FOR THE W126 METRIC

The four air quality spatial field techniques were evaluated to determine which method was most appropriate for generating spatial fields of W126 for recent air quality data, air quality data adjusted to meet the existing O_3 standard, and air quality data further adjusted to meet the potential alternative W126 standards of 15 ppm-hrs, 11 ppm-hrs, and 7 ppm-hrs. Section 3.1 describes the evaluation of these techniques for recent air quality data, and section 3.2 describes the evaluation of these techniques for the various adjusted air quality scenarios.

4A.3.1 RECENT AIR QUALITY DATA

The evaluation based on recent air quality data was designed to assess the relative ability of each spatial field technique to reproduce monitored W126 concentrations. For the ambient monitoring data, the W126 metric was calculated for all monitors in the contiguous U.S. with complete data for 2007 based on the initial dataset and the data completeness criteria described in Appendix 4a of the 2nd draft Health REA. For the photochemical modeling data, the W126 metric was calculated from hourly O_3 concentrations based on a CMAQ simulation with a 12 km gridded domain covering the contiguous U.S., and 2007 emissions and meteorology inputs (EPA, 2012b).

Cross-validation is a method commonly used to evaluate the ability of statistical models to make accurate predictions. In a cross-validation analysis, the data are split into two subsets, the "calibration" subset, and the "validation" subset. The calibration subset is used to "fit" the model, usually by estimating parameters which establish a relationship between the variable of interest and one or more dependent variables. The resulting model fit is then applied to the dependent variable(s) in the validation subset, and the predictive ability of the model is assessed by how accurately it is able to reproduce the variable of interest in the validation subset.

The evaluation used a systematic "4-fold" cross-validation scheme based on the CMAQ model grid. The CMAQ model grid was divided into four groups, or "folds", so that each 2x2 block of 12 km grid cells had one member in each fold. Figure 4A-3 shows an example of the resulting four folds with O_3 monitor locations for the area surrounding southern Lake Michigan. Four cross-validations were performed using VNA, eVNA, and DS to predict W126 values at monitored locations. The calibration subset in the first cross-validation consisted of the monitors in folds 2, 3, and 4 as shown in Figure 4A-3 (blue dots), while the validation subset consisted of the monitors in fold 1 (red dots). The remaining cross-validations were performed in a similar manner, with three of the four folds used as the calibration subset and the final fold used as the validation subset. Thus, each monitor was included in the validation subset exactly once, resulting in a validation dataset with observed W126 values paired with VNA, eVNA, and DS predictions of those values at the monitor locations. The CMAQ predictions were simply the modeled W126 values for the 12 km grid cells containing O_3 monitors.

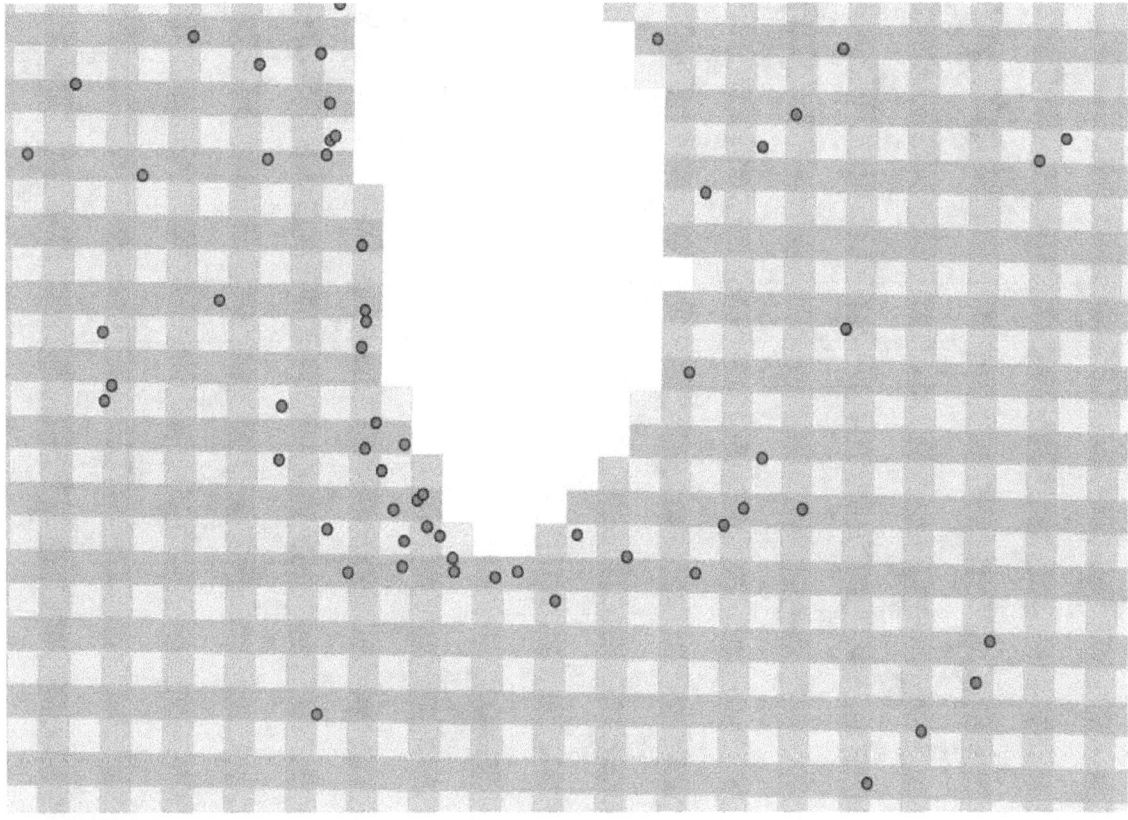

4 Fold Validation
- Fold 1 for validation
- Folds 2; 3; 4 for calibaration

Figure 4A-3 **Example of the "4-fold" cross-validation scheme used in the evaluation of the air quality spatial field techniques for the southern Lake Michigan area**

The cross-validation predictions based on the four air quality spatial field techniques were compared with the observed W126 values based on the ambient data. The comparison focused on three performance metrics: 1) the root mean squared error (RMSE), 2) the coefficient of variation (R^2), and 3) the mean bias (MB). The results of these comparisons are shown in Figure 4A-4, and Table 4A-1 contains a summary of the three performance metrics for each technique.

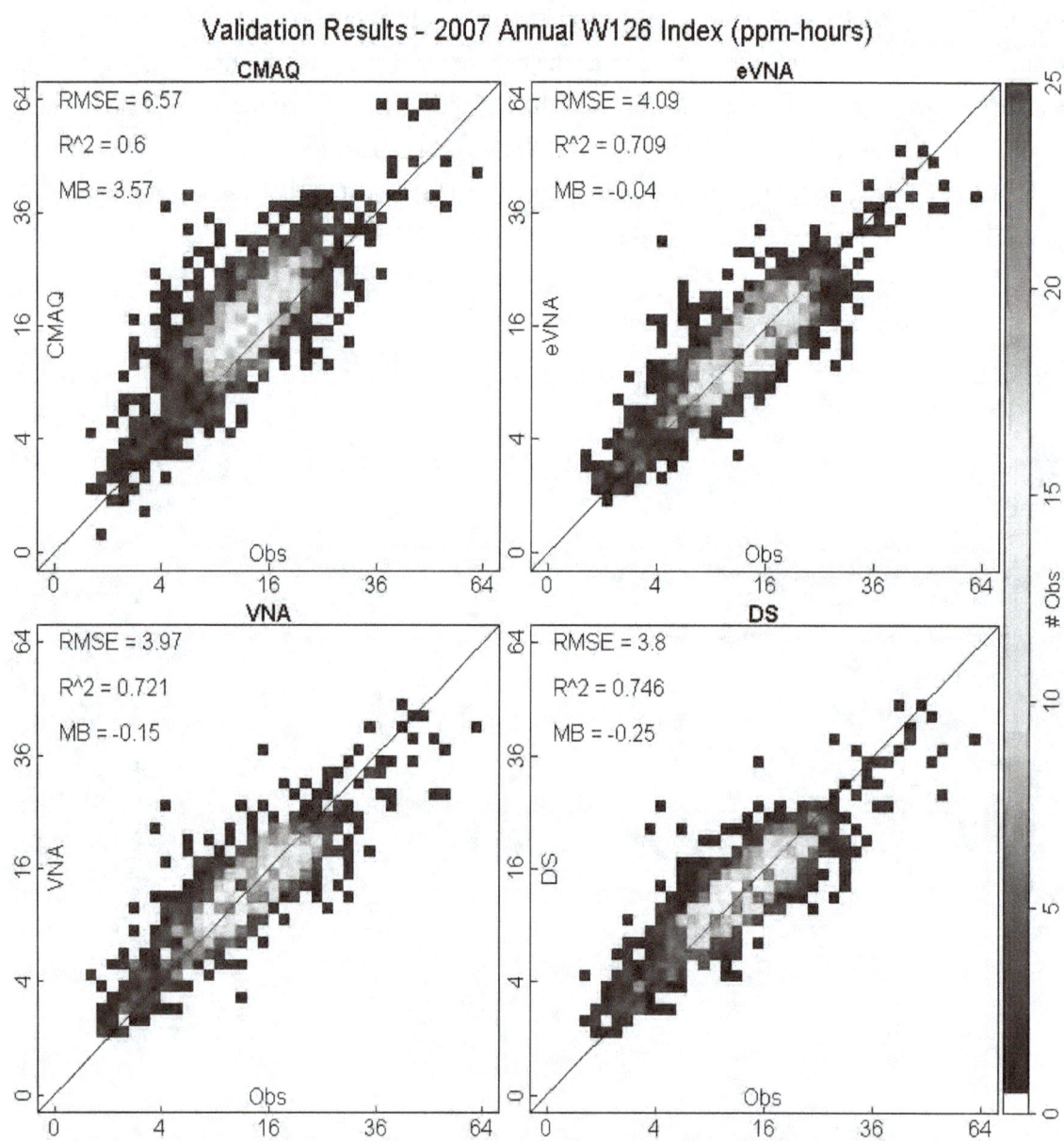

Figure 4A-4   Cross-validation results for the 2007 annual W126 concentrations

Performance Metric	VNA	CMAQ	eVNA	DS
RMSE	3.97	6.57	4.09	3.80
R^2	0.721	0.600	0.709	0.746

| | MB | -0.15 | 3.57 | -0.04 | -0.25 |

Table 4A-1   Summary of the cross-validation performance metrics for W126

The cross-validation results clearly showed that VNA, eVNA, and DS more accurately predict monitored W126 concentrations than the CMAQ model. The scatter plots and the mean bias statistics indicated that both eVNA and DS were effective at reducing the amount of bias present in the modeled concentrations. The differences between VNA, eVNA, and DS were much smaller. Although the performance metrics indicated that DS had the highest R^2 and the lowest RMSE of those three techniques, DS also had the largest absolute mean bias, and none of the differences were statistically significant.

4A.3.2 ADJUSTED AIR QUALITY DATA

As described in Chapter 4, the air quality monitoring data were adjusted using HDDM based on domain-wide reductions in U.S. anthropogenic NOx emissions, so that in each of the 9 NOAA climate regions, the highest monitor just met the existing standard, and the alternative W126-based standards of 15 ppm-hrs, 11 ppm-hrs, and 7 ppm-hrs. Figure 4A-5 shows a map of the 9 NOAA climate regions for reference. Table 4A-2 shows the percent reduction in nationwide anthropogenic NOx emissions that were used to reach the existing and alternative standards in each of the 9 regions. In a few cases, all monitors in the region met one or more of the alternative standards based on 2006-2008 observations, and thus there was no need for model-based adjustments. These cases are represented by values of "0%" in Table 4A-2. Finally, Figure 4A-6 shows the 2006-2008 average W126 values in monitored locations based on the air quality monitoring data adjusted to meet the existing standard and the alternative W126-based standards with levels of 15 ppm-hrs, 11 ppm-hrs, and 7 ppm-hrs.

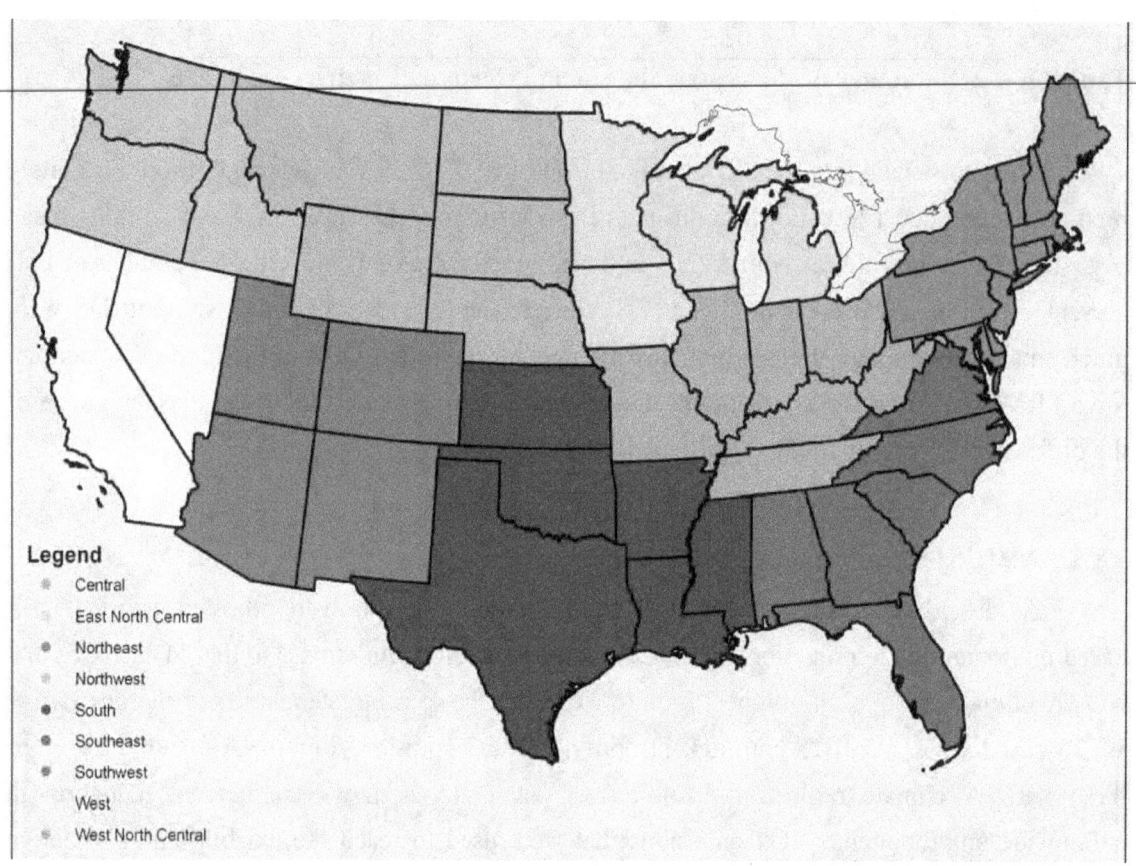

Figure 4A-5 NOAA climate regions used in the model-based air quality adjustments

Region	75 ppb	15 ppm-hrs	11 ppm-hrs	7 ppm-hrs
Central	48%	14%	58%	70%
East North Central	65%	0%	23%	61%
Northeast	96%	36%	51%	81%
Northwest	51%	0%	0%	0%
South	54%	44%	56%	66%
Southeast	64%	14%	38%	58%
Southwest	55%	67%	85%	90%
West	90%	91%	93%	95%
West North Central	23%	0%	6%	39%

Table 4A-2 Percent reductions in U.S. anthropogenic NOx emissions used to reach existing and alternative standard the in nine climate regions.

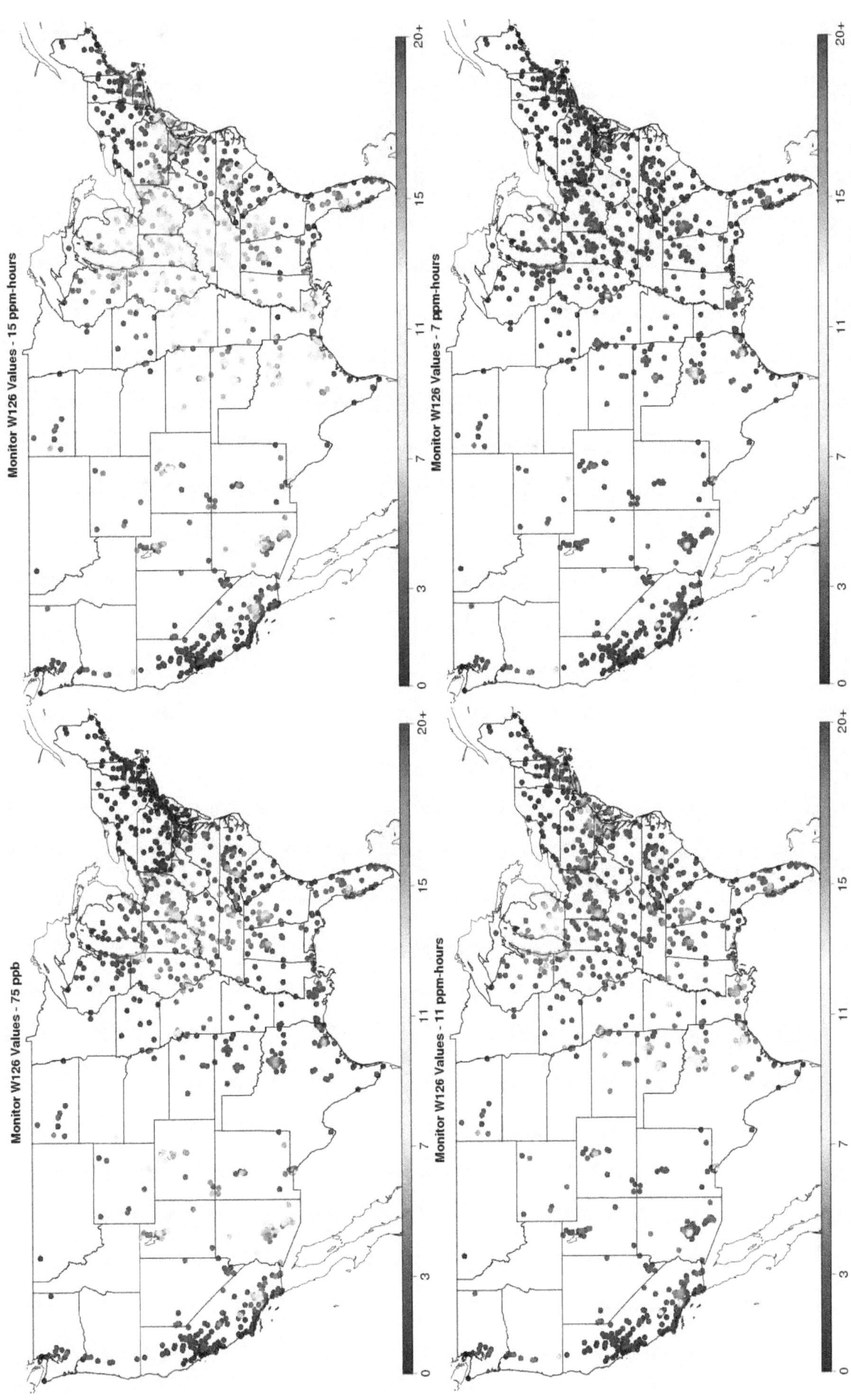

Figure 4A-6 Monitored 2006-2008 average W126 concentrations adjusted to meet the existing standard (top left), and the alternative W126-based standards of 15 ppm-hrs (top right), 11 ppm-hrs (bottom left), and 7 ppm-hrs (bottom right)

4A-15

The model-based adjustment technique was developed to adjust hourly O_3 concentrations in monitored locations. However, gridded modeled air quality concentrations are required as inputs to the eVNA and DS techniques. Thus, for the adjusted air quality scenarios there was a need to create adjusted model W126 fields to "fuse" with the adjusted monitor W126 values. To accomplish this, the modeled hourly O_3 concentrations were adjusted based on the grid- and hour-specific HDDM sensitivities combined with the percent emissions perturbations shown in Table 4A-2 (note that the NOx reductions listed in Table 4A-2 were determined to meet the targeted standards at monitor locations only and would not guarantee levels below the standard in unmonitored areas). This resulted in gridded spatial fields of adjusted hourly O_3 concentrations for April-October 2007, which were then aggregated to create gridded spatial fields of W126 index values. The method used to adjust the modeled concentrations differed from the procedure used to adjust the monitored concentrations in three important respects:

1) The starting (unadjusted) concentrations were different. As shown in Figure 4A-7, modest differences in hourly measured and modeled O_3 concentration data can lead to substantial differences in the W126 metric. In addition, the different time periods represented by the monitoring data (2006-2008) and the modeled data (2007) added to the differences in the starting concentrations. Finally, unlike the monitoring data, the modeled concentrations do not have any missing values. Thus, while adjustments were made to the observed W126 index values to account for time periods with missing data, the model estimates are based on complete, continuous hourly O_3 concentration data.

2) For the adjustments to the monitored concentrations, relationships were derived between the HDDM sensitivities and the monitored O_3 concentrations for each monitoring site, hour-of-the-day, and season. The sensitivities were then applied to the observed hourly O_3 concentrations at each monitoring site based on the linear relationship between the HDDM sensitivities and monitored O_3 concentrations. This was meant to account for differences in monitored and modeled concentrations at specific times and locations, and to allow application of HDDM sensitivities to monitored concentrations from unmodeled years (e.g., 2006 and 2008). For the adjustments to the modeled concentration data, the model-predicted HDDM sensitivities were applied directly to the modeled concentrations by pairing them spatially and temporally (i.e., on a grid-cell and hour-specific basis). In cases where the monitored hourly O_3 concentrations and their respective modeled values were quite different, or where ozone response was atypical, the sensitivities applied to the monitored concentrations may differ substantially from the sensitivities applied to the modeled concentrations.

3) For the adjustments to the monitored concentrations, floors based on the 5th percentile values were applied to the regression relationships for each specific monitor, hour, and

season. This prevented estimates of negative sensitivities that were derived from modeled conditions with low hourly O_3 due to NOx titration from being inaccurately applied to situations where the monitored concentrations were not titrated (e.g., different years). Since the adjusted model concentrations were derived directly from the HDDM sensitivities, and not values based on regression relationships, there was no need to apply floor values to those sensitivities.

Figure 4A-8 shows the resulting adjusted CMAQ model surfaces for W126 based on the HDDM adjustments for the existing standard, and the alternative W126-based standards with levels of 15 ppm-hrs, 11 ppm-hrs, and 7 ppm-hrs.

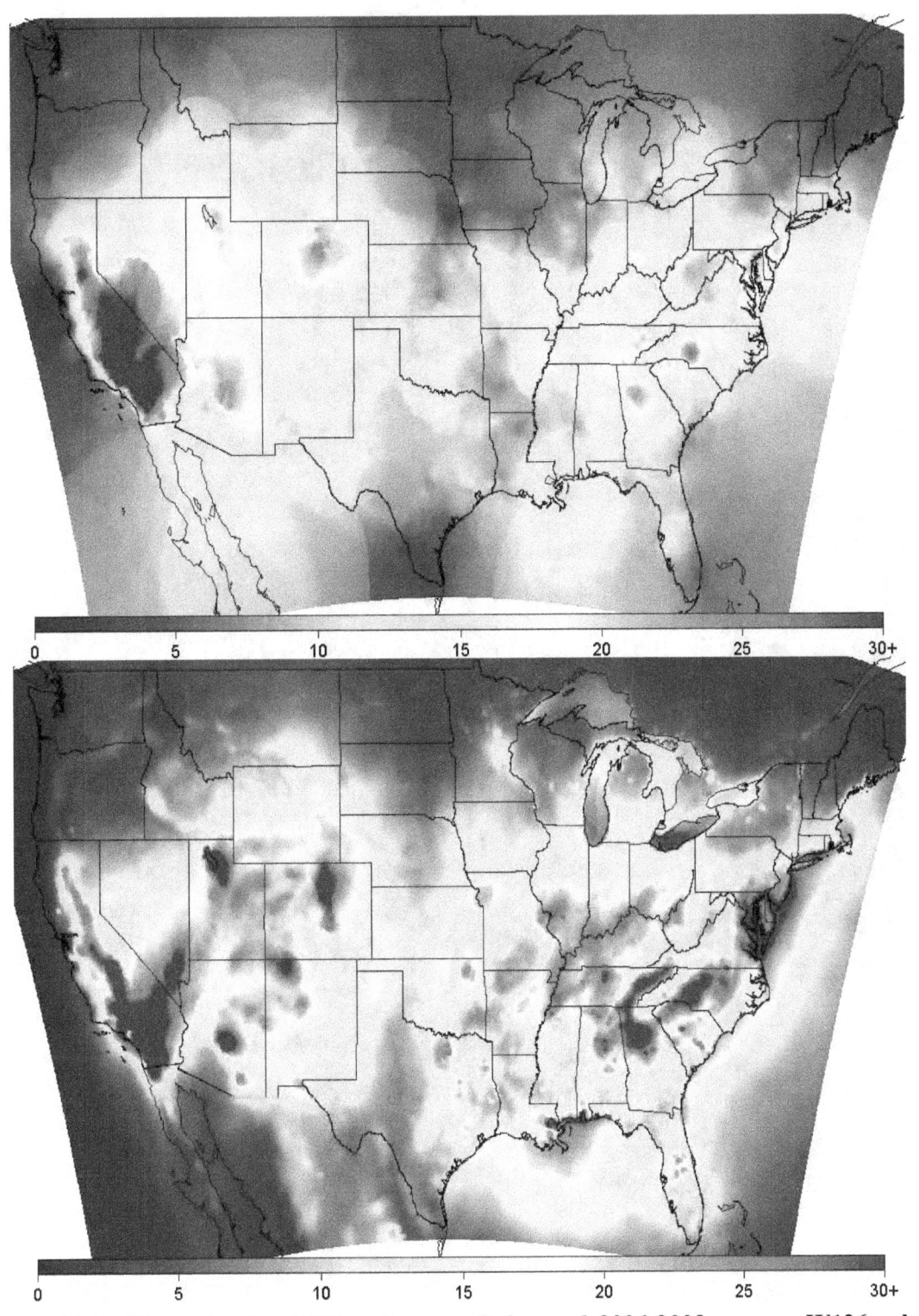

Figure 4A-7   Maps showing VNA estimates of observed 2006-2008 average W126 values
(top), and base CMAQ estimates of 2007 annual W126 values (bottom)

CMAQ W126 Values - 75 ppb

CMAQ W126 Values - 15 ppm-hours

CMAQ W126 Values - 11 ppm-hours

CMAQ W126 Values - 7 ppm-hours

4A-20

Figure 4A-8 CMAQ model surfaces of W126 adjusted using HDDM based on emissions reductions used to meet the existing standard (top left), and the alternative W126-based standards of 15 ppm-hrs (top right), 11 ppm-hrs (bottom left), and 7 ppm-hrs (bottom right) at all monitoring locations.

As shown in Figure 4A-8, the large amount of variability in the emissions reductions used to meet the various standards across the nine regions resulted in very sharp spatial gradients of W126 along regional boundaries in the adjusted model surfaces in some cases. For example, a 67% reduction in U.S. anthropogenic NOx emissions was used to meet a 15 ppm-hr standard in the Southwest region, while no adjustments were used in the West North Central region, which resulted in a sharp change in W126 concentrations near the Colorado/Wyoming border. These disparities were much less apparent in the adjusted monitor W126 values in Figure 4A-6 due in part to the scarcity of monitors in those locations in contrast to a continuous modeled surface. In addition, the unadjusted modeled W126 concentrations were substantially higher than the corresponding unadjusted monitored values in many locations (see Figure 4A-7), resulting in adjusted model surfaces having W126 values that were much higher than the respective adjusted monitor W126 values. The differences in adjusted W126 concentration were partially due to the different time periods represented by the monitored and modeled concentrations, and partially due to the differences in the adjustment methodology as explained above.

Thus, we determined that in this situation, it was not appropriate to apply data fusion methods due to the magnitude of the differences in the "monitored" and "modeled" W126 values. On this basis, we determined that VNA was the most appropriate method to use for the adjusted W126 spatial fields since it is the only technique which relies on only monitored concentrations. We also determined that VNA was the most appropriate method to use for the W126 spatial fields based on recent air quality, based on the comparable performance of VNA to the other techniques in the cross-validation, and for the purpose of eliminating any uncertainties associated with comparing risk results based on air quality inputs created using different techniques.

4A.4 AIR QUALITY INPUTS TO THE WELFARE RISK AND EXPOSURE ASSESSMENT

This section presents the set of spatial fields used as air quality inputs to the Welfare Risk and Exposure Assessments. For the biomass loss analyses presented in Chapter 6, the air quality inputs were VNA spatial fields of 2006-2008 average W126 concentrations for observed air quality, air quality adjusted to meet the existing O_3 standard, and air quality adjusted to meet potential alternative W126-based standards of 15 ppm-hrs, 11 ppm-hrs, and 7 ppm-hrs. For the foliar injury analyses presented in Chapter 7, the air quality inputs were VNA spatial fields of the observed annual W126 values for 2006-2010. All VNA fields were evaluated over a 12 km x 12 km gridded CMAQ domain covering the continental U.S., with estimates taken at the center of each grid cell.

4A.4.1 **AIR QUALITY INPUTS TO THE BIOMASS LOSS ANALYSES (CHAPTER 6)**

Figure 4A-9 shows the VNA surfaces of the 2006-2008 average W126 based on air quality data adjusted to meet the existing standard, and the alternative W126-based standards of 15 ppm-hrs, 11 ppm-hrs, and 7 ppm-hrs. Recall from Chapter 4 that in order to assess the changes in welfare-related impacts that could result from meeting a W126-based standard in addition to the existing standard, the final VNA spatial fields for the alternative W126-based standards were created using air quality adjusted to meet the existing standard as a starting point. The adjusted monitoring data used as inputs to these spatial fields were spliced together by region, based on which standard (either the existing standard or the relevant W126-based standard) was the "controlling" standard in each region (i.e., which standard used a greater reduction in U.S. anthropogenic NOx emissions in order to meet it). For example, the final VNA surface for meeting the existing standard AND 15 ppm-hrs used monitoring data adjusted to meet the 15 ppm-hr standard in the Southwest and West regions, and monitoring data adjusted to meet the existing standard in all other regions. A national VNA surface was created using the spliced-together monitor values resulting in more gentle gradients than would have been produced if VNA surfaces were first created by region and then spliced together. Figure 4A-10 shows maps of the final VNA surfaces used in the biomass loss analyses in Chapter 6. Note that Figure 4A-9 shows air quality which does not first adjust to meet the existing standard before adjusting to meet the alternative standards while Figure 4A-10 shows air quality adjusted using the approach described above to meet the alternative W126-based standards *after* first meeting the current standard. The top left panels showing W126 values based on air quality adjusted to meet the current standard are identical in the two figures. When air quality data were adjusted to meet the existing standard, there were only 5 monitors in the U.S. with W126 values above 15 ppm-hrs, and only 17 monitors with W126 values above 11 ppm-hrs. All of these monitors were located in urban areas.

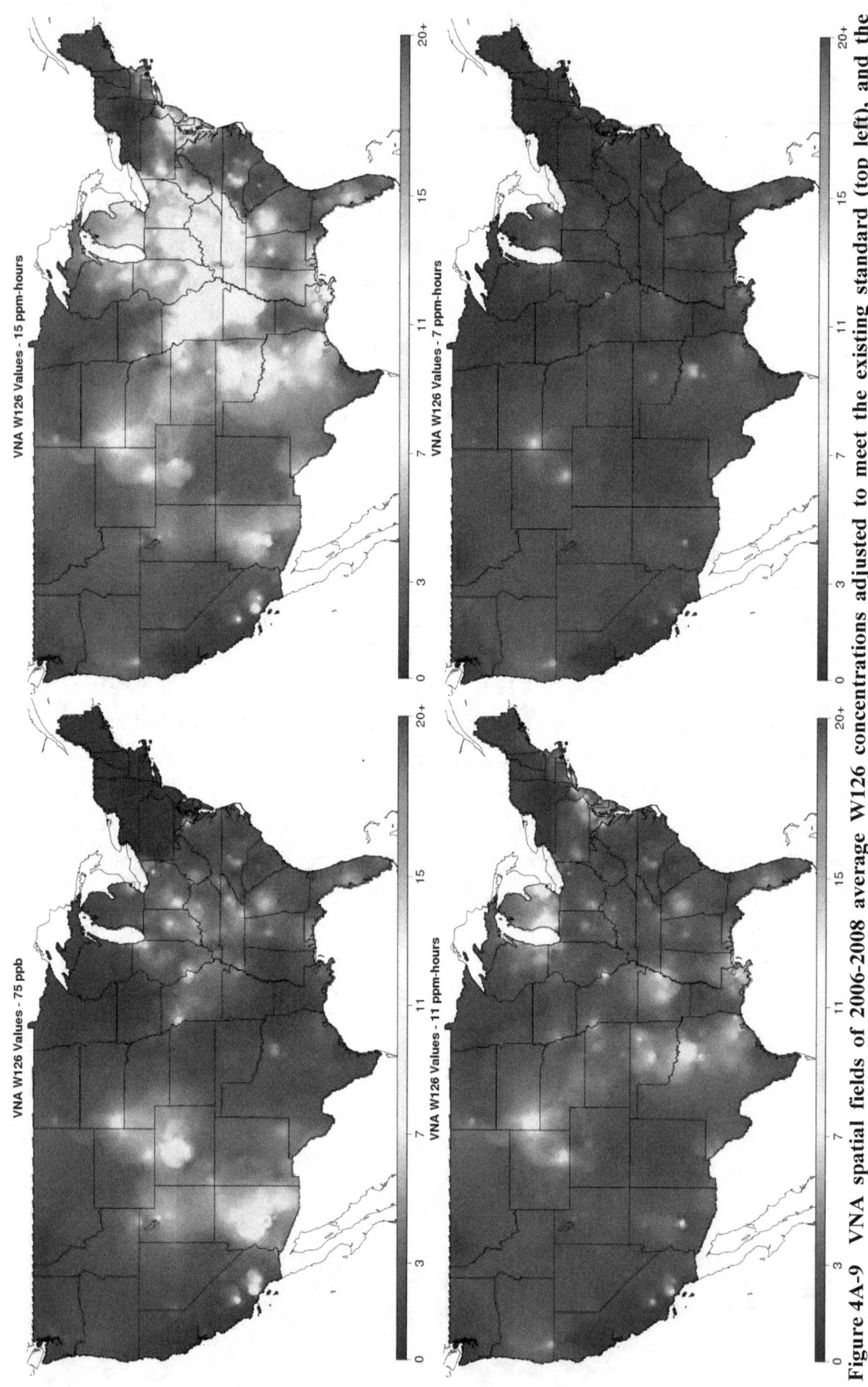

Figure 4A-9 VNA spatial fields of 2006-2008 average W126 concentrations adjusted to meet the existing standard (top left), and the alternative W126-based standards of 15 ppm-hrs (top right), 11 ppm-hrs (bottom left), and 7 ppm-hrs (bottom right)

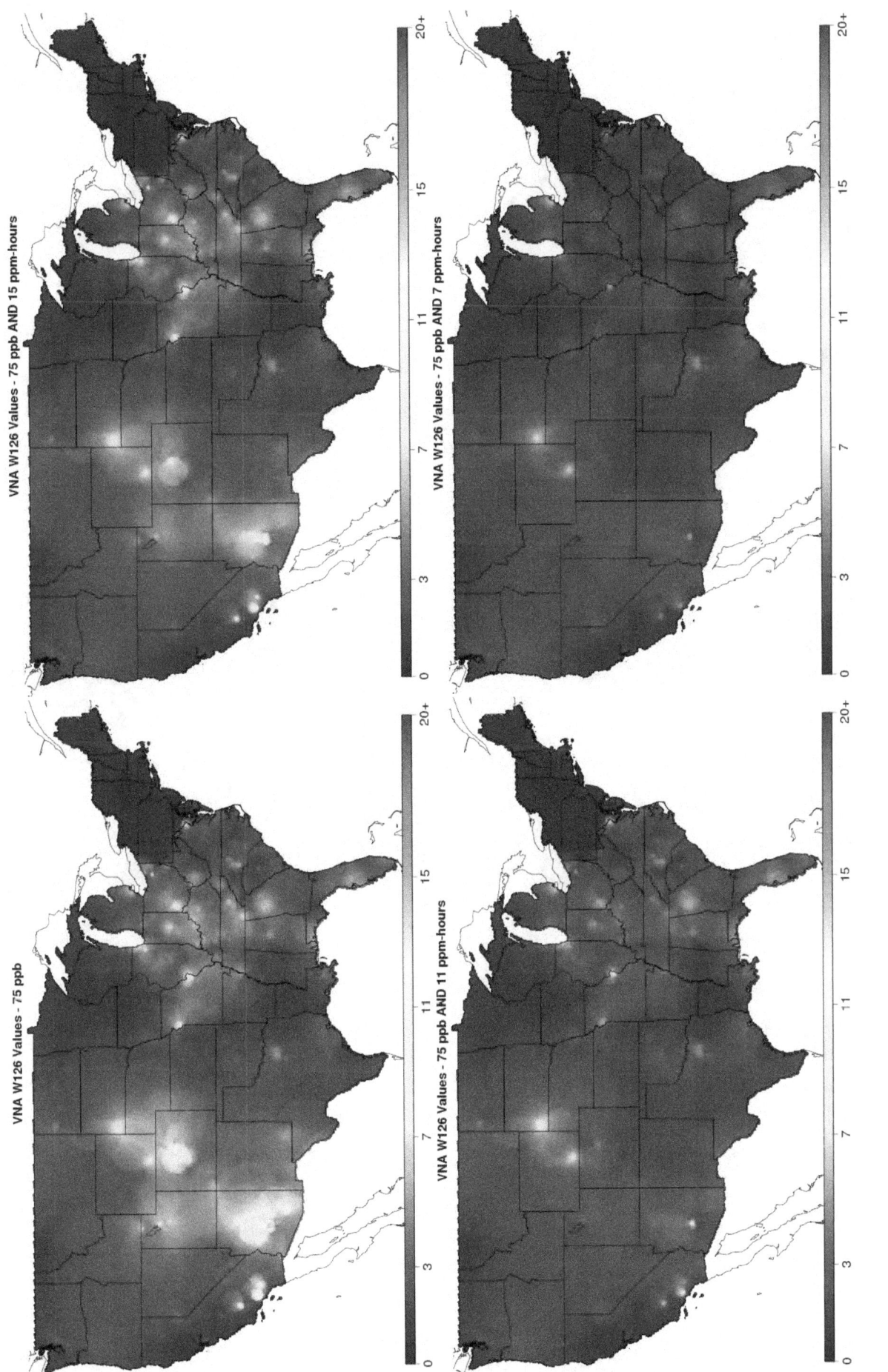

4A-25

Figure 4A-10 VNA spatial fields of 2006-2008 average W126 concentrations adjusted to meet the existing standard (top left), and then further adjusted to meet the alternative W126-based standards of 15 ppm-hrs (top right), 11 ppm-hrs (bottom left), and 7 ppm-hrs (bottom right)

4A-26

4A.4.2 **AIR QUALITY INPUTS TO THE FOLIAR INJURY ANALYSES (CHAPTER 7)**

For the foliar injury analyses presented in Chapter 7, we used VNA to create spatial fields of the observed annual W126 values for 2006-2010, which are shown in Figure 4A-11 through Figure 4A-15. Figure 4A-16 shows the empirical distributions of these spatial fields. There was a substantial amount of inter-annual variability in the W126 spatial fields, with median VNA estimates ranging from about 5.5 ppm-hrs in 2009 to about 11 ppm-hrs in 2006.

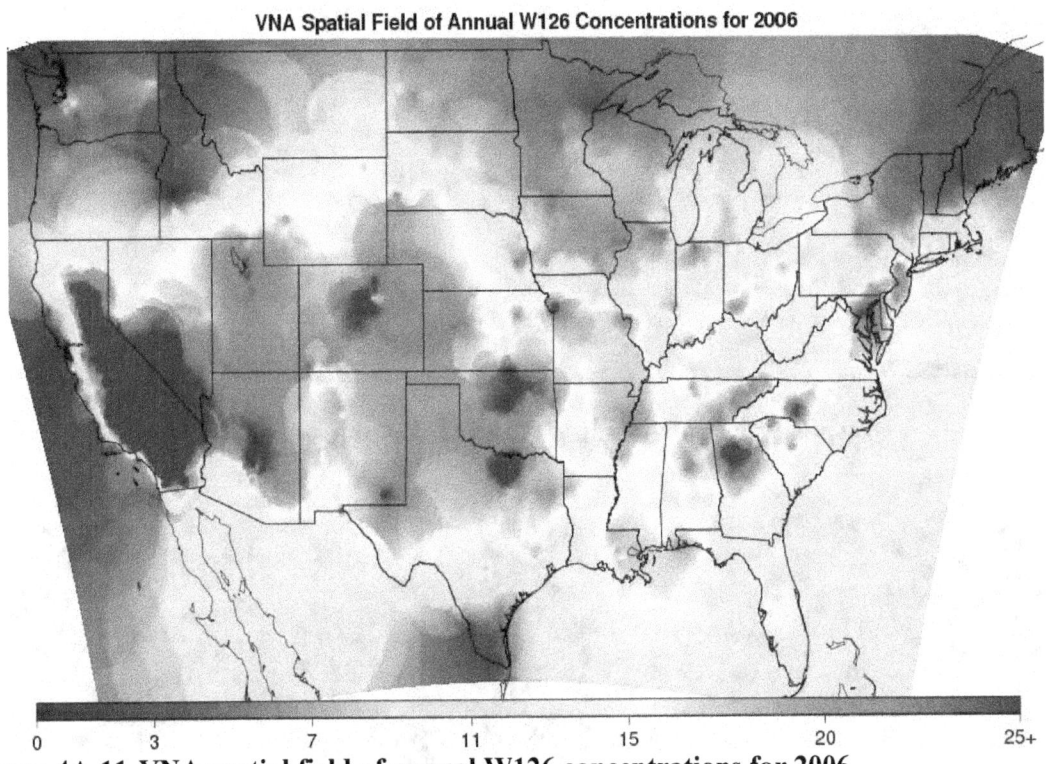

Figure 4A-11 VNA spatial field of annual W126 concentrations for 2006

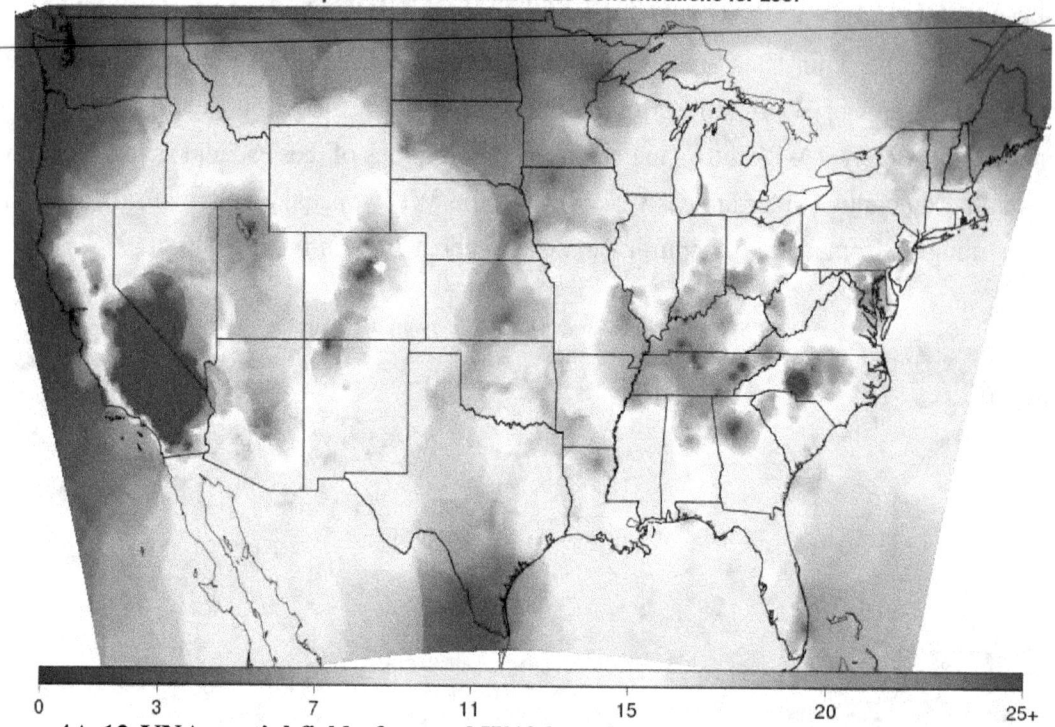

VNA Spatial Field of Annual W126 Concentrations for 2007

Figure 4A-12 VNA spatial field of annual W126 concentrations for 2007

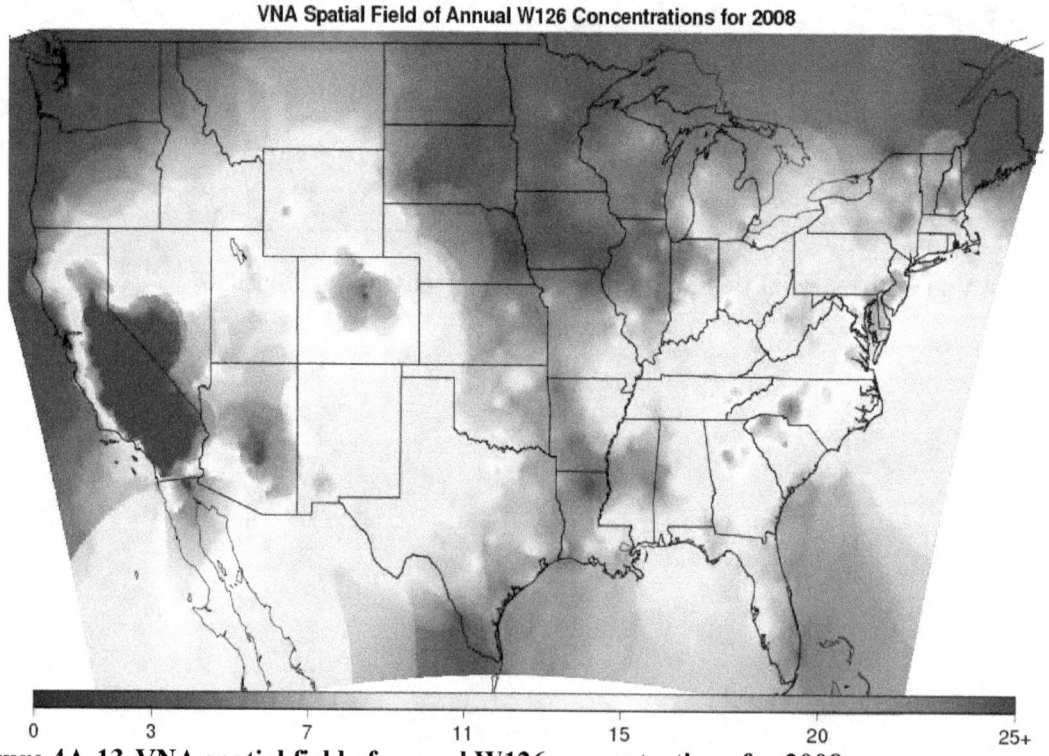

VNA Spatial Field of Annual W126 Concentrations for 2008

Figure 4A-13 VNA spatial field of annual W126 concentrations for 2008

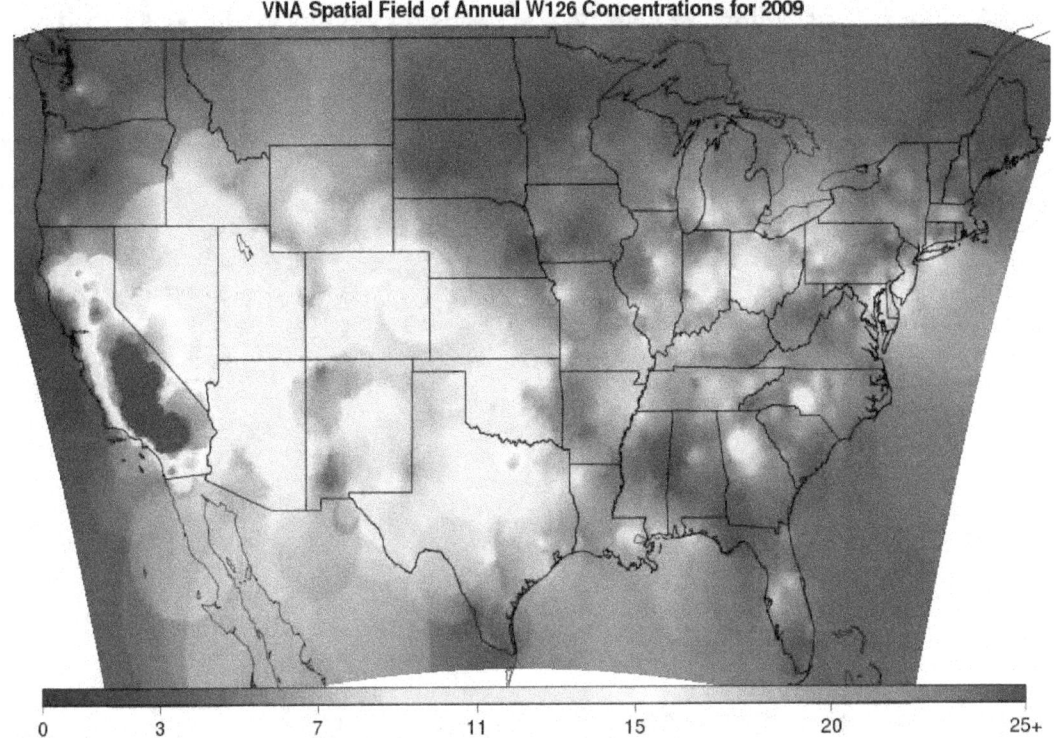

VNA Spatial Field of Annual W126 Concentrations for 2009

| 0 | 3 | 7 | 11 | 15 | 20 | 25+ |

Figure 4A-14 VNA spatial field of annual W126 concentrations for 2009

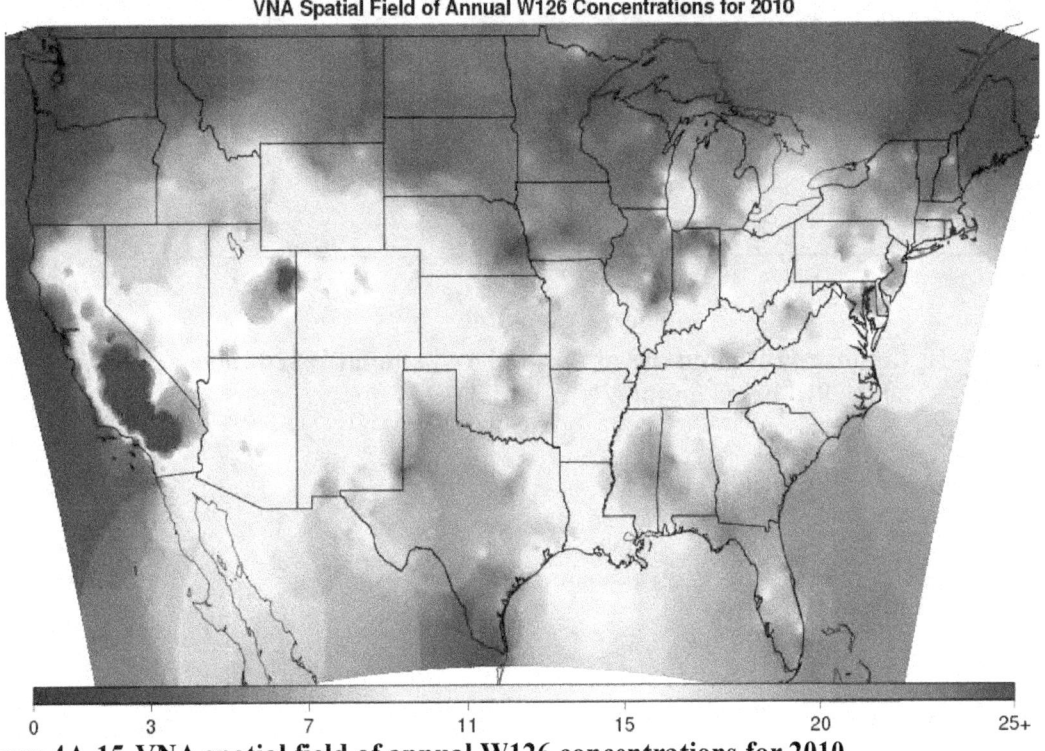

VNA Spatial Field of Annual W126 Concentrations for 2010

| 0 | 3 | 7 | 11 | 15 | 20 | 25+ |

Figure 4A-15 VNA spatial field of annual W126 concentrations for 2010

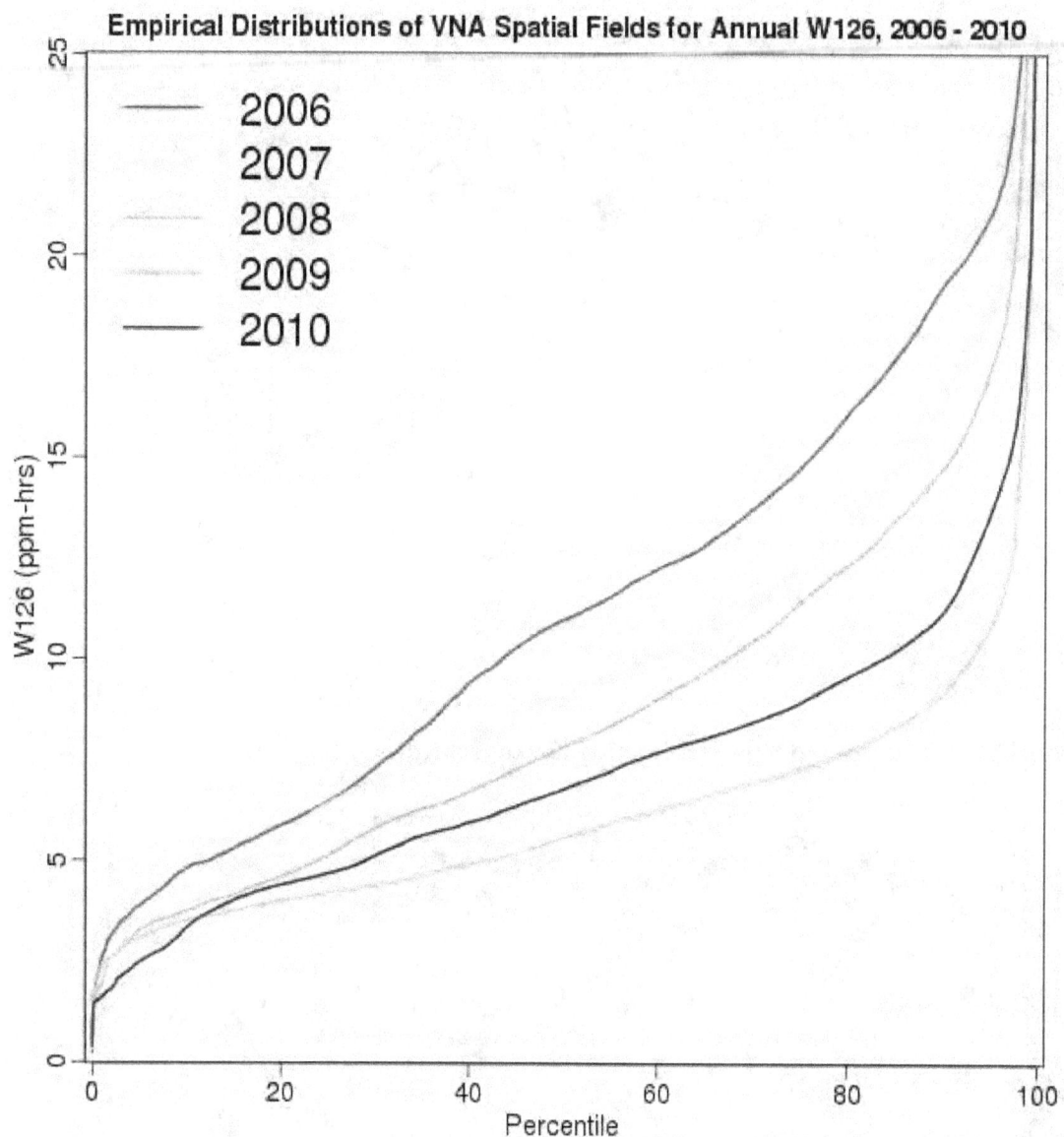

Figure 4A-16 Empirical distributions of the observed annual W126 concentrations for 2006-2010 based on the VNA spatial fields

4A.5 REFERENCES

Berrocal, V.J., Gelfand, A.E., Holland, D.M. (2009). A Spatio-Temporal Downscaler for Output from Numerical Models. Journal of Agricultural, Biological, and Environmental Statistics, 15(2), 176–197.

Berrocal, V.J., Gelfand, A.E., Holland, D.M. (2010). A Bivariate Space-Time Downscaler Under Space and Time Misalignment. Annals of Applied Statistics, 4(4), 1942-1975.

Berrocal, V.J., Gelfand, A.E., Holland, D.M. (2011). Space-Time Data Fusion Under Error in Computer Model Output: An Application to Modeling Air Quality. Biometrics, 68(3), 837-848.

Chen, J., Zhao, R., Li, Z. (2004). Voronoi-based k-order neighbor relations for spatial analysis. ISPRS J Photogrammetry Remote Sensing, 59(1-2), 60-72.

Gold, C. (1997). Voronoi methods in GIS. In: Algorithmic Foundation of Geographic Information Systems (va Kereveld M., Nievergelt, J., Roos, T., Widmayer, P., eds). Lecture Notes in Computer Sicnece, Vol 1340. Berlin: Springer-Verlag, 21-35.

Hall, E., Eyth, A., Phillips, S. (2012). Hierarchical Bayesian Model (HBM)-Derived Estimates of Air Quality for 2007: Annual Report. EPA/600/R-12/538. Available on the Internet at: http://www.epa.gov/heasd/sources/projects/CDC/AnnualReports/2007_HBM.pdf

McMillan, N.J., Holland, D.M., Morara, M., Feng, J. (2009). Combining Numerical Model Output and Particulate Data using Bayesian Space-Time Modeling. Environmetrics, Vol. 21, 48-65.

R Core Team (2012). R: A language and environment for statistical computing. R Foundation for Statistical Computing, Vienna, Austria. http://www.R-project.org/.

Timin B, Wesson K, Thurman J. (2010). Application of Model and Ambient Data Fusion Techniques to Predict Current and Future Year $PM_{2.5}$ Concentrations in Unmonitored Areas. Pp. 175-179 in Steyn DG, Rao St (eds). Air Pollution Modeling and Its Application XX. Netherlands: Springer.

Turner, R. (2012). deldir: Delaunay Triangulation and Dirichlet (Voronoi) Tessellation. R package version 0.0-19. http://CRAN.R-project.org/package=deldir

U.S. Environmental Protection Agency. (2012a). Health Risk and Exposure Assessment for Ozone, First External Review Draft. Available on the Internet at: http://www.epa.gov/ttn/naaqs/standards/ozone/s_o3_2008_isa.html

U.S. Environmental Protection Agency. (2012b). Air Quality Modeling Technical Support Document for the Regulatory Impact Analysis for the Revisions to the National Ambient Air Quality Standards for Particulate Matter. Available on the Internet at: http://www.epa.gov/ttn/naaqs/standards/pm/data/201212aqm.pdf

Wells, B., Wesson, K., Jenkins, S. (2012). Analysis of Recent U.S. Ozone Air Quality Data to Support the O3 NAAQS Review and Quadratic Rollback Simulations to Support the First Draft of the Risk and Exposure Assessment. Available on the Internet at: http://www.epa.gov/ttn/naaqs/standards/ozone/s_o3_td.html

APPENDIX 5A: LARGER MAPS OF

FIRE THREAT AND BASAL AREA LOSS

O₃ Levels at Recent Conditions in Areas where Fire Threat is Moderate to Severe

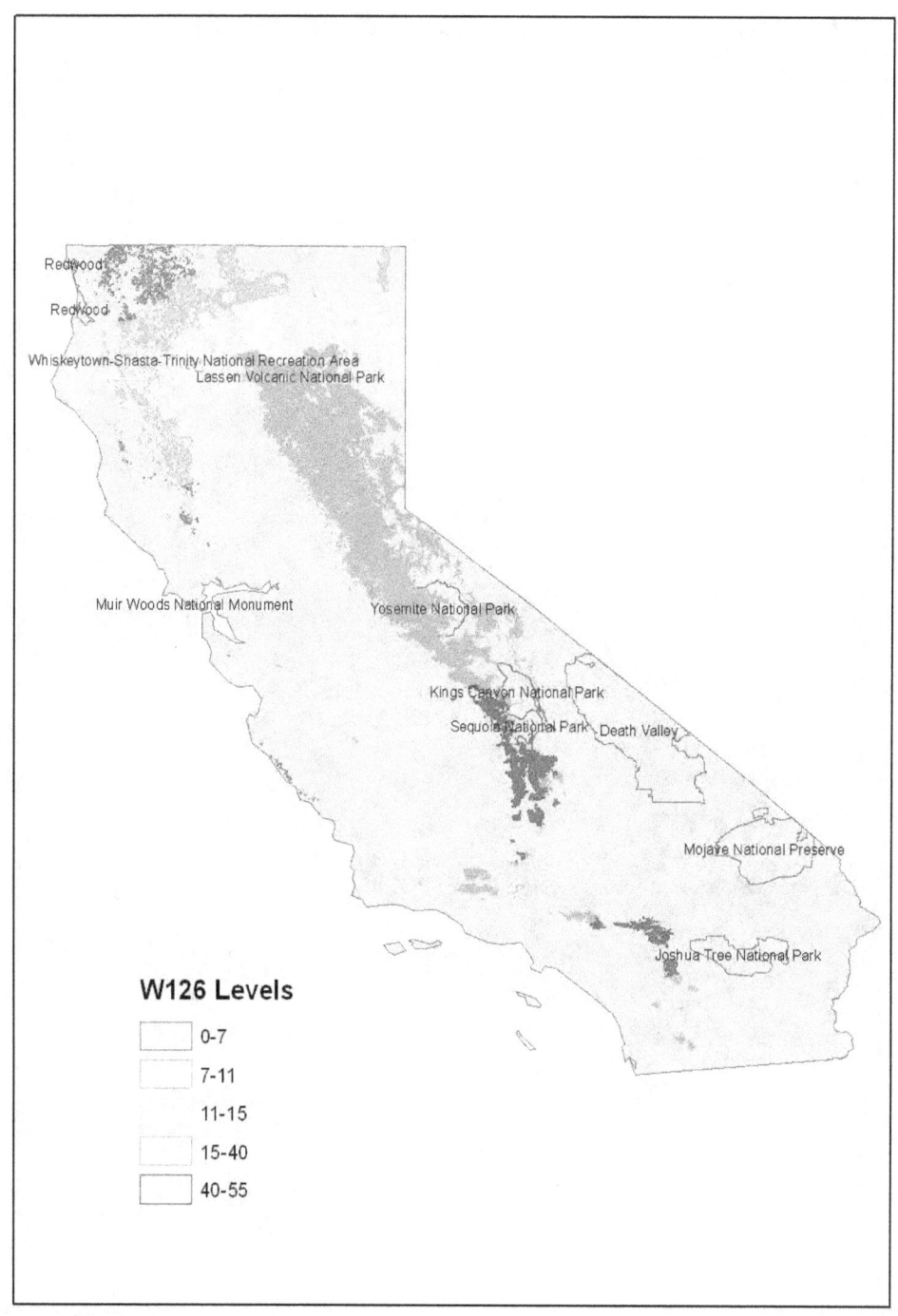

W126 Levels

0-7
7-11
11-15
15-40
40-55

O₃ Levels after Adjusting Air Quality to Just Meeting the Existing Standard in Areas where Fire Threat is Moderate to Severe

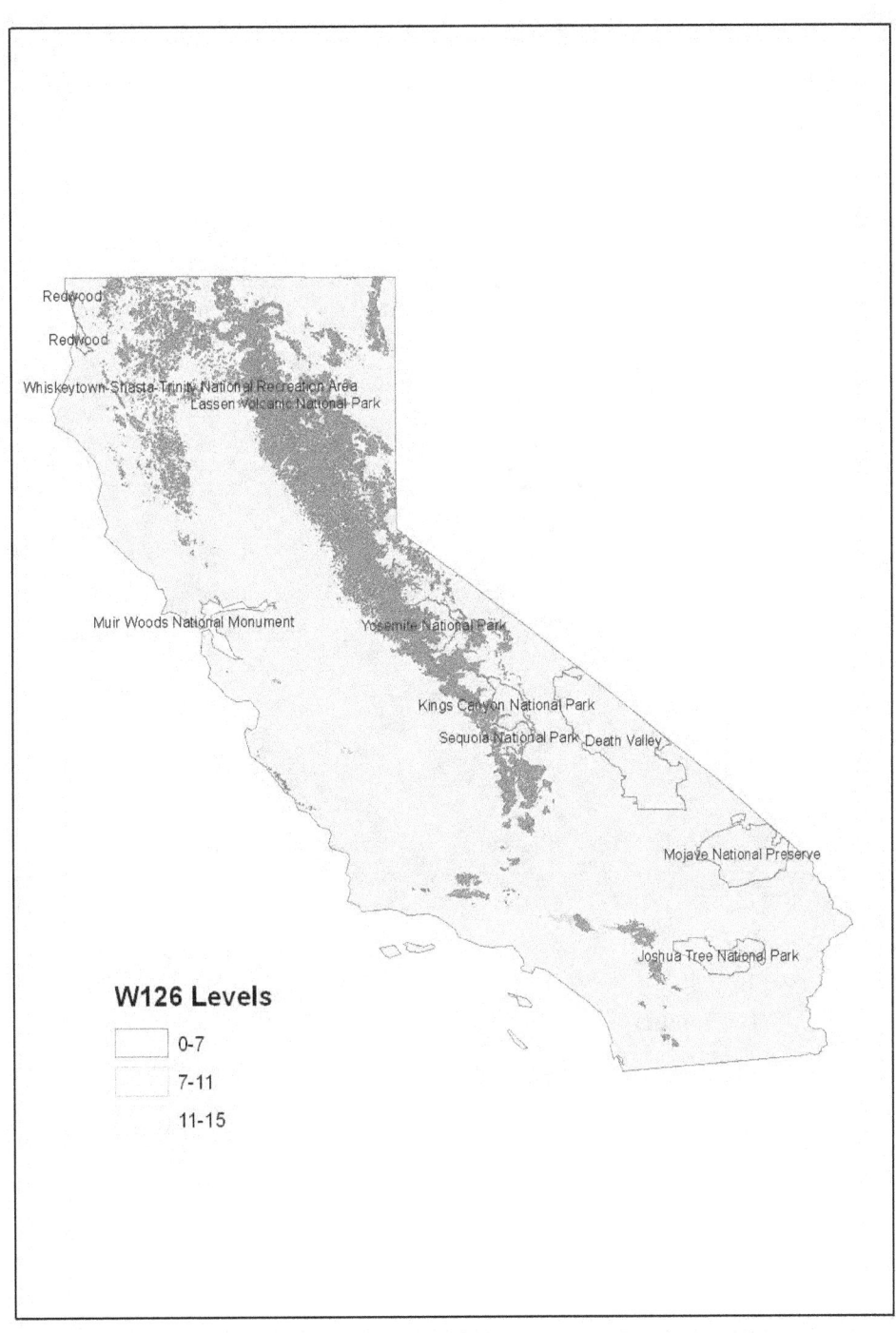

O₃ Levels after Adjusting Air Quality to Just Meeting an Alternate W126 Standard of 15 ppm-hrs in Areas where Fire Threat is Moderate to Severe

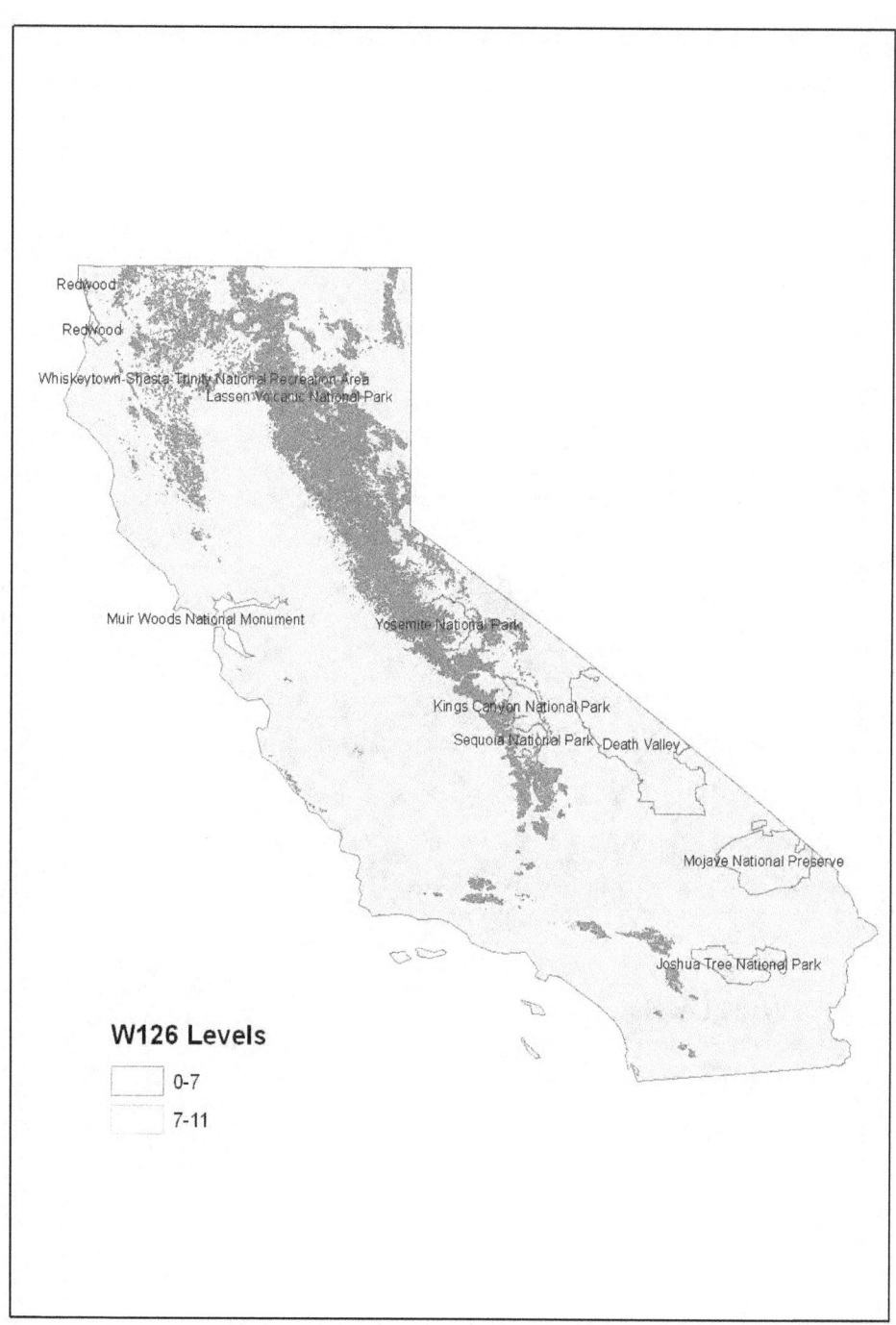

O₃ Levels after Adjusting Air Quality to Just Meeting Alternate W126 Standards of 11 and 7 ppm-hrs in Areas where Fire Threat is Moderate to Severe

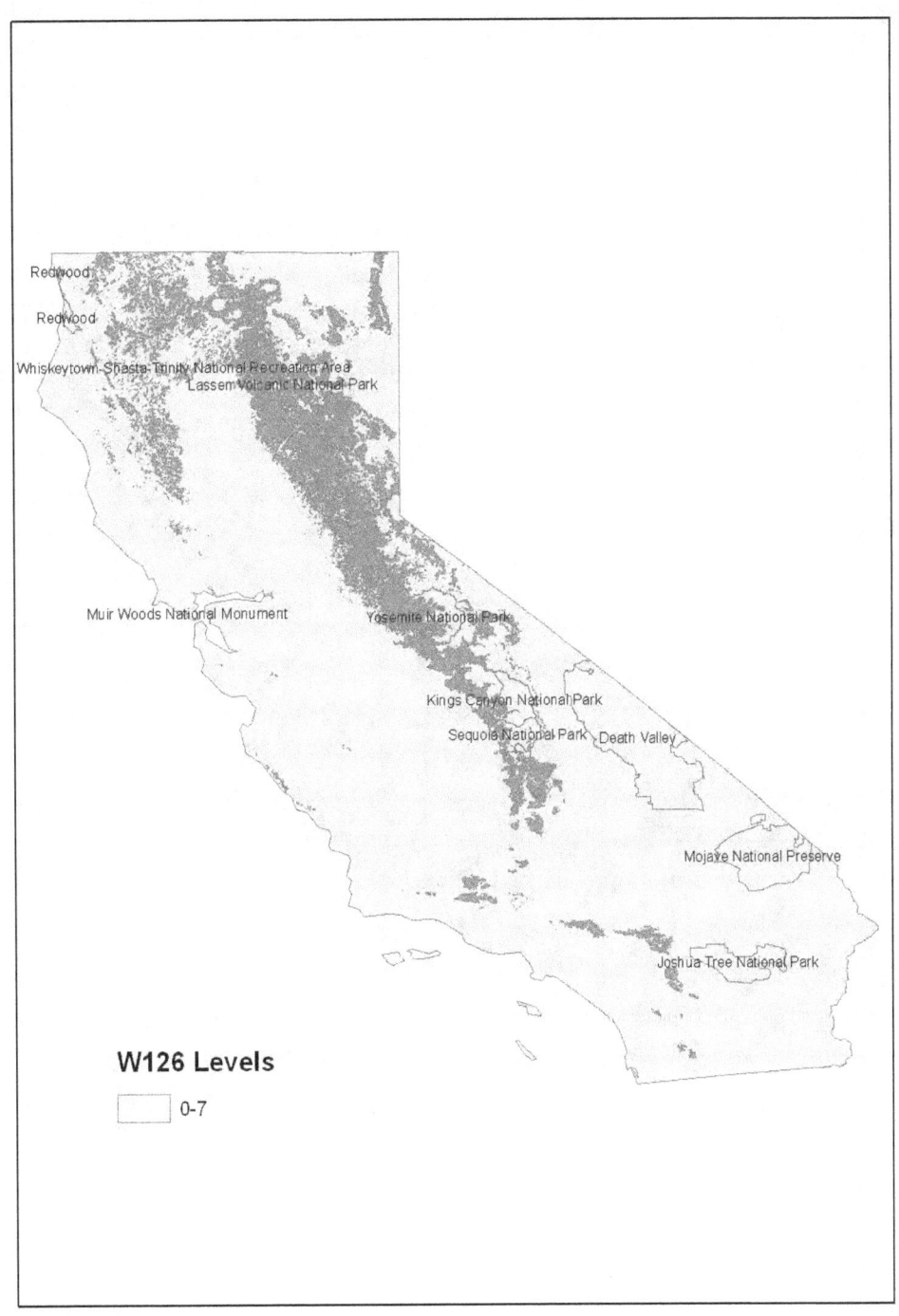

O₃ Levels at Recent Conditions in Areas Considered 'At Risk' of High Basal Area Loss
(>25% Loss)

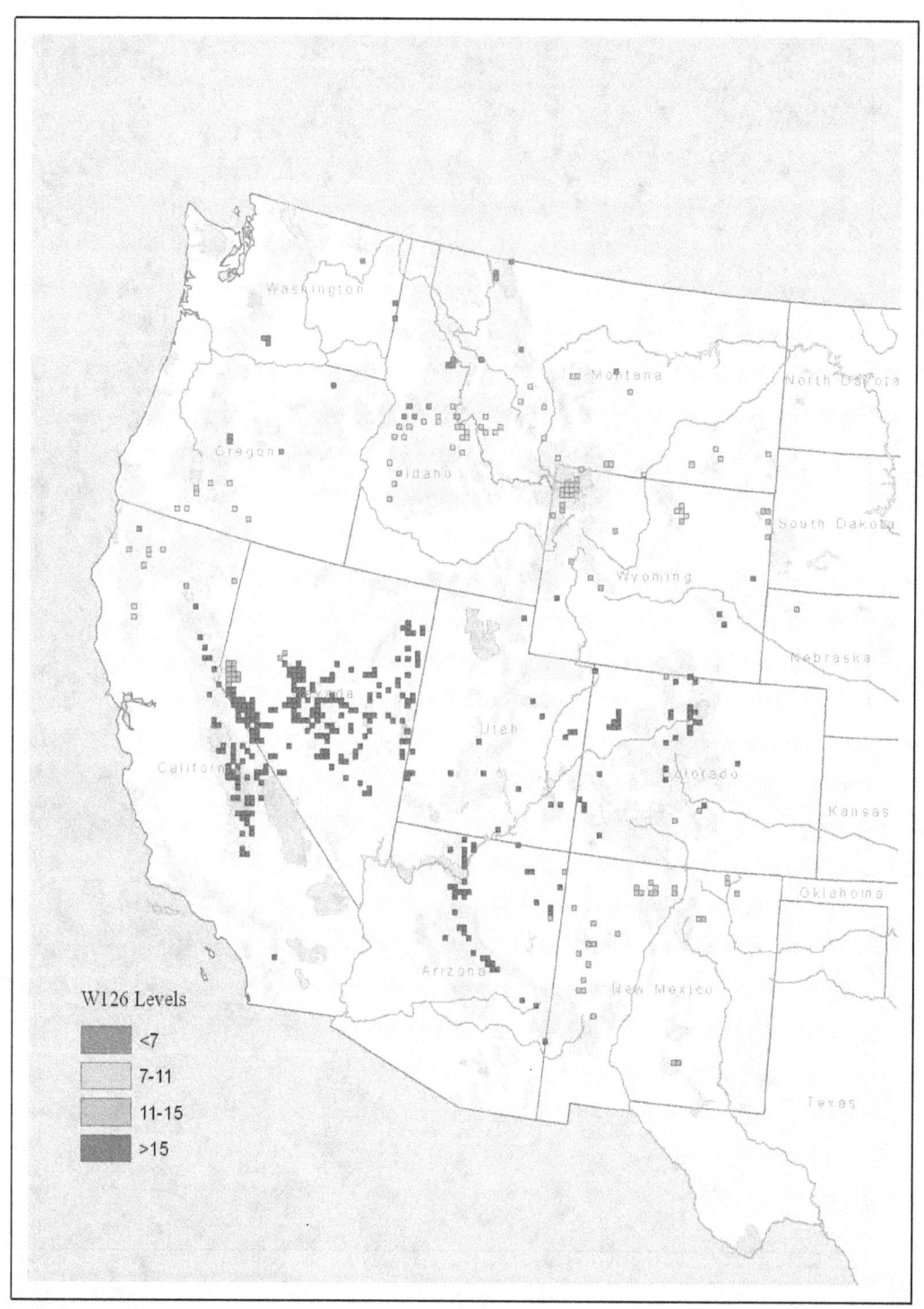

O₃ Levels after Adjusting Air Quality to Just Meeting the Existing Standard in Areas Considered 'At Risk' of High Basal Area Loss (>25% Loss)

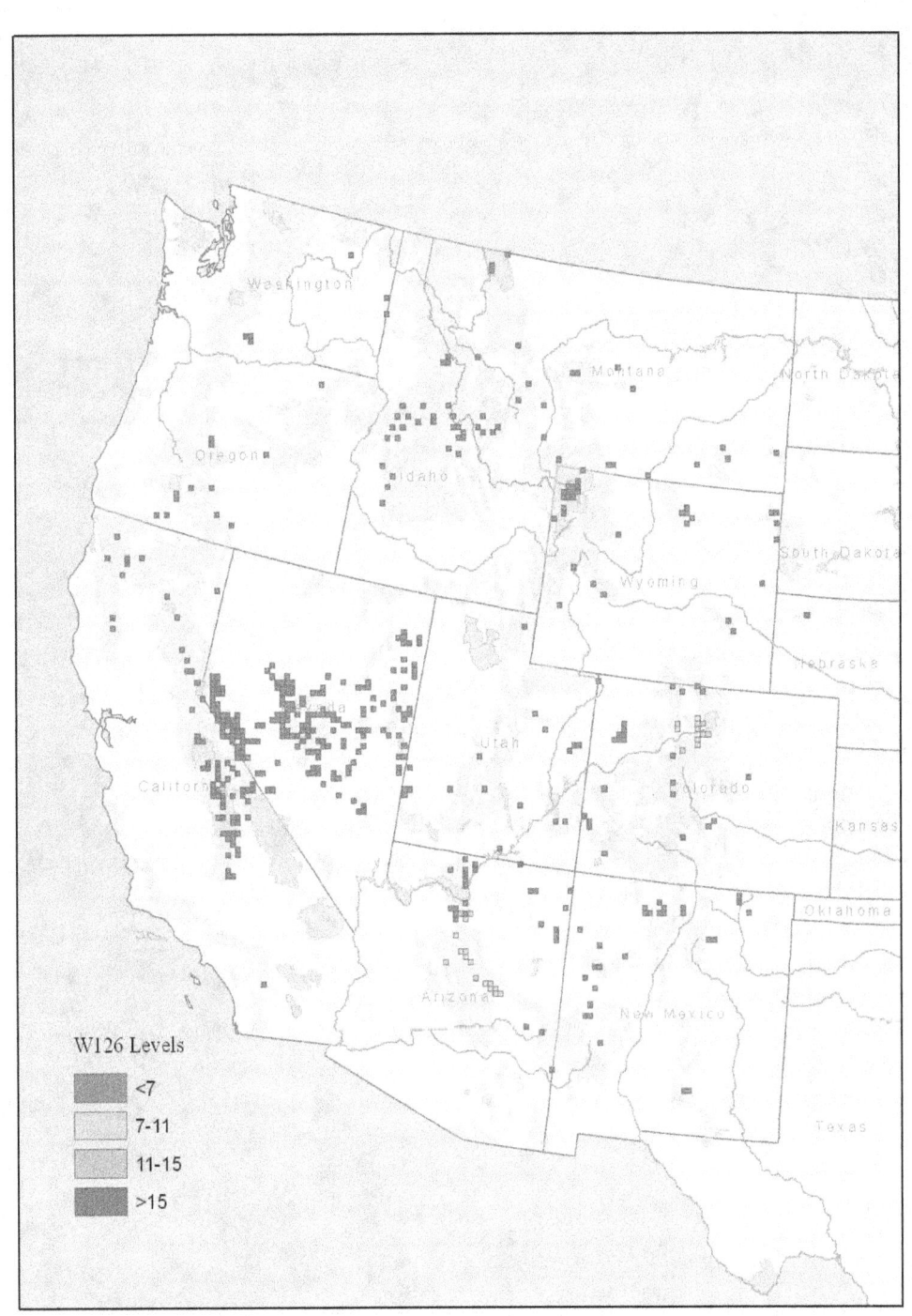

O₃ Levels after Adjusting Air Quality to Just Meeting an Alternate W126 Standard of 15 ppm-hrs in Areas Considered 'At Risk' of High Basal Area Loss (>25% Loss)

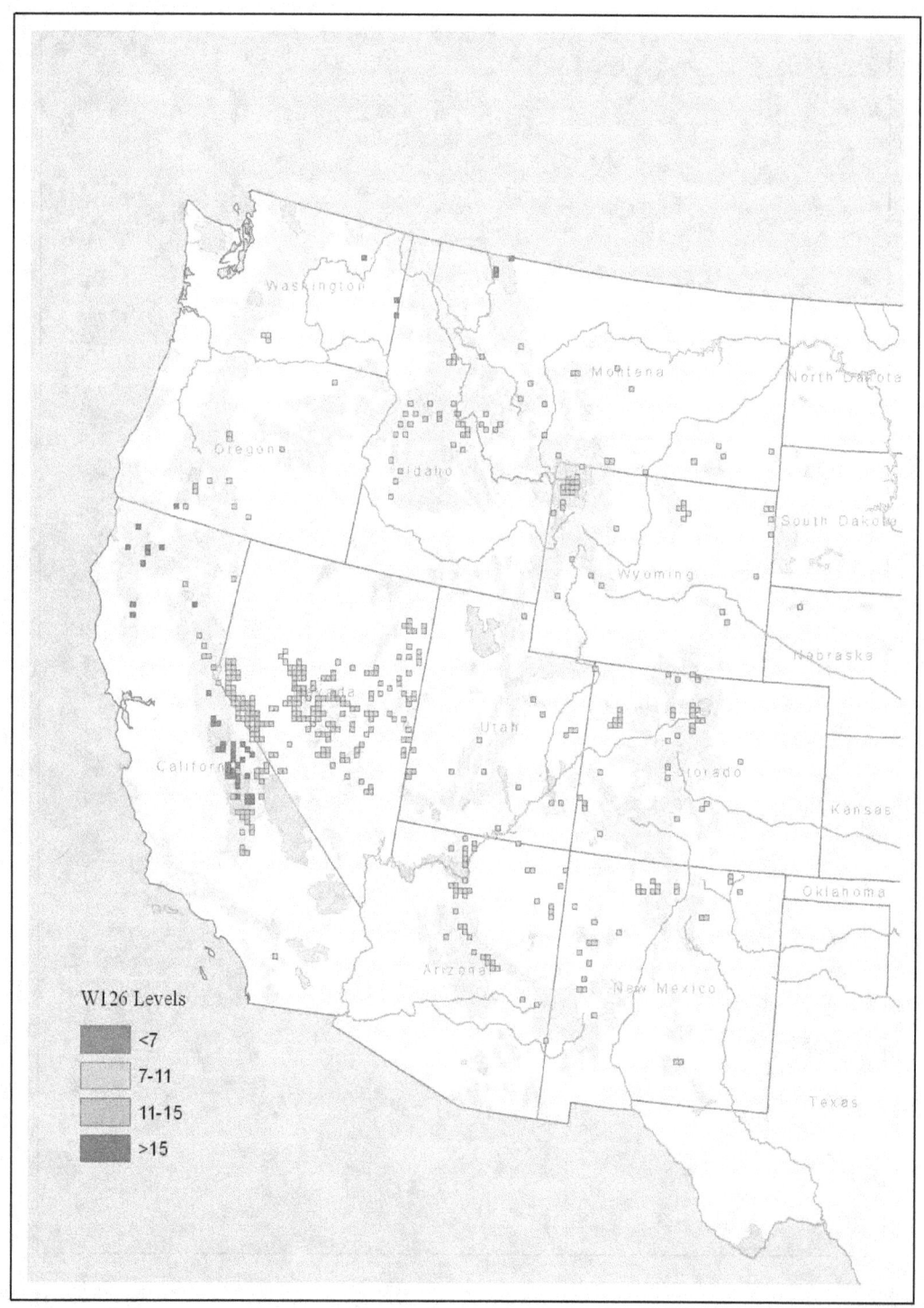

O₃ Levels after Adjusting Air Quality to Just Meeting Alternate W126 Standards of 11 and 7 ppm-hrs in Areas Considered 'At Risk' of High Basal Area Loss (>25% Loss)

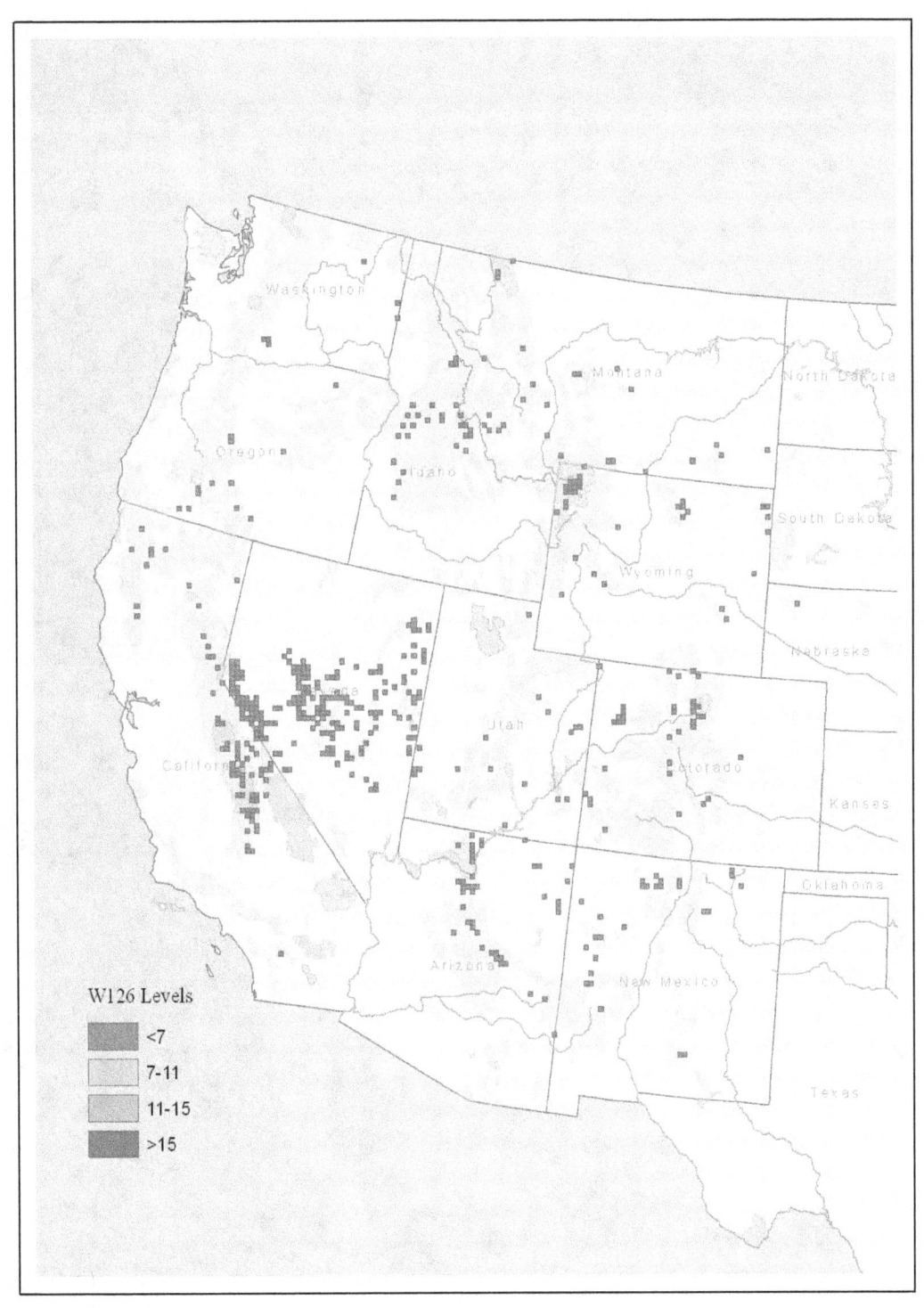

O₃ Levels at Recent Conditions in Areas Considered 'At Risk' of High Basal Area Loss (>25% Loss)

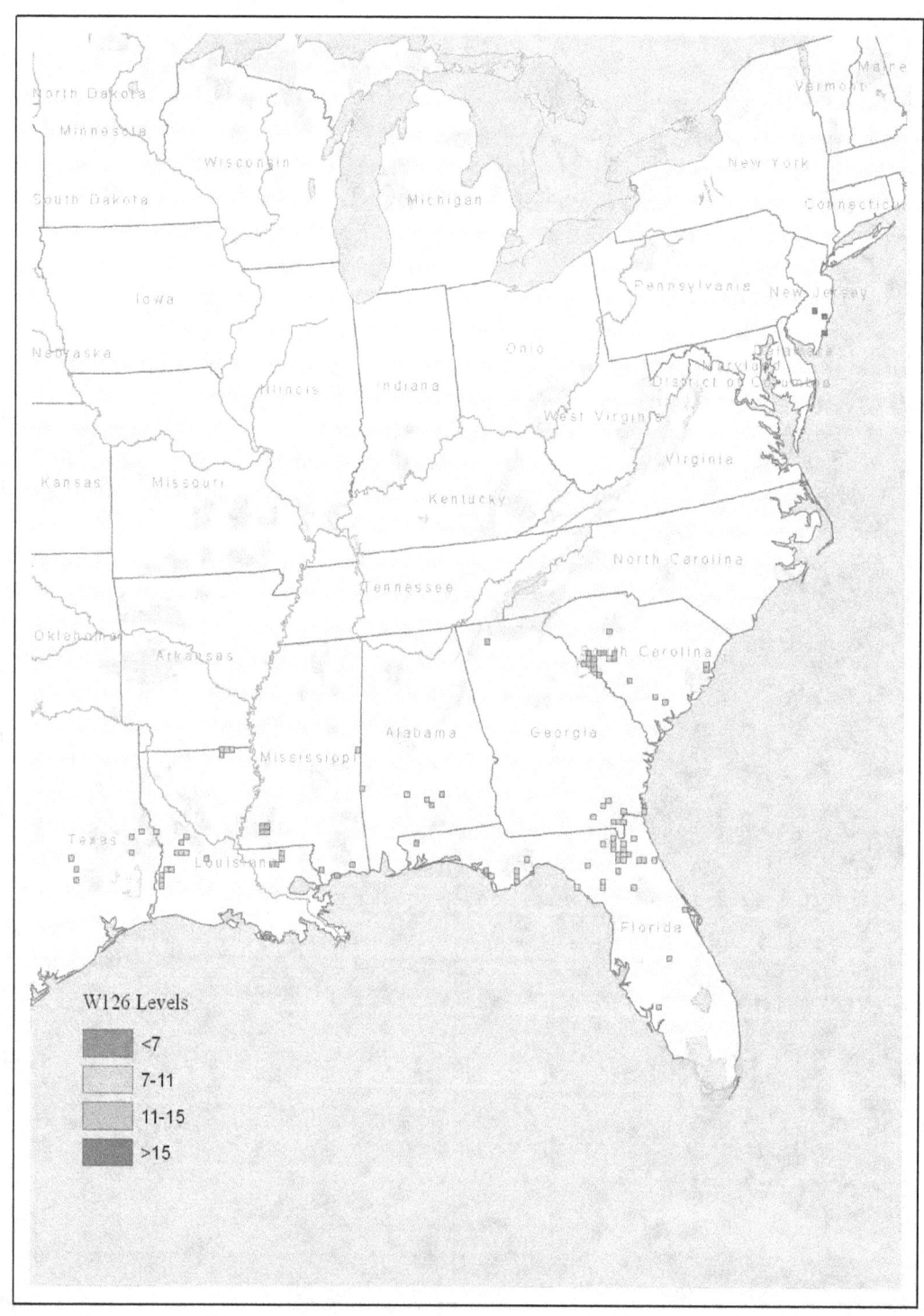

O₃ Levels after Adjusting Air Quality to Just Meeting the Existing Standard and Alternate W126 Standards in Areas Considered 'At Risk' of High Basal Area Loss

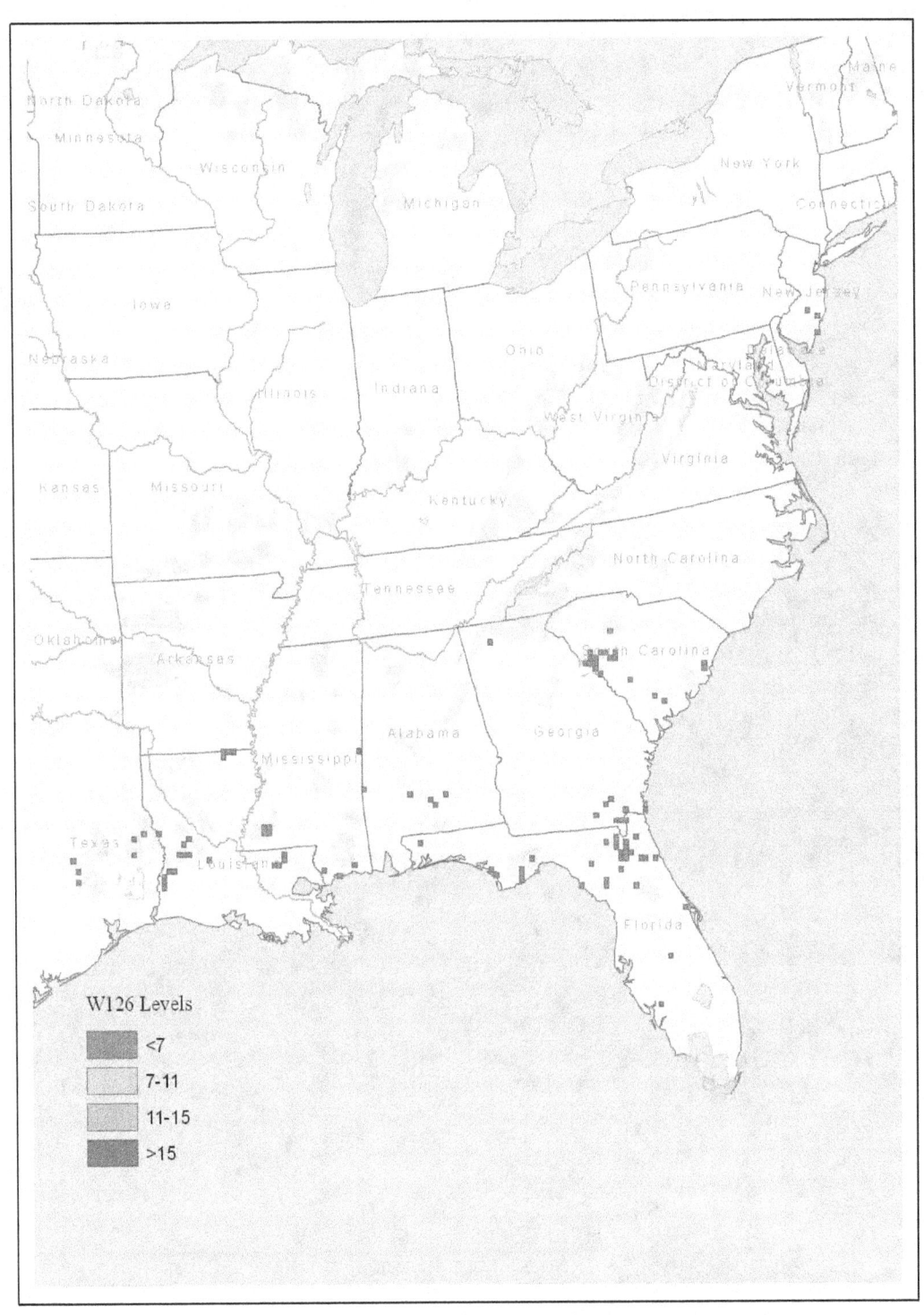

APPENDIX 6A:

MAPS OF INDIVIDUAL TREE SPECIES

6A.1   Discussion

This appendix includes summary maps and figures of relative biomass loss (RBL) as discussed in Section 6.2.1.3 for each of the 12 tree species included in Chapter 6. Data are included for Red Maple (*Acer rubrum*), Sugar Maple (*Acer saccharum*), Red Alder (*Alnus rubra*), Tulip Poplar (*Liriodendron tulipifera*), Ponderosa Pine (*Pinus Ponderosa*), Eastern White Pine (*Pinus strobus*), Loblolly Pine (*Pinus taeda*), Virginia Pine (*Pinus virginiana*), Eastern Cottonwood (*Populus deltoides*), Quaking Aspen (*Populus tremuloides*), Black Cherry (*Prusnus serotina*), and Douglas Fir (*Pseudotsuga menzeiesii*).

For each species there are five maps of RBL. The first is under recent O$_3$ conditions (2006 to 2008) and the following four show RBL under four additional air quality scenarios (75 ppb, 15 ppm-hrs, 11 ppm-hrs, and 7 ppm-hrs,. In addition to the maps, we include histograms showing the RBL distribution under recent O$_3$ conditions and under the four additional air quality scenarios and the proportion RBL (relative to the 75 ppb scenario). Note that in the final panel of histograms, the 75 ppb scenario is by definition 1, so it is not included. For the eastern species, the 75 ppb scenario was controlling below 15 ppm-hrs (i.e., the O$_3$ levels are the same for those two air quality scenarios in the eastern U.S.), so for those species, the 15 ppm-hrs scenario is not included because it is also by definition 1.

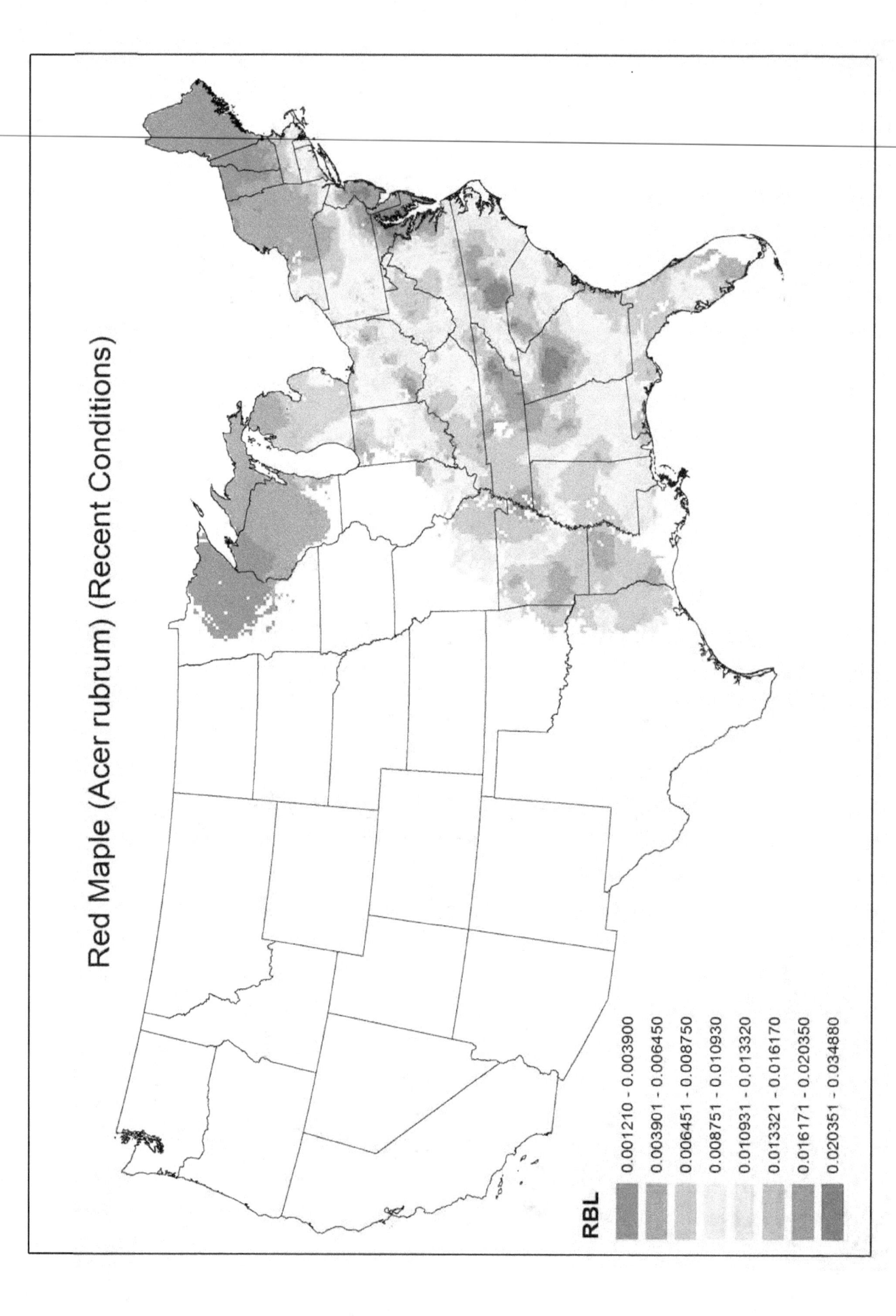

Red Maple (Acer rubrum) (Recent Conditions)

RBL

0.001210 - 0.003900
0.003901 - 0.006450
0.006451 - 0.008750
0.008751 - 0.010930
0.010931 - 0.013320
0.013321 - 0.016170
0.016171 - 0.020350
0.020351 - 0.034880

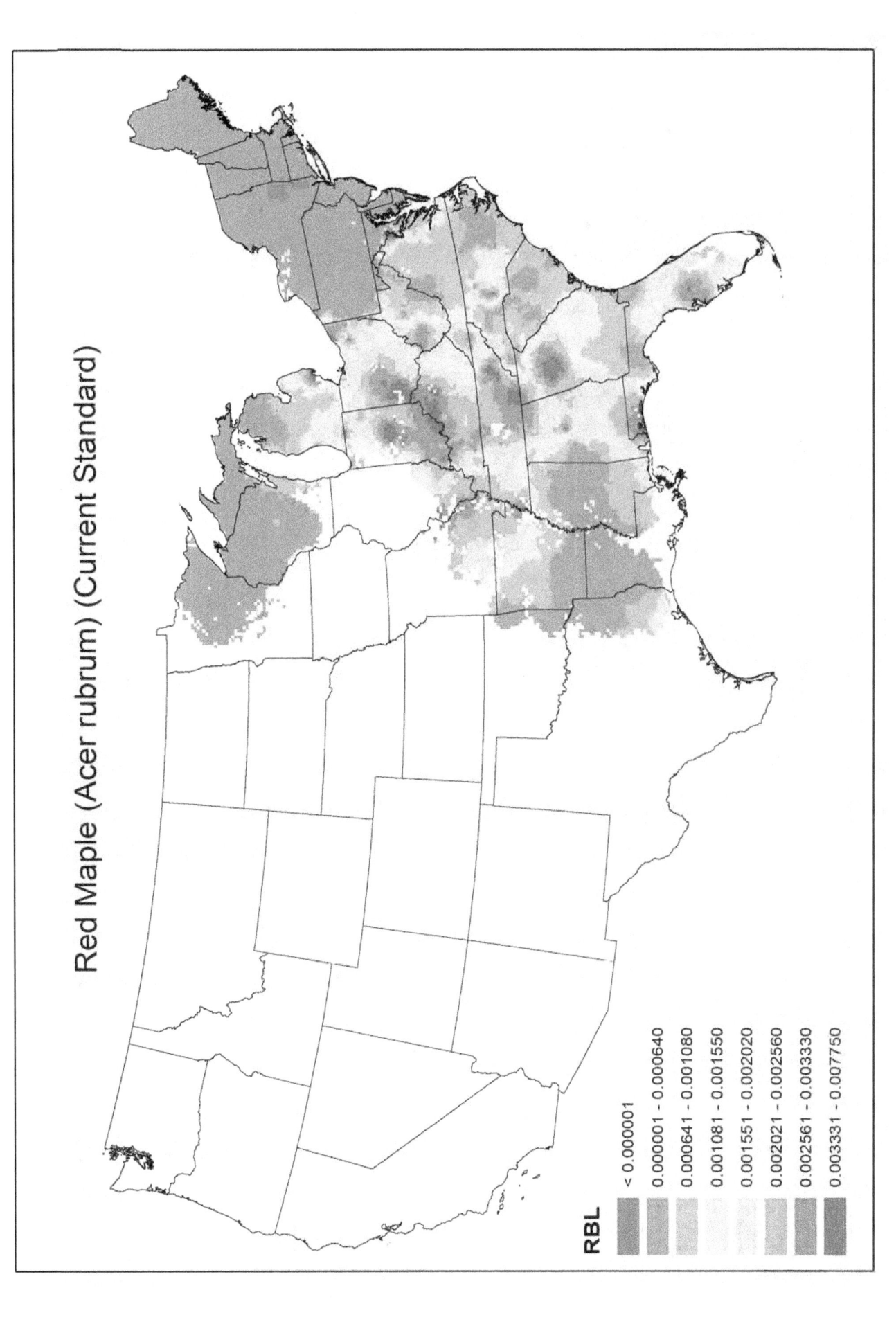

Red Maple (Acer rubrum) (Current Standard)

RBL

< 0.000001
0.000001 - 0.000640
0.000641 - 0.001080
0.001081 - 0.001550
0.001551 - 0.002020
0.002021 - 0.002560
0.002561 - 0.003330
0.003331 - 0.007750

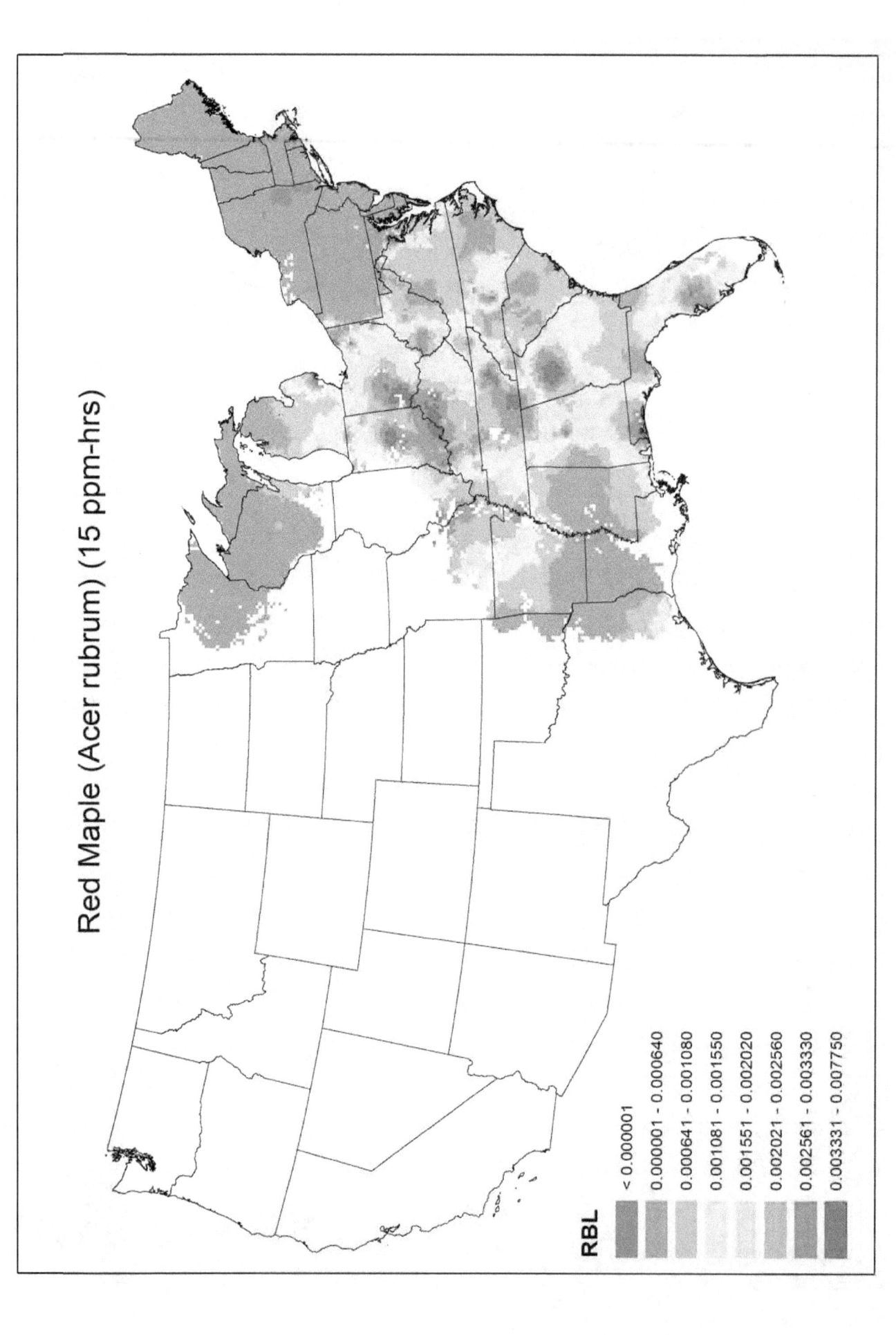

Red Maple (Acer rubrum) (15 ppm-hrs)

RBL
< 0.000001
0.000001 - 0.000640
0.000641 - 0.001080
0.001081 - 0.001550
0.001551 - 0.002020
0.002021 - 0.002560
0.002561 - 0.003330
0.003331 - 0.007750

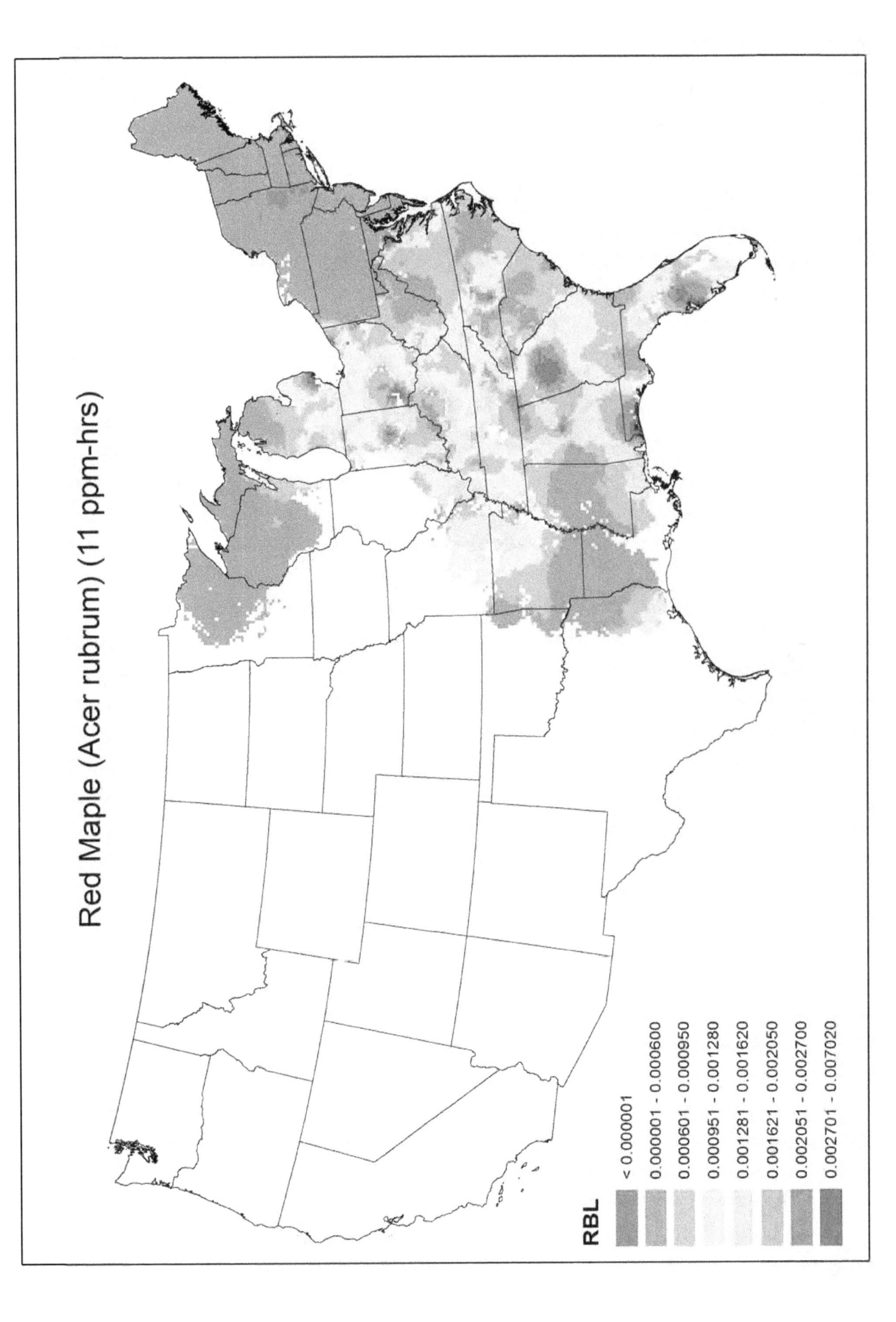

Red Maple (Acer rubrum) (11 ppm-hrs)

RBL
< 0.000001
0.000001 - 0.000600
0.000601 - 0.000950
0.000951 - 0.001280
0.001281 - 0.001620
0.001621 - 0.002050
0.002051 - 0.002700
0.002701 - 0.007020

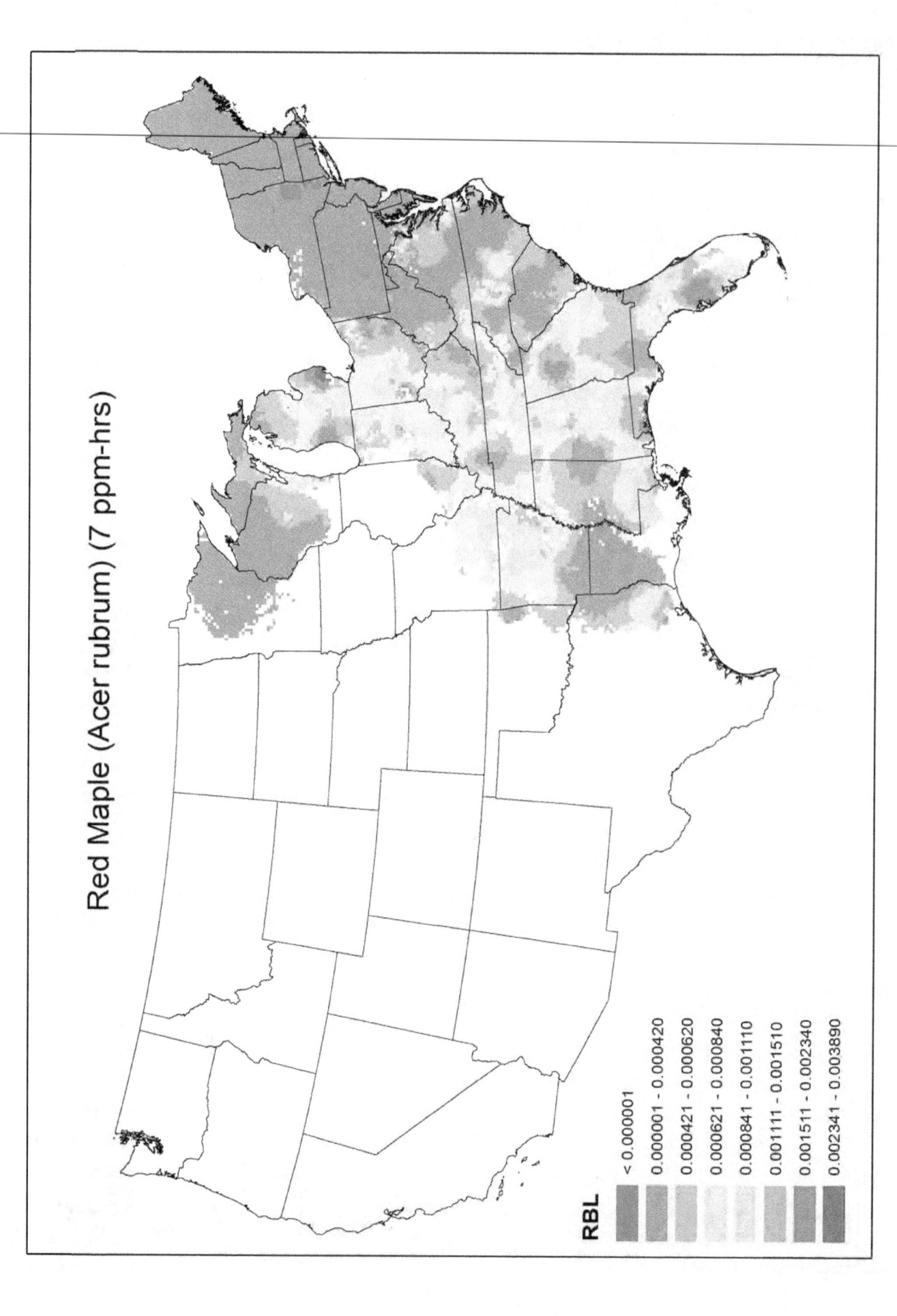

Red Maple (Acer rubrum) (7 ppm-hrs)

RBL

< 0.000001
0.000001 - 0.000420
0.000421 - 0.000620
0.000621 - 0.000840
0.000841 - 0.001110
0.001111 - 0.001510
0.001511 - 0.002340
0.002341 - 0.003890

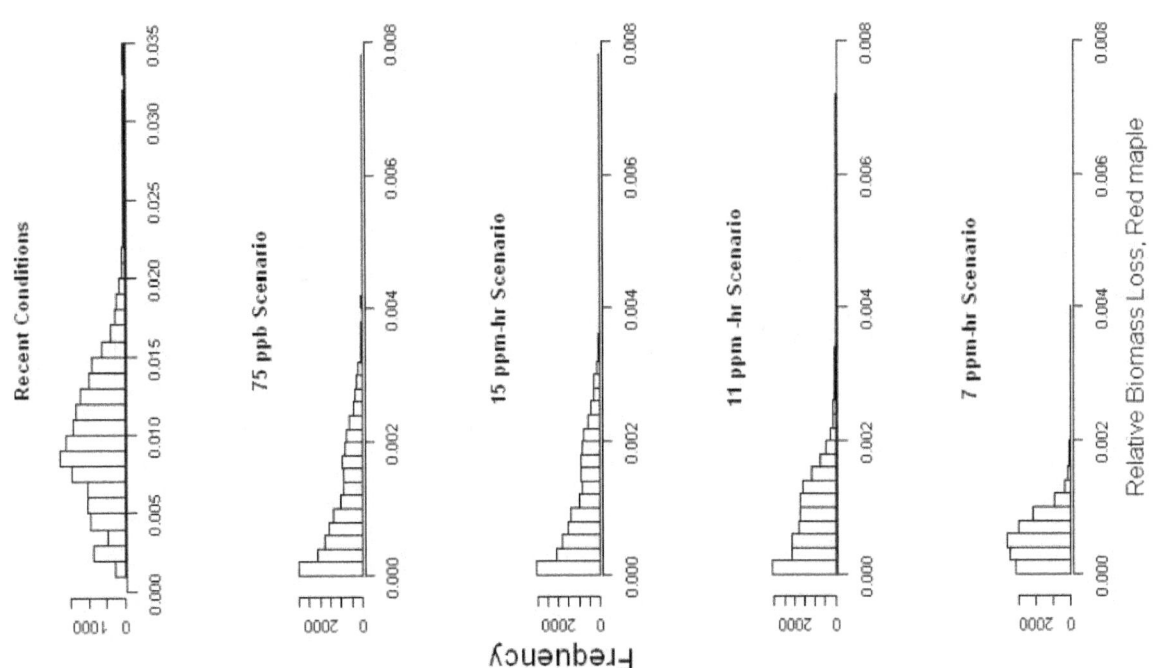

7

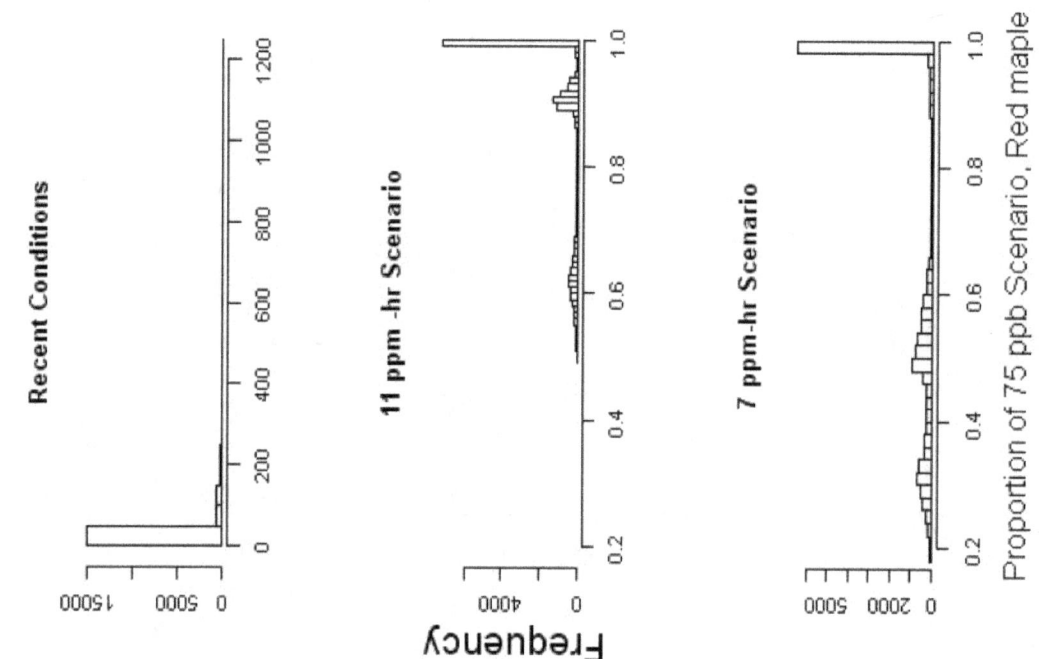

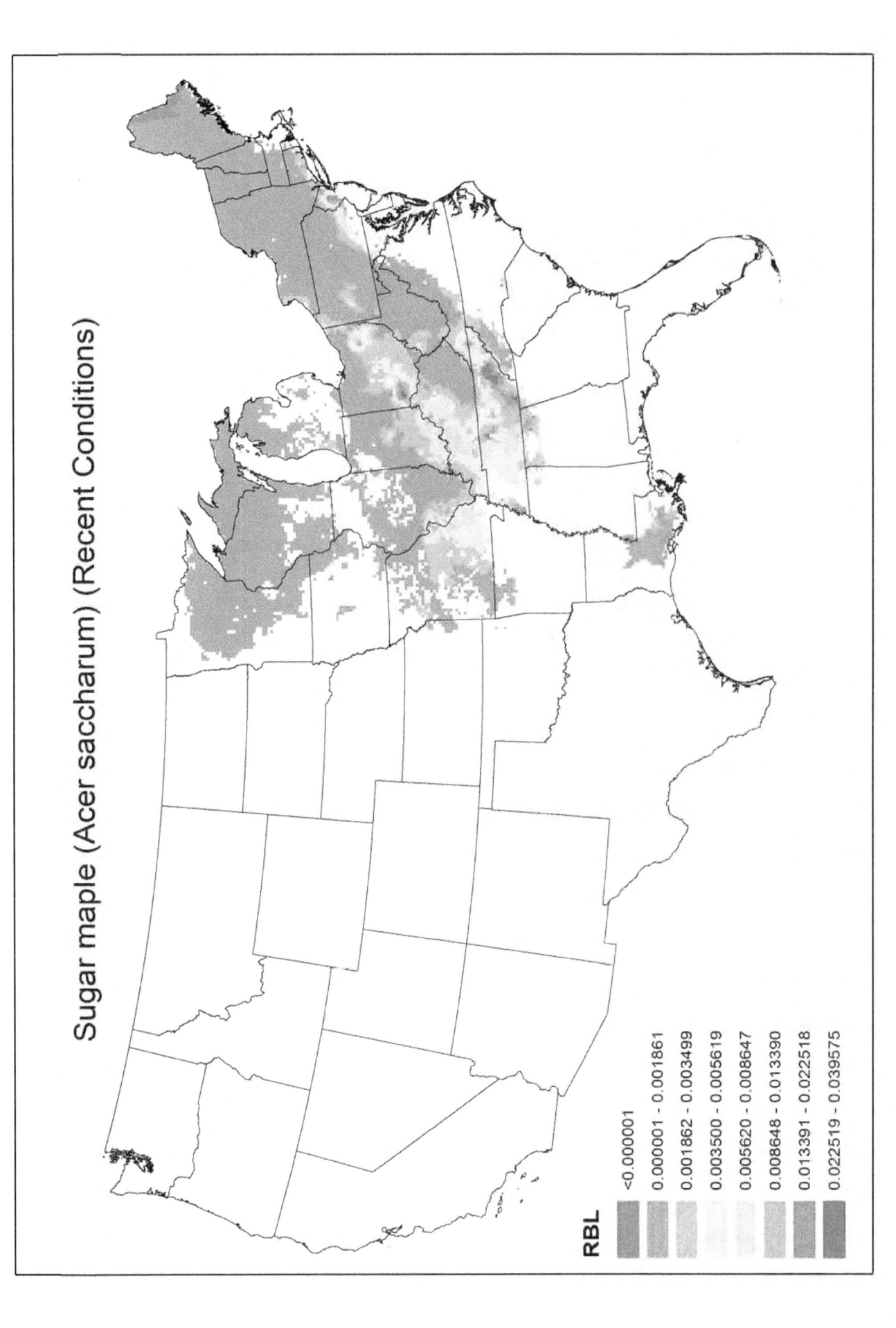

Sugar maple (Acer saccharum) (Recent Conditions)

RBL

- <0.000001
- 0.000001 - 0.001861
- 0.001862 - 0.003499
- 0.003500 - 0.005619
- 0.005620 - 0.008647
- 0.008648 - 0.013390
- 0.013391 - 0.022518
- 0.022519 - 0.039575

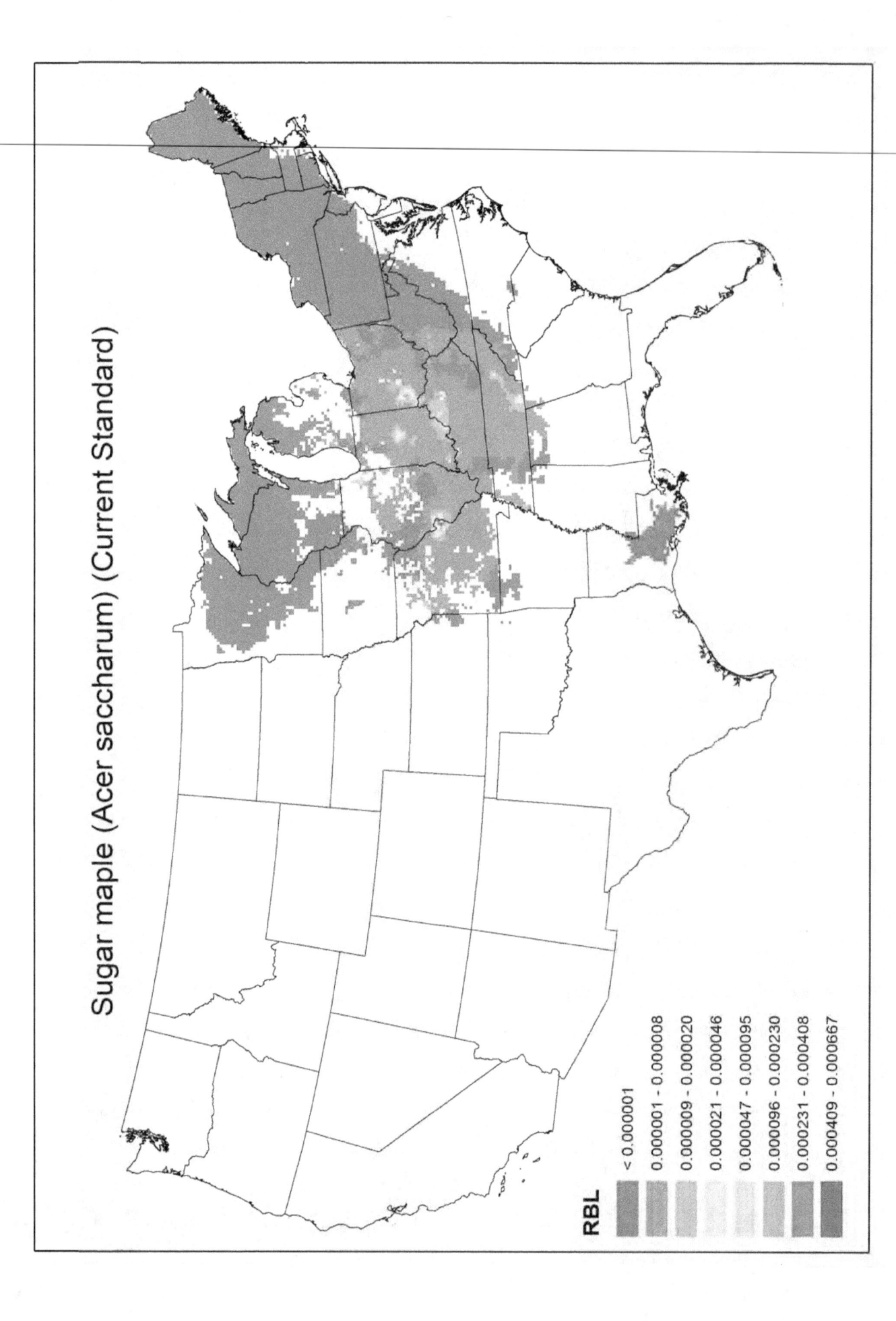

Sugar maple (Acer saccharum) (Current Standard)

RBL

< 0.000001
0.000001 - 0.000008
0.000009 - 0.000020
0.000021 - 0.000046
0.000047 - 0.000095
0.000096 - 0.000230
0.000231 - 0.000408
0.000409 - 0.000667

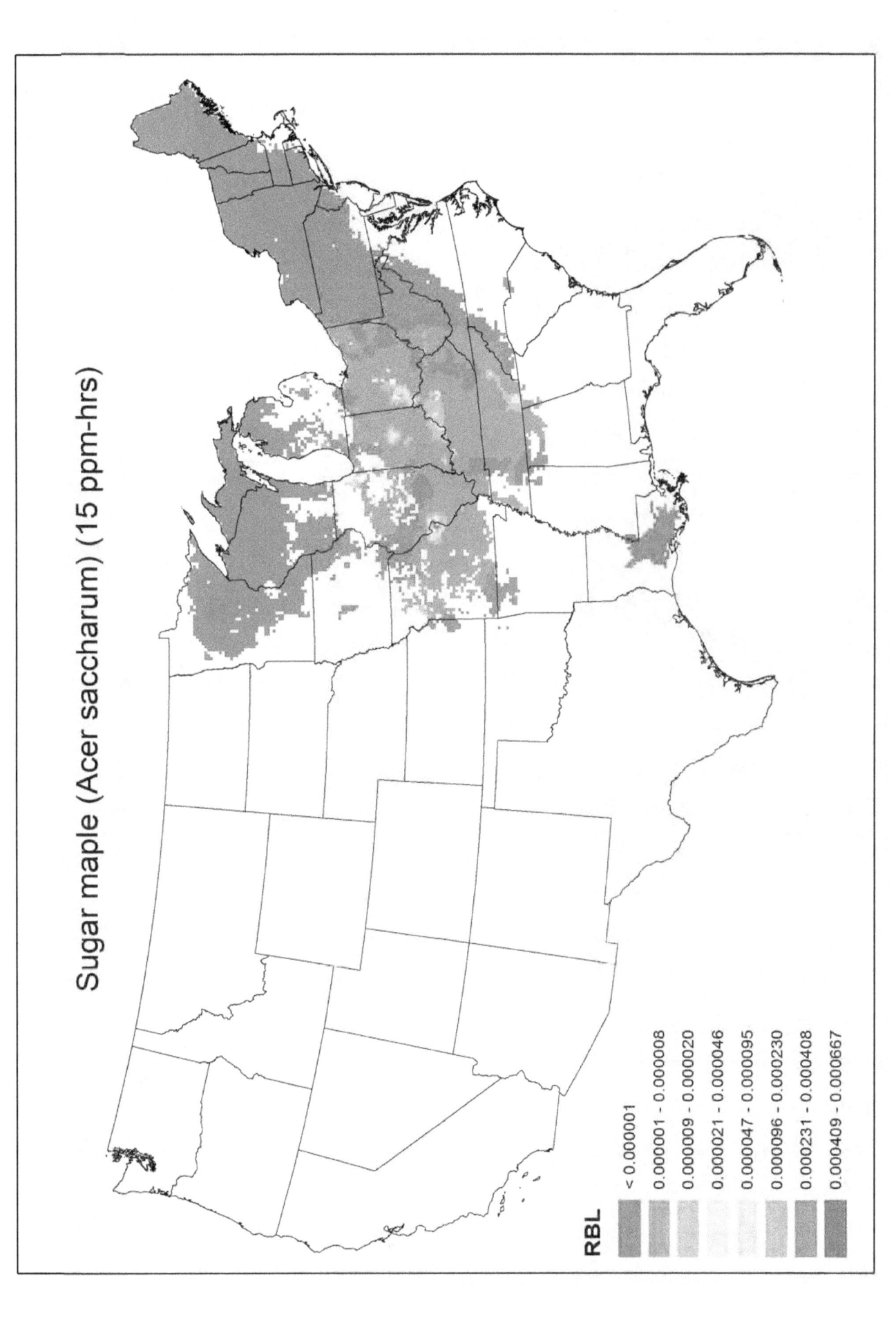

Sugar maple (Acer saccharum) (15 ppm-hrs)

RBL

< 0.000001
0.000001 - 0.000008
0.000009 - 0.000020
0.000021 - 0.000046
0.000047 - 0.000095
0.000096 - 0.000230
0.000231 - 0.000408
0.000409 - 0.000667

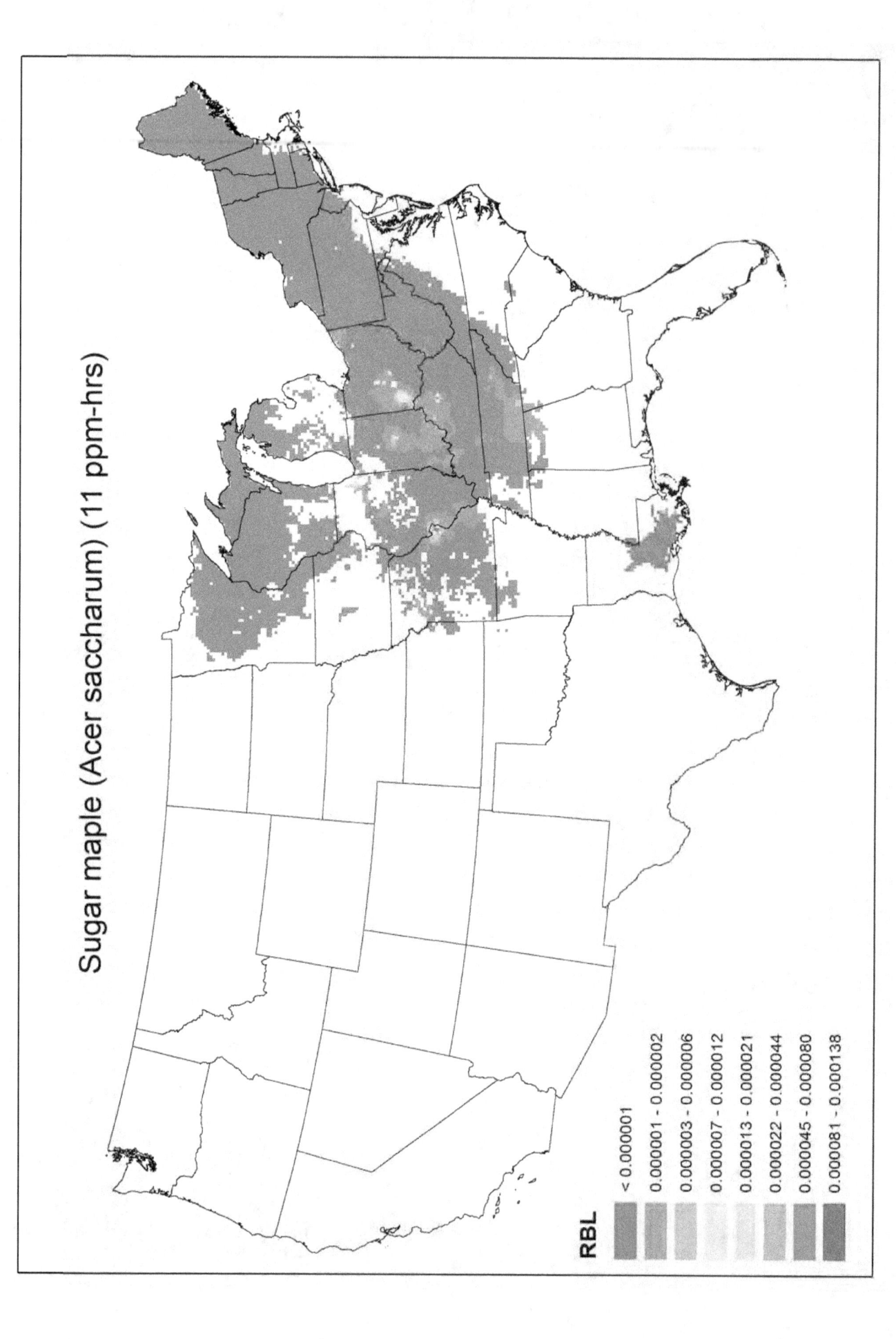

Sugar maple (Acer saccharum) (11 ppm-hrs)

RBL

< 0.000001	
0.000001 - 0.000002	
0.000003 - 0.000006	
0.000007 - 0.000012	
0.000013 - 0.000021	
0.000022 - 0.000044	
0.000045 - 0.000080	
0.000081 - 0.000138	

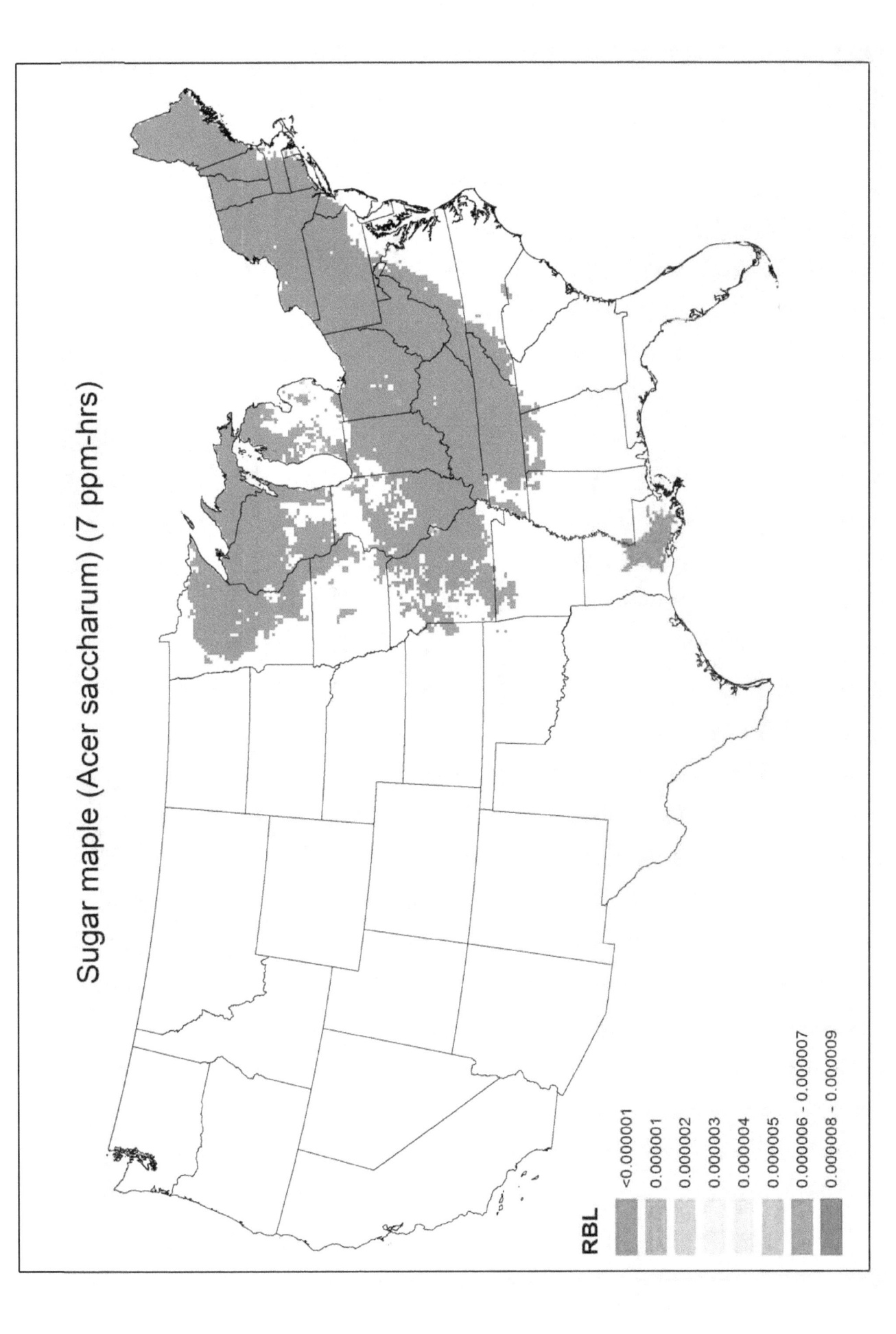

Sugar maple (Acer saccharum) (7 ppm-hrs)

RBL
<0.000001
0.000001
0.000002
0.000003
0.000004
0.000005
0.000006 - 0.000007
0.000008 - 0.000009

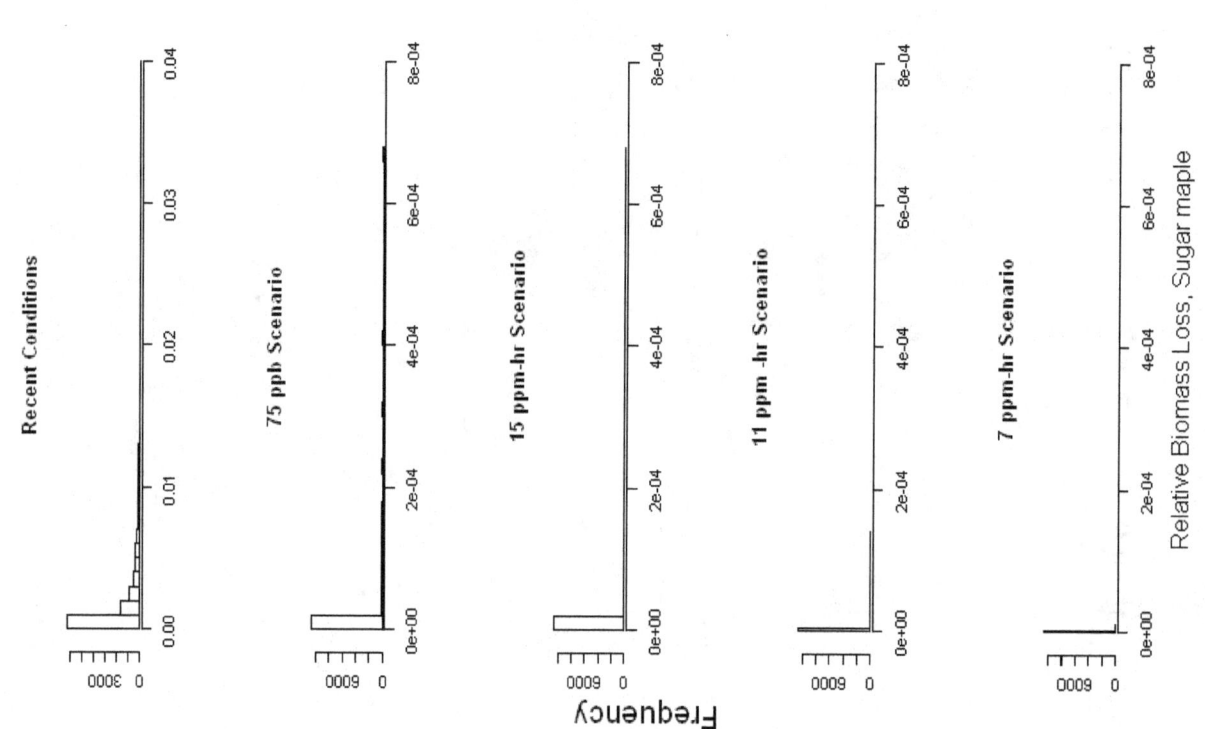

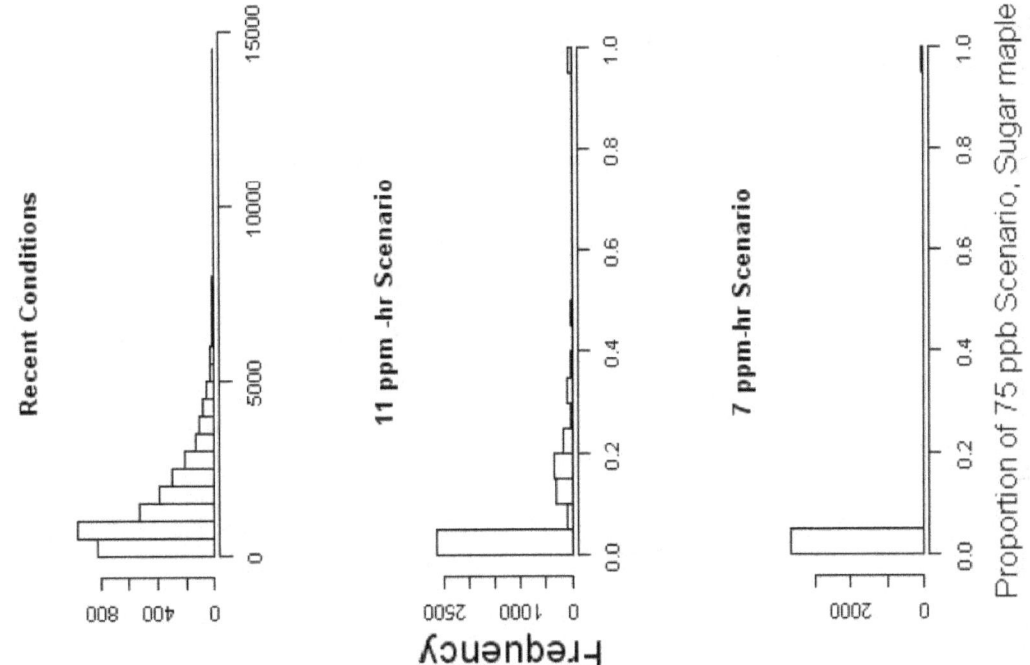

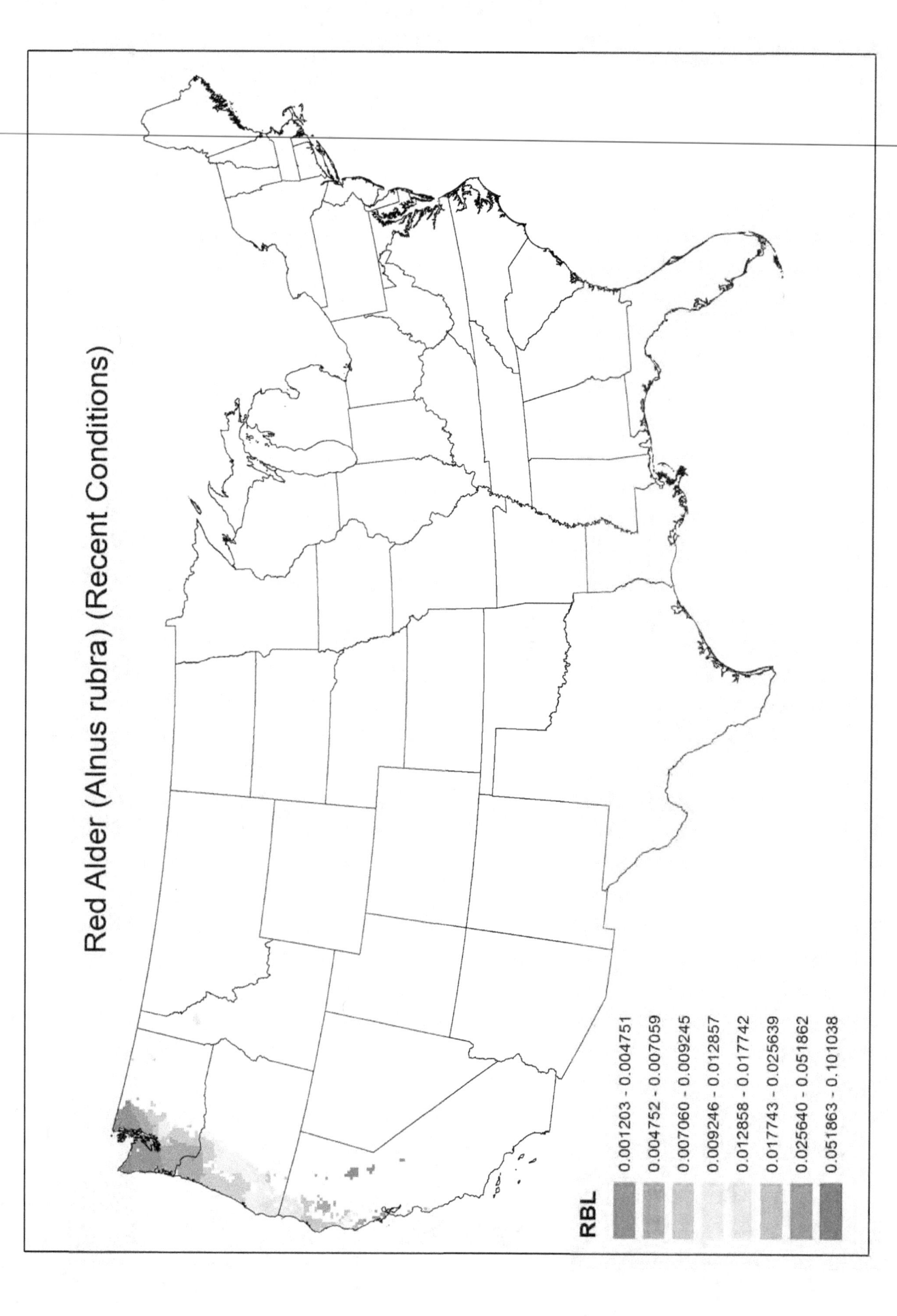

Red Alder (Alnus rubra) (Recent Conditions)

RBL

0.001203 - 0.004751
0.004752 - 0.007059
0.007060 - 0.009245
0.009246 - 0.012857
0.012858 - 0.017742
0.017743 - 0.025639
0.025640 - 0.051862
0.051863 - 0.101038

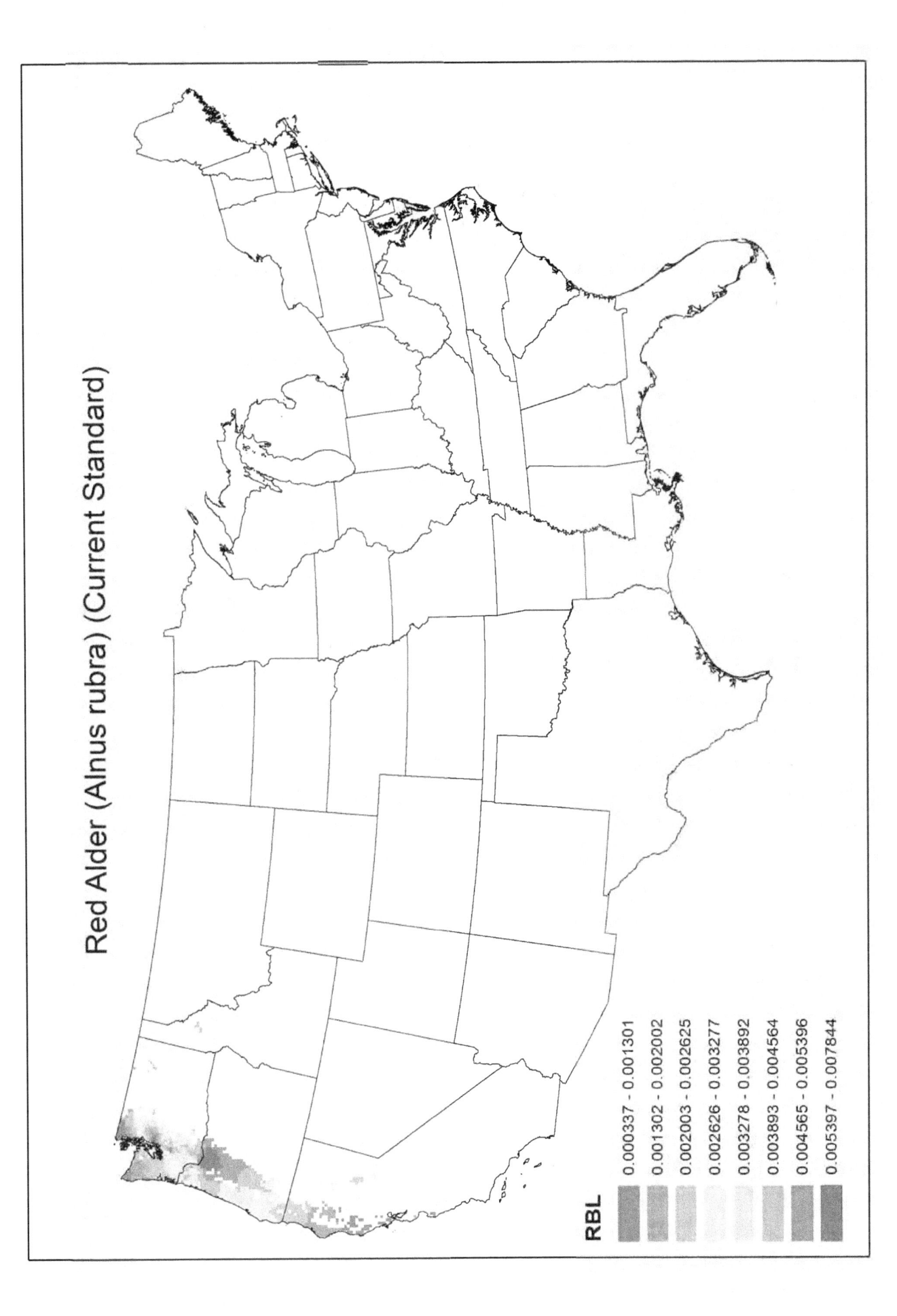

Red Alder (Alnus rubra) (Current Standard)

RBL

0.000337 - 0.001301
0.001302 - 0.002002
0.002003 - 0.002625
0.002626 - 0.003277
0.003278 - 0.003892
0.003893 - 0.004564
0.004565 - 0.005396
0.005397 - 0.007844

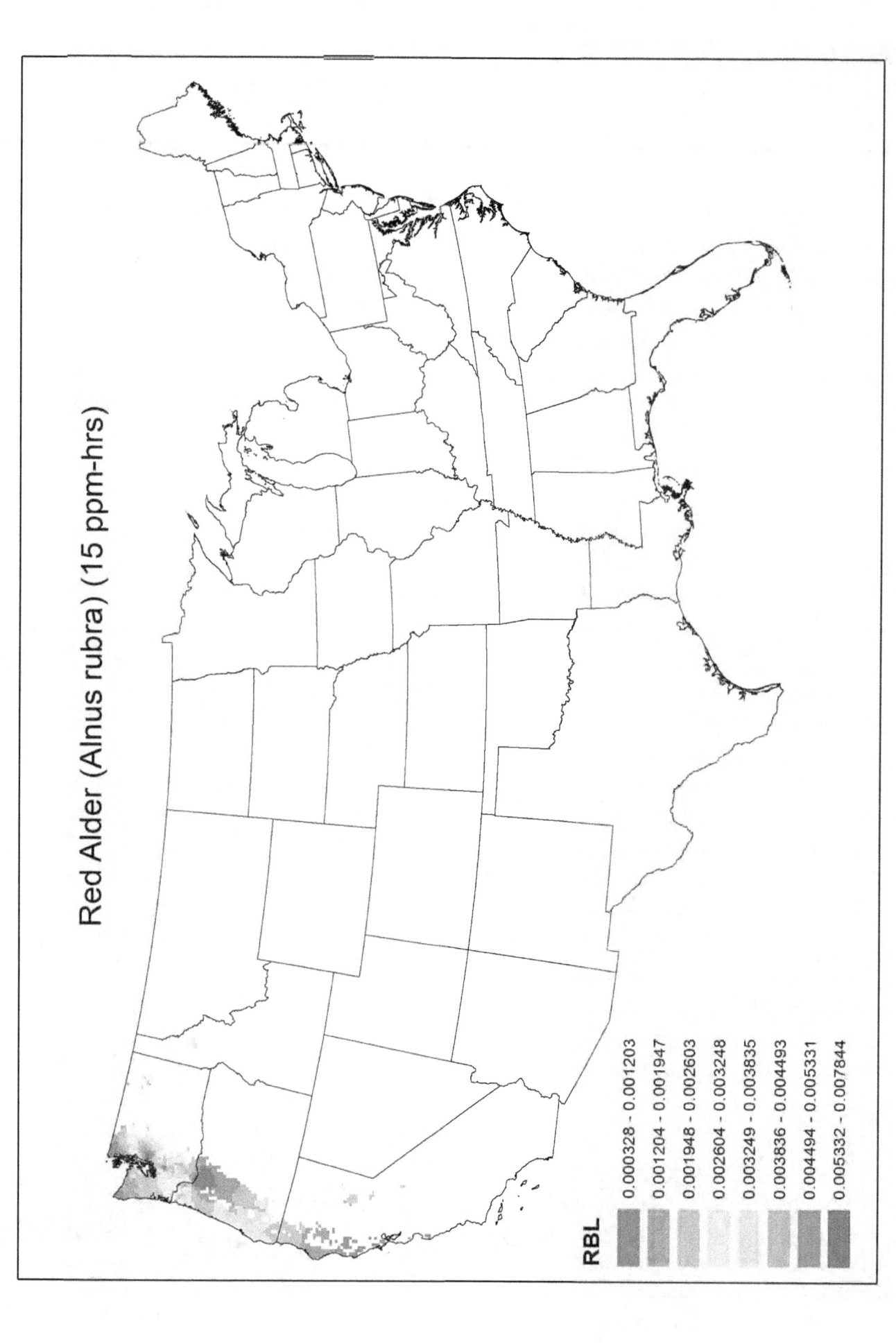

Red Alder (Alnus rubra) (15 ppm-hrs)

RBL

0.000328 - 0.001203
0.001204 - 0.001947
0.001948 - 0.002603
0.002604 - 0.003248
0.003249 - 0.003835
0.003836 - 0.004493
0.004494 - 0.005331
0.005332 - 0.007844

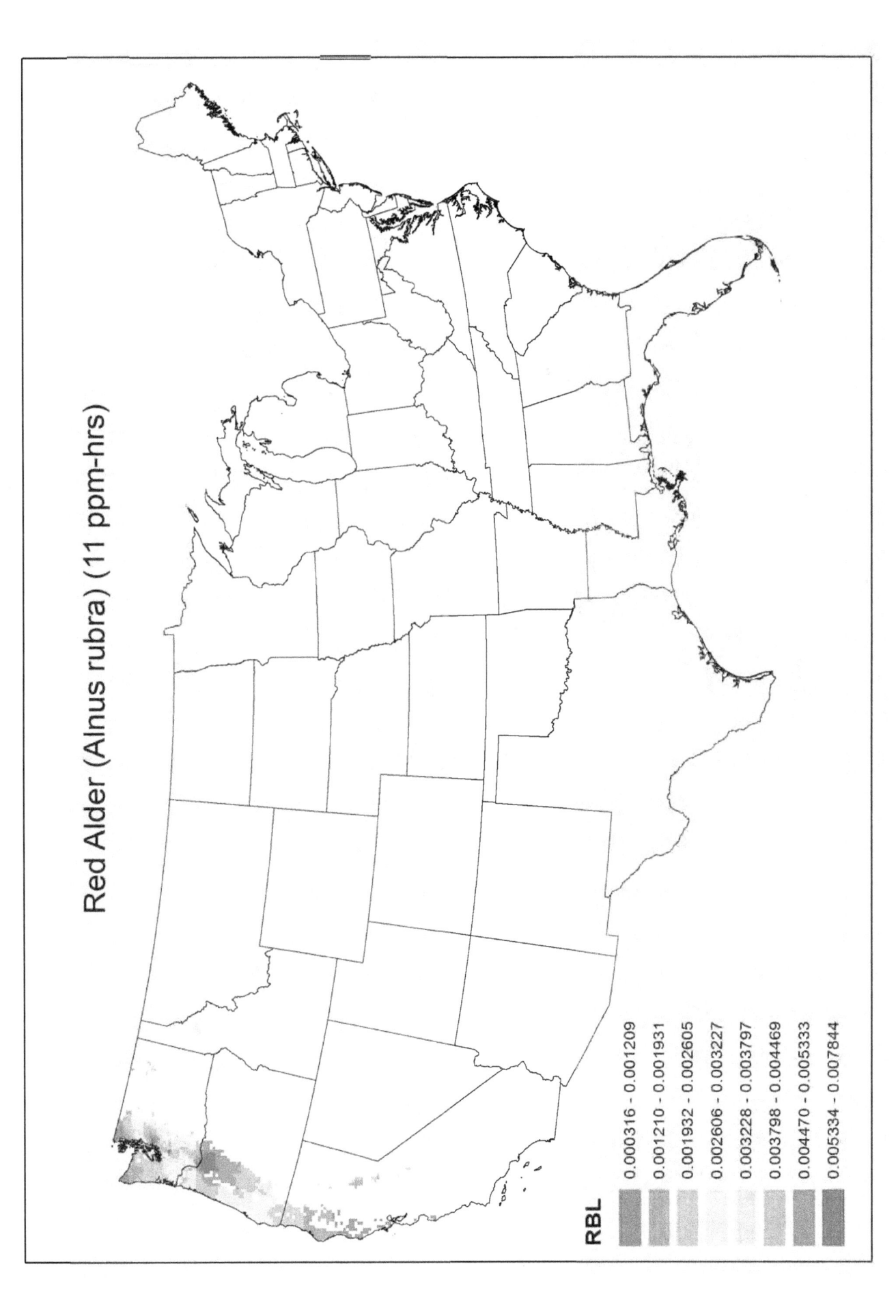

Red Alder (Alnus rubra) (11 ppm-hrs)

RBL

	0.000316 - 0.001209
	0.001210 - 0.001931
	0.001932 - 0.002605
	0.002606 - 0.003227
	0.003228 - 0.003797
	0.003798 - 0.004469
	0.004470 - 0.005333
	0.005334 - 0.007844

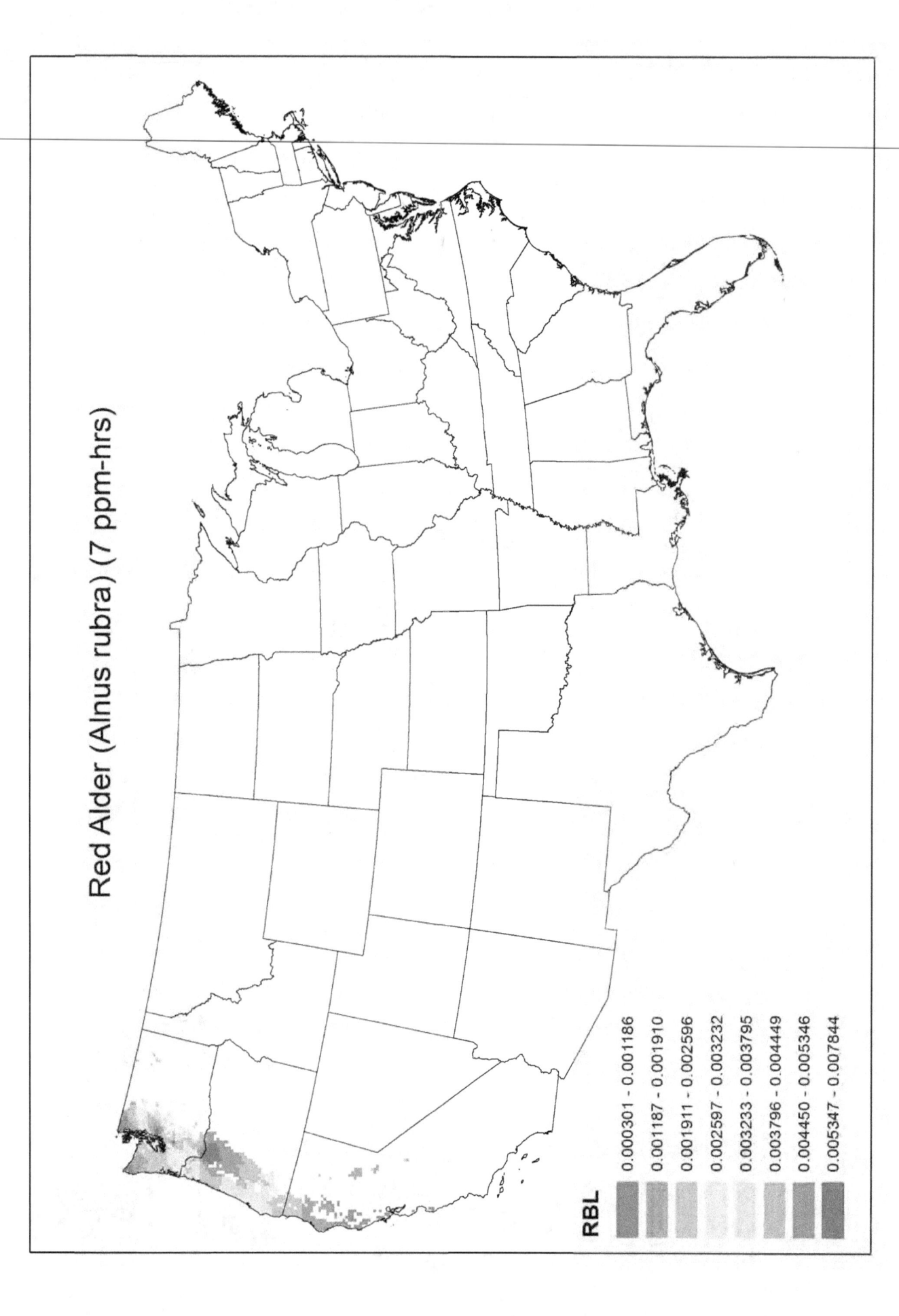

Red Alder (Alnus rubra) (7 ppm-hrs)

RBL
0.000301 - 0.001186
0.001187 - 0.001910
0.001911 - 0.002596
0.002597 - 0.003232
0.003233 - 0.003795
0.003796 - 0.004449
0.004450 - 0.005346
0.005347 - 0.007844

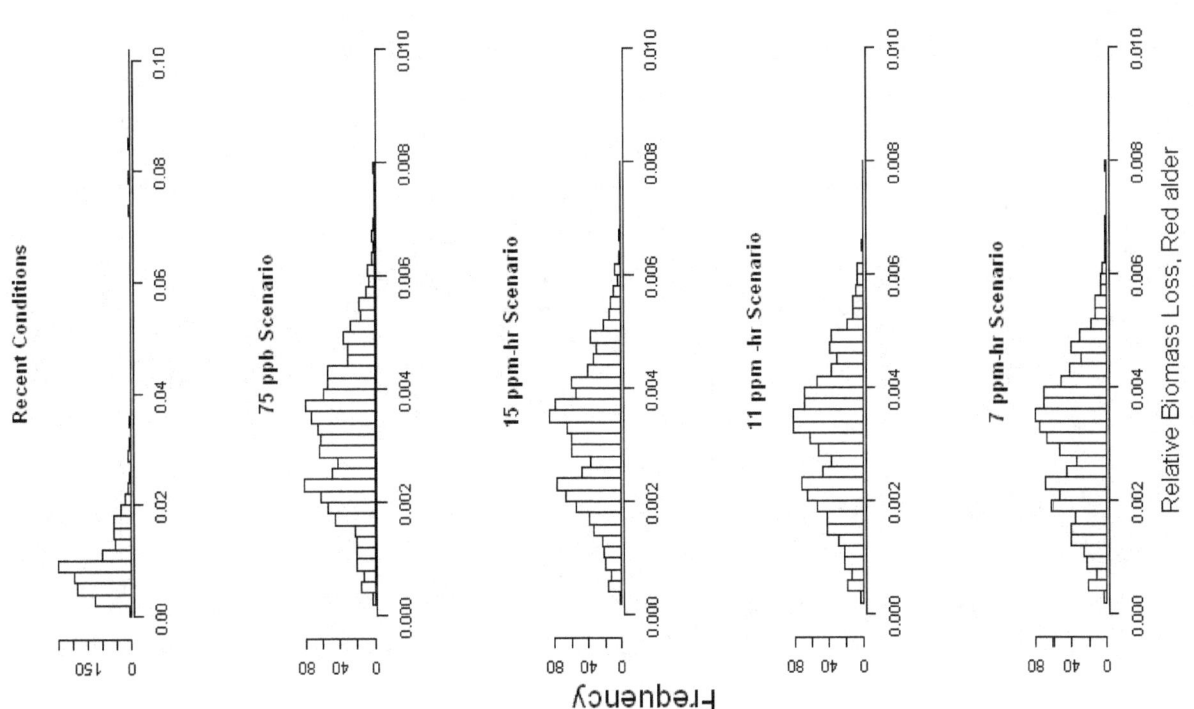

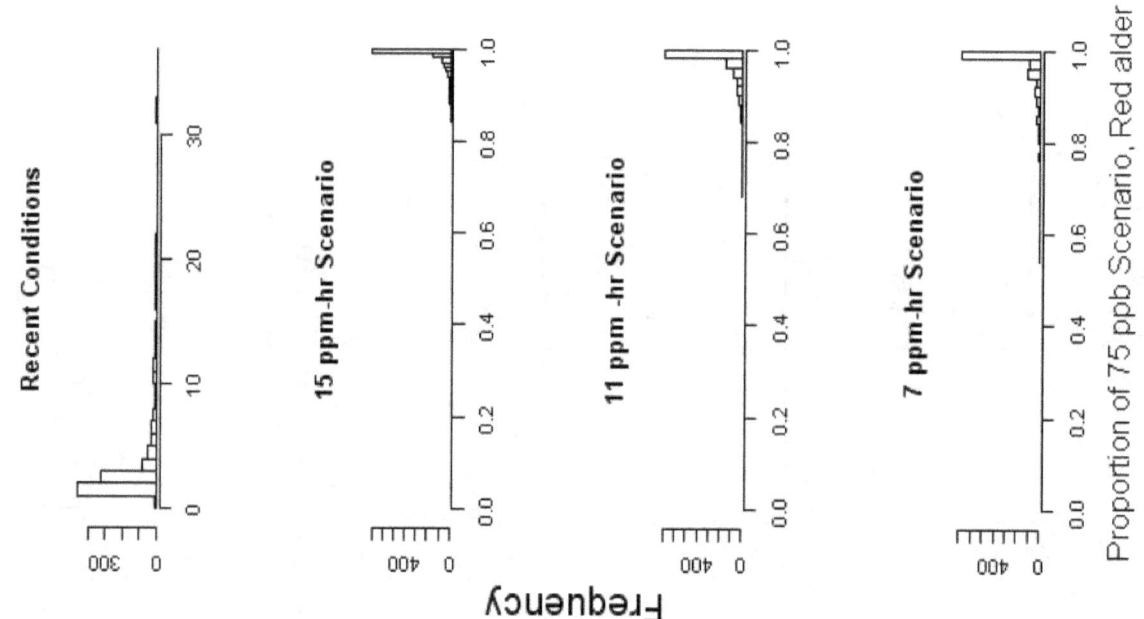

Proportion of 75 ppb Scenario, Red alder

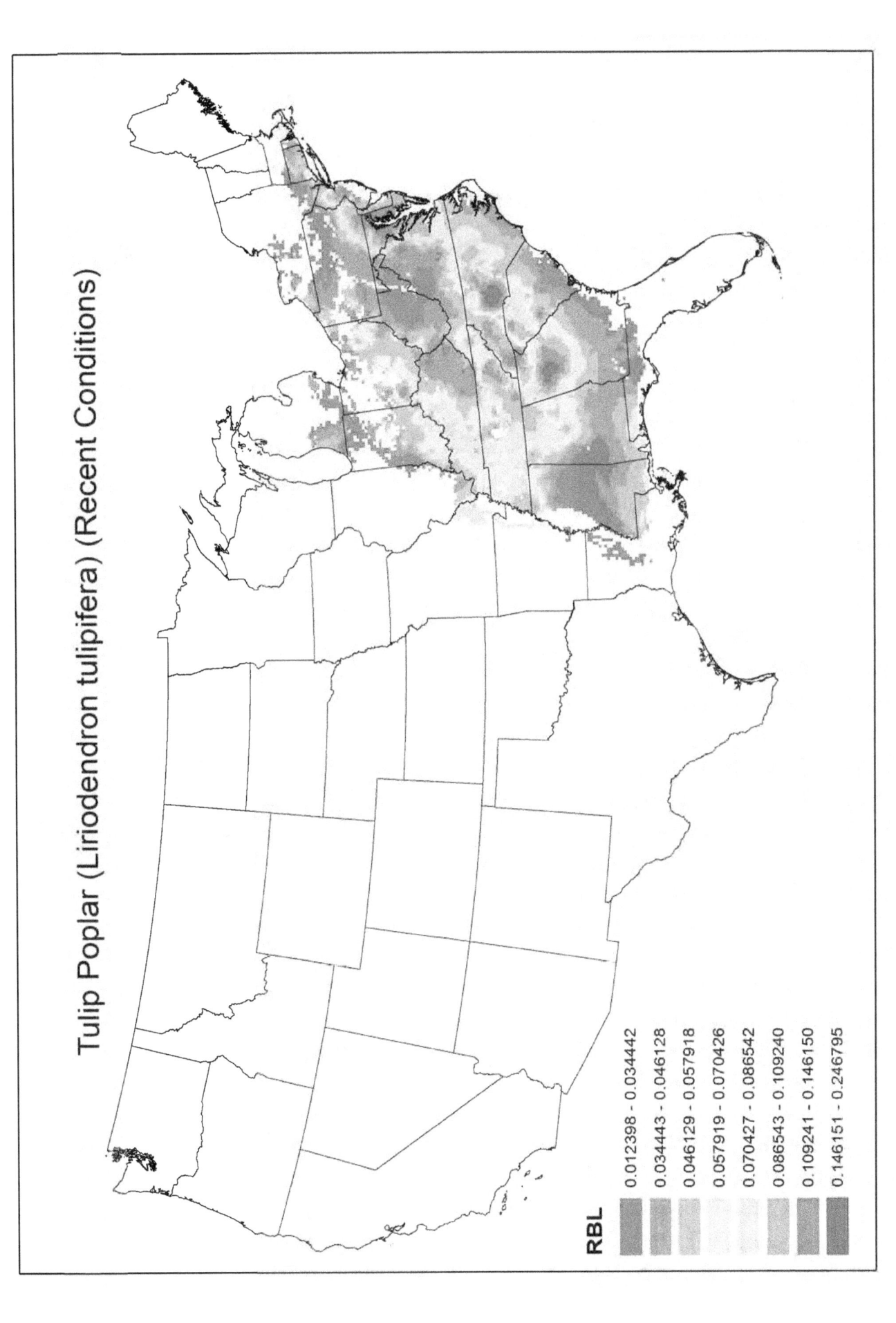

Tulip Poplar (Liriodendron tulipifera) (Recent Conditions)

RBL

0.012398 - 0.034442
0.034443 - 0.046128
0.046129 - 0.057918
0.057919 - 0.070426
0.070427 - 0.086542
0.086543 - 0.109240
0.109241 - 0.146150
0.146151 - 0.246795

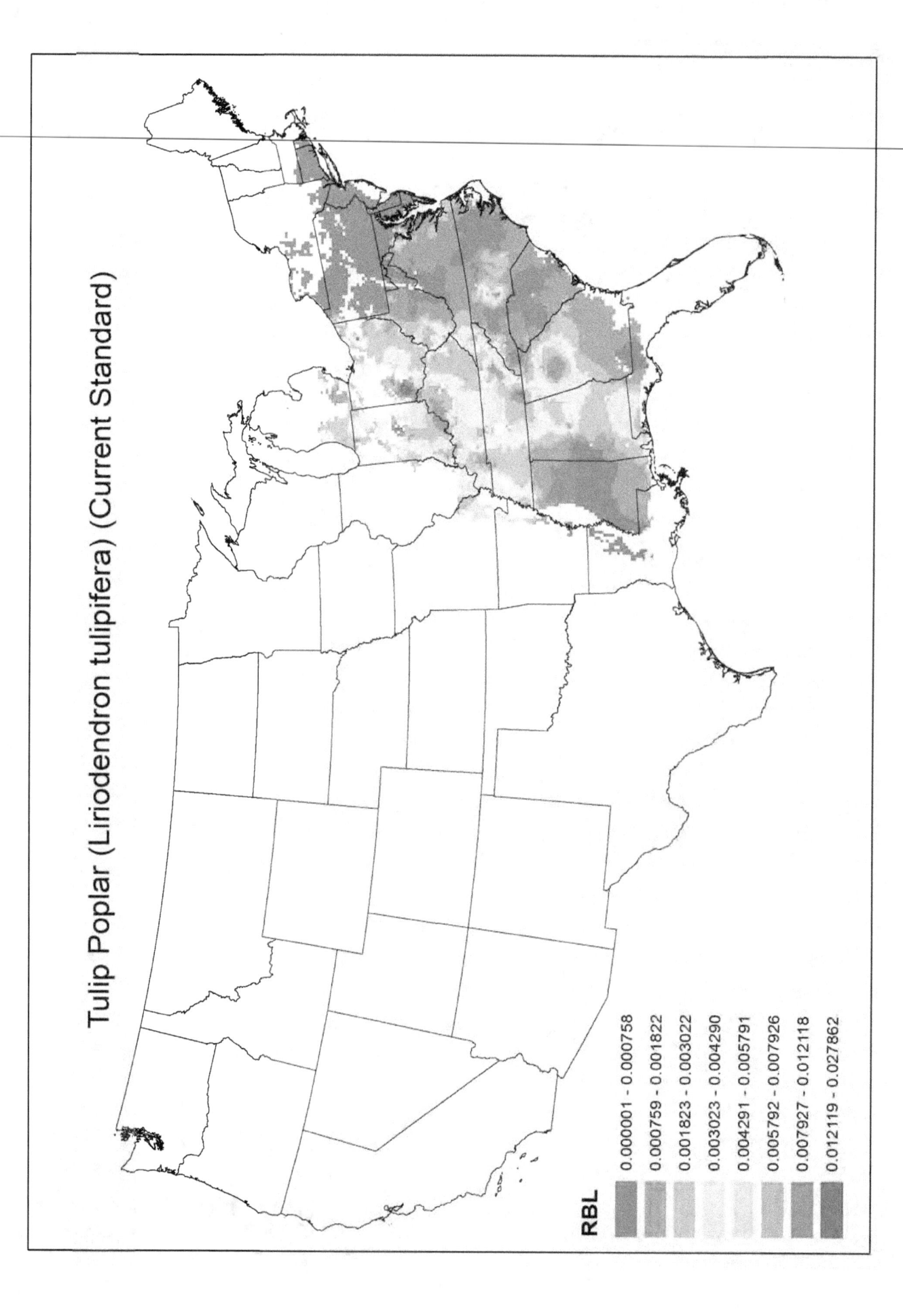

Tulip Poplar (Liriodendron tulipifera) (Current Standard)

RBL

- 0.000001 - 0.000758
- 0.000759 - 0.001822
- 0.001823 - 0.003022
- 0.003023 - 0.004290
- 0.004291 - 0.005791
- 0.005792 - 0.007926
- 0.007927 - 0.012118
- 0.012119 - 0.027862

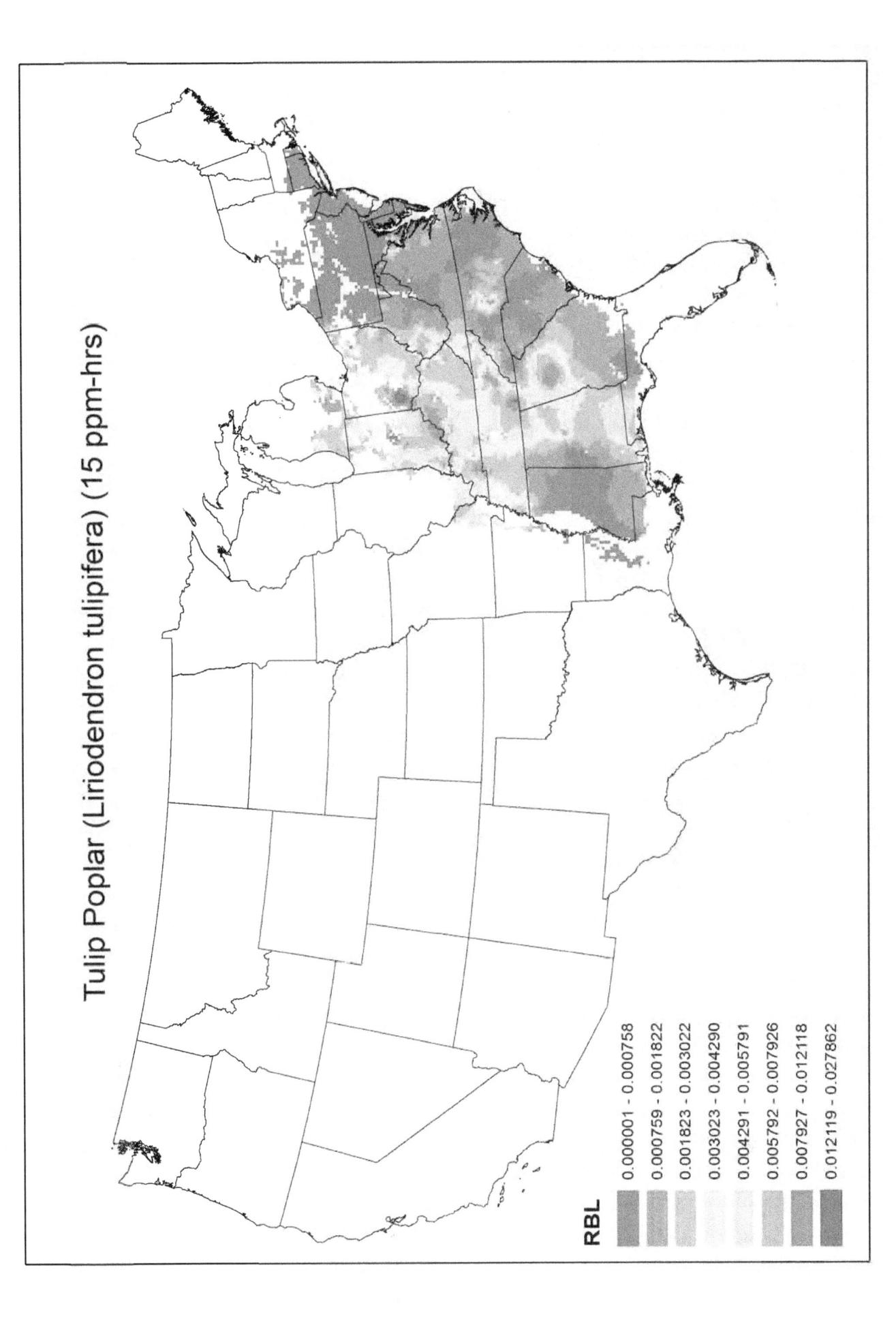

Tulip Poplar (Liriodendron tulipifera) (15 ppm-hrs)

RBL

0.000001 - 0.000758
0.000759 - 0.001822
0.001823 - 0.003022
0.003023 - 0.004290
0.004291 - 0.005791
0.005792 - 0.007926
0.007927 - 0.012118
0.012119 - 0.027862

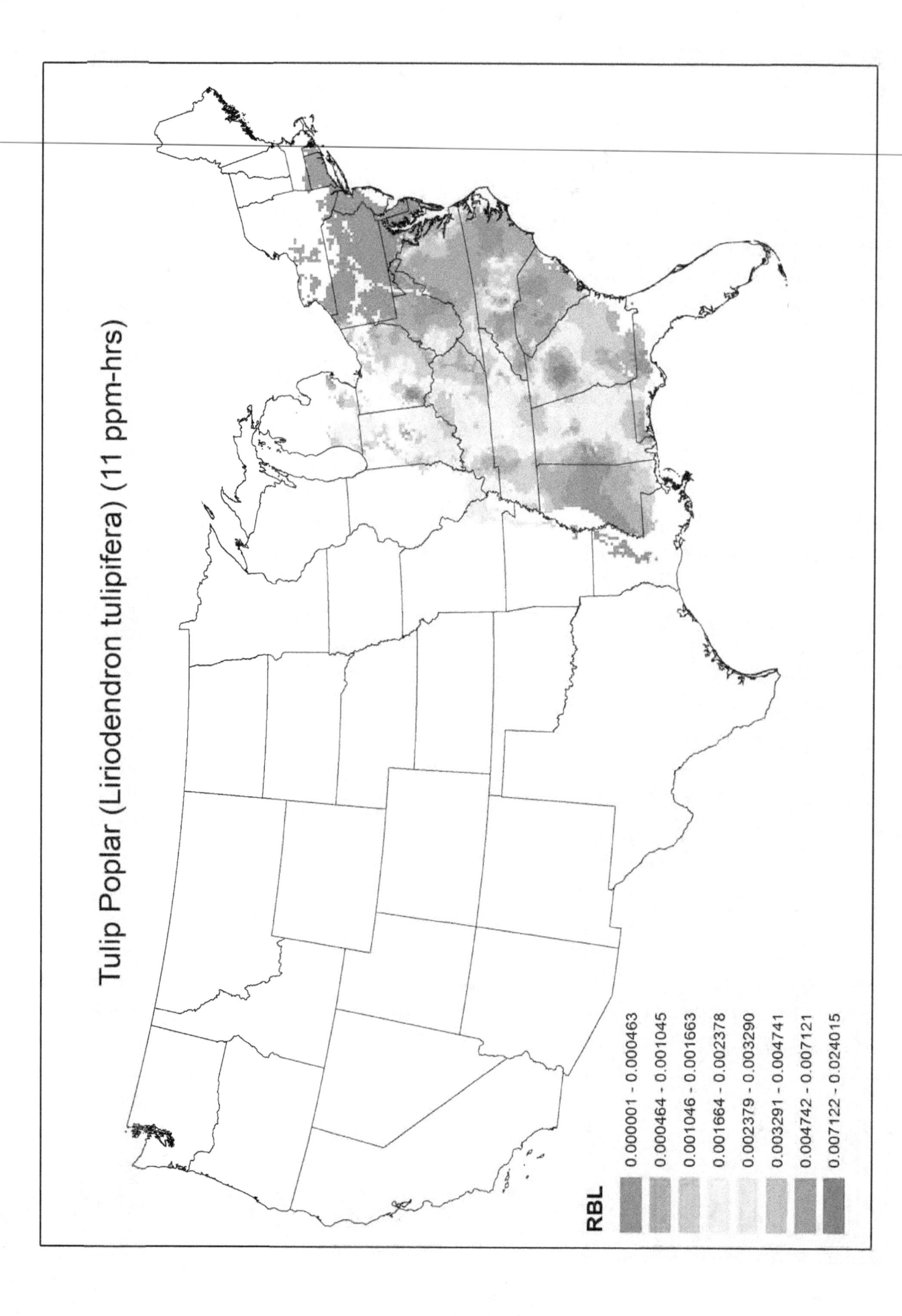

Tulip Poplar (Liriodendron tulipifera) (11 ppm-hrs)

RBL

0.000001 - 0.000463
0.000464 - 0.001045
0.001046 - 0.001663
0.001664 - 0.002378
0.002379 - 0.003290
0.003291 - 0.004741
0.004742 - 0.007121
0.007122 - 0.024015

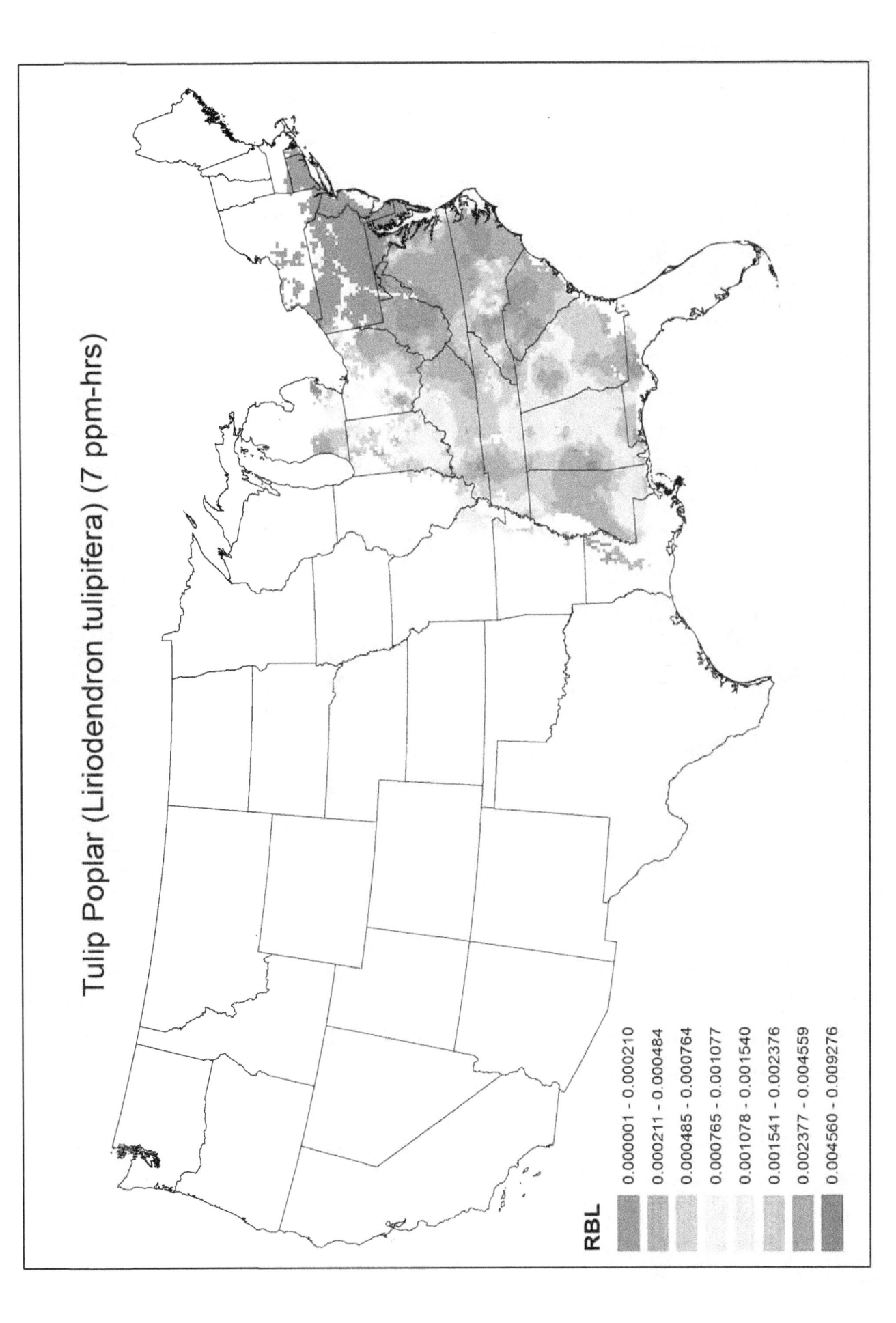

Tulip Poplar (Liriodendron tulipifera) (7 ppm-hrs)

RBL

0.000001 - 0.000210
0.000211 - 0.000484
0.000485 - 0.000764
0.000765 - 0.001077
0.001078 - 0.001540
0.001541 - 0.002376
0.002377 - 0.004559
0.004560 - 0.009276

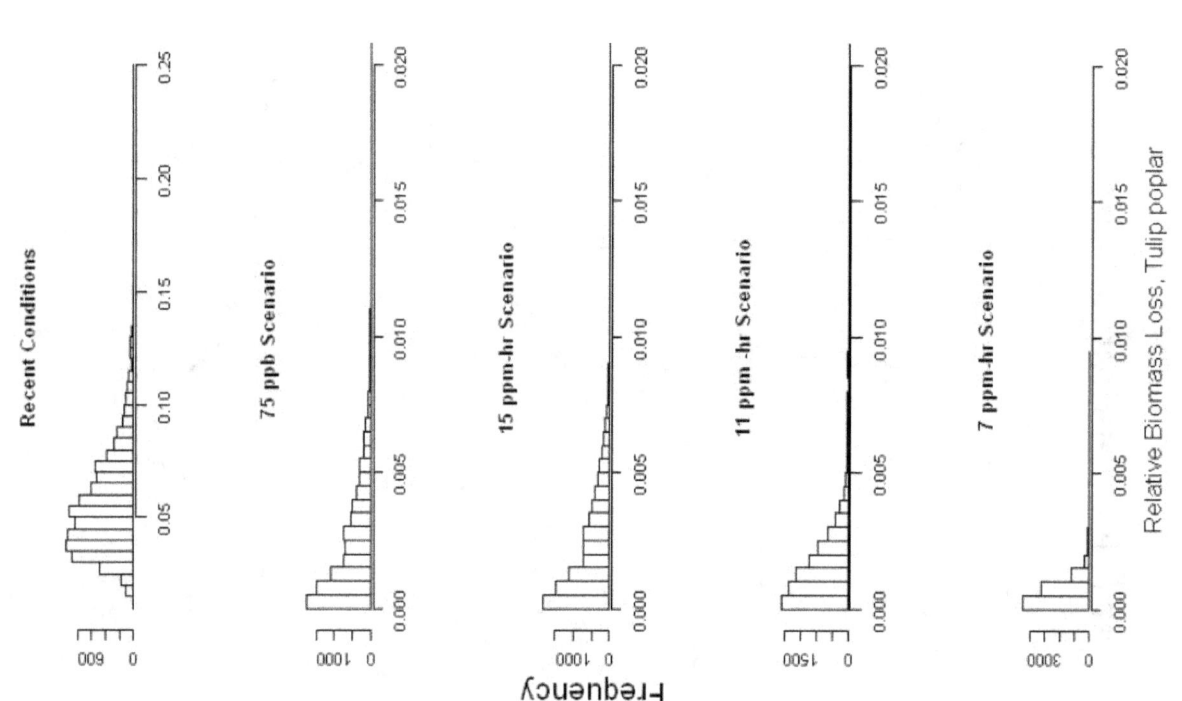

Relative Biomass Loss, Tulip poplar

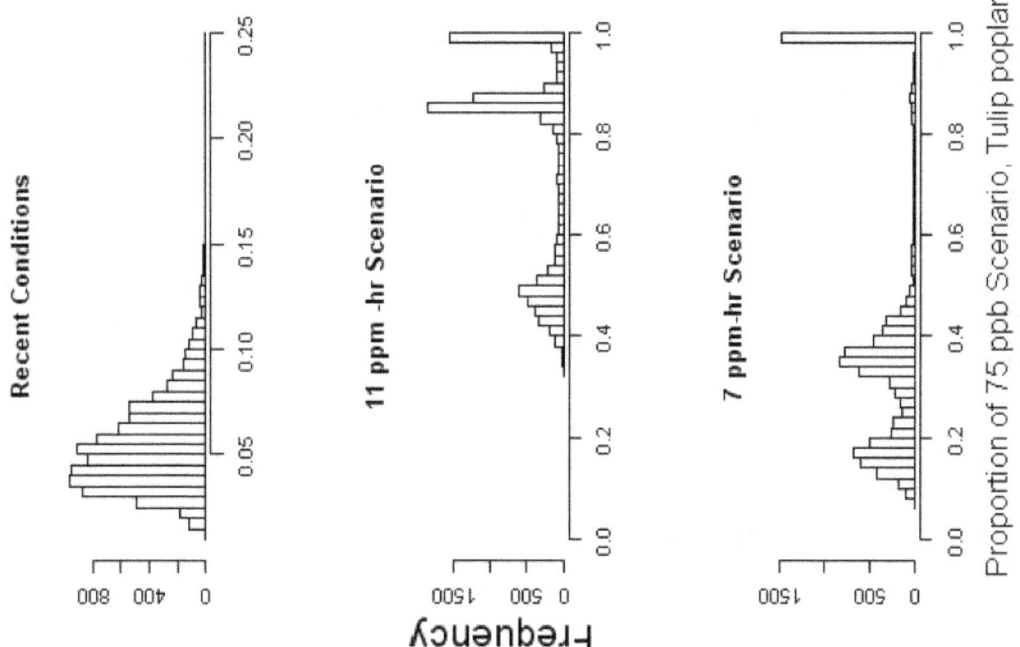

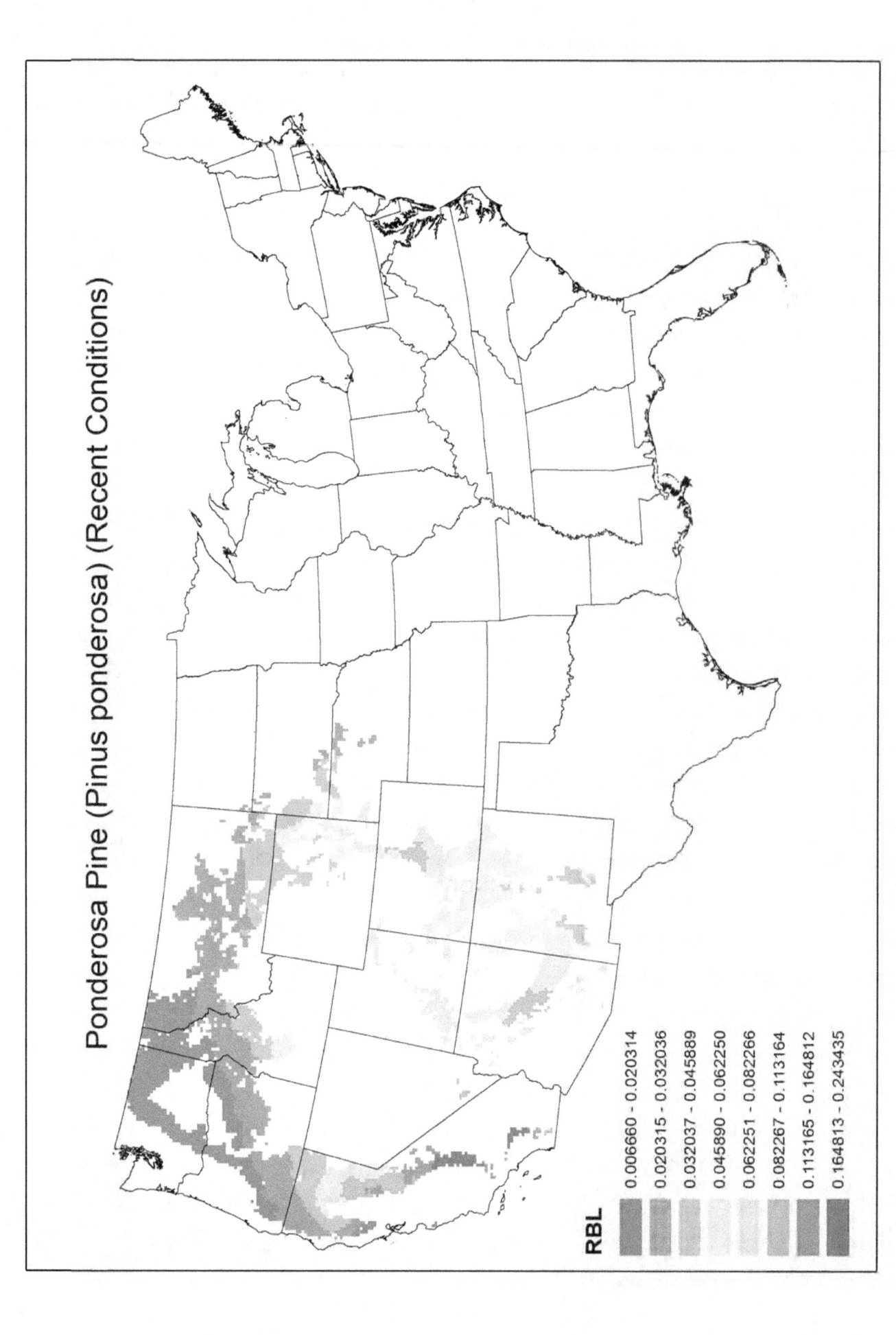

Ponderosa Pine (Pinus ponderosa) (Recent Conditions)

RBL
0.006660 - 0.020314
0.020315 - 0.032036
0.032037 - 0.045889
0.045890 - 0.062250
0.062251 - 0.082266
0.082267 - 0.113164
0.113165 - 0.164812
0.164813 - 0.243435

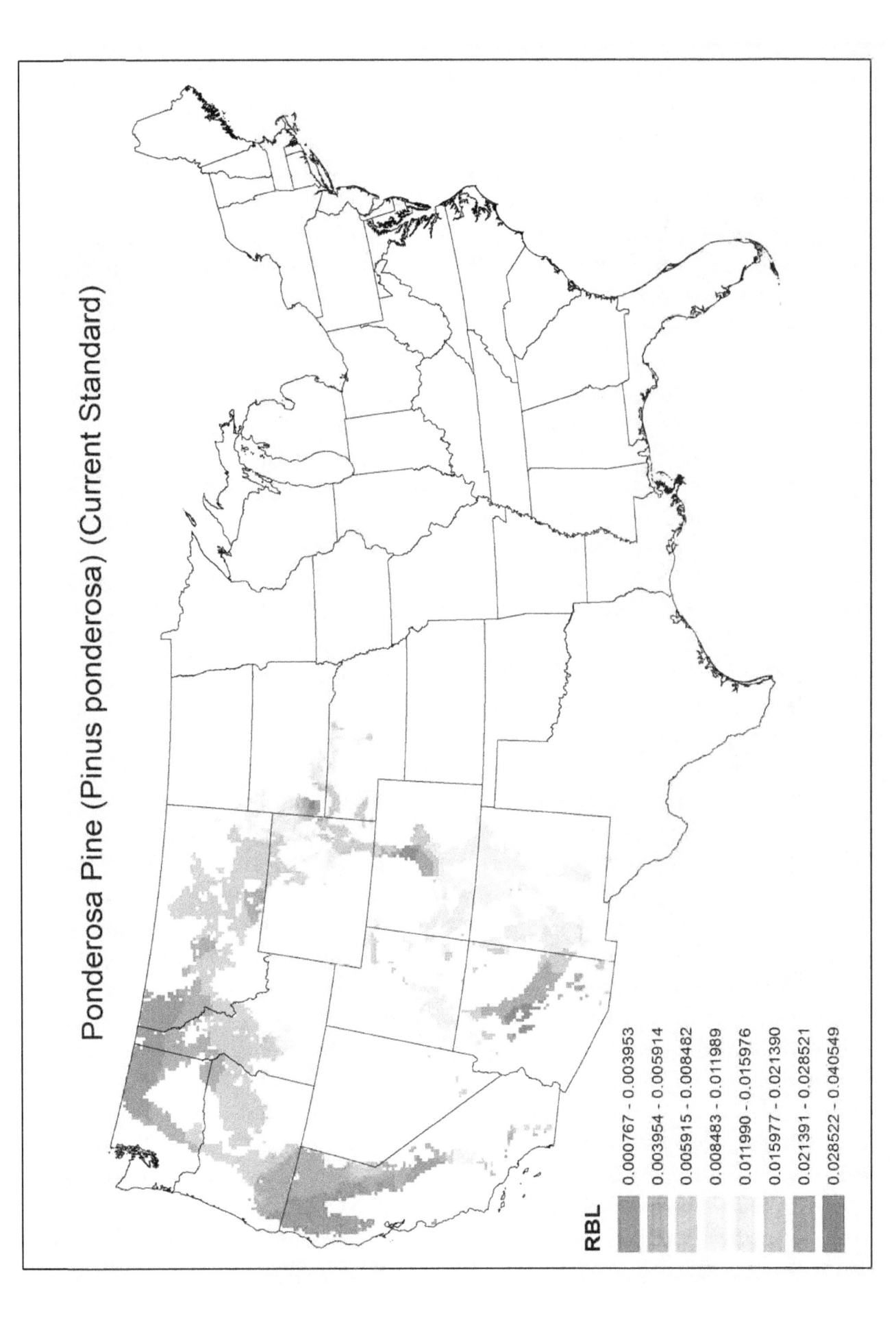

Ponderosa Pine (Pinus ponderosa) (Current Standard)

RBL

0.000767 - 0.003953
0.003954 - 0.005914
0.005915 - 0.008482
0.008483 - 0.011989
0.011990 - 0.015976
0.015977 - 0.021390
0.021391 - 0.028521
0.028522 - 0.040549

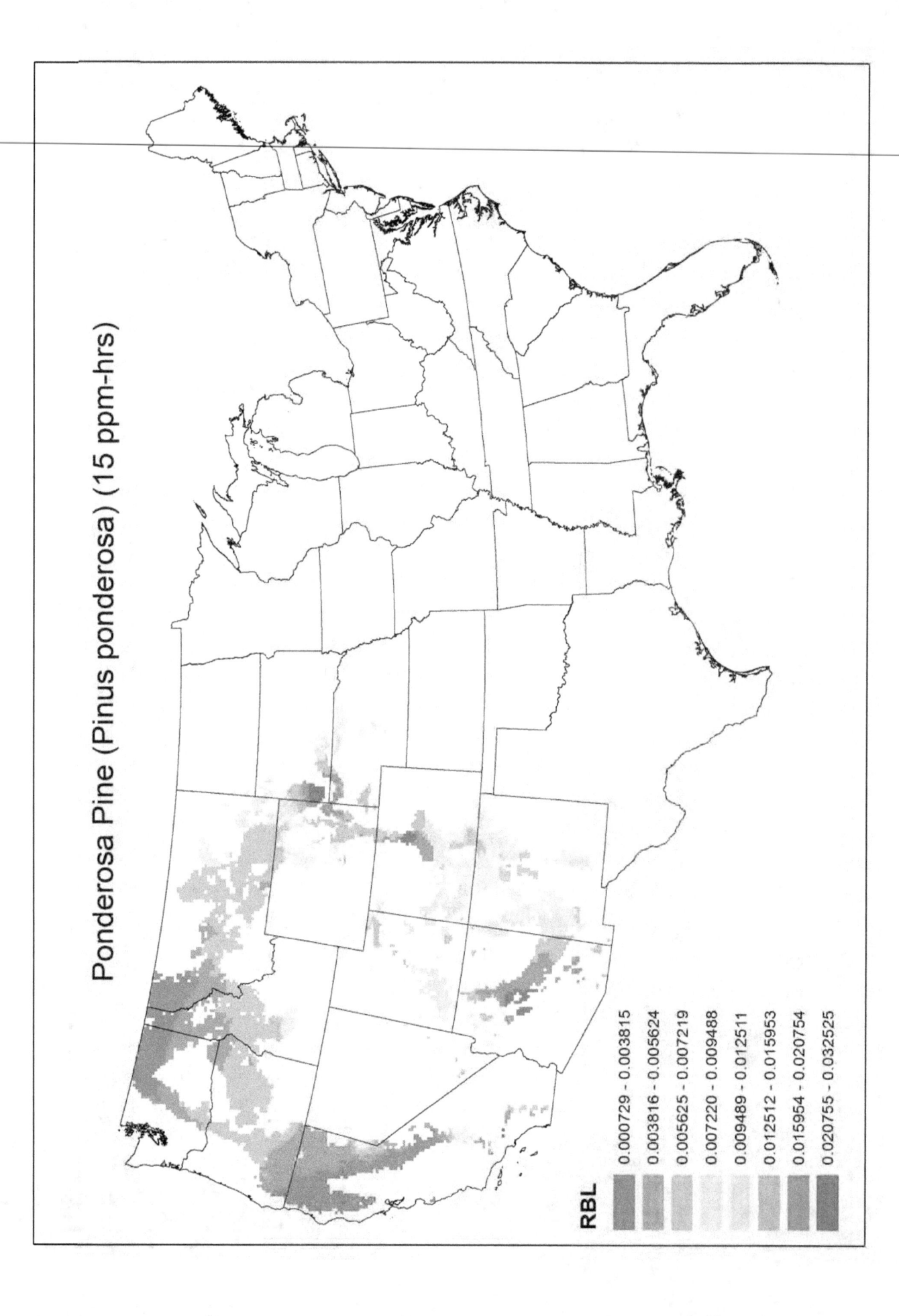

Ponderosa Pine (Pinus ponderosa) (15 ppm-hrs)

RBL

- 0.000729 - 0.003815
- 0.003816 - 0.005624
- 0.005625 - 0.007219
- 0.007220 - 0.009488
- 0.009489 - 0.012511
- 0.012512 - 0.015953
- 0.015954 - 0.020754
- 0.020755 - 0.032525

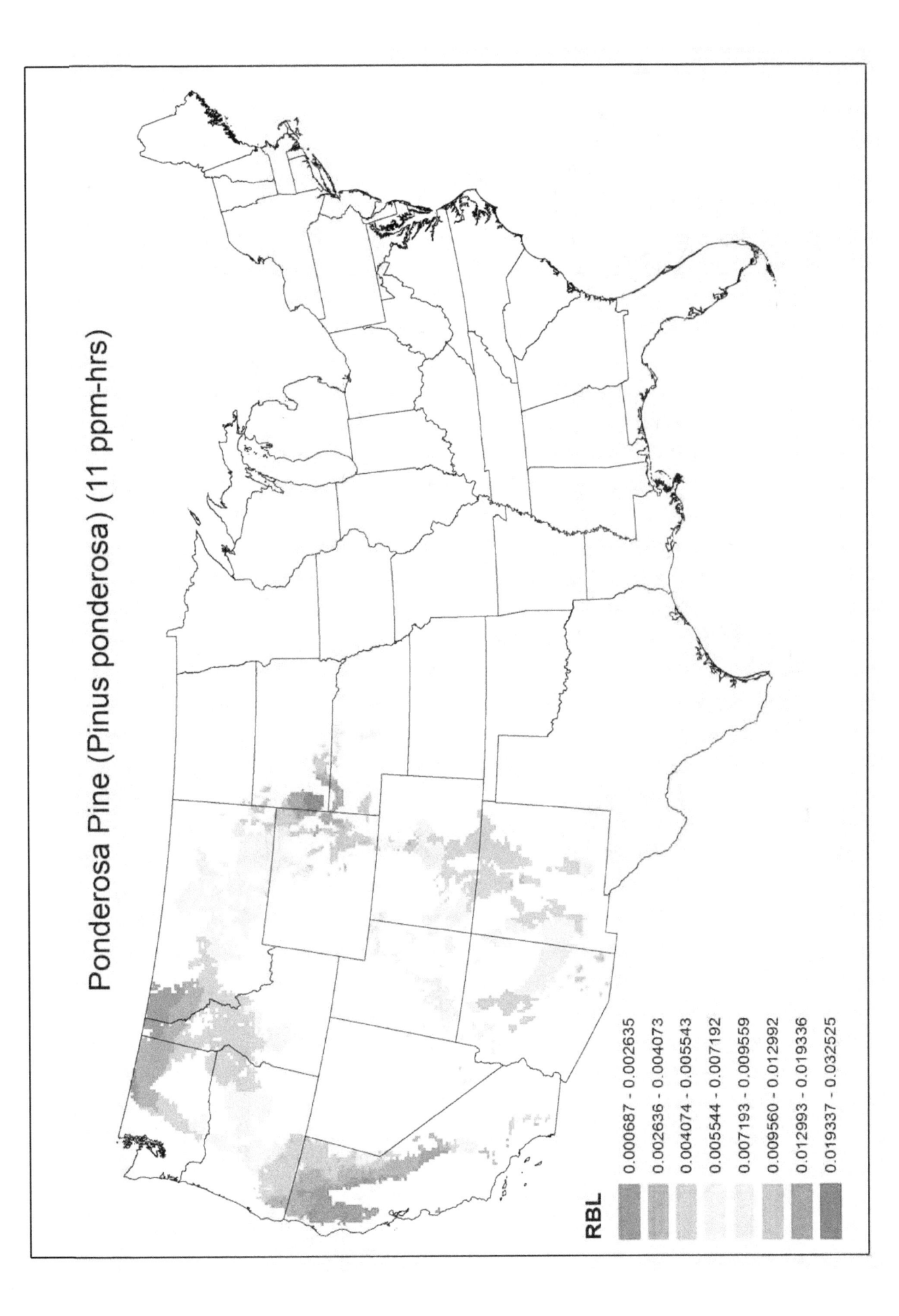

Ponderosa Pine (Pinus ponderosa) (11 ppm-hrs)

RBL

- 0.000687 - 0.002635
- 0.002636 - 0.004073
- 0.004074 - 0.005543
- 0.005544 - 0.007192
- 0.007193 - 0.009559
- 0.009560 - 0.012992
- 0.012993 - 0.019336
- 0.019337 - 0.032525

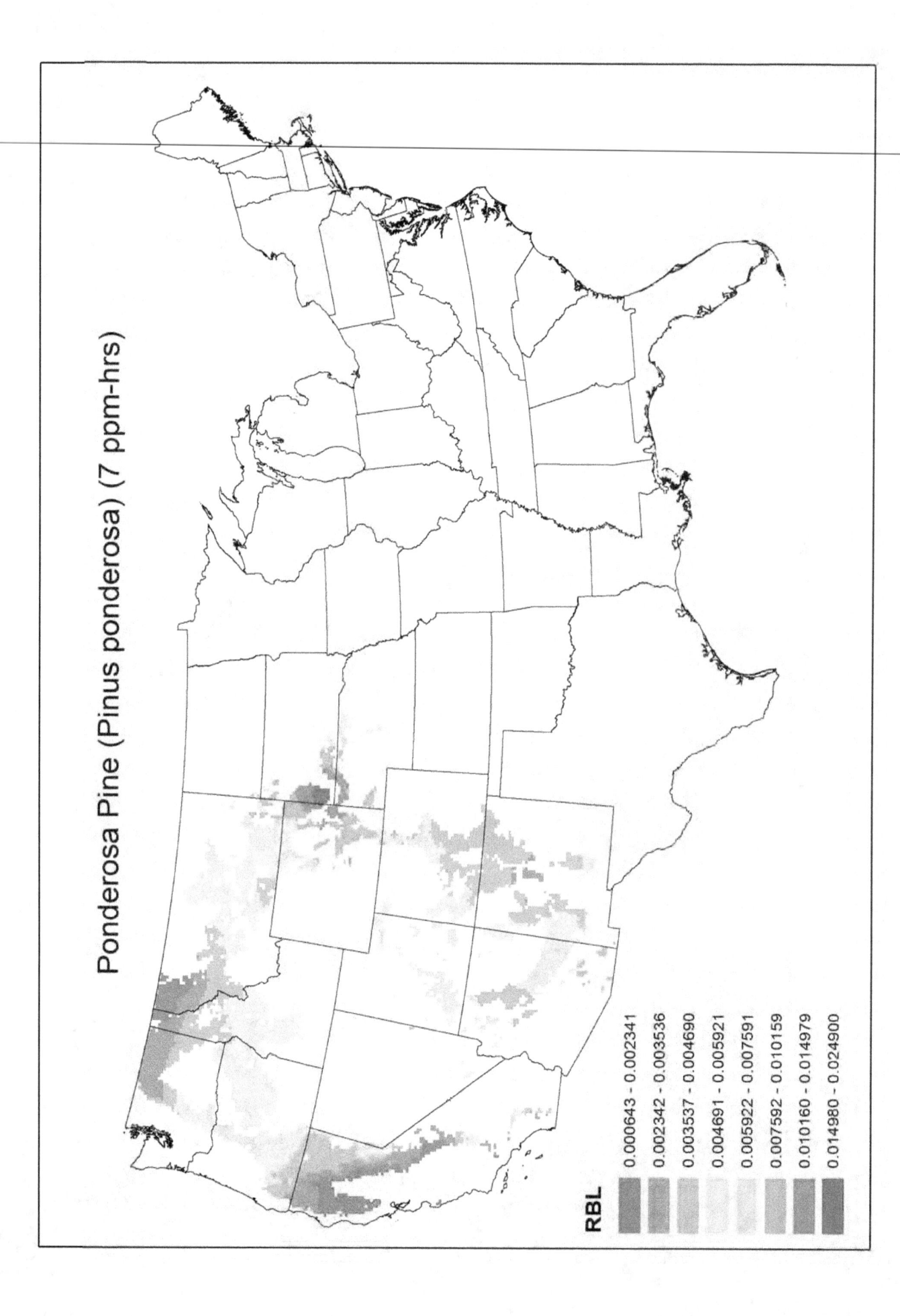

Ponderosa Pine (Pinus ponderosa) (7 ppm-hrs)

RBL

0.000643 - 0.002341
0.002342 - 0.003536
0.003537 - 0.004690
0.004691 - 0.005921
0.005922 - 0.007591
0.007592 - 0.010159
0.010160 - 0.014979
0.014980 - 0.024900

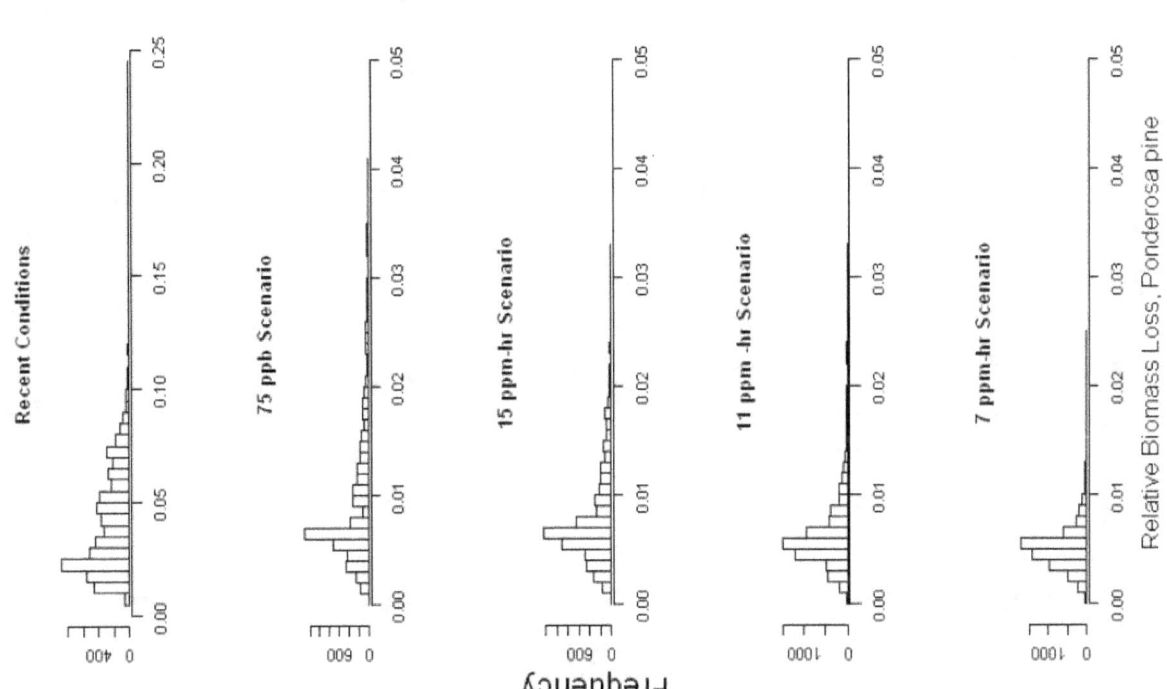

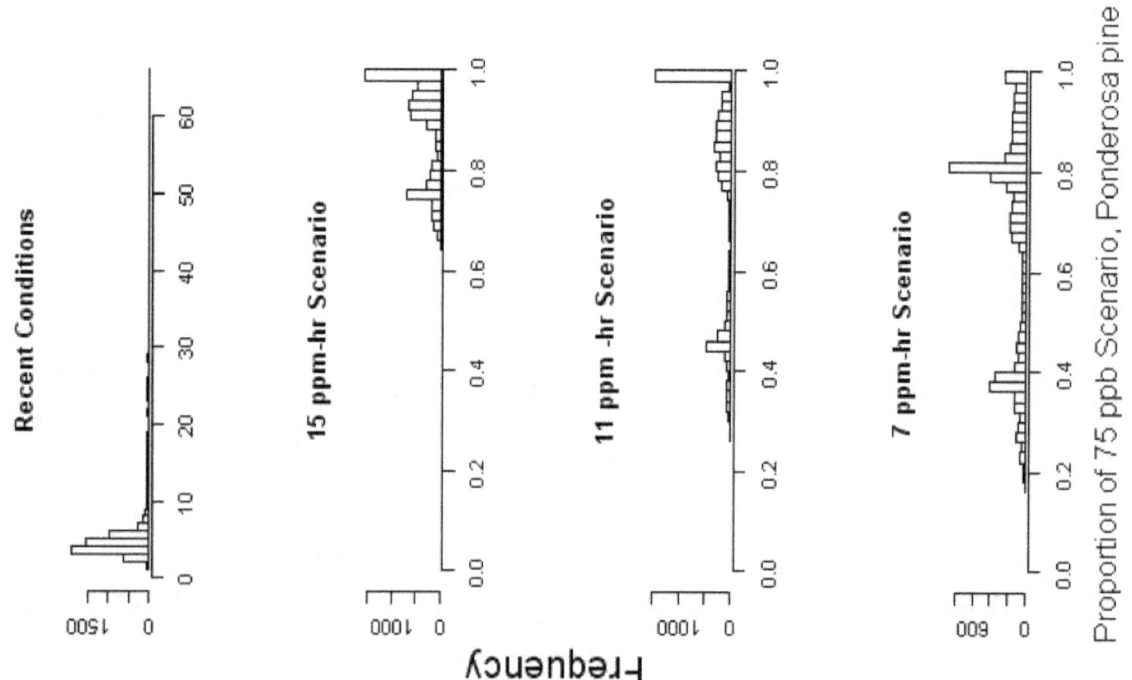

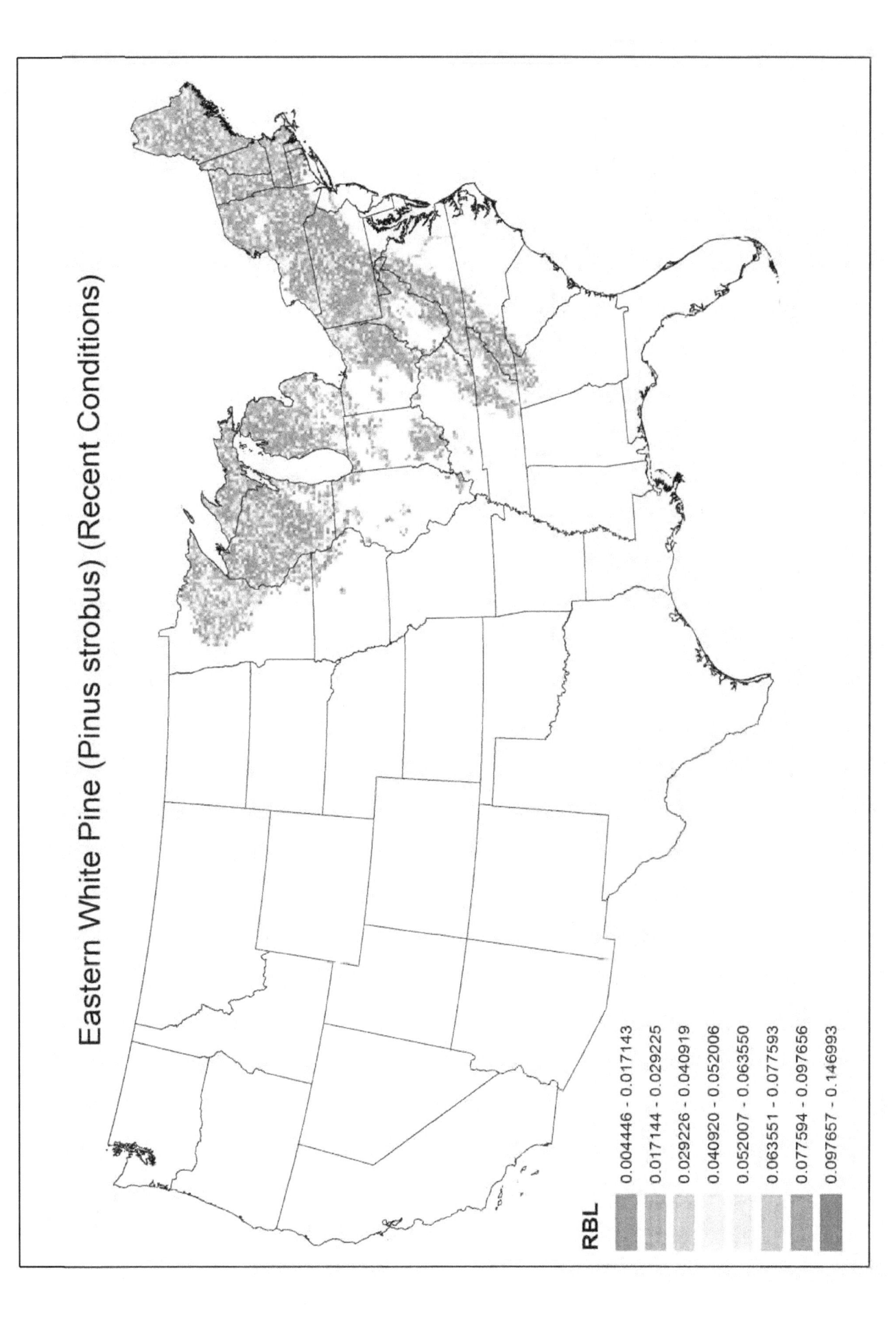

Eastern White Pine (Pinus strobus) (Recent Conditions)

RBL

0.004446 - 0.017143
0.017144 - 0.029225
0.029226 - 0.040919
0.040920 - 0.052006
0.052007 - 0.063550
0.063551 - 0.077593
0.077594 - 0.097656
0.097657 - 0.146993

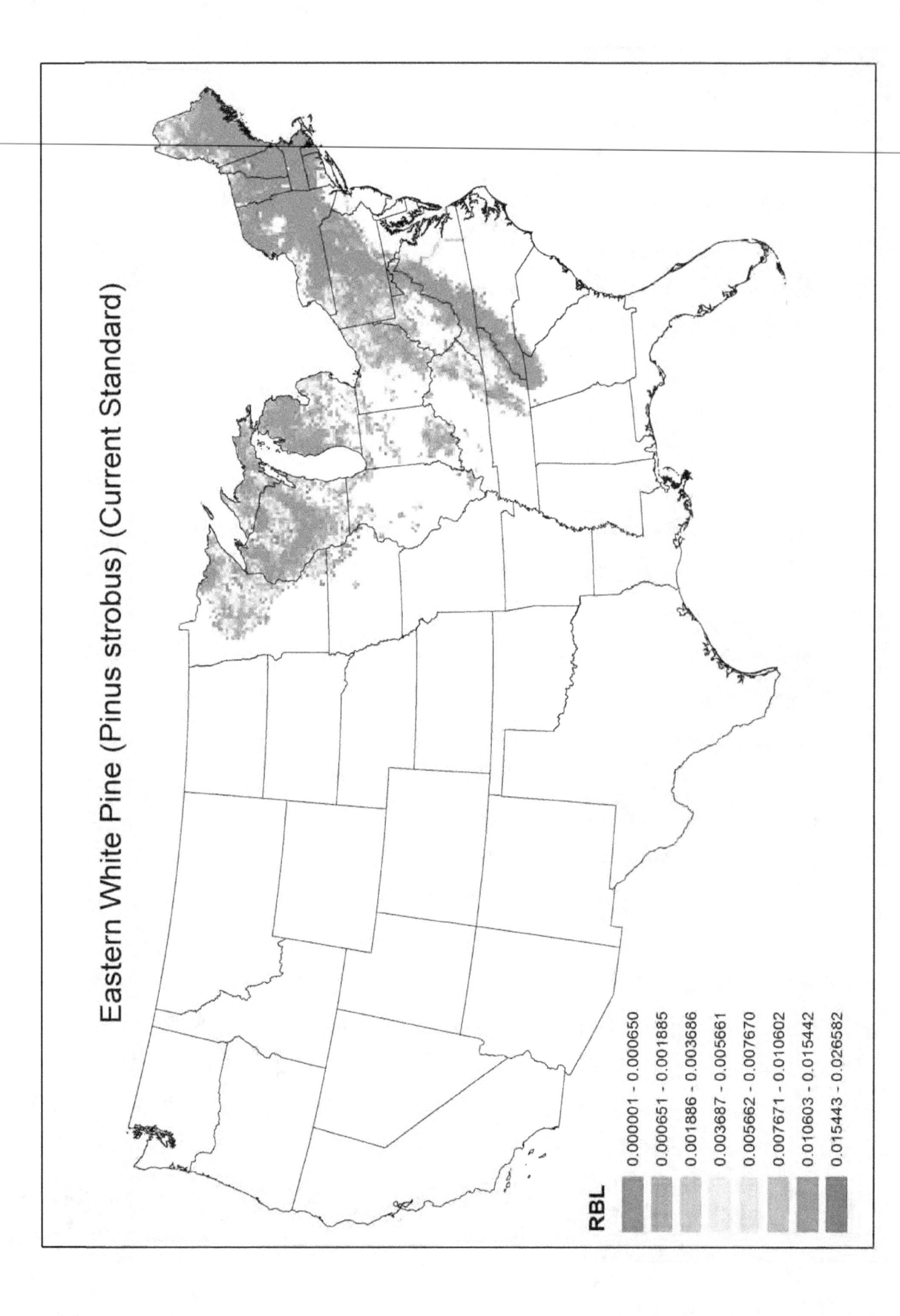

Eastern White Pine (Pinus strobus) (Current Standard)

RBL

- 0.000001 - 0.000650
- 0.000651 - 0.001885
- 0.001886 - 0.003686
- 0.003687 - 0.005661
- 0.005662 - 0.007670
- 0.007671 - 0.010602
- 0.010603 - 0.015442
- 0.015443 - 0.026582

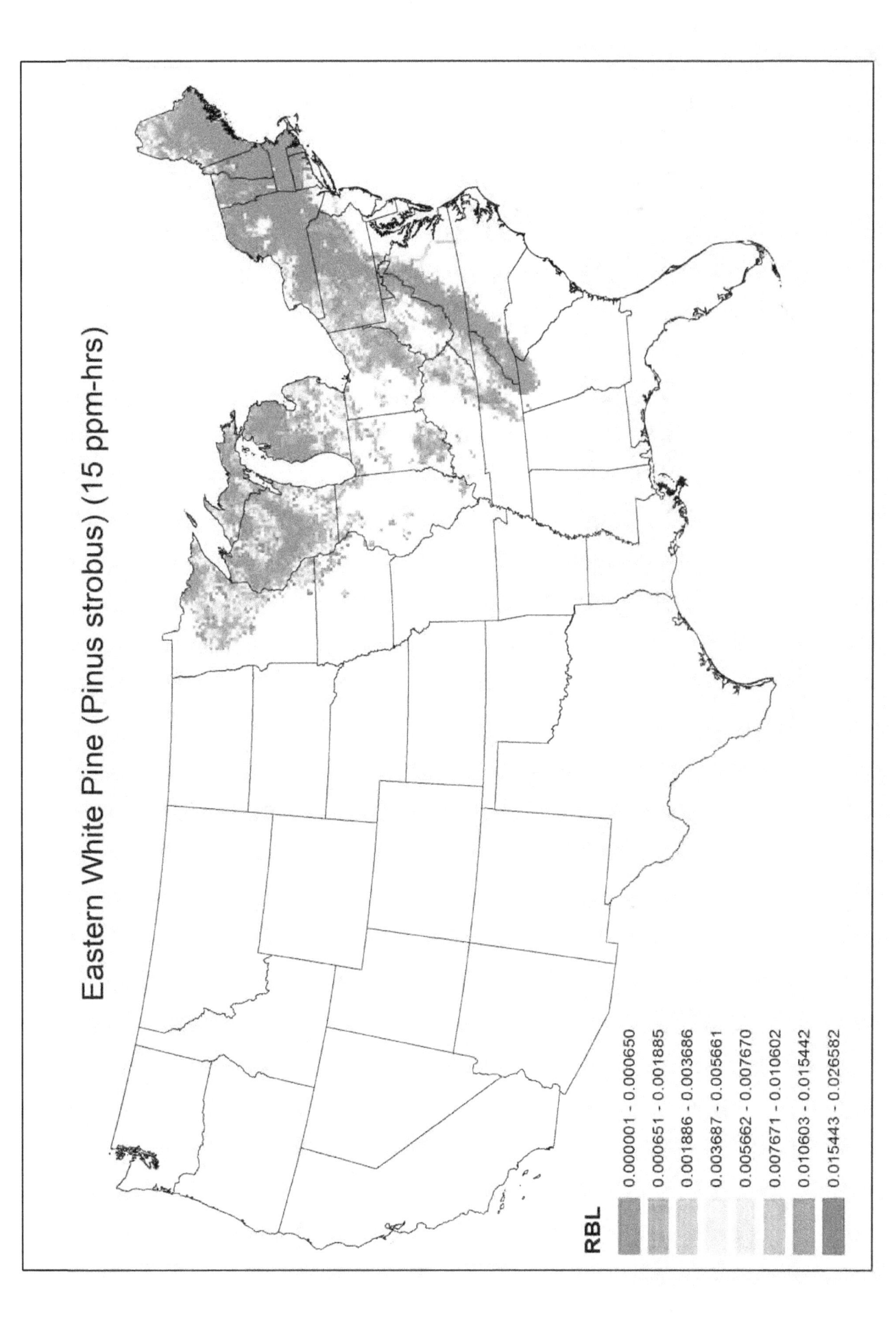

Eastern White Pine (Pinus strobus) (15 ppm-hrs)

RBL

0.000001 - 0.000650
0.000651 - 0.001885
0.001886 - 0.003686
0.003687 - 0.005661
0.005662 - 0.007670
0.007671 - 0.010602
0.010603 - 0.015442
0.015443 - 0.026582

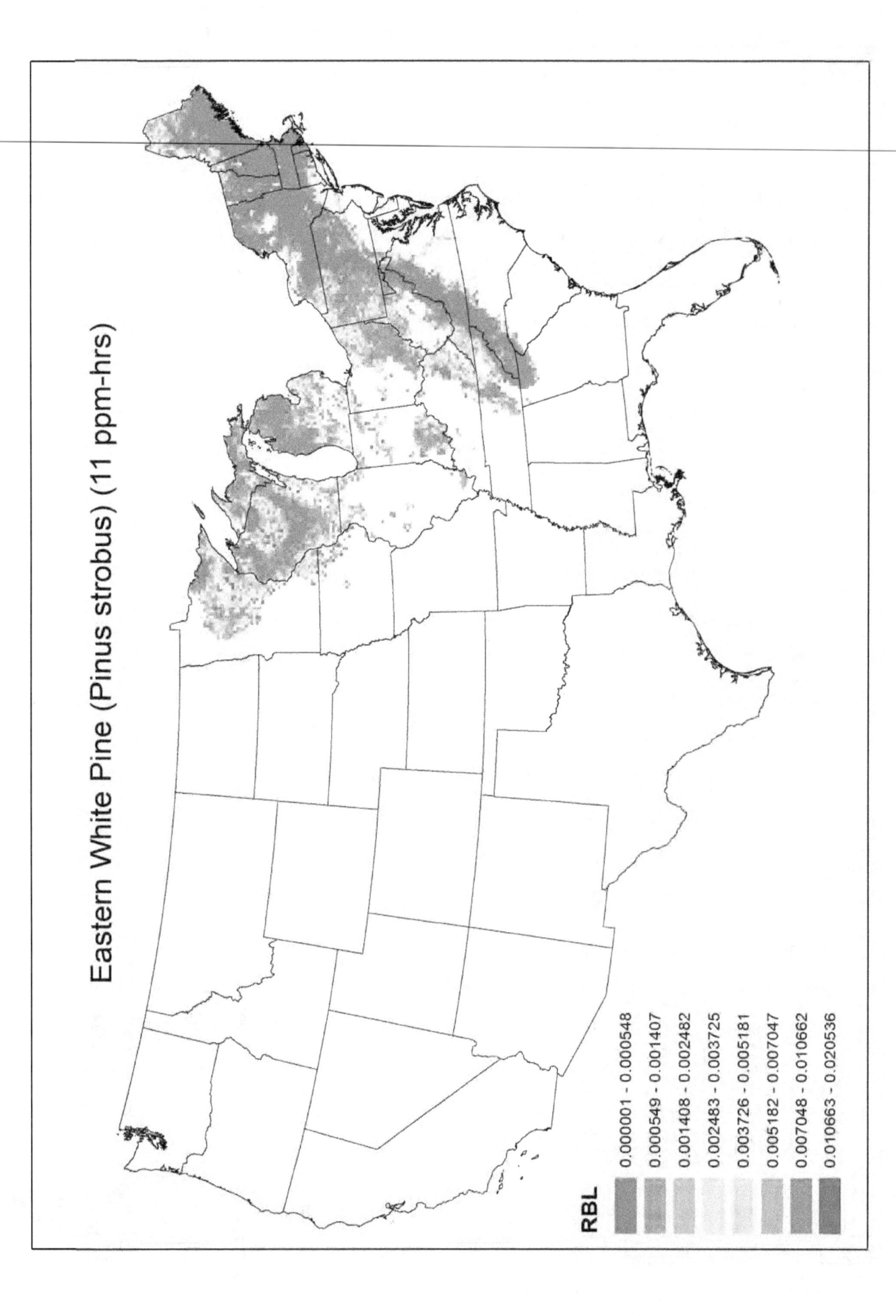

Eastern White Pine (Pinus strobus) (11 ppm-hrs)

RBL

- 0.000001 - 0.000548
- 0.000549 - 0.001407
- 0.001408 - 0.002482
- 0.002483 - 0.003725
- 0.003726 - 0.005181
- 0.005182 - 0.007047
- 0.007048 - 0.010662
- 0.010663 - 0.020536

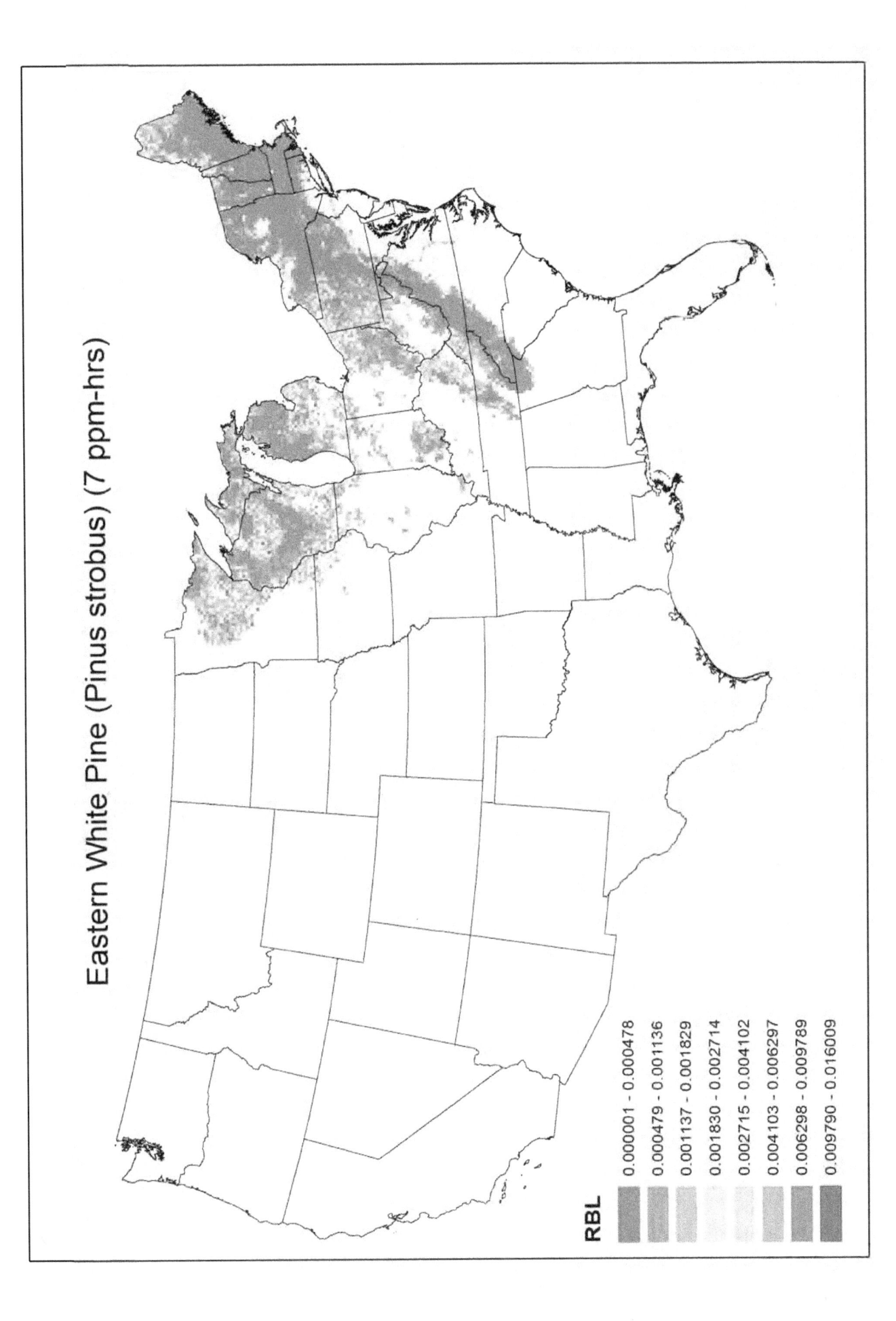

Eastern White Pine (Pinus strobus) (7 ppm-hrs)

RBL
0.000001 - 0.000478
0.000479 - 0.001136
0.001137 - 0.001829
0.001830 - 0.002714
0.002715 - 0.004102
0.004103 - 0.006297
0.006298 - 0.009789
0.009790 - 0.016009

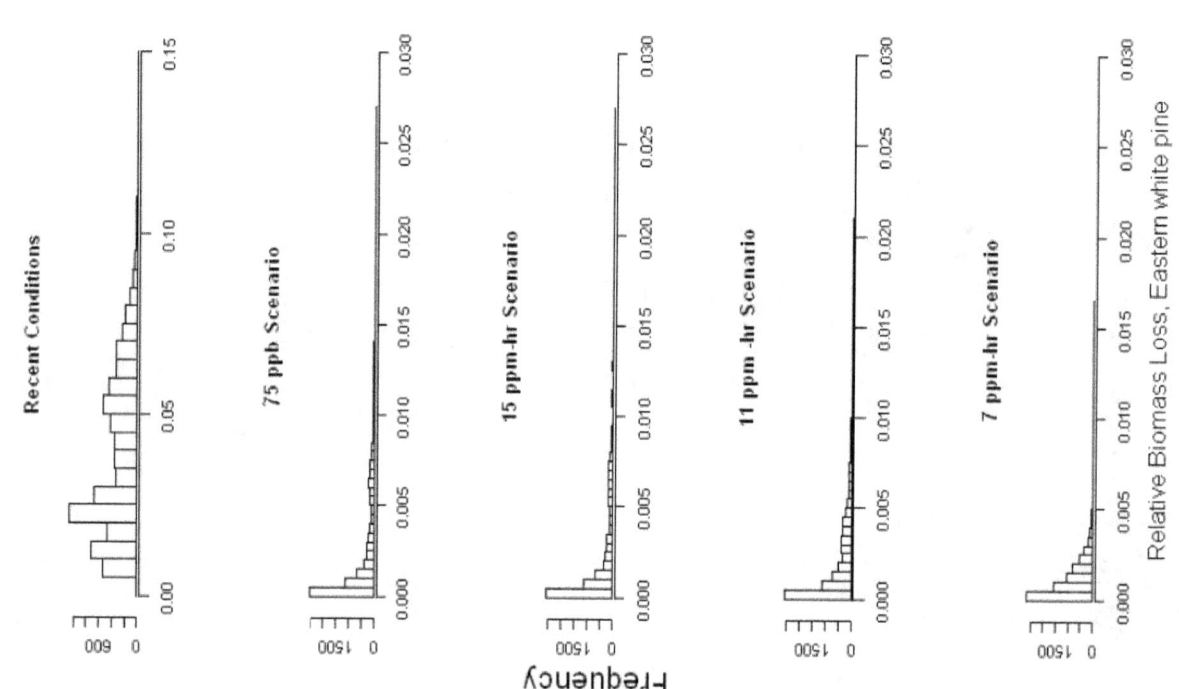

Relative Biomass Loss, Eastern white pine

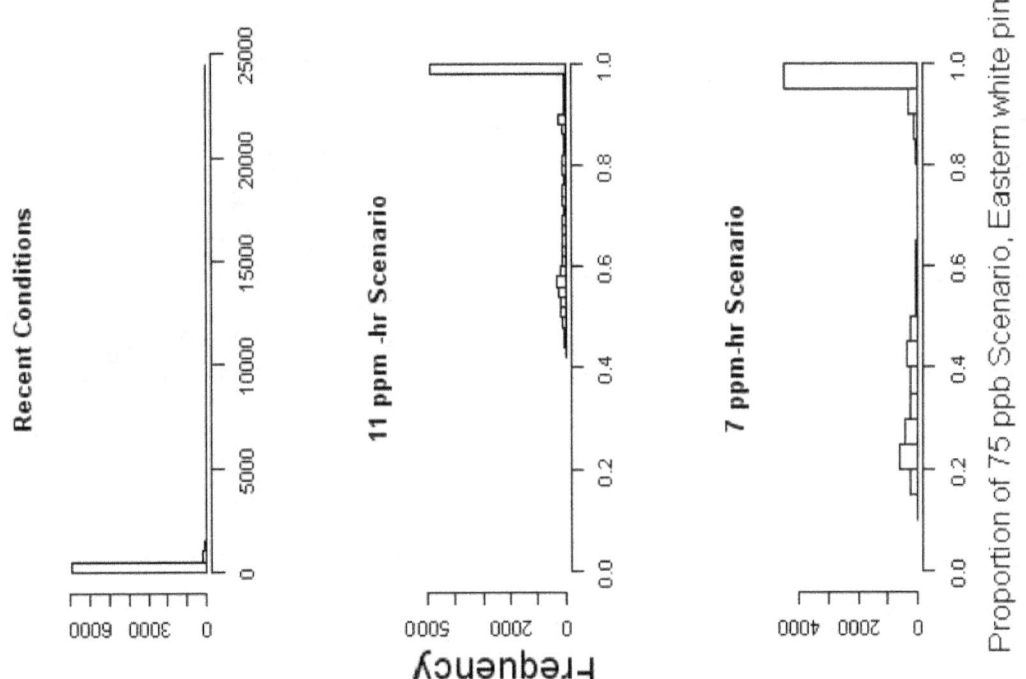

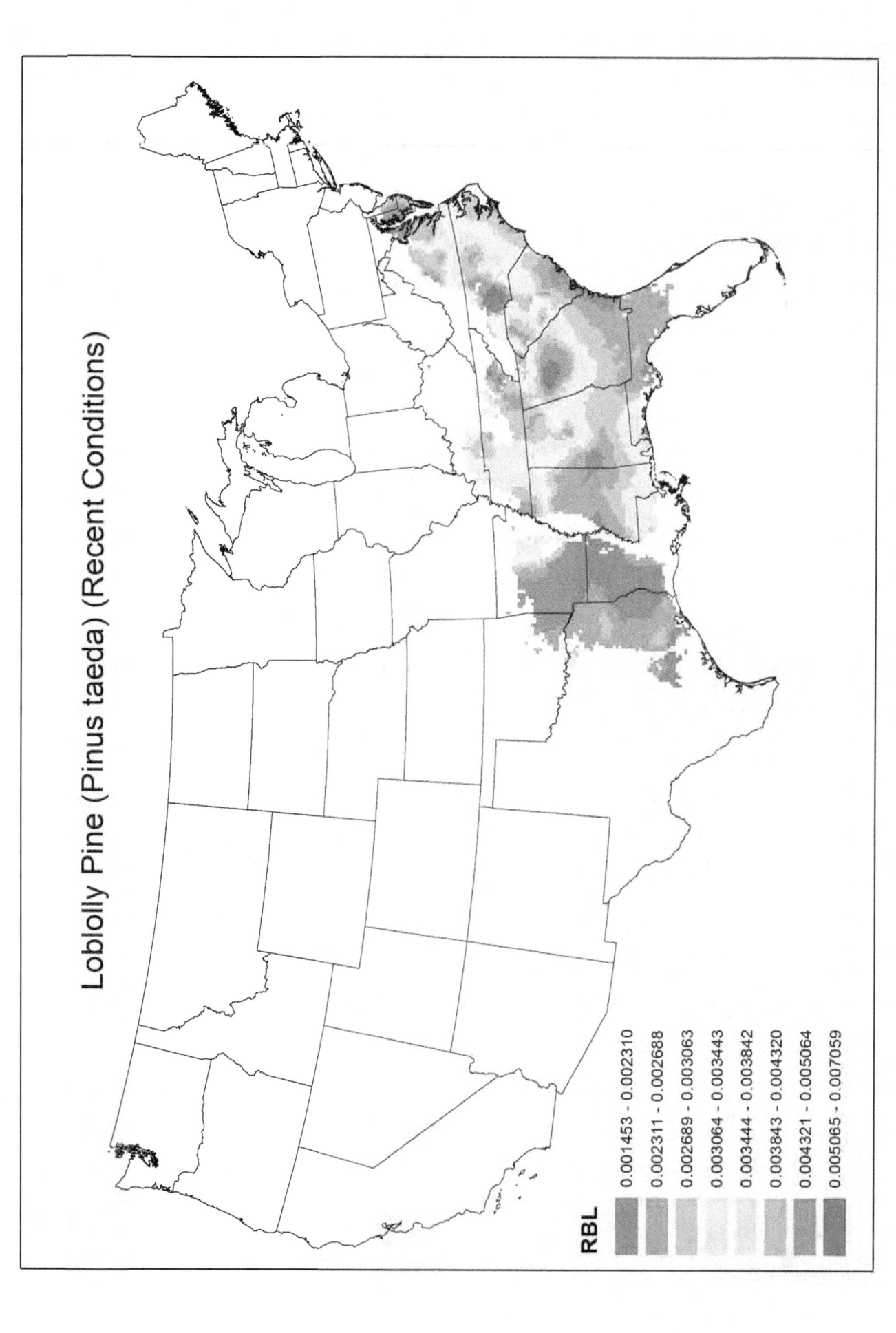

Loblolly Pine (Pinus taeda) (Recent Conditions)

RBL

0.001453 - 0.002310
0.002311 - 0.002688
0.002689 - 0.003063
0.003064 - 0.003443
0.003444 - 0.003842
0.003843 - 0.004320
0.004321 - 0.005064
0.005065 - 0.007059

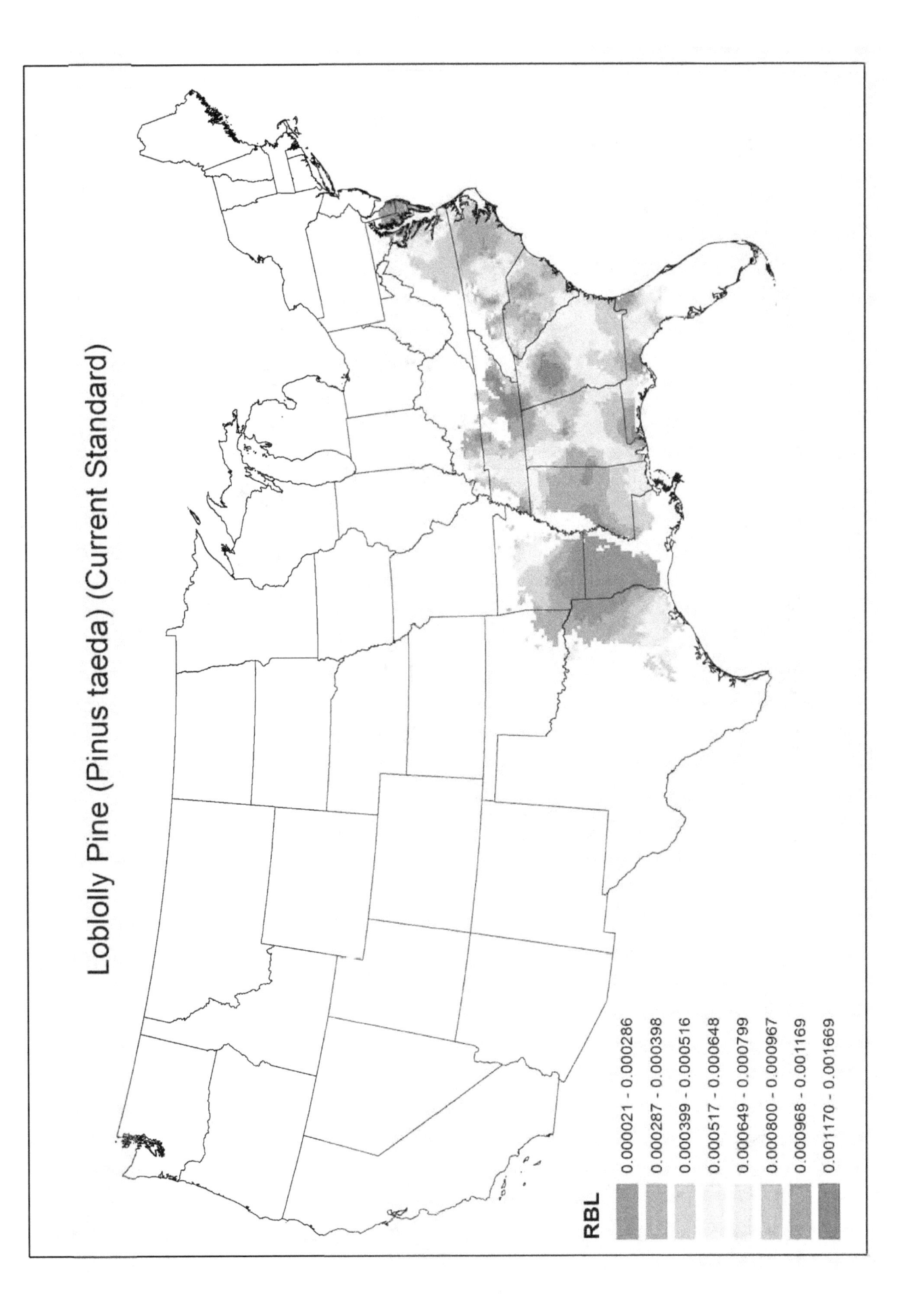

Loblolly Pine (Pinus taeda) (Current Standard)

RBL

0.000021 - 0.000286
0.000287 - 0.000398
0.000399 - 0.000516
0.000517 - 0.000648
0.000649 - 0.000799
0.000800 - 0.000967
0.000968 - 0.001169
0.001170 - 0.001669

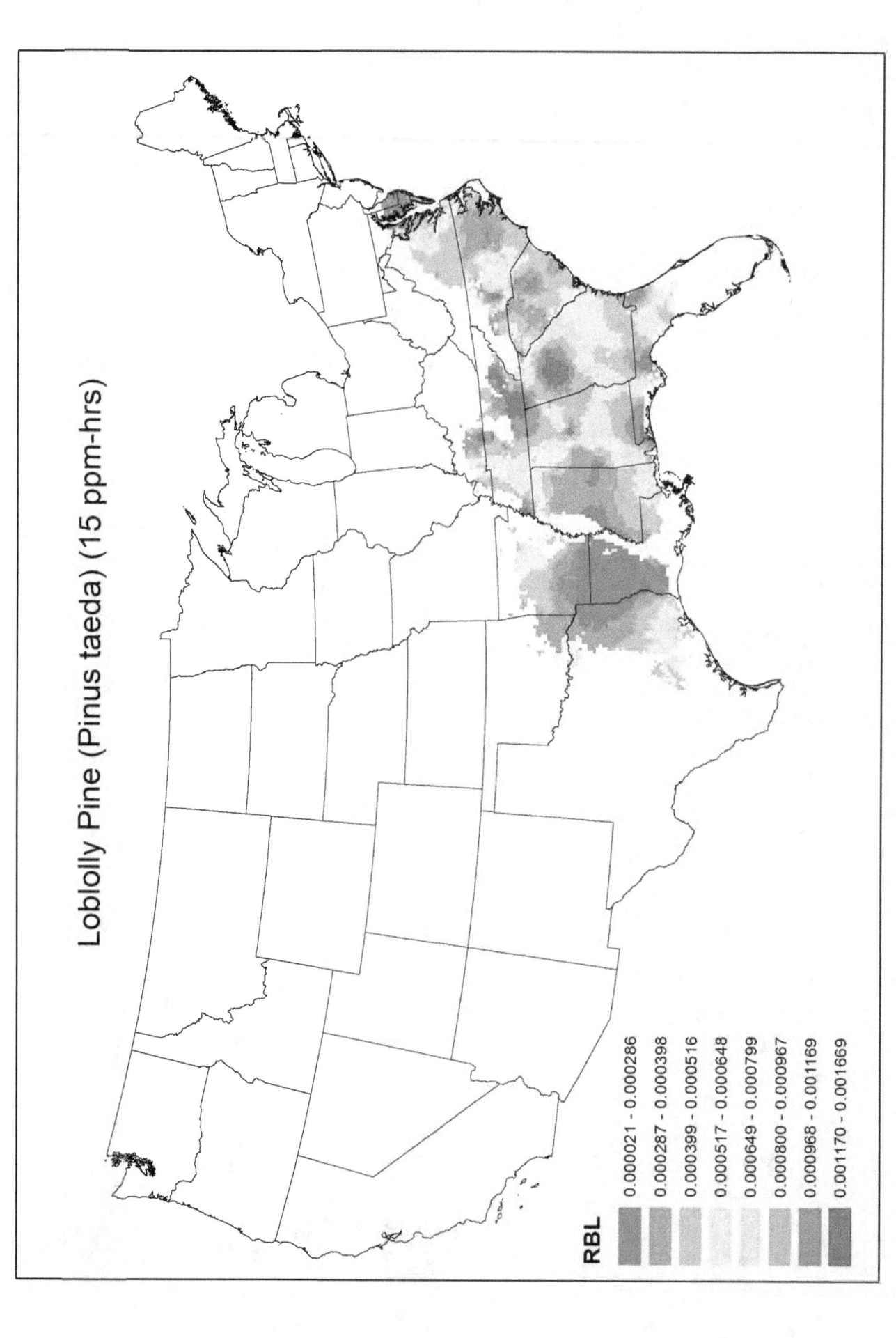

Loblolly Pine (Pinus taeda) (15 ppm-hrs)

RBL

0.000021 - 0.000286
0.000287 - 0.000398
0.000399 - 0.000516
0.000517 - 0.000648
0.000649 - 0.000799
0.000800 - 0.000967
0.000968 - 0.001169
0.001170 - 0.001669

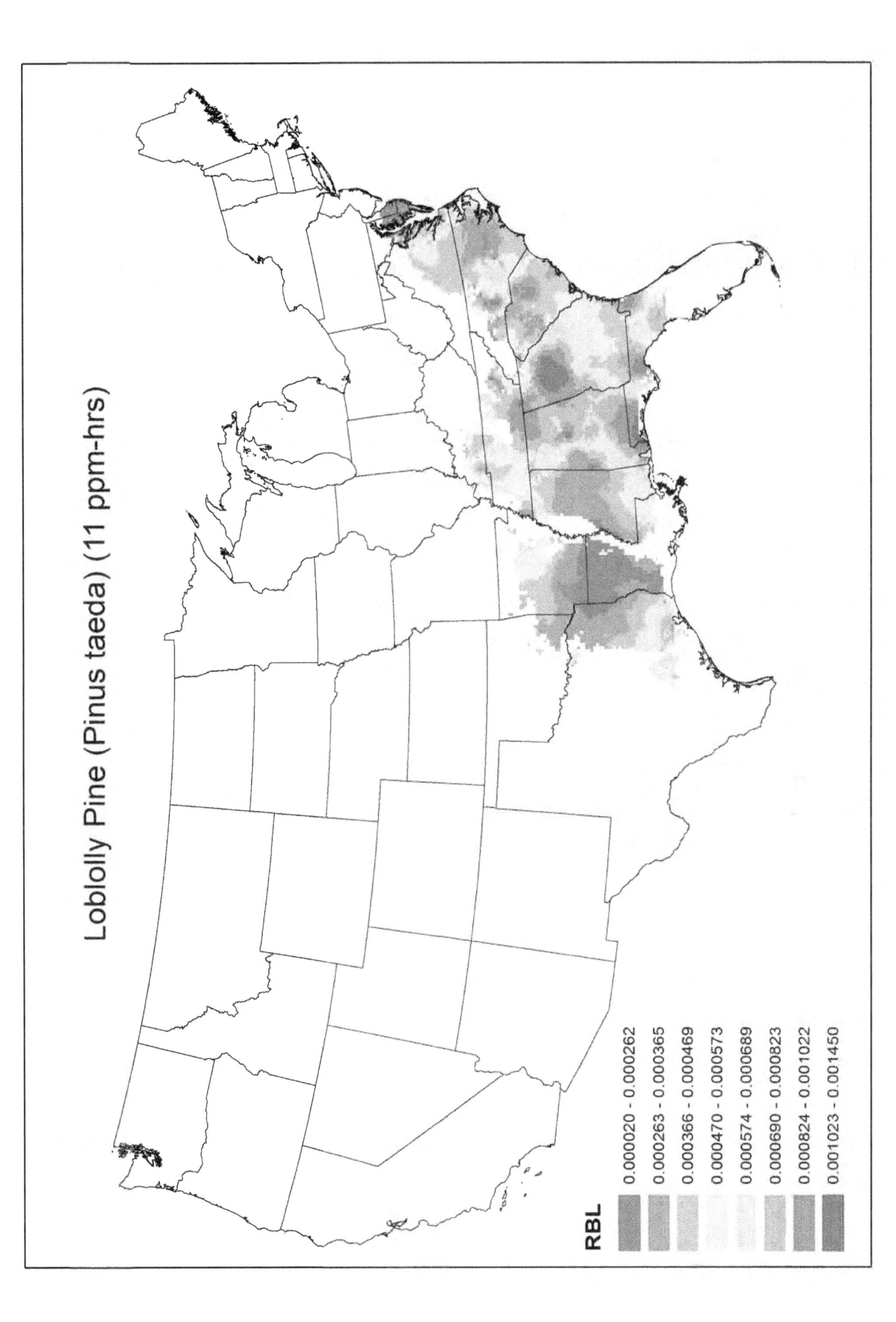

Loblolly Pine (Pinus taeda) (11 ppm-hrs)

RBL

- 0.000020 - 0.000262
- 0.000263 - 0.000365
- 0.000366 - 0.000469
- 0.000470 - 0.000573
- 0.000574 - 0.000689
- 0.000690 - 0.000823
- 0.000824 - 0.001022
- 0.001023 - 0.001450

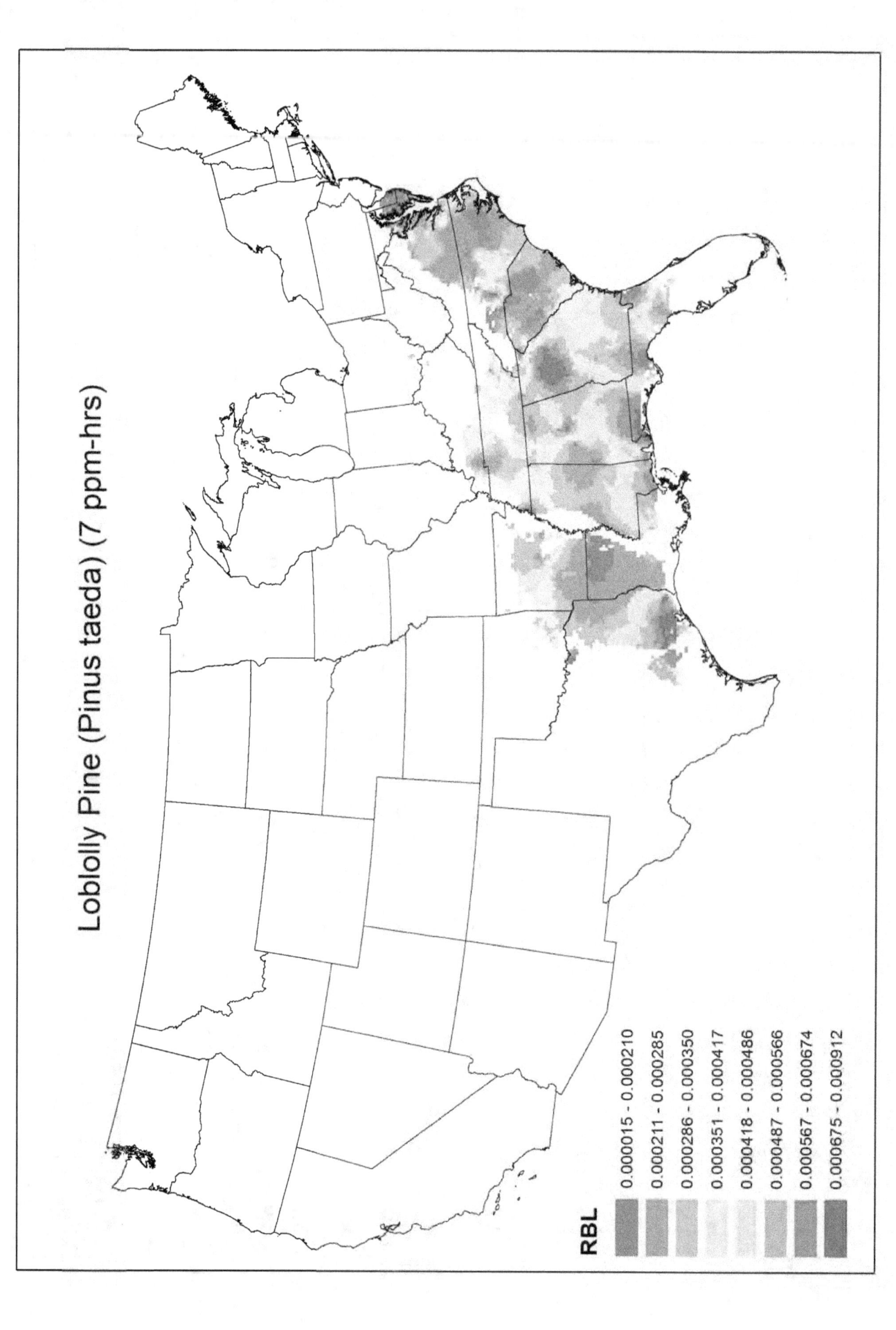

Loblolly Pine (Pinus taeda) (7 ppm-hrs)

RBL

0.000015 - 0.000210
0.000211 - 0.000285
0.000286 - 0.000350
0.000351 - 0.000417
0.000418 - 0.000486
0.000487 - 0.000566
0.000567 - 0.000674
0.000675 - 0.000912

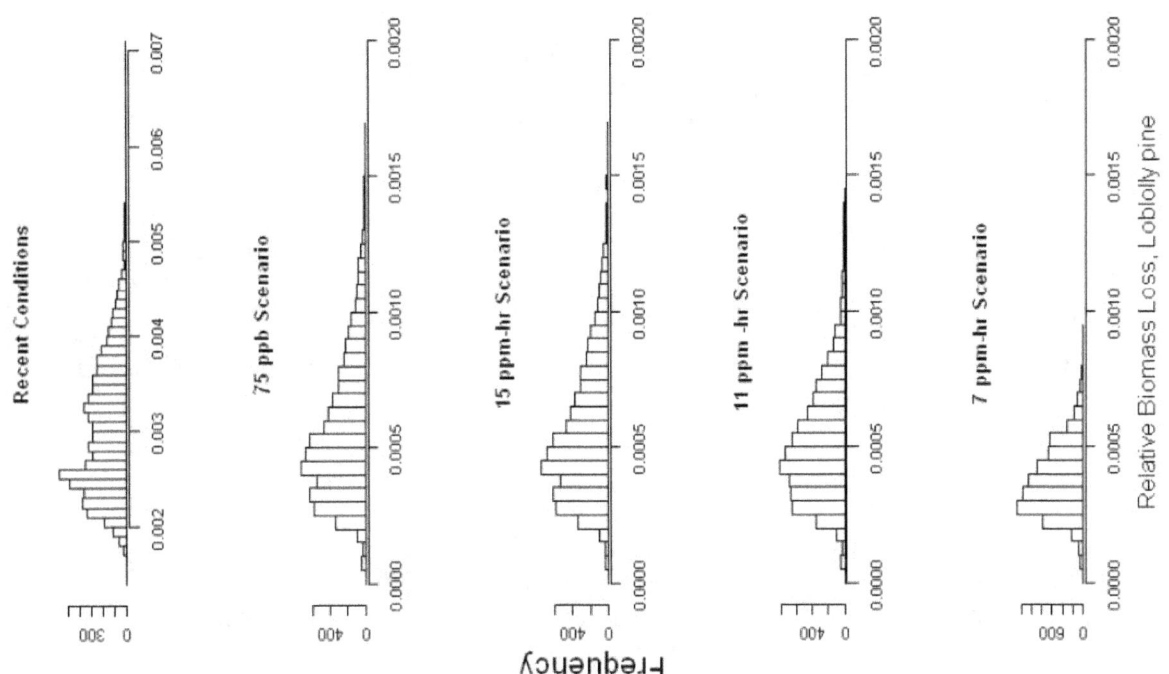

Relative Biomass Loss, Loblolly pine

Frequency

Recent Conditions

75 ppb Scenario

15 ppm-hr Scenario

11 ppm -hr Scenario

7 ppm-hr Scenario

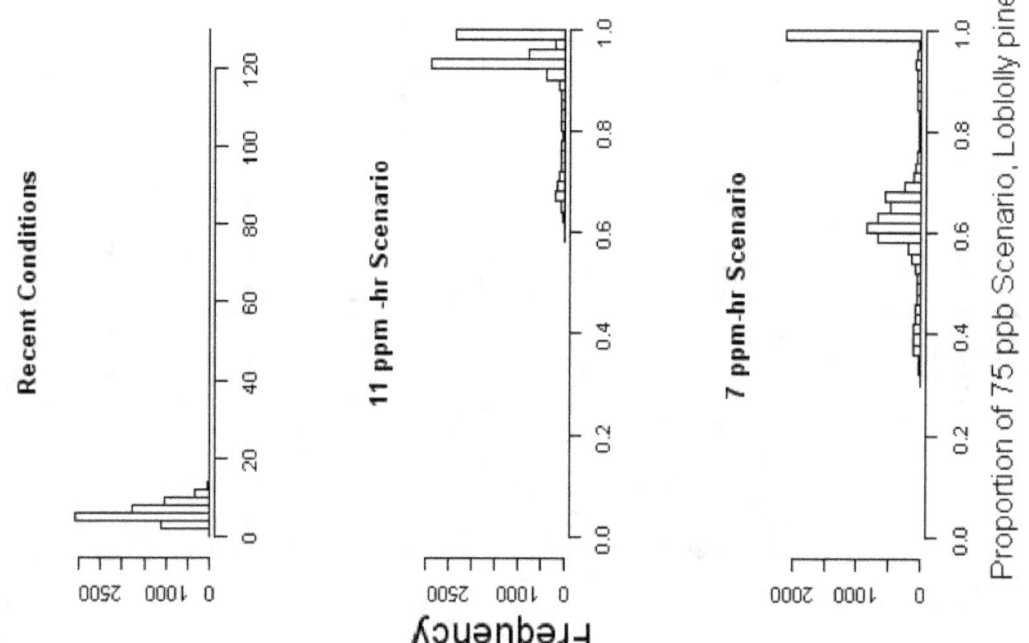

Recent Conditions

11 ppm -hr Scenario

7 ppm-hr Scenario

Frequency

Proportion of 75 ppb Scenario, Loblolly pine

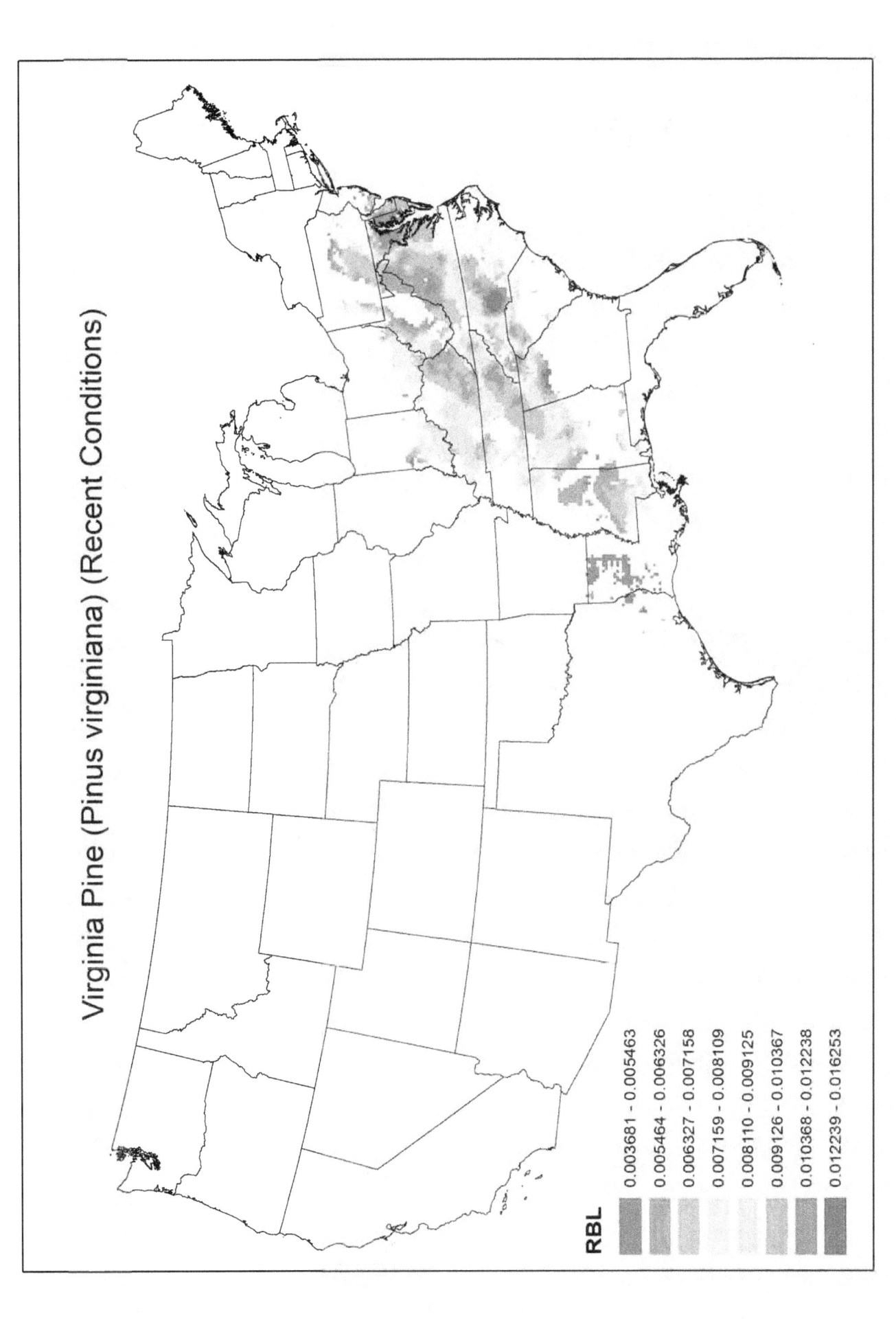

Virginia Pine (Pinus virginiana) (Recent Conditions)

RBL

0.003681 - 0.005463
0.005464 - 0.006326
0.006327 - 0.007158
0.007159 - 0.008109
0.008110 - 0.009125
0.009126 - 0.010367
0.010368 - 0.012238
0.012239 - 0.016253

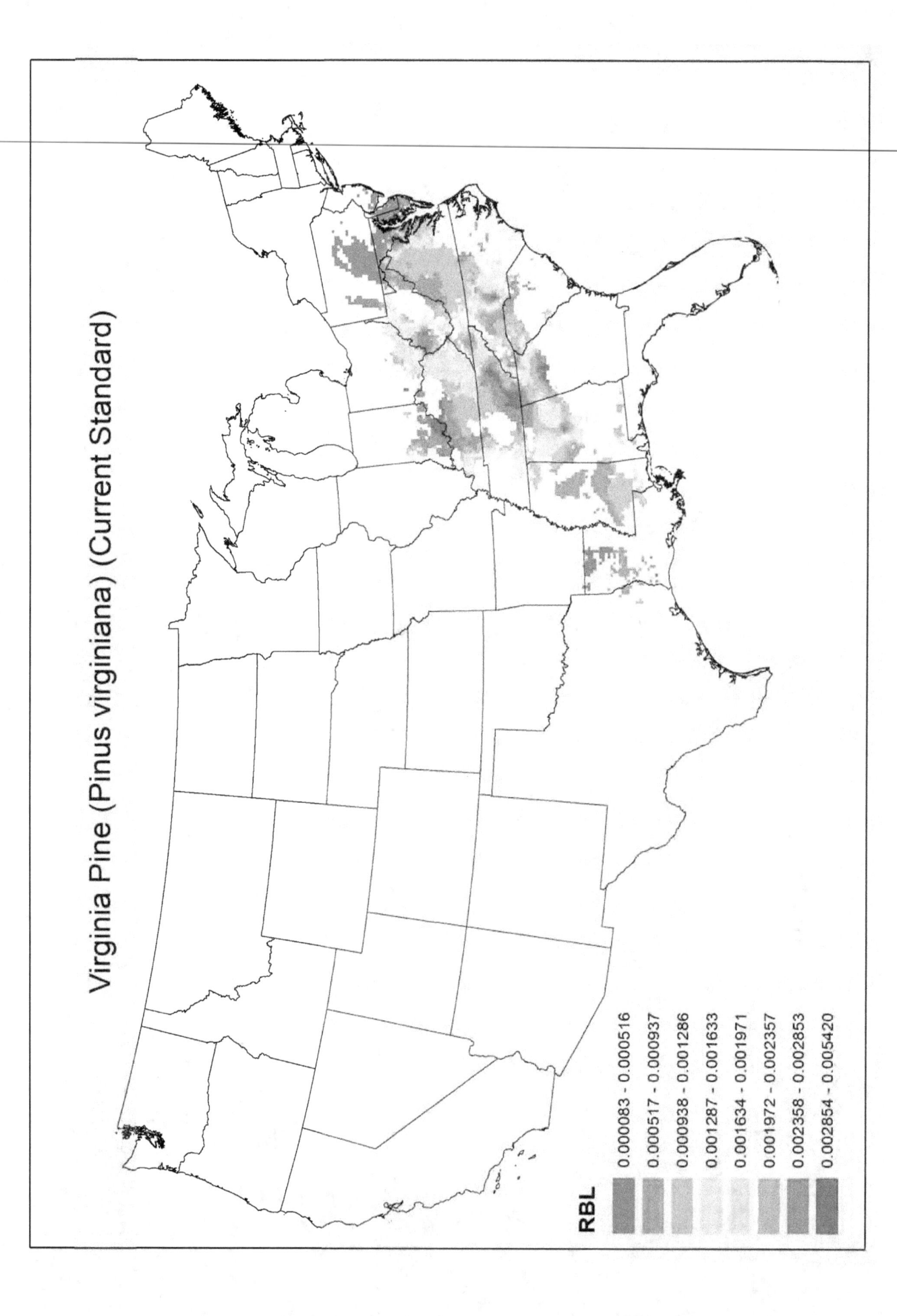

Virginia Pine (Pinus virginiana) (Current Standard)

RBL

- 0.000083 - 0.000516
- 0.000517 - 0.000937
- 0.000938 - 0.001286
- 0.001287 - 0.001633
- 0.001634 - 0.001971
- 0.001972 - 0.002357
- 0.002358 - 0.002853
- 0.002854 - 0.005420

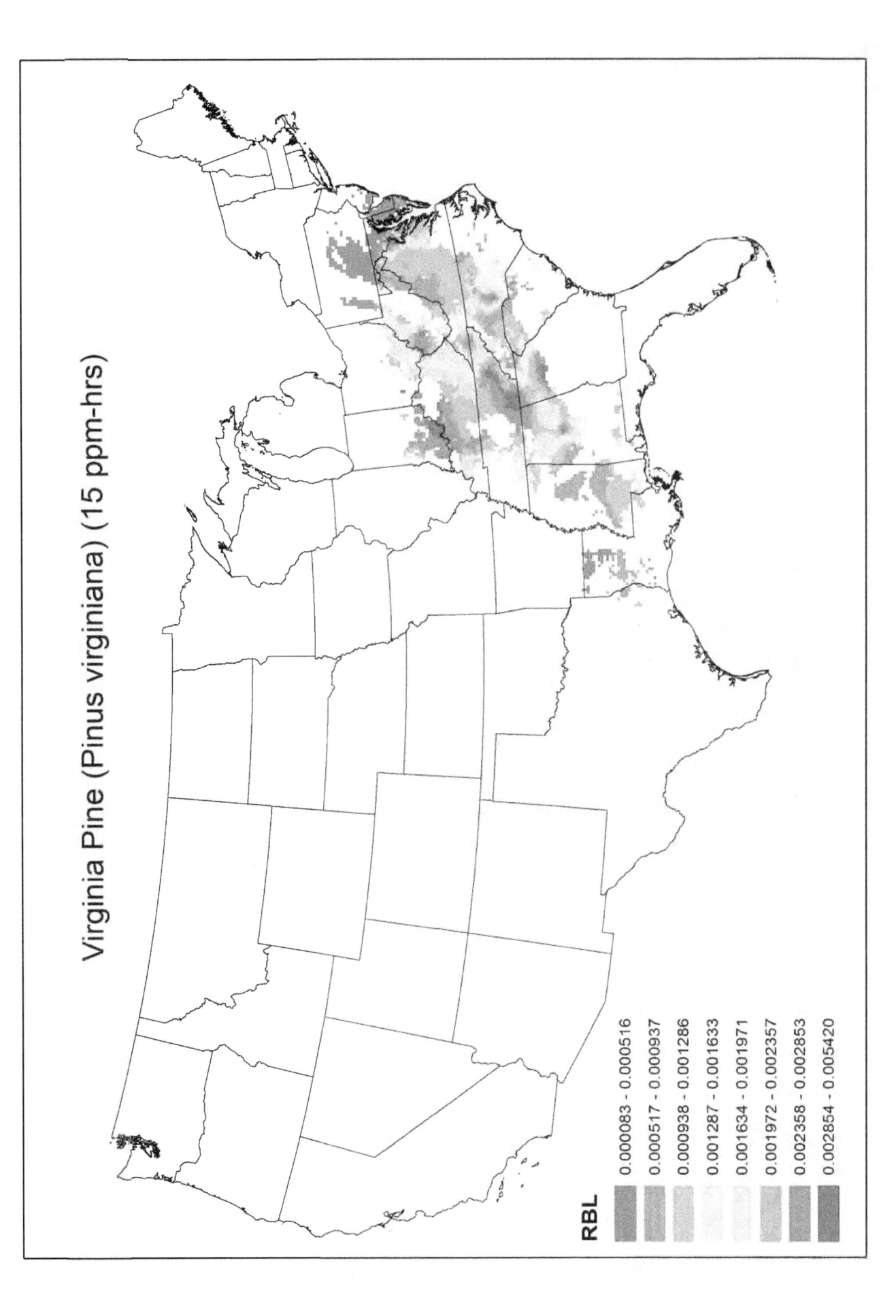

Virginia Pine (Pinus virginiana) (15 ppm-hrs)

RBL

0.000083 - 0.000516
0.000517 - 0.000937
0.000938 - 0.001286
0.001287 - 0.001633
0.001634 - 0.001971
0.001972 - 0.002357
0.002358 - 0.002853
0.002854 - 0.005420

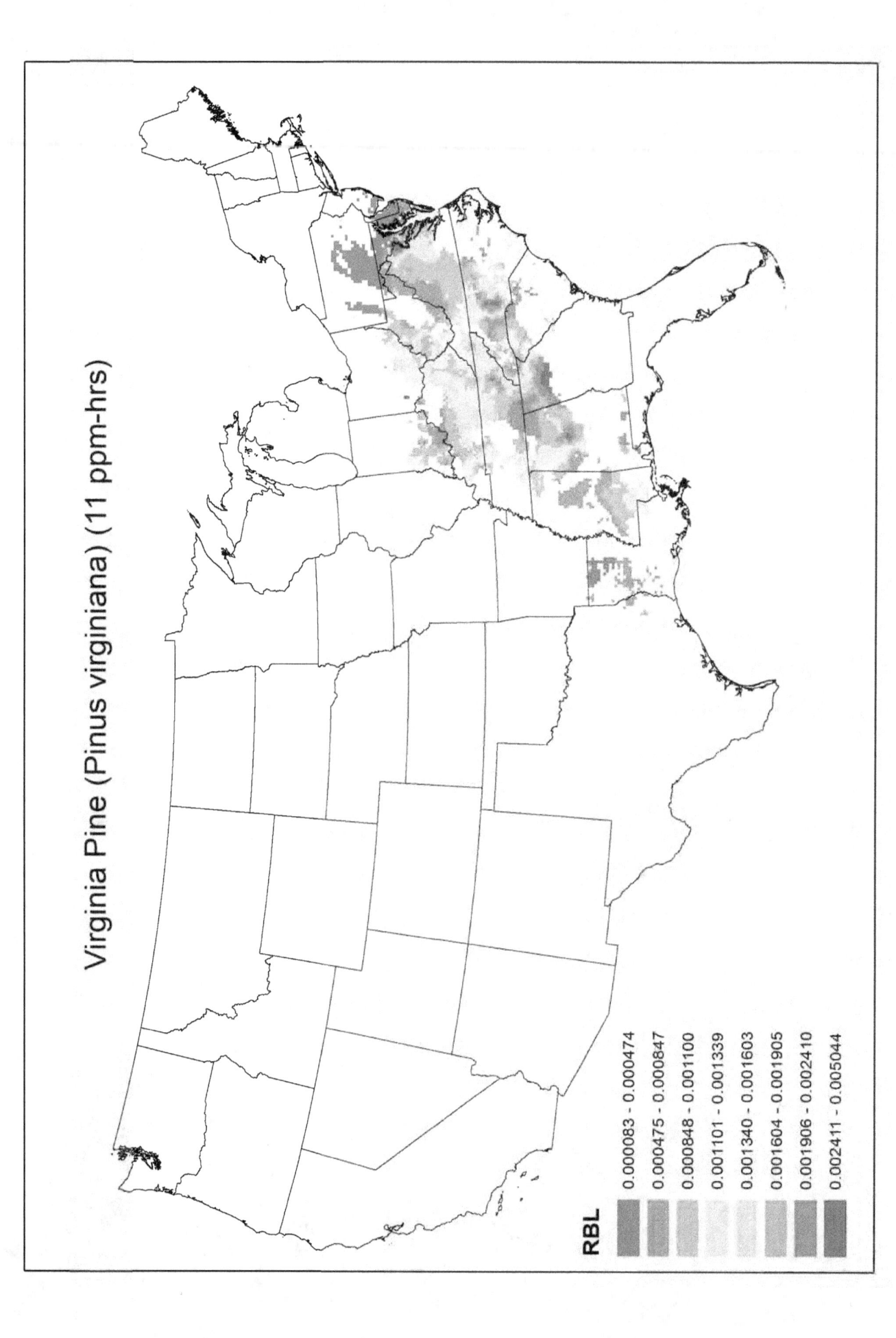

Virginia Pine (Pinus virginiana) (11 ppm-hrs)

RBL
0.000083 - 0.000474
0.000475 - 0.000847
0.000848 - 0.001100
0.001101 - 0.001339
0.001340 - 0.001603
0.001604 - 0.001905
0.001906 - 0.002410
0.002411 - 0.005044

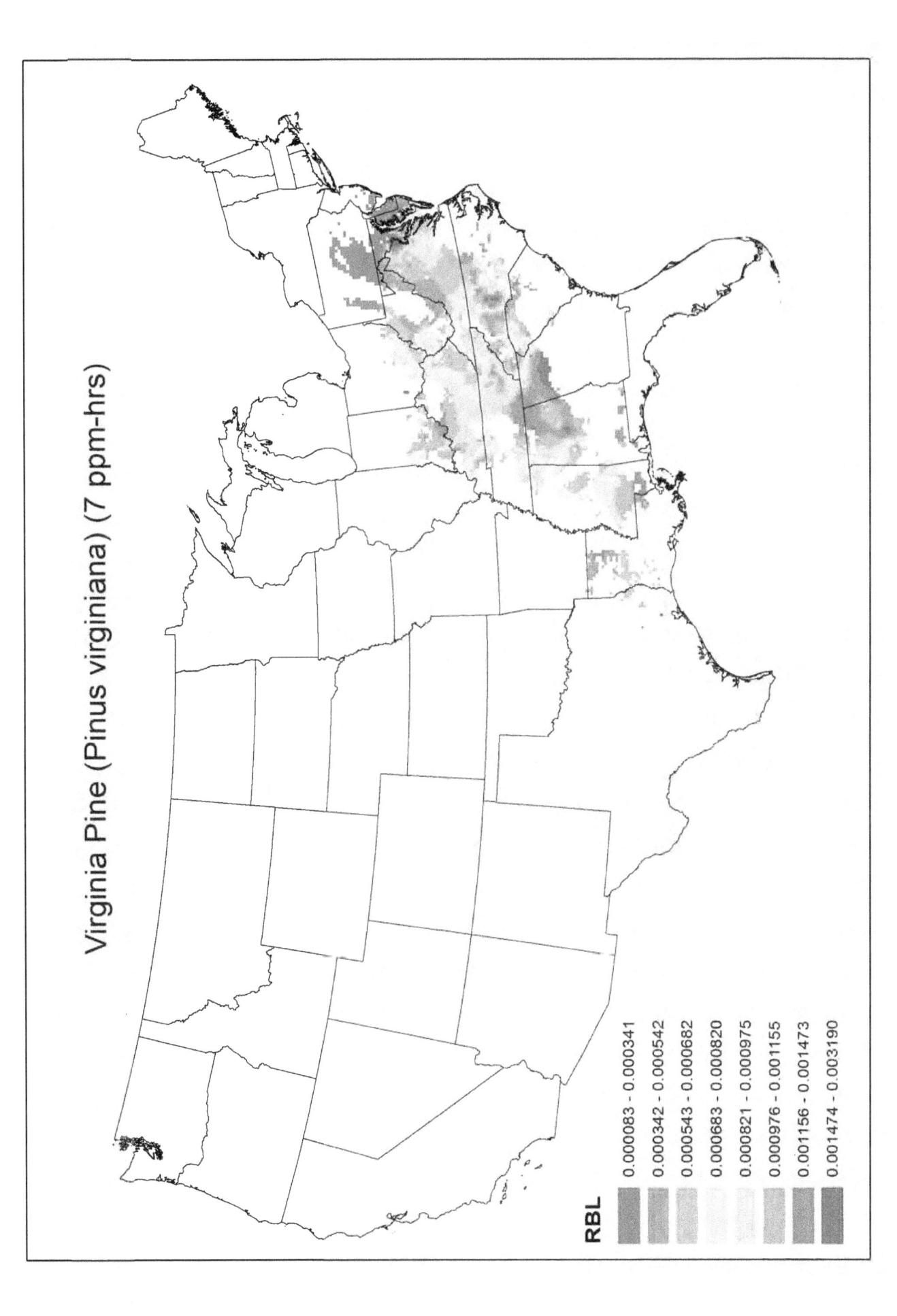

Virginia Pine (Pinus virginiana) (7 ppm-hrs)

RBL

- 0.000083 - 0.000341
- 0.000342 - 0.000542
- 0.000543 - 0.000682
- 0.000683 - 0.000820
- 0.000821 - 0.000975
- 0.000976 - 0.001155
- 0.001156 - 0.001473
- 0.001474 - 0.003190

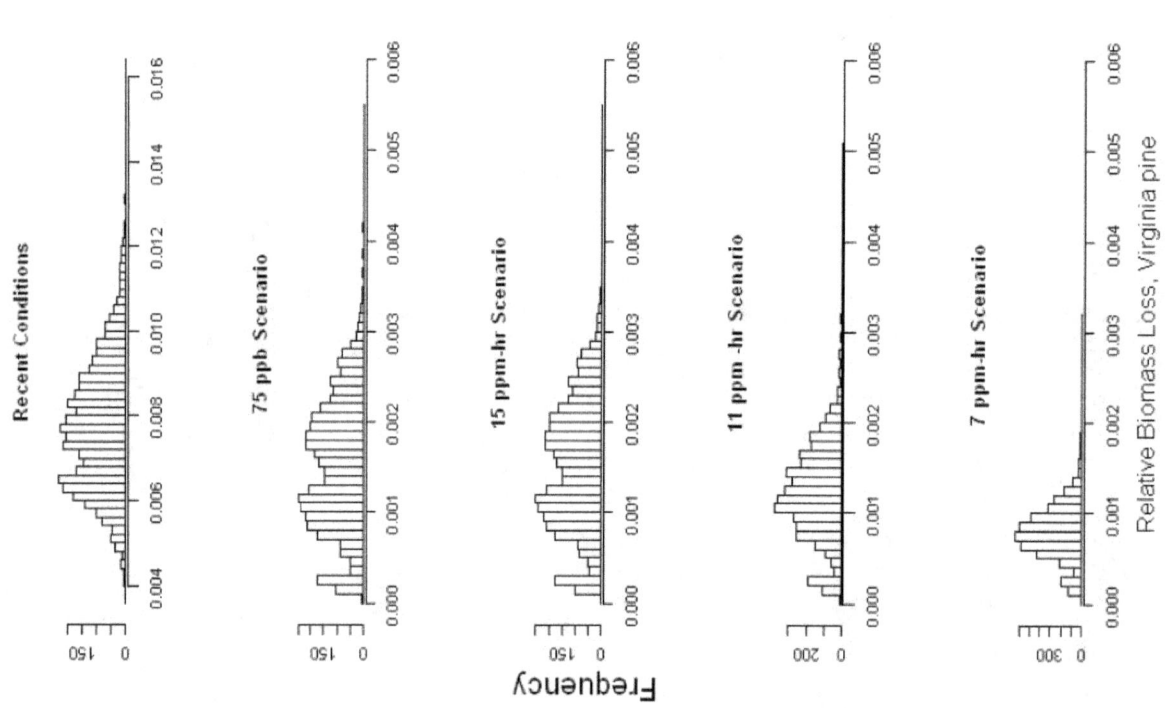

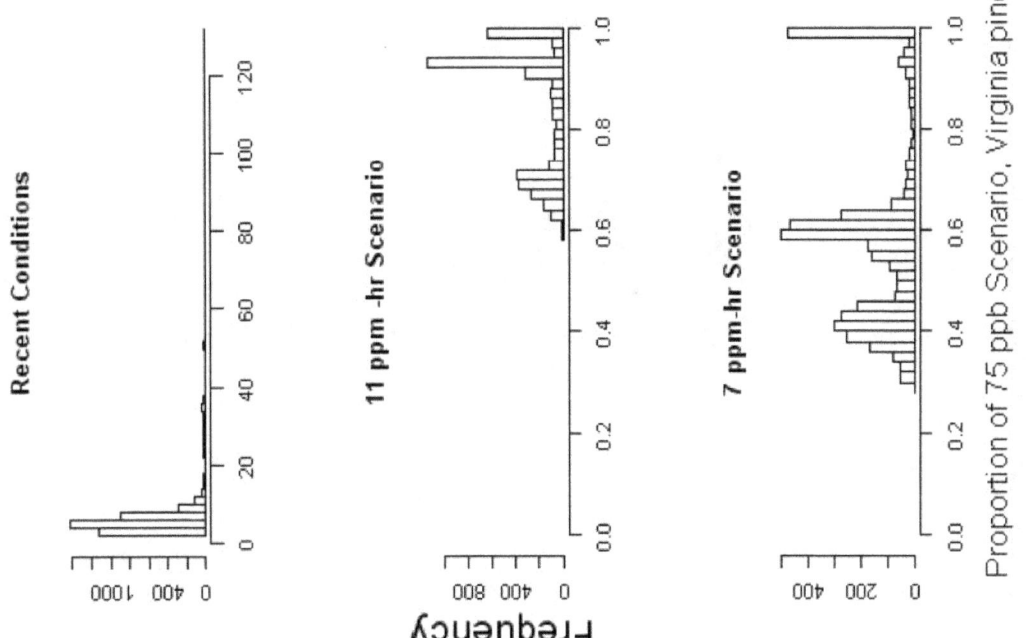

Recent Conditions

11 ppm -hr Scenario

7 ppm-hr Scenario

Frequency

Proportion of 75 ppb Scenario, Virginia pine

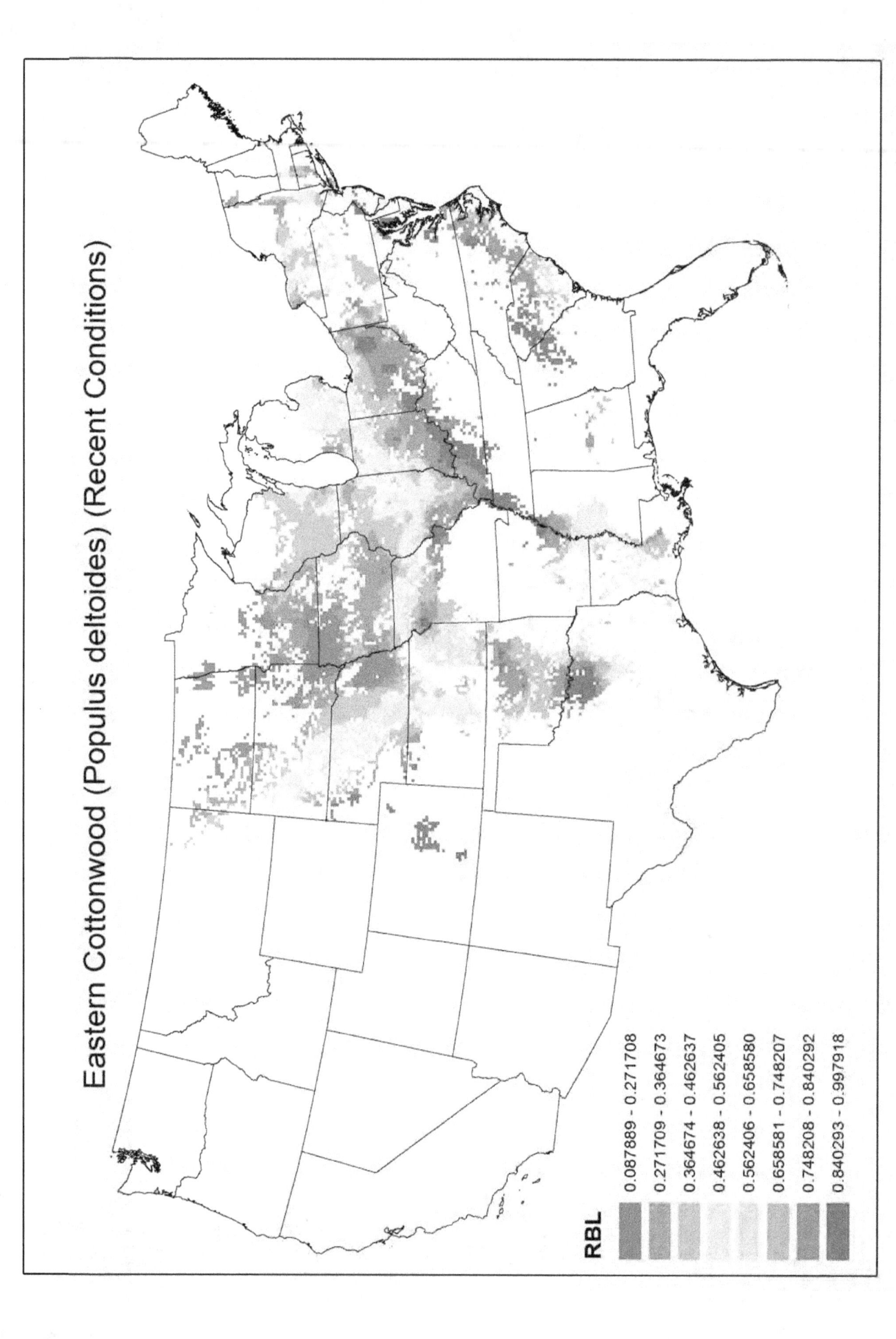

Eastern Cottonwood (Populus deltoides) (Recent Conditions)

RBL

0.087889 - 0.271708
0.271709 - 0.364673
0.364674 - 0.462637
0.462638 - 0.562405
0.562406 - 0.658580
0.658581 - 0.748207
0.748208 - 0.840292
0.840293 - 0.997918

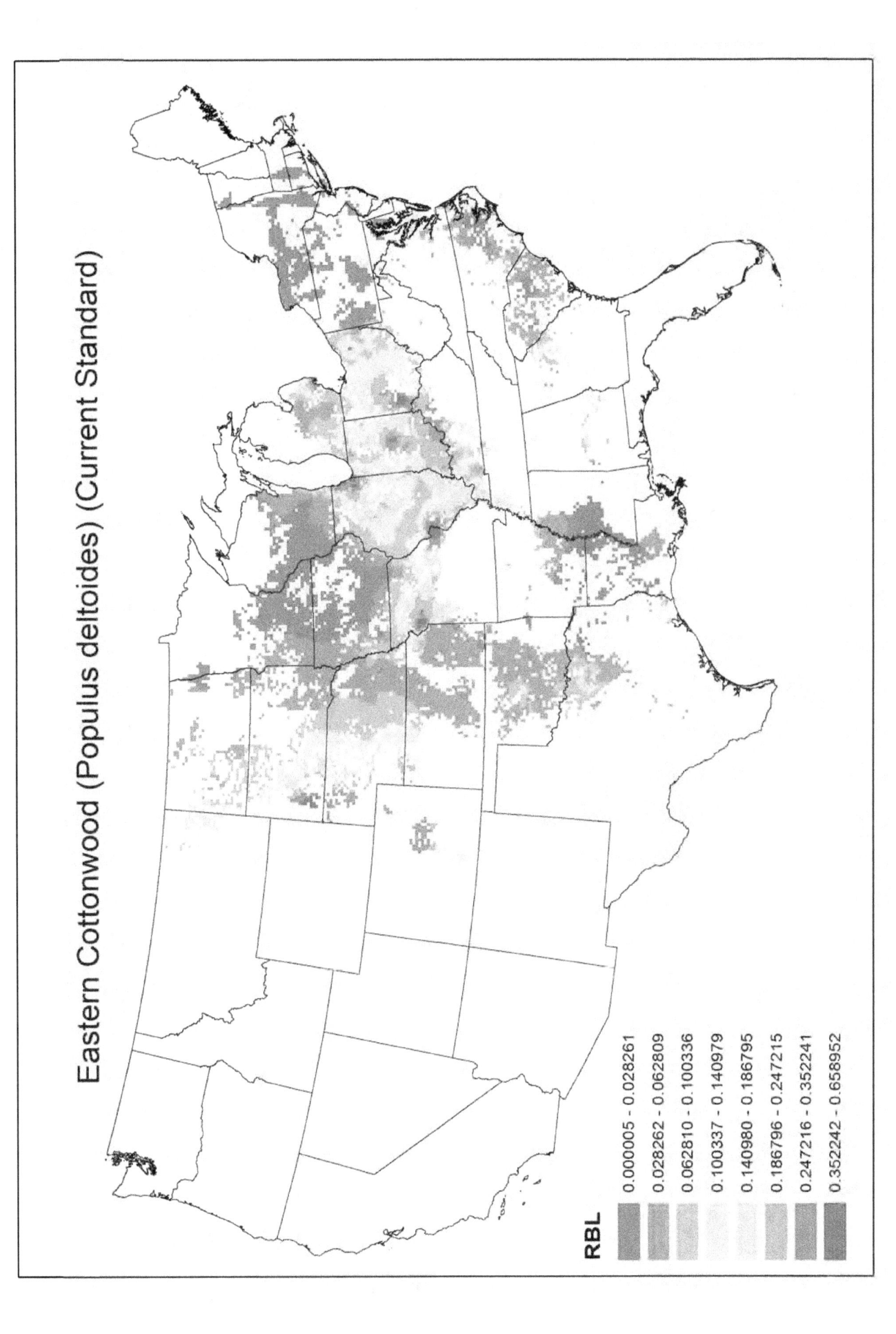

Eastern Cottonwood (Populus deltoides) (Current Standard)

RBL

- 0.000005 - 0.028261
- 0.028262 - 0.062809
- 0.062810 - 0.100336
- 0.100337 - 0.140979
- 0.140980 - 0.186795
- 0.186796 - 0.247215
- 0.247216 - 0.352241
- 0.352242 - 0.658952

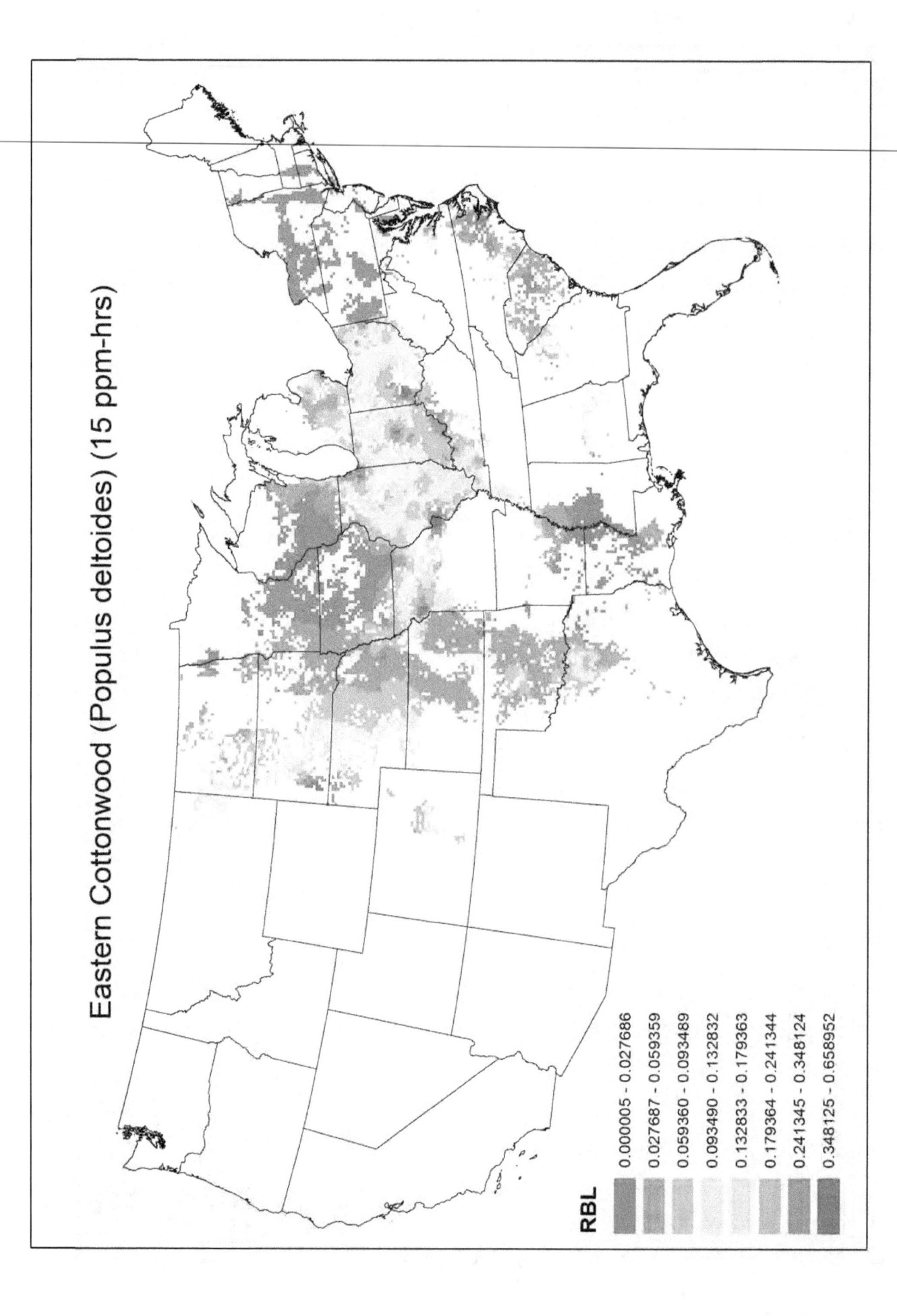

Eastern Cottonwood (Populus deltoides) (15 ppm-hrs)

RBL

0.000005 - 0.027686
0.027687 - 0.059359
0.059360 - 0.093489
0.093490 - 0.132832
0.132833 - 0.179363
0.179364 - 0.241344
0.241345 - 0.348124
0.348125 - 0.658952

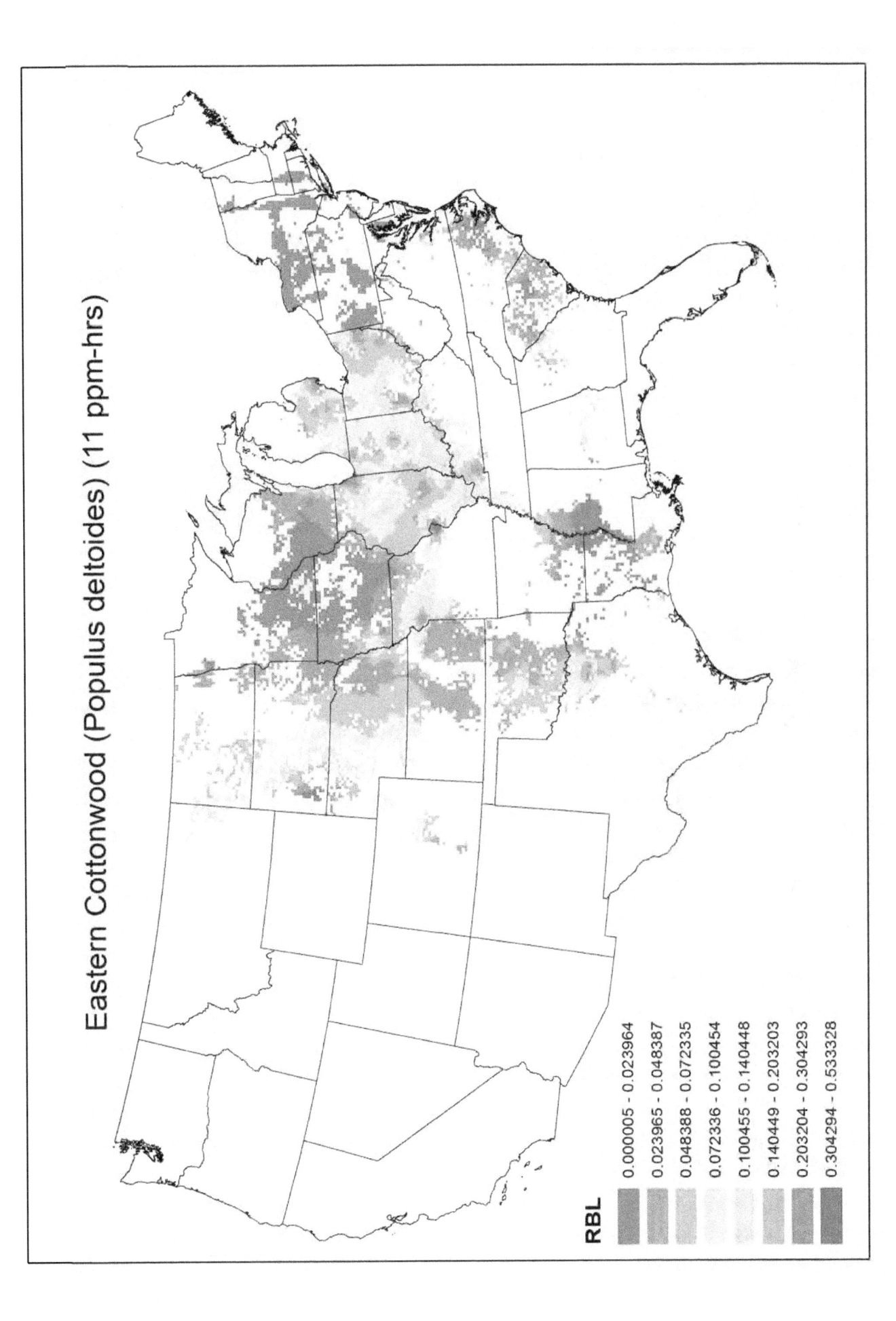

Eastern Cottonwood (Populus deltoides) (11 ppm-hrs)

RBL

0.000005 - 0.023964
0.023965 - 0.048387
0.048388 - 0.072335
0.072336 - 0.100454
0.100455 - 0.140448
0.140449 - 0.203203
0.203204 - 0.304293
0.304294 - 0.533328

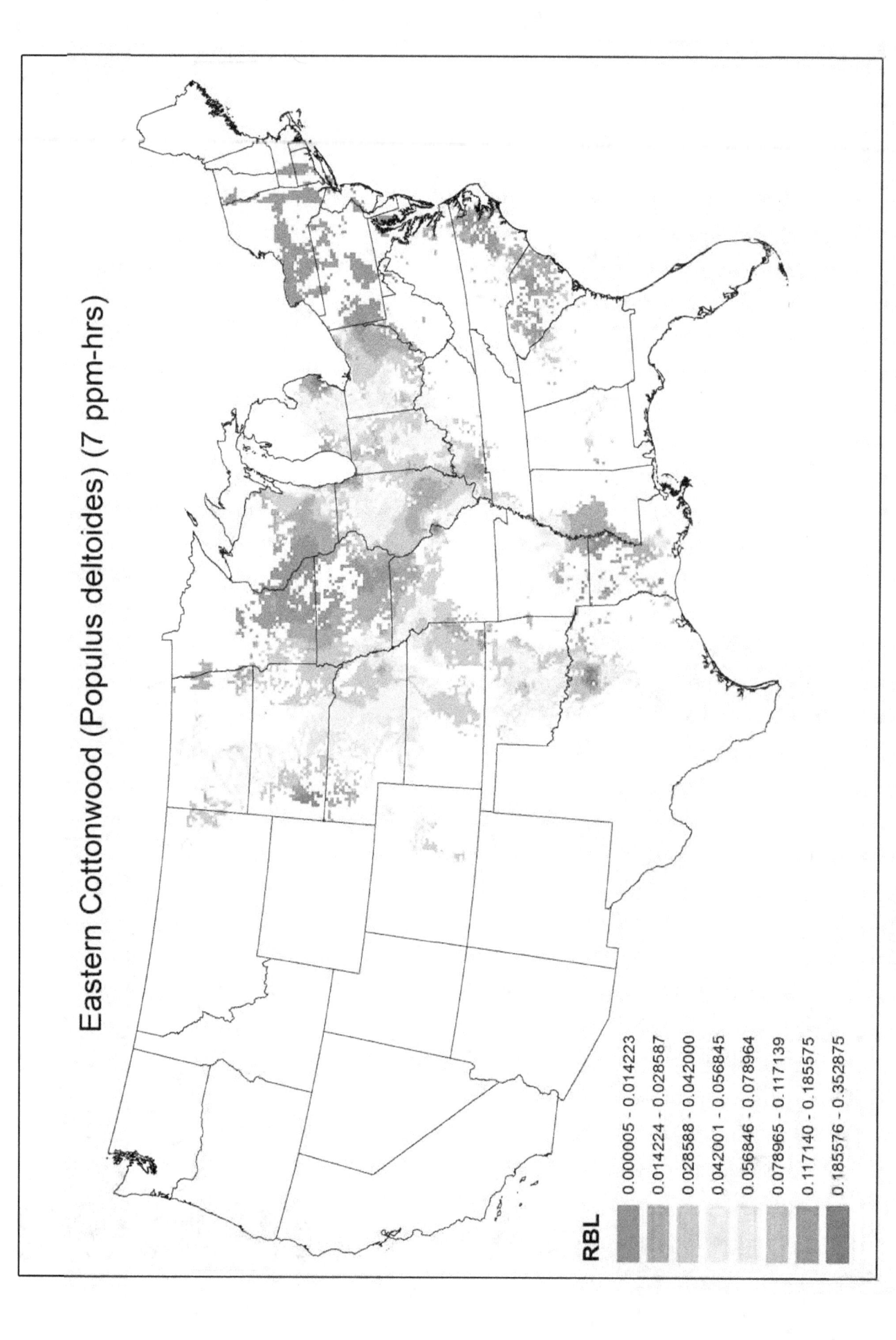

Eastern Cottonwood (Populus deltoides) (7 ppm-hrs)

RBL

	0.000005 - 0.014223
	0.014224 - 0.028587
	0.028588 - 0.042000
	0.042001 - 0.056845
	0.056846 - 0.078964
	0.078965 - 0.117139
	0.117140 - 0.185575
	0.185576 - 0.352875

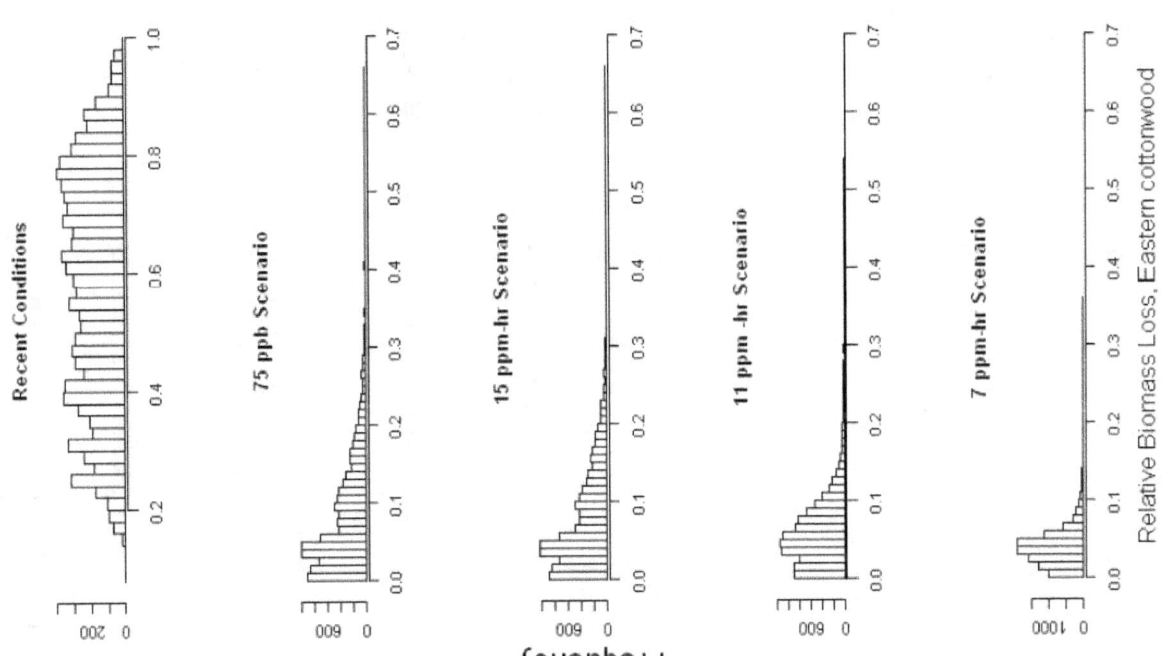

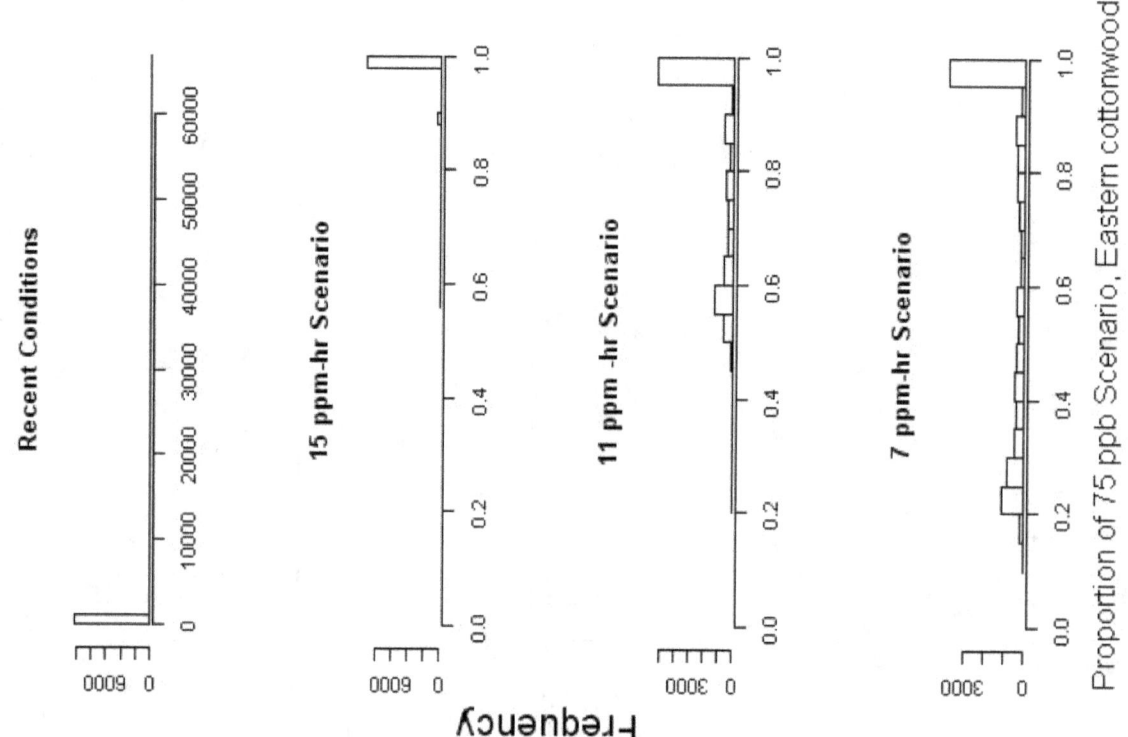

Proportion of 75 ppb Scenario, Eastern cottonwood

Quaking Aspen (Populus tremuloides) (Recent Conditions)

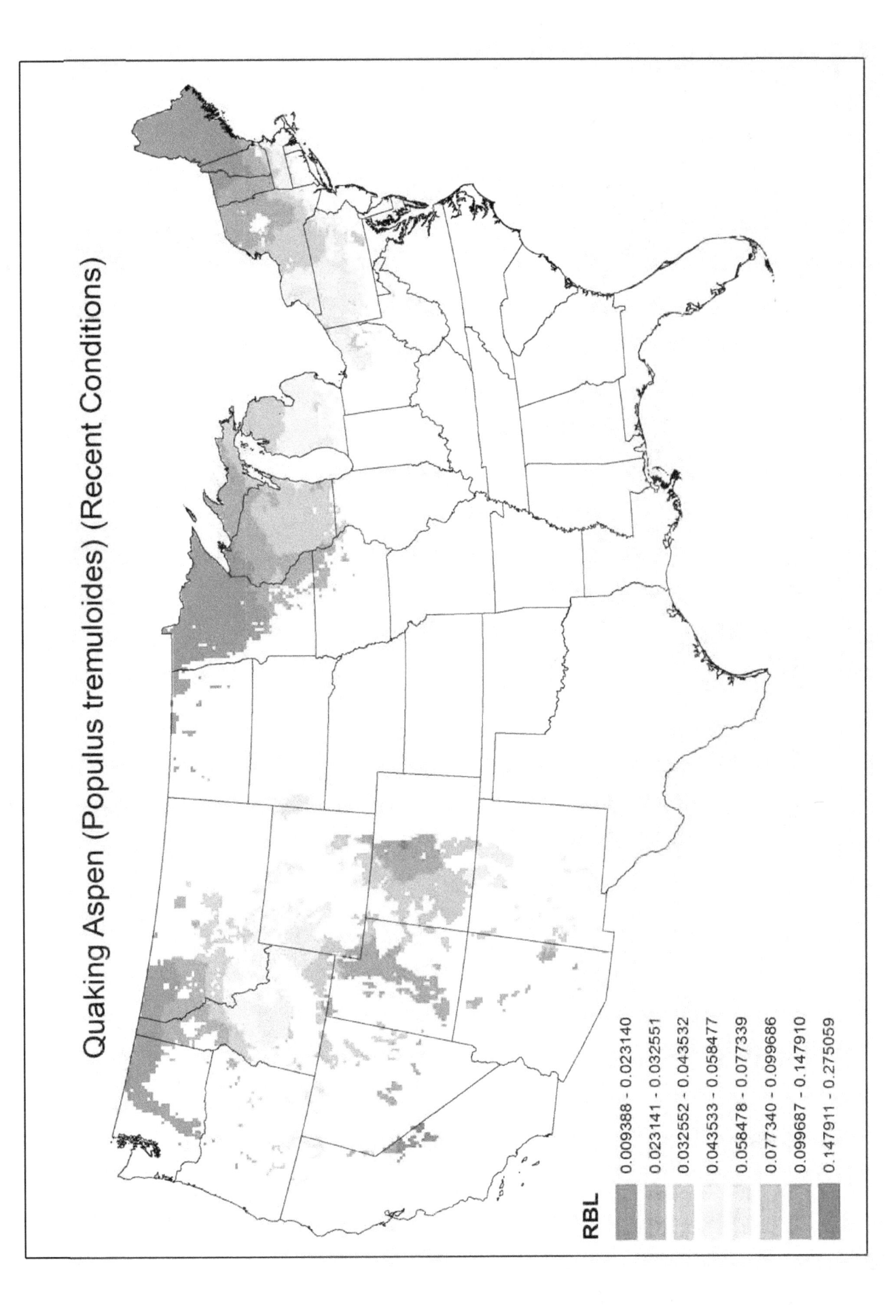

RBL

0.009388 - 0.023140
0.023141 - 0.032551
0.032552 - 0.043532
0.043533 - 0.058477
0.058478 - 0.077339
0.077340 - 0.099686
0.099687 - 0.147910
0.147911 - 0.275059

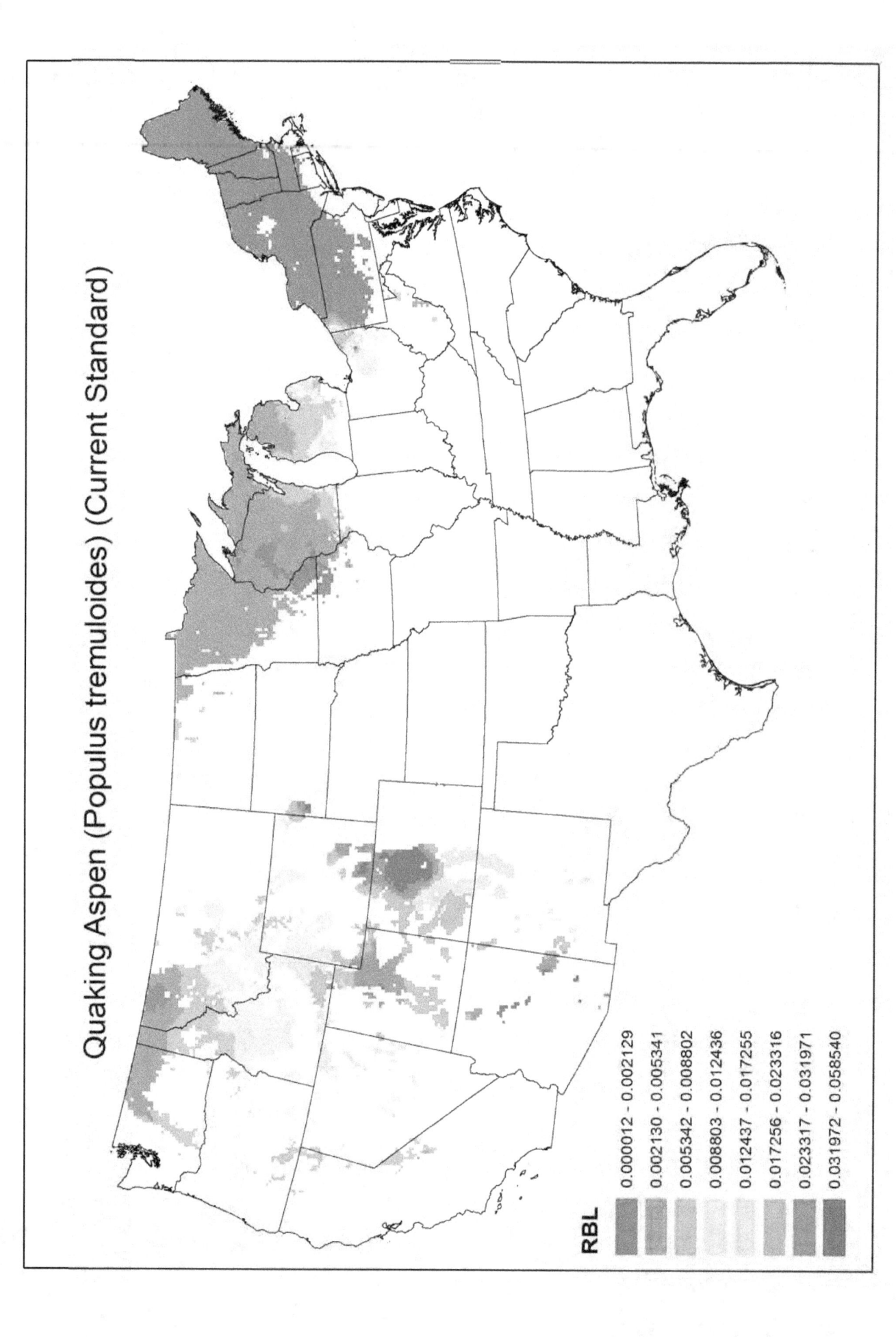

Quaking Aspen (Populus tremuloides) (Current Standard)

RBL

0.000012 - 0.002129
0.002130 - 0.005341
0.005342 - 0.008802
0.008803 - 0.012436
0.012437 - 0.017255
0.017256 - 0.023316
0.023317 - 0.031971
0.031972 - 0.058540

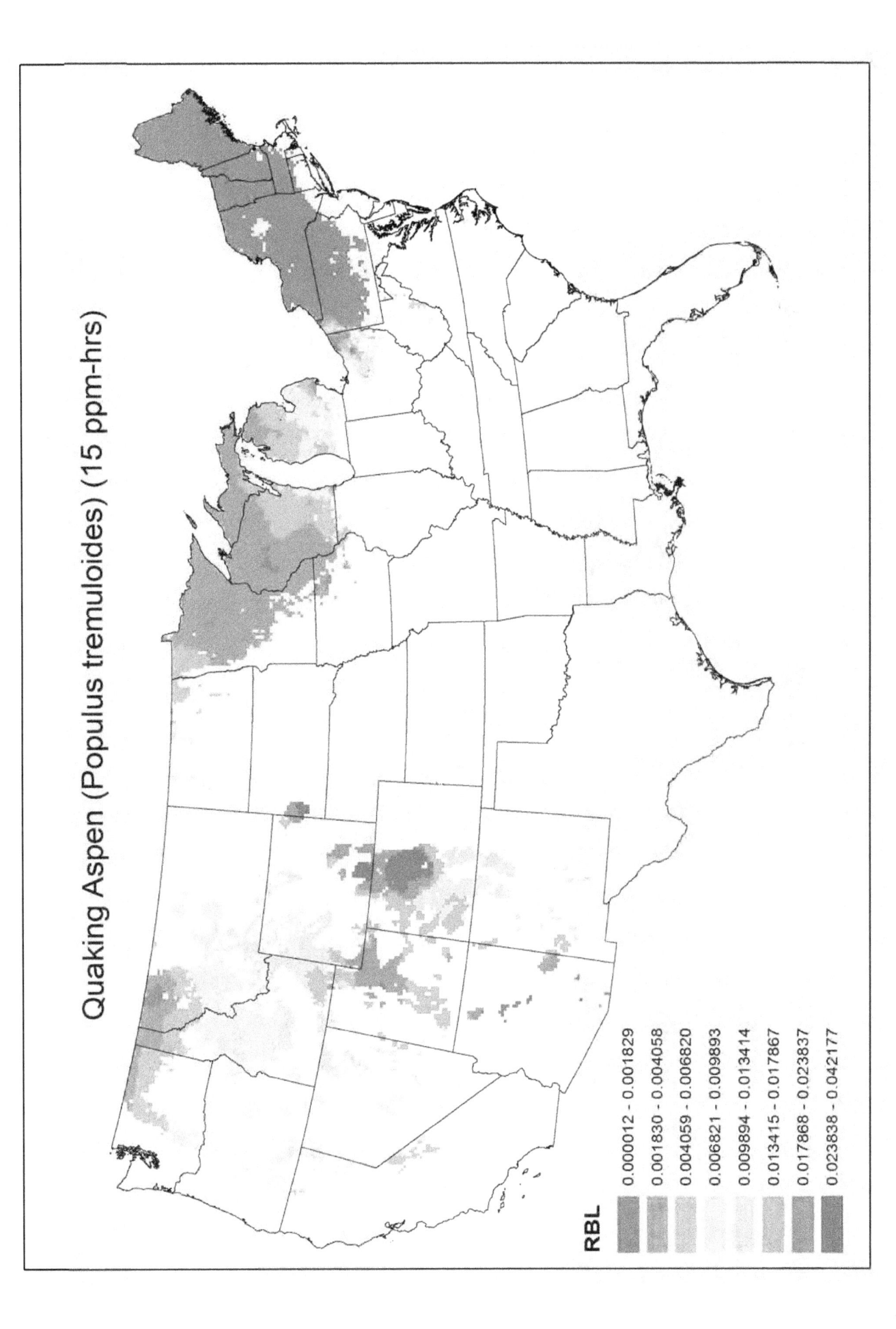

Quaking Aspen (Populus tremuloides) (15 ppm-hrs)

RBL

0.000012 - 0.001829
0.001830 - 0.004058
0.004059 - 0.006820
0.006821 - 0.009893
0.009894 - 0.013414
0.013415 - 0.017867
0.017868 - 0.023837
0.023838 - 0.042177

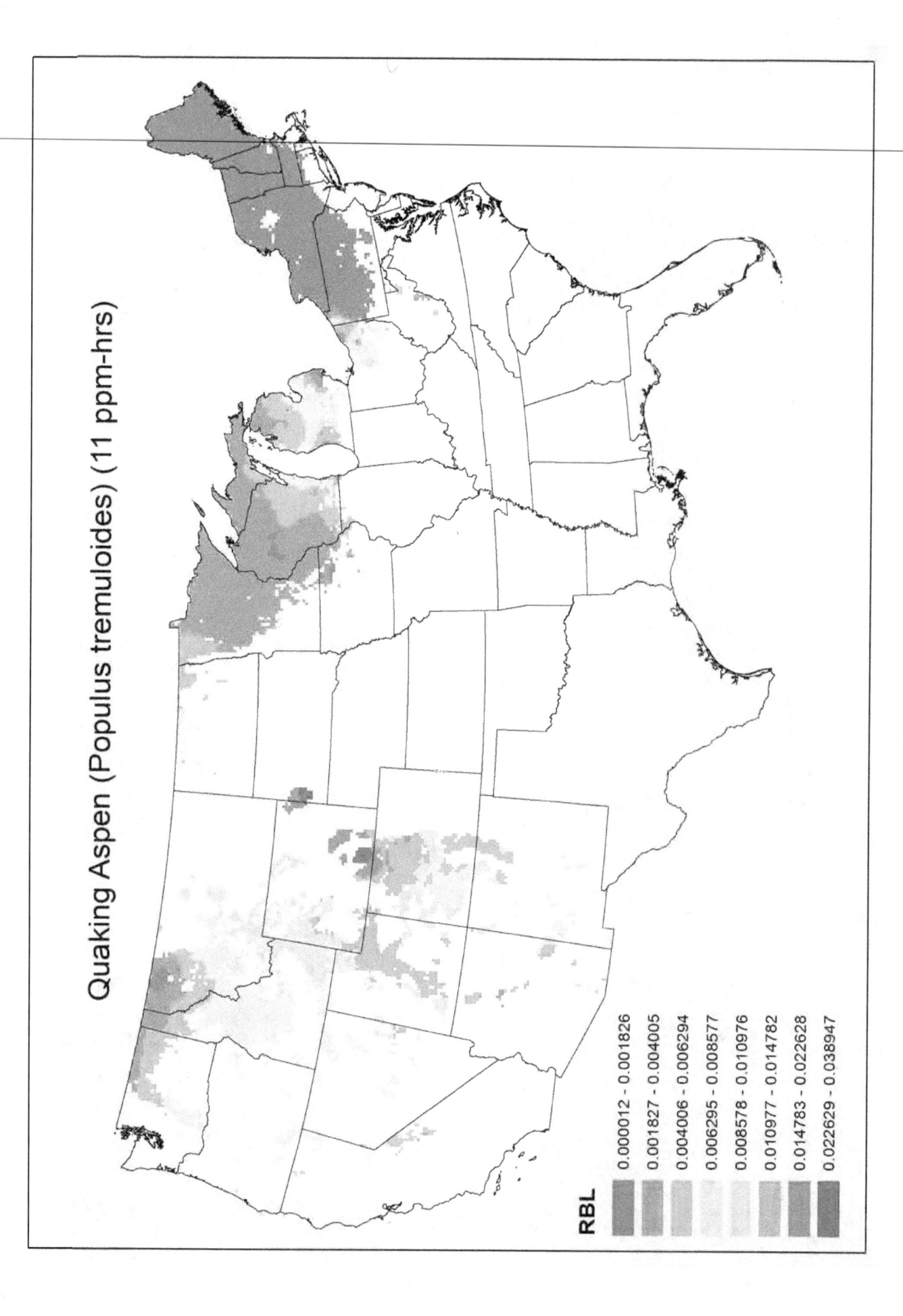

Quaking Aspen (Populus tremuloides) (11 ppm-hrs)

RBL

- 0.000012 - 0.001826
- 0.001827 - 0.004005
- 0.004006 - 0.006294
- 0.006295 - 0.008577
- 0.008578 - 0.010976
- 0.010977 - 0.014782
- 0.014783 - 0.022628
- 0.022629 - 0.038947

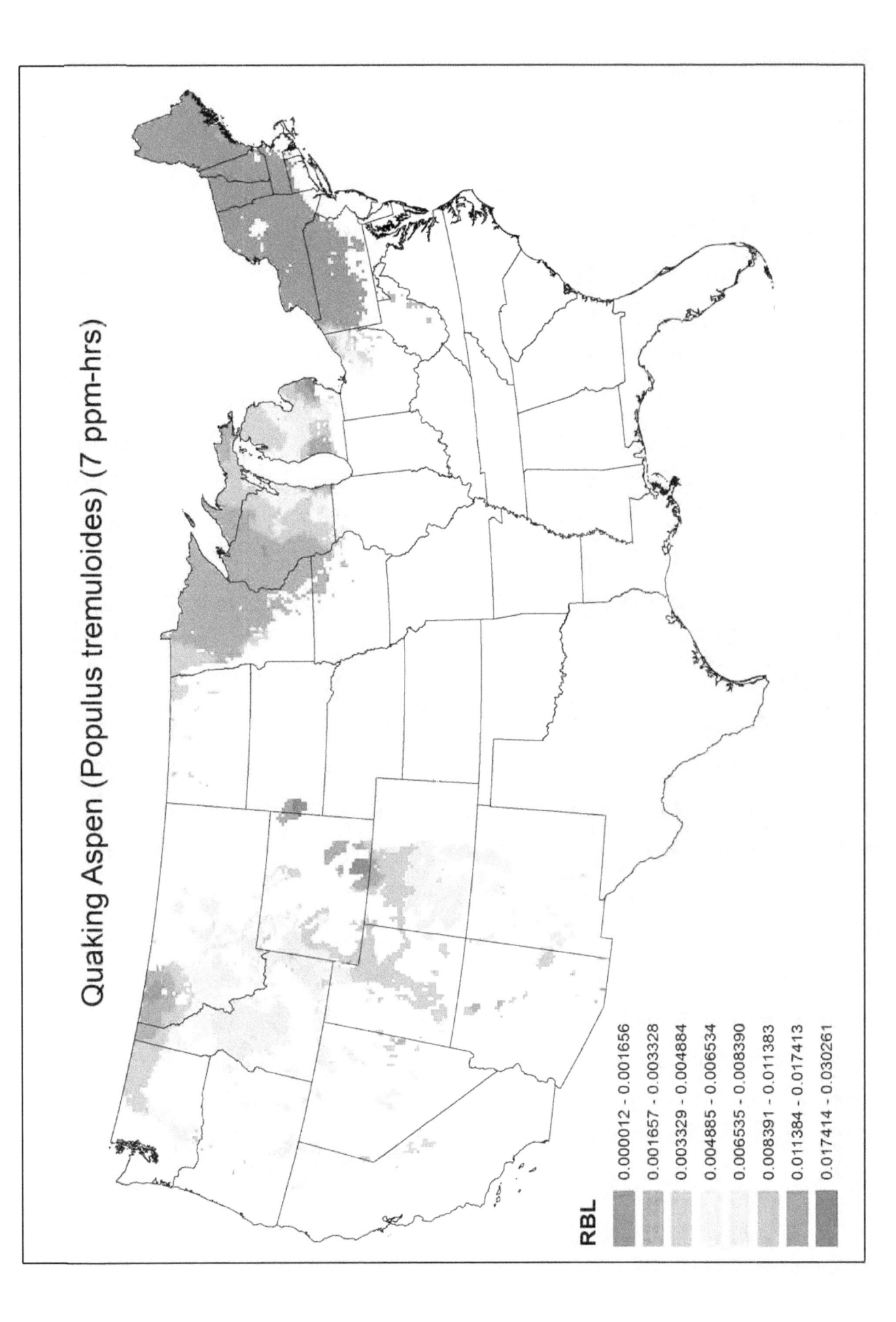

Quaking Aspen (Populus tremuloides) (7 ppm-hrs)

RBL
0.000012 - 0.001656
0.001657 - 0.003328
0.003329 - 0.004884
0.004885 - 0.006534
0.006535 - 0.008390
0.008391 - 0.011383
0.011384 - 0.017413
0.017414 - 0.030261

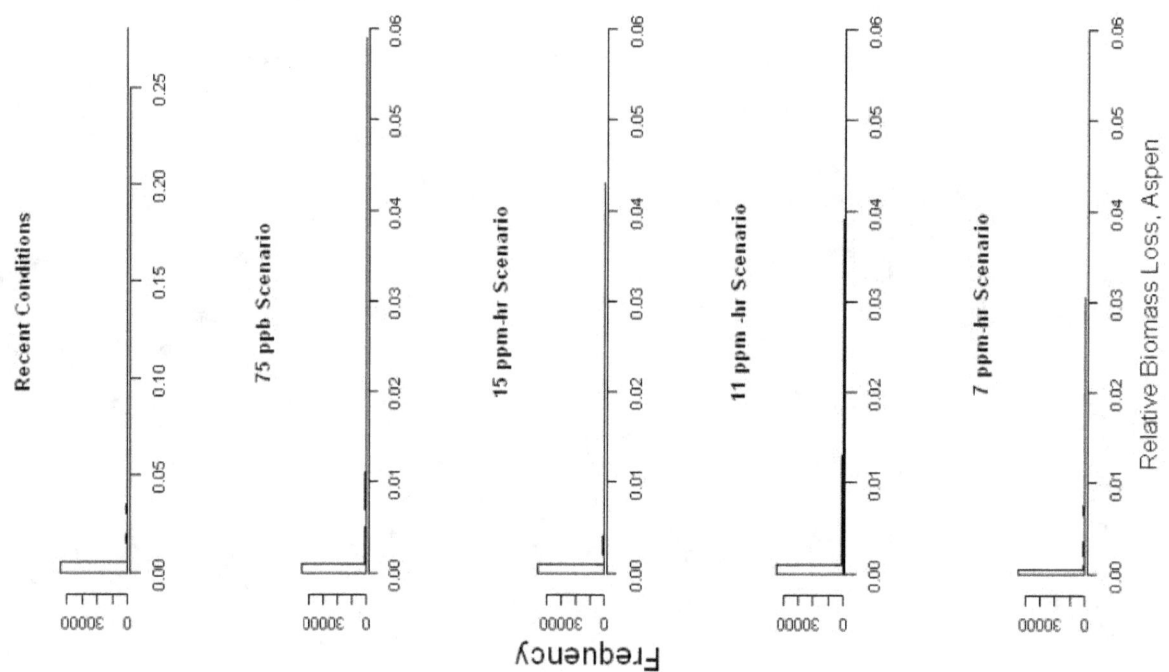

Relative Biomass Loss, Aspen

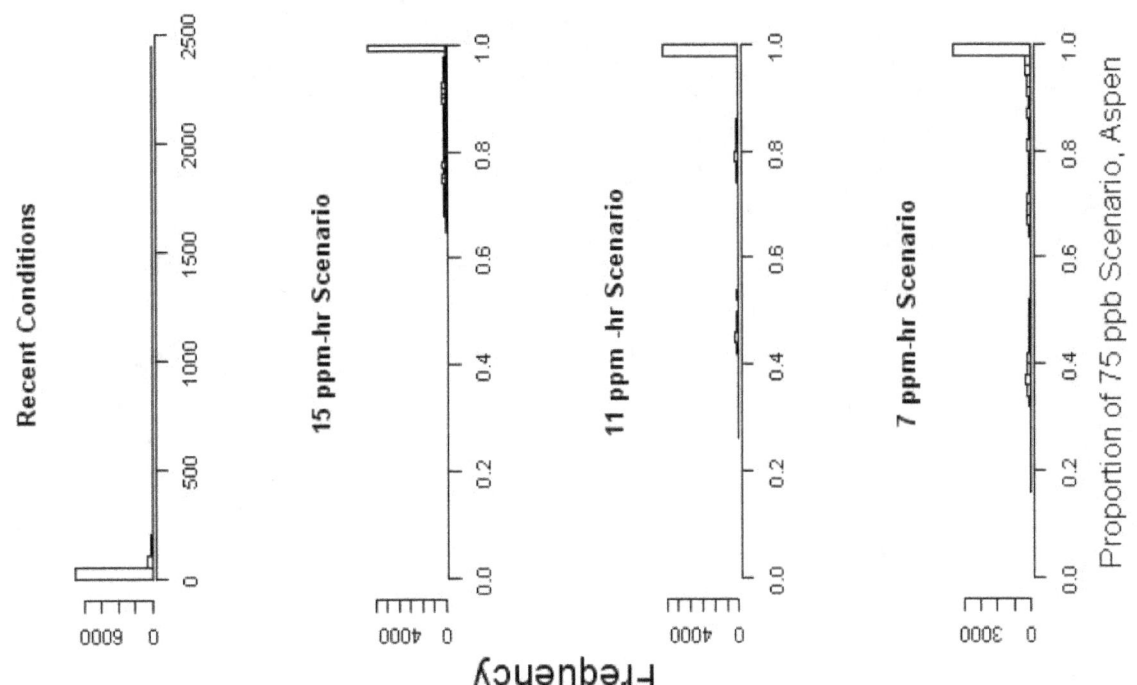

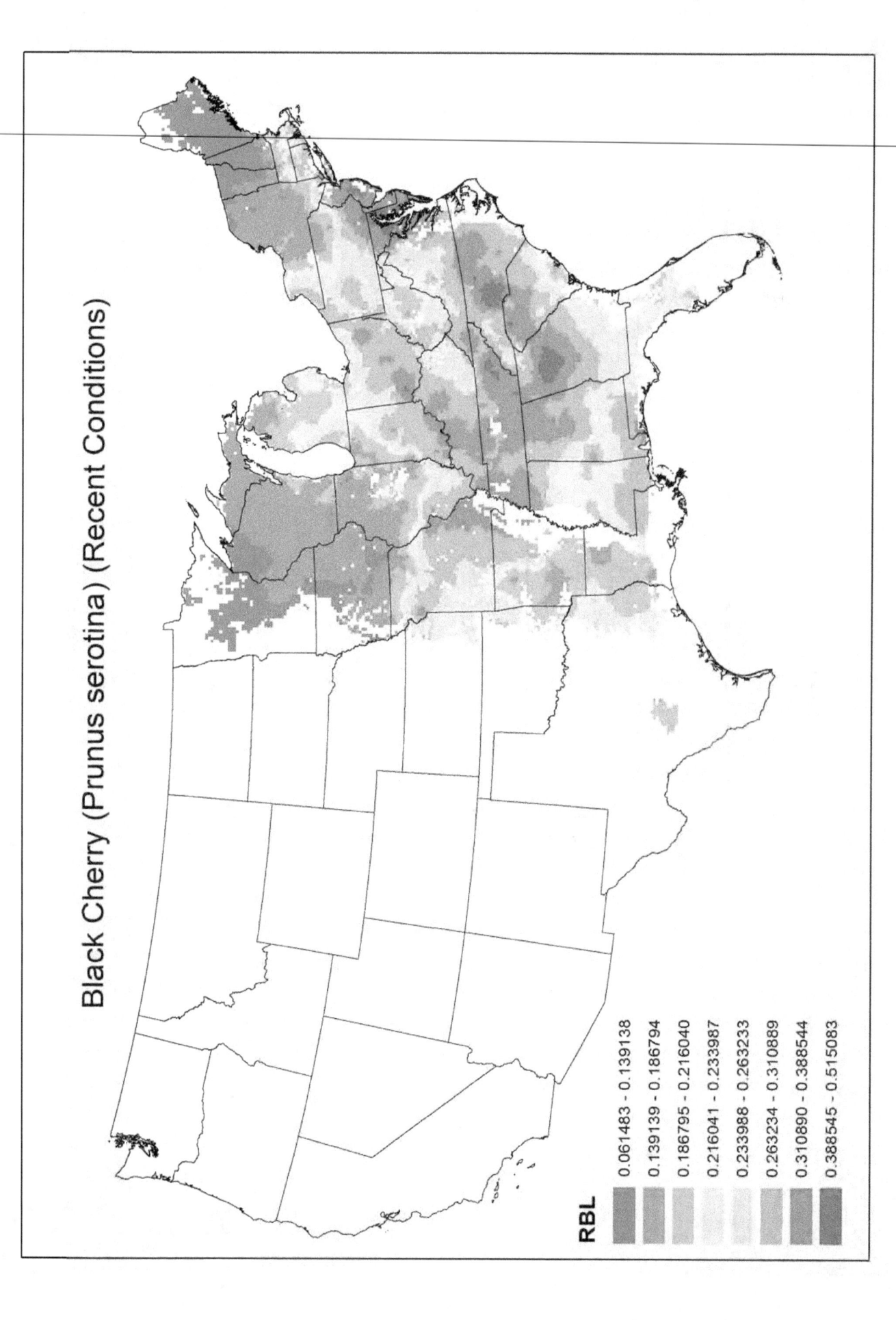

Black Cherry (Prunus serotina) (Recent Conditions)

RBL

0.061483 - 0.139138
0.139139 - 0.186794
0.186795 - 0.216040
0.216041 - 0.233987
0.233988 - 0.263233
0.263234 - 0.310889
0.310890 - 0.388544
0.388545 - 0.515083

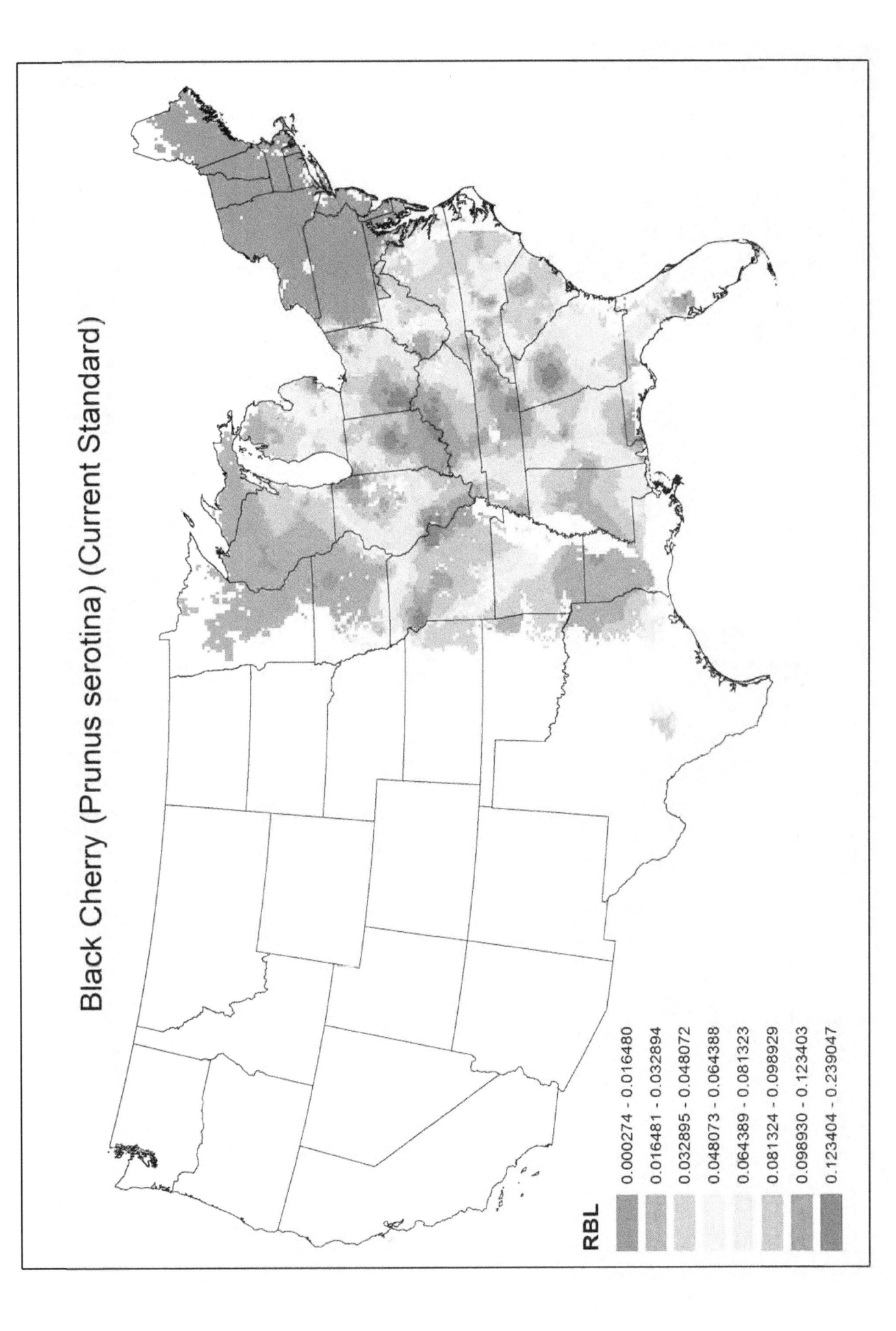

Black Cherry (Prunus serotina) (Current Standard)

RBL

0.000274 - 0.016480
0.016481 - 0.032894
0.032895 - 0.048072
0.048073 - 0.064388
0.064389 - 0.081323
0.081324 - 0.098929
0.098930 - 0.123403
0.123404 - 0.239047

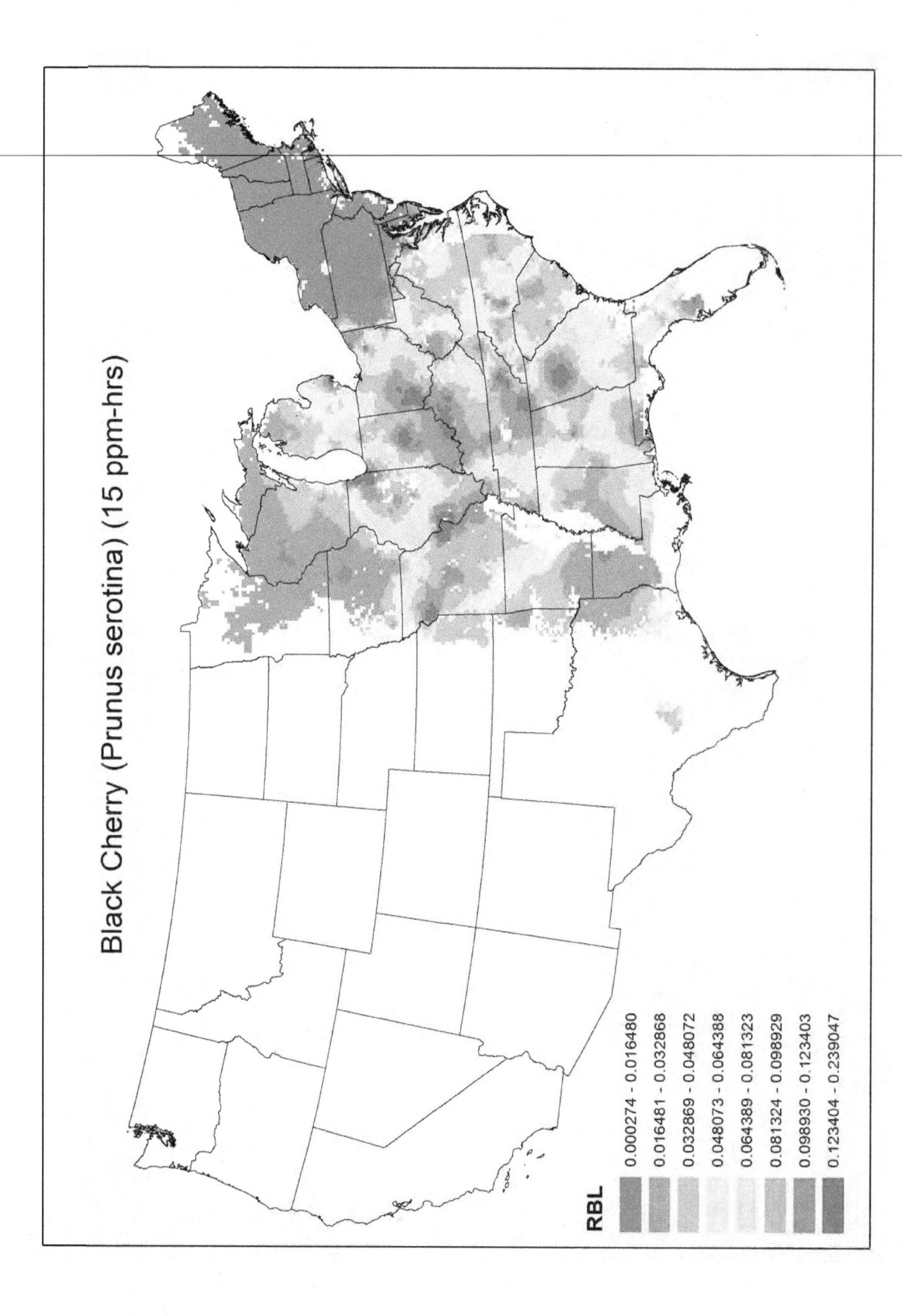

Black Cherry (Prunus serotina) (15 ppm-hrs)

RBL

0.000274 - 0.016480
0.016481 - 0.032868
0.032869 - 0.048072
0.048073 - 0.064388
0.064389 - 0.081323
0.081324 - 0.098929
0.098930 - 0.123403
0.123404 - 0.239047

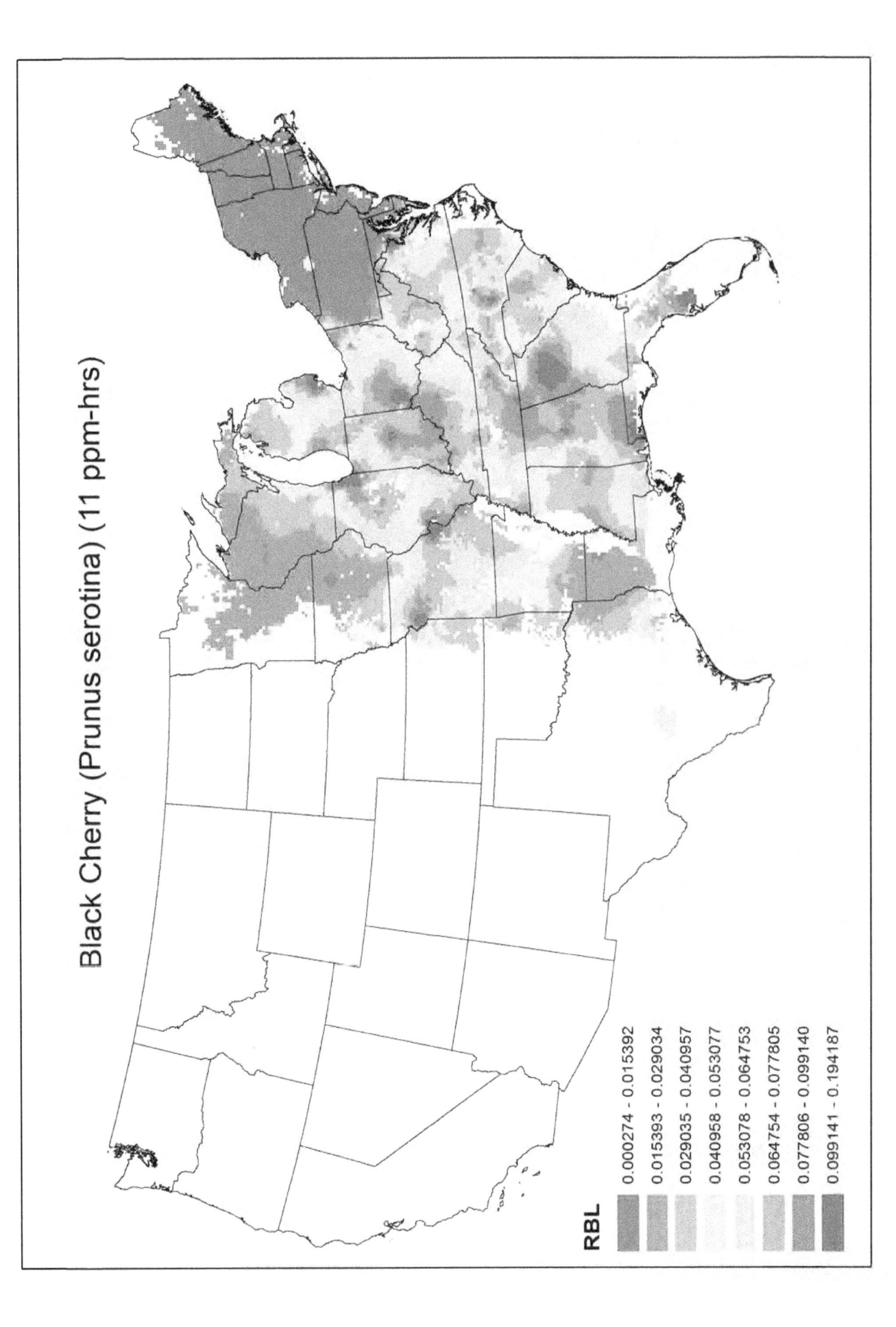

Black Cherry (Prunus serotina) (11 ppm-hrs)

RBL

0.000274 - 0.015392
0.015393 - 0.029034
0.029035 - 0.040957
0.040958 - 0.053077
0.053078 - 0.064753
0.064754 - 0.077805
0.077806 - 0.099140
0.099141 - 0.194187

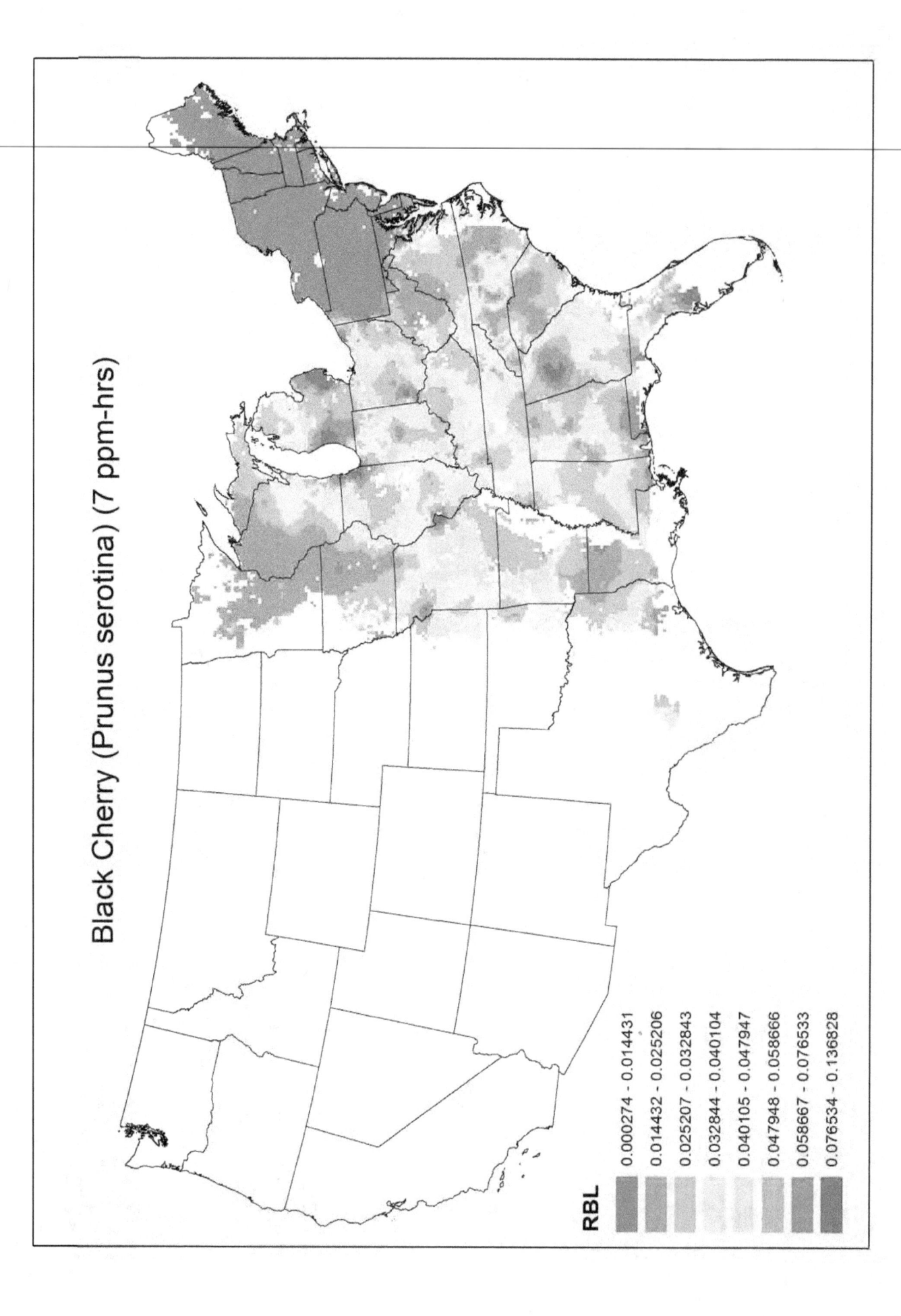

Black Cherry (Prunus serotina) (7 ppm-hrs)

RBL

- 0.000274 - 0.014431
- 0.014432 - 0.025206
- 0.025207 - 0.032843
- 0.032844 - 0.040104
- 0.040105 - 0.047947
- 0.047948 - 0.058666
- 0.058667 - 0.076533
- 0.076534 - 0.136828

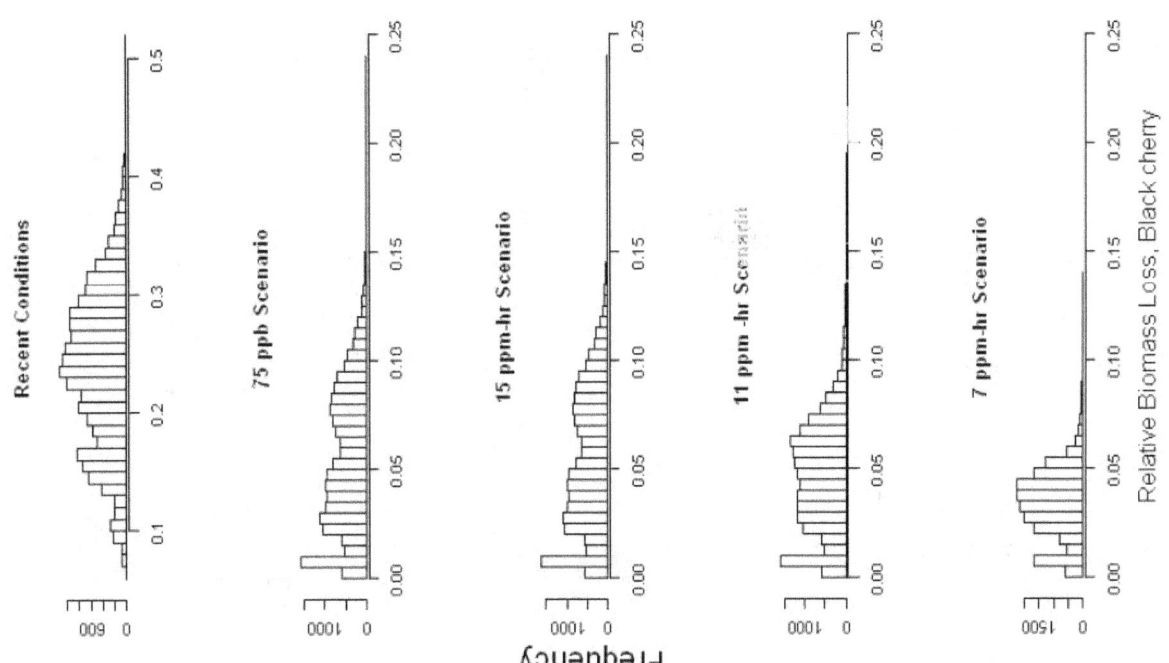

Recent Conditions

75 ppb Scenario

15 ppm-hr Scenario

11 ppm -hr Scenario

7 ppm-hr Scenario

Frequency

Relative Biomass Loss, Black cherry

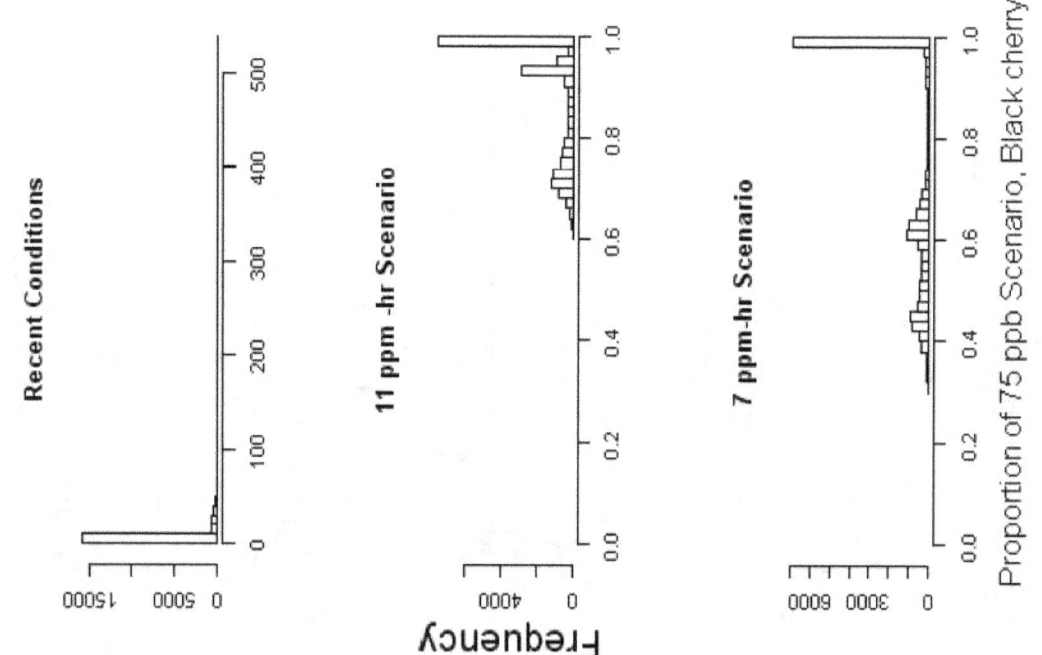

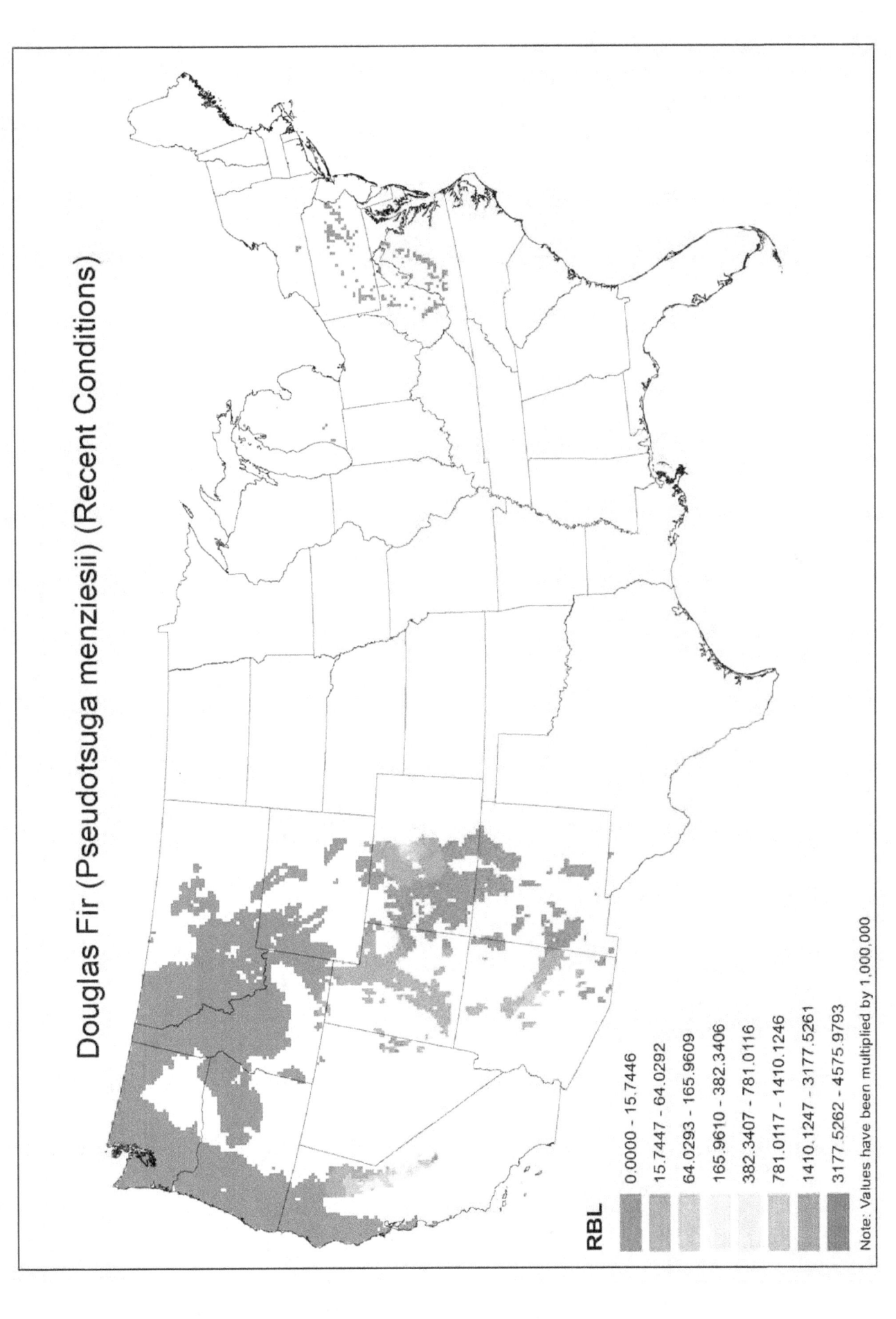

Douglas Fir (Pseudotsuga menziesii) (Recent Conditions)

RBL

- 0.0000 - 15.7446
- 15.7447 - 64.0292
- 64.0293 - 165.9609
- 165.9610 - 382.3406
- 382.3407 - 781.0116
- 781.0117 - 1410.1246
- 1410.1247 - 3177.5261
- 3177.5262 - 4575.9793

Note: Values have been multiplied by 1,000,000

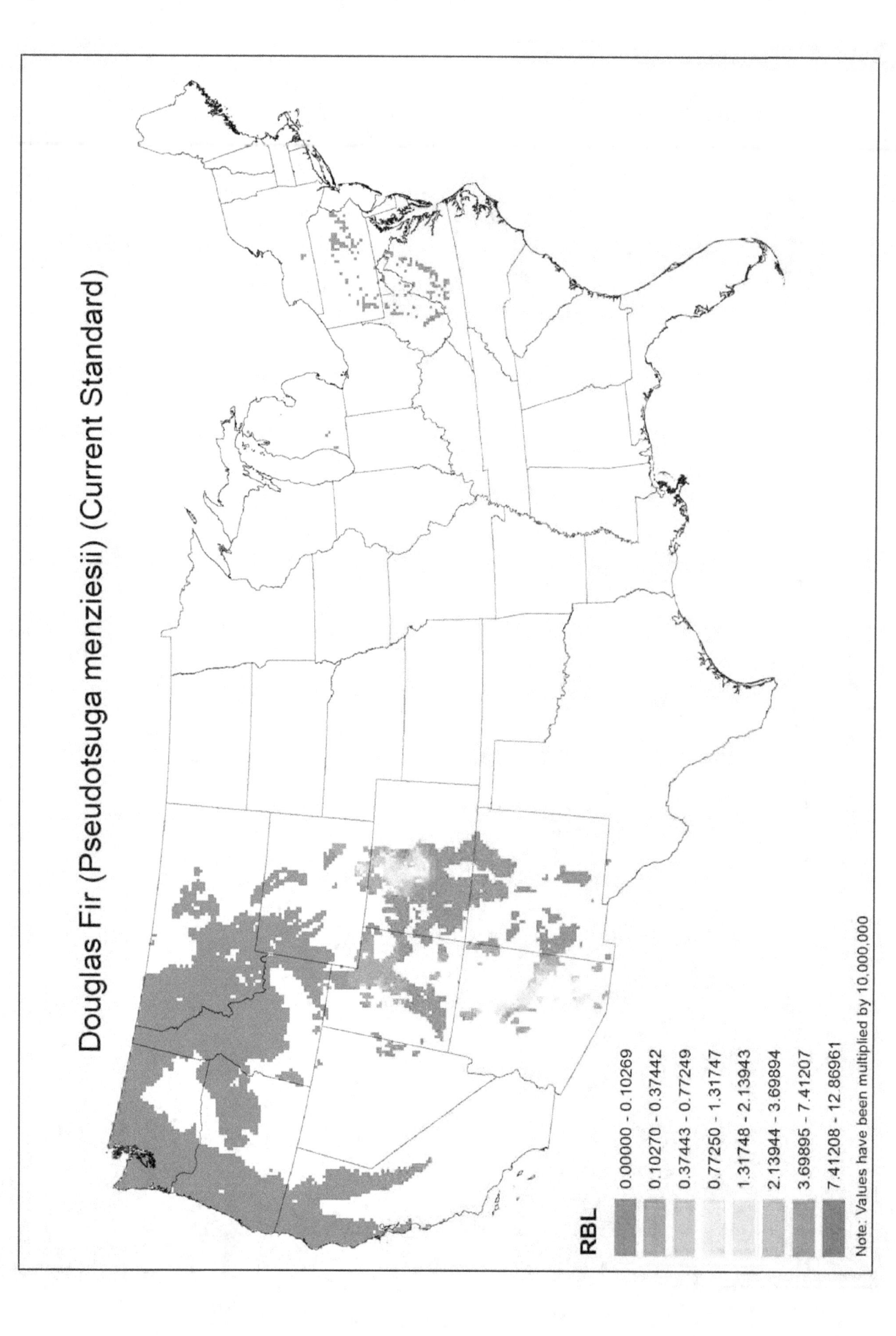

Douglas Fir (Pseudotsuga menziesii) (Current Standard)

RBL

0.00000 - 0.10269
0.10270 - 0.37442
0.37443 - 0.77249
0.77250 - 1.31747
1.31748 - 2.13943
2.13944 - 3.69894
3.69895 - 7.41207
7.41208 - 12.86961

Note: Values have been multiplied by 10,000,000

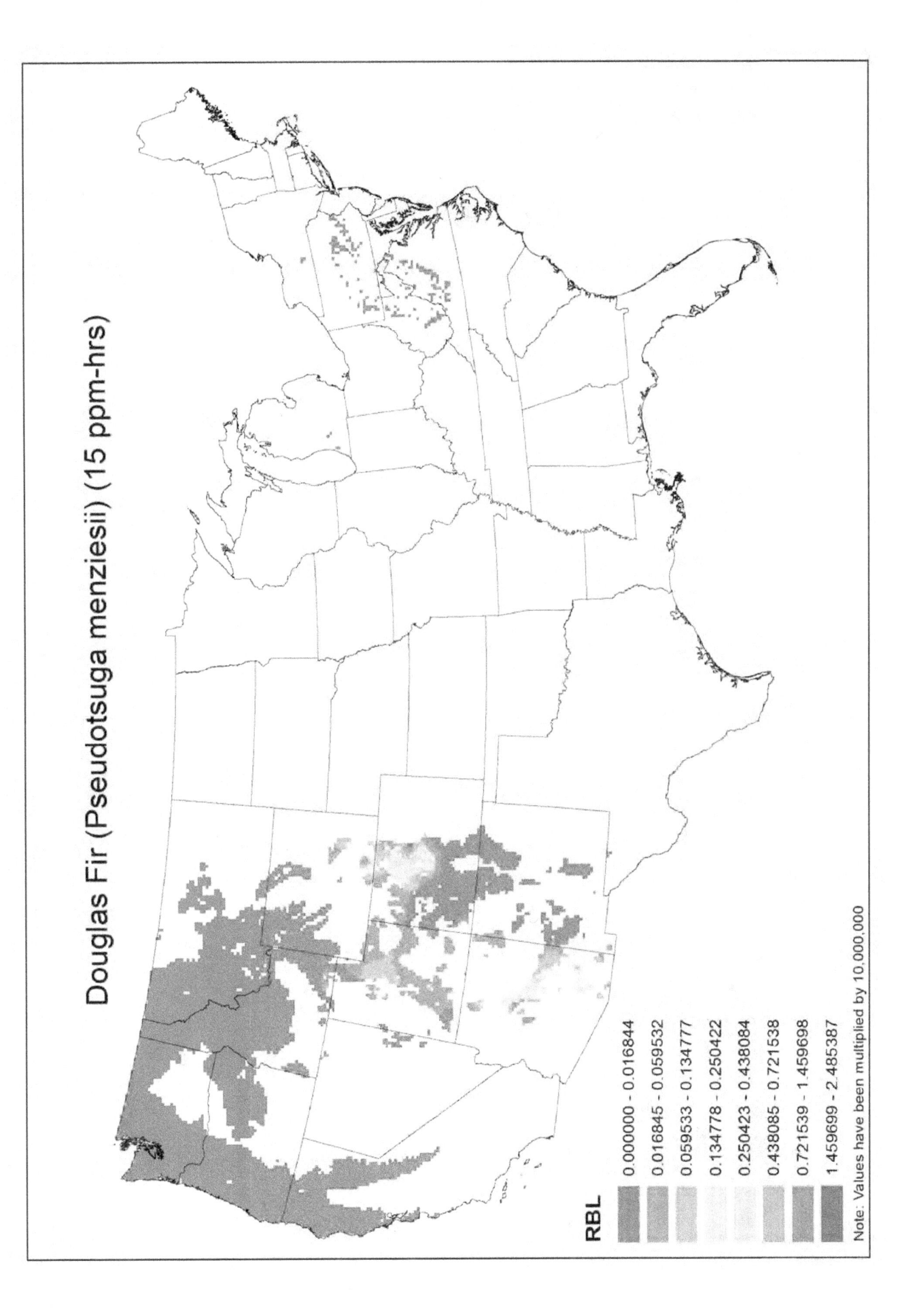

Douglas Fir (Pseudotsuga menziesii) (15 ppm-hrs)

RBL

0.000000 - 0.016844
0.016845 - 0.059532
0.059533 - 0.134777
0.134778 - 0.250422
0.250423 - 0.438084
0.438085 - 0.721538
0.721539 - 1.459698
1.459699 - 2.485387

Note: Values have been multiplied by 10,000,000

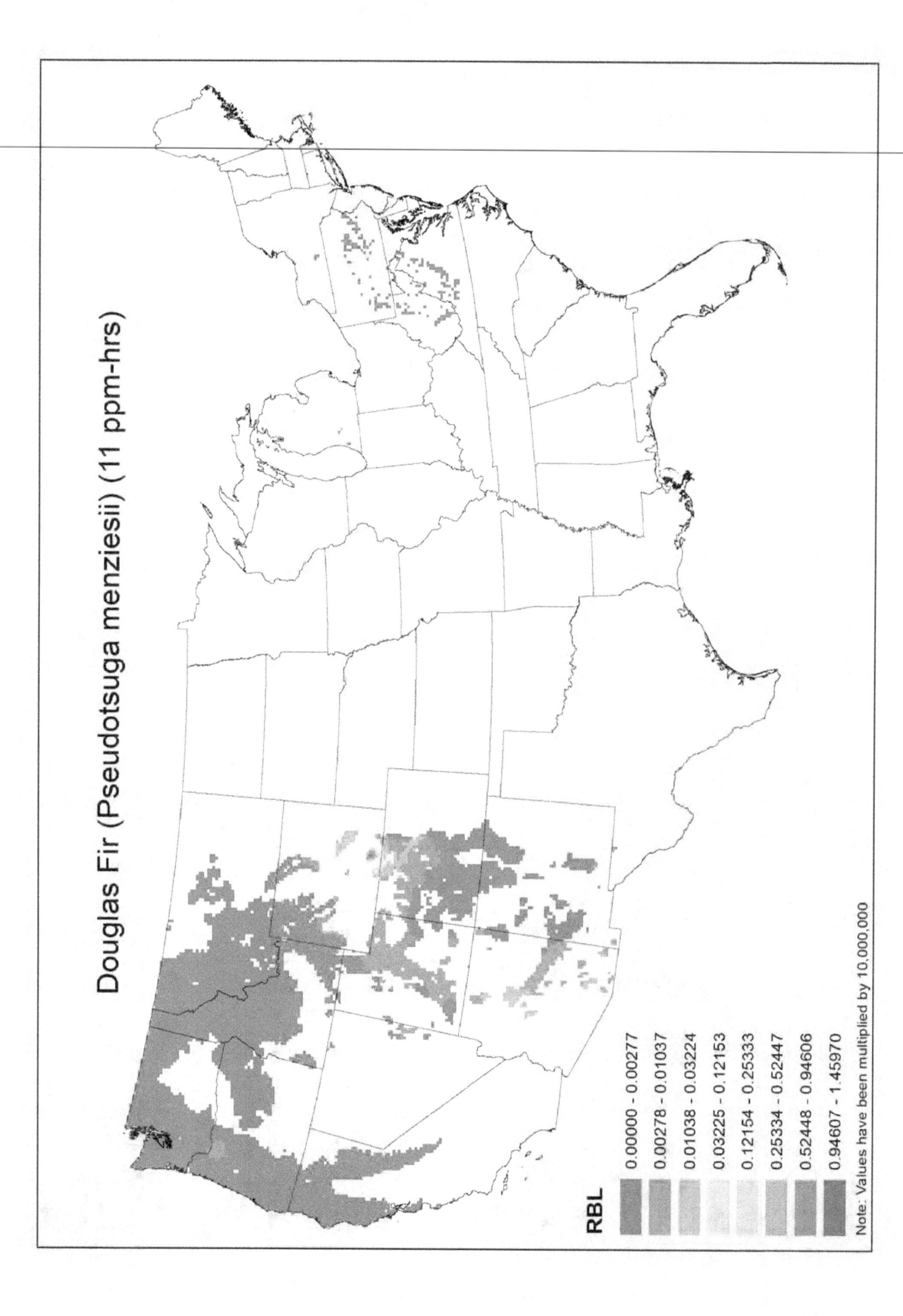

Douglas Fir (Pseudotsuga menziesii) (11 ppm-hrs)

RBL

0.00000 - 0.00277
0.00278 - 0.01037
0.01038 - 0.03224
0.03225 - 0.12153
0.12154 - 0.25333
0.25334 - 0.52447
0.52448 - 0.94606
0.94607 - 1.45970

Note: Values have been multiplied by 10,000,000

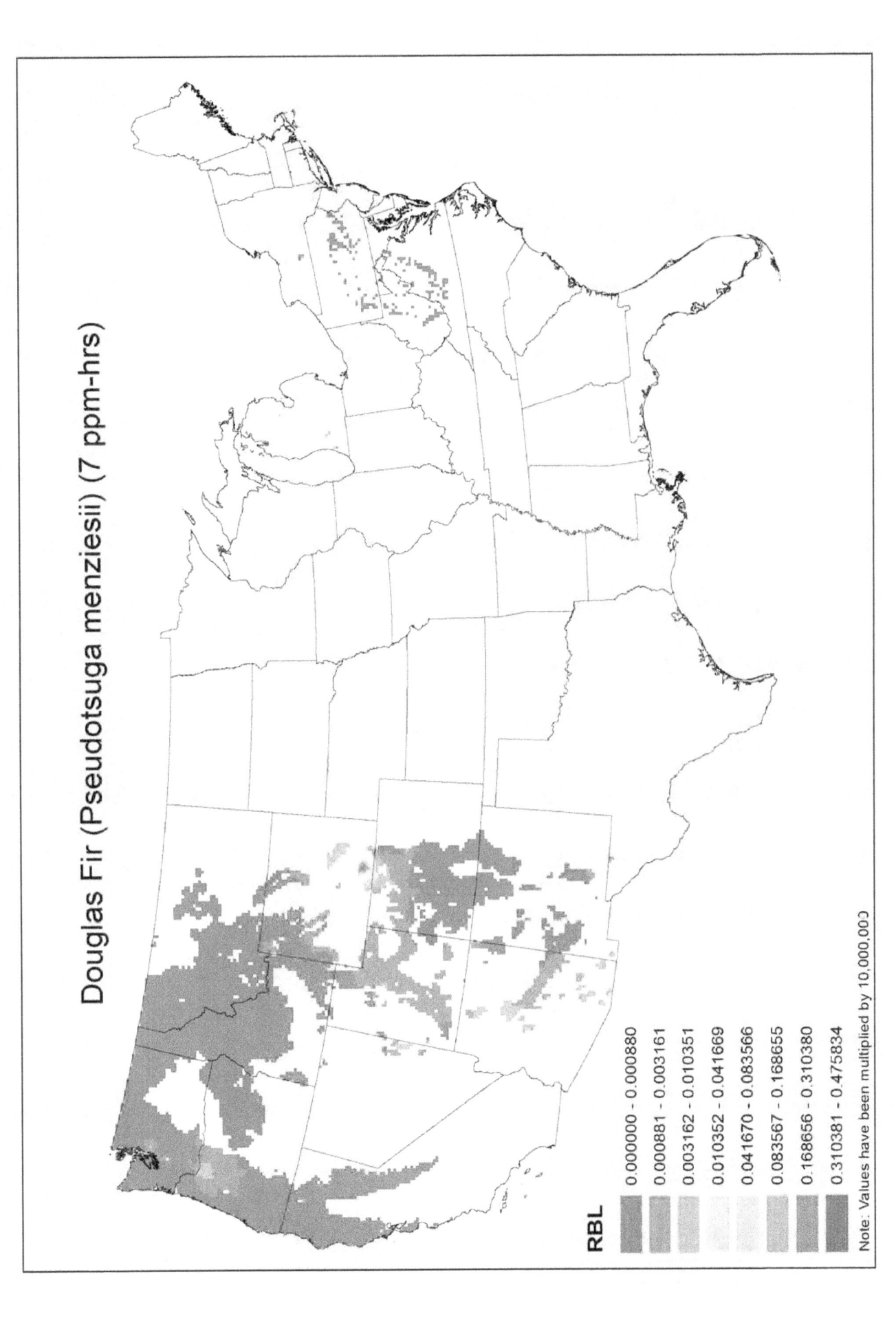

Douglas Fir (Pseudotsuga menziesii) (7 ppm-hrs)

RBL

0.000000 - 0.000880
0.000881 - 0.003161
0.003162 - 0.010351
0.010352 - 0.041669
0.041670 - 0.083566
0.083567 - 0.168655
0.168656 - 0.310380
0.310381 - 0.475834

Note: Values have been multiplied by 10,000,000.

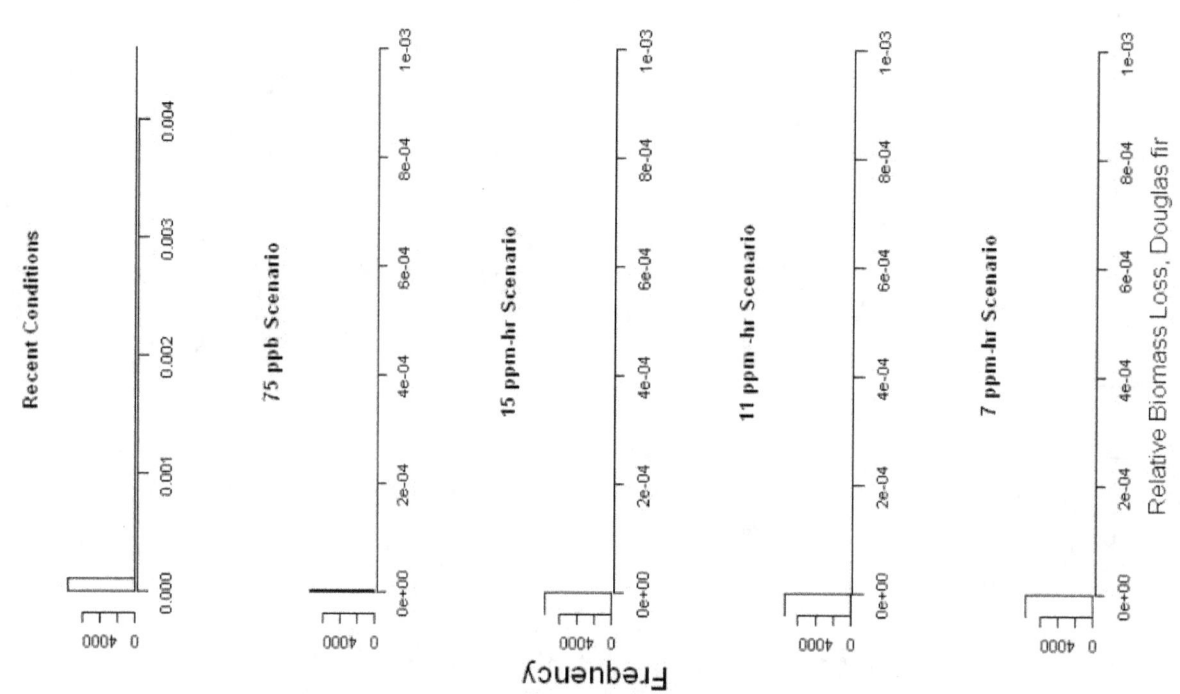

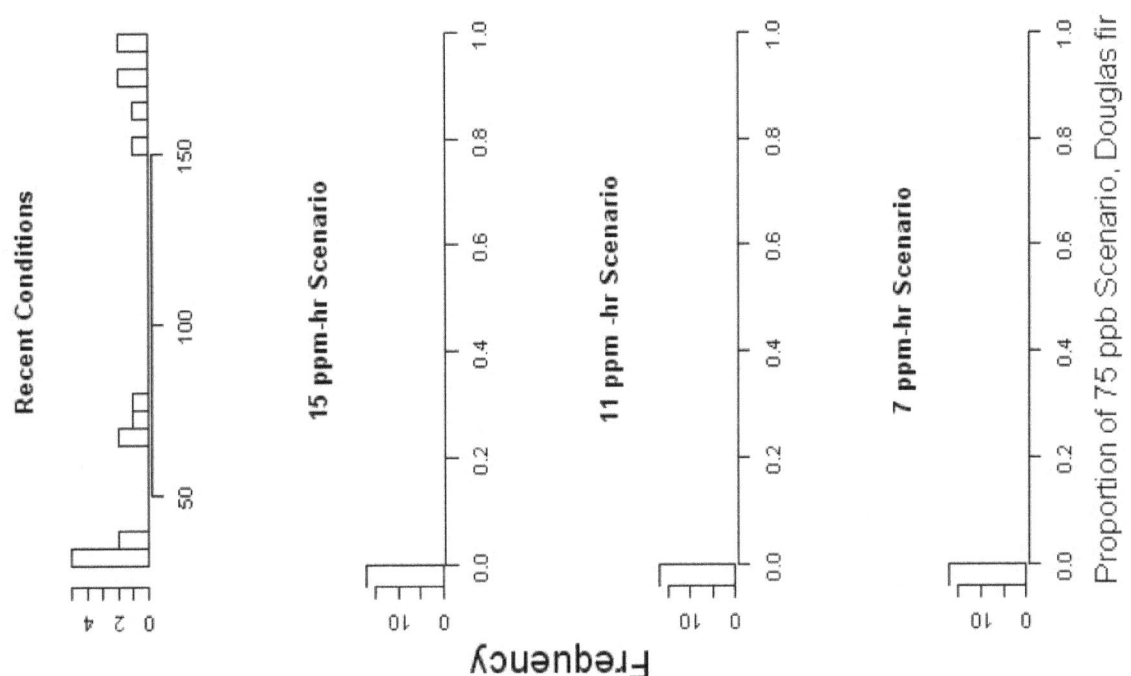

Recent Conditions

15 ppm-hr Scenario

11 ppm-hr Scenario

7 ppm-hr Scenario

Frequency

Proportion of 75 ppb Scenario, Douglas fir

July 2013

Assessment of the Impacts of Alternative Ozone Concentrations on the U.S. Forest and Agriculture Sectors

Revised Draft Report

Prepared for

Christine Davis
U.S. Environmental Protection Agency
Office of Air Quality Planning and Standards
Research Triangle Park, NC 27711

Prepared by

Robert Beach, PhD
Yuquan Zhang, PhD
Viola Glenn, MS
Ross Loomis, MA
RTI International
3040 Cornwallis Road
Research Triangle Park, NC 27709

RTI Project Number 0212979.002.016

RTI Project Number
0212979.002.016

Assessment of the Impacts of Alternative Ozone Concentrations on the U.S. Forest and Agriculture Sectors

Revised Draft Report

July 2013

Prepared for

Christine Davis
U.S. Environmental Protection Agency
Office of Air Quality Planning and Standards
Research Triangle Park, NC 27711

Prepared by

Robert Beach, PhD
Yuquan Zhang, PhD
Viola Glenn, MS
Ross Loomis, MA
RTI International
3040 Cornwallis Road
Research Triangle Park, NC 27709

CONTENTS

Appendixes

LIST OF FIGURES

LIST OF TABLES

SECTION 1
INTRODUCTION

Ground-level ozone has a number of negative impacts on human health and ecosystems that increase with ambient ozone concentrations. One important category of impacts is damage to plants that results in reduced growth rates, leading to lower productivity for agricultural crops and for the trees used to produce forestry products. As one component of the risk and ecosystem services impacts assessment of the effects of ozone, RTI International is working with the U.S. Environmental Protection Agency (EPA) to examine the potential impacts on agriculture and forests. We are examining the potential forest and agricultural market responses under alternative ambient ozone concentrations, as well as the associated effects on consumer and producer welfare. To adequately investigate the dynamic effects of policies affecting the forestry and agricultural sectors, we need an analytical framework that can simulate the time path of market and environmental impacts. The model we are using to simulate market outcomes under alternative ozone concentrations is the Forest and Agricultural Sector Optimization Model with Greenhouse Gases (FASOMGHG).

FASOMGHG is a dynamic, nonlinear programming model of the U.S. forest and agricultural sectors. Although public timberland is not explicitly modeled (the focus of the model is on private decision-maker responses to changing incentives), FASOMGHG includes an exogenous timber supply from public forestlands. Harvests from public forestlands are included in the model but are treated as exogenously determined by the government. The model solves a constrained dynamic optimization problem that maximizes the net present value of the sum of producer and consumer surplus across the two sectors over time. The model is constrained such that total production is equal to total consumption, technical input/output relationships hold, and total land use must remain constant. FASOMGHG simulates the allocation of land over time to competing activities in both the forest and agricultural sectors and the associated impacts on commodity markets. In addition, the model simulates environmental impacts resulting from changing land allocation and production practices, including detailed accounting for changes in net greenhouse gas (GHG) emissions. The model was developed to evaluate the welfare and market impacts of policies that influence land allocation and alter production activities within these sectors. FASOMGHG has been used in numerous studies to examine such issues as the potential impacts of GHG mitigation policy, climate change, timber harvest policy on public lands, federal farm programs, bioenergy production, changes in ozone levels, and other policies affecting the forest and agricultural sectors.

The comprehensive sectoral coverage provided by FASOMGHG offers several advantages for analysis of policies affecting the forest and agricultural sectors. Because the model accounts for land competition among forestry, crop production, and livestock production (pasture) and landowner responses to changing relative prices, FASOMGHG provides a more complete assessment of the net market impacts associated with a policy than models that focus only on direct policy impacts on an individual commodity or subset of alternative land uses. Using FASOMGHG enables determination of secondary impacts, such as crop switching, movements between cropland and pasture, movements between forestland and agricultural land, and changes in equilibrium quantities of forest and agricultural commodities due to changes in relative commodity prices. FASOMGHG also captures changes in the livestock market due to changes in feed costs and pasture rents, as well as changes in U.S. exports and imports of major agricultural commodities. In addition, the model accounts for changes in the primary agricultural GHGs (carbon dioxide [CO_2], methane [CH_4], and nitrous oxide [N_2O]) from the majority of emitting agricultural activities and tracks carbon sequestration and carbon losses over time. The intertemporal dynamics of the economic and biophysical systems allow for an accounting of environmental impacts over time and by region. This approach allows for a more complete quantification of net impacts, providing additional insights into the important environmental and economic impacts in these sectors.

FASOMGHG simulates a dynamic baseline and changes from that baseline in response to changes in public policy or other factors affecting these sectors. For instance, the model is often used to evaluate the joint economic and biophysical effects of GHG mitigation and bioenergy scenarios in U.S. forestry and agriculture. The model has also been used for previous studies of ozone and climate impacts on forests and agriculture. The primary data required for simulations of the impacts of changing ambient ozone concentrations are regionally disaggregated productivity effects of these concentrations for each crop and forest type included within FASOMGHG. These values are incorporated as shifts in the model production functions. Because of changes in the relative returns available for alternative land uses, landowners will alter their land use, crop mix, production practices, and other factors, moving to a new equilibrium.

In the remainder of this report, we provide an overview of FASOMGHG (Section 2), describe the methodology we used to calculate productivity effects associated with alternative ozone concentrations (Section 3), present the model inputs used to represent ozone impacts in our scenarios (Section 4), summarize the results of our analyses (Section 5), and show the distribution of welfare effects on agricultural producers (Section 6). The calculations of impacts in the main body of the report are based on comparison with current primary standards for ambient ozone concentrations. Appendixes A, B, and C present the ozone impacts on crop yields, model results, and welfare impacts, respectively, for our scenarios relative to current conditions for ambient ozone concentrations.

SECTION 2
OVERVIEW OF THE FASOMGHG MODEL

FASOMGHG[1] combines component models of agricultural crop and livestock production, renewable fuels production, livestock feeding, agricultural processing, log production, forest processing, carbon sequestration, GHG emissions, wood product markets, agricultural markets, GHG payments, and land use to systematically capture the rich mix of biophysical and economic processes that will determine the technical, economic, and environmental implications of changes in policies. FASOMGHG covers private timberlands (along with an exogenously determined timber supply from public forestlands[2]) and all agricultural activity across the conterminous ("lower 48") United States, broken into 11 market regions. Finally, FASOMGHG tracks approximately 80 forest product categories and more than 2,000 production possibilities for field crops, livestock, and renewable energy feedstocks.

FASOMGHG assumes intertemporal optimizing behavior by economic agents. For instance, the decision to continue growing a stand of timber rather than harvesting it now is based on a comparison of the net present value of timber harvest from a future period versus the net present value of harvesting now and replanting (or not replanting and shifting the land to agricultural use). Similarly, landowners make a decision to keep their land in agriculture versus afforestation based on a comparison of the net present value of returns in agriculture and forestry. Land can also move between cropland and pasture, depending on relative returns. This process establishes a land price equilibrium across the sectors (reflecting productivity in alternative uses and land conversion costs) and, given the land base interaction, a link between contemporaneous commodity prices in the two sectors as well.

The model solution portrays simultaneous multiperiod, multicommodity, multifactor market equilibria, typically over 60 to 100 years on a 5-year time-step basis, when running the combined forest-agriculture version of the model. Results yield a dynamic simulation of prices, production, management, consumption, GHG effects, and other environmental and economic indicators within these sectors under each scenario defined in the model run.

The key endogenous variables in FASOMGHG include

---

[1] See Adams et al. (2005), Beach et al. (2010), and Beach and McCarl (2010) for more detailed documentation of FASOMGHG.

[2] In the scenarios modeled for this draft report, we assumed that timber supply from public forestlands remains constant under all scenarios. However, we may revisit this assumption in the future to examine the potential effects of reduced ozone concentrations on public forests and timber supply from public lands.

- commodity and factor prices;

- production, consumption, and export and import quantities;

- land use allocations between sectors;

- management strategy adoption;

- resource use;

- economic welfare measures;

- producer and consumer surplus;

- transfer payments;

- net welfare effects; and

- environmental impact indicators, such as

  - GHG emission/sequestration of CO_2, CH_4, and N_2O and
  - total nitrogen and phosphorous applications.

Additional details on the model and key characteristics are provided in the following subsections.

Brief History and Previous Applications

The current version of FASOMGHG reflects numerous model enhancements that have been made over time, dating back to the first version of the Agricultural Sector Model (ASM) (Baumes, 1978). Since the initial version of ASM, the model has undergone many changes, including improvements for pesticide analysis by Burton (1982), as reported in Burton and Martin (1987), and a number of model additions to enable more detailed environmental and resource analyses. ASM has been used for analyses of renewable fuels dating back to the late 1970s and 1980s (Tyner et al., 1979; Chattin, 1982; Hickenbotham, 1987). In addition, ASM was applied to study ozone impacts (Hamilton, 1985; Adams, Hamilton, and McCarl, 1984), acid rain (Adams, Callaway, and McCarl, 1986), soil conservation policy (Chang et al., 1994), global climate change impacts (Adams et al., 1988, 1990, 1999, 2001; McCarl, 1999; Reilly et al., 2000, 2002), and GHG mitigation (Adams et al., 1993; McCarl and Schneider, 2001).

One of the drivers behind integrating ASM with forest-sector models to create the initial Forest and Agricultural Sector Optimization Model (FASOM) was an ASM study examining issues regarding joint forestry and agricultural GHG mitigation (Adams et al., 1993). Attempts to reconcile forestry production possibilities with the static single-year equilibrium representation in ASM led to the recognition that the model did not adequately reflect the dynamic issues associated with land allocation between forestry and agriculture. Thus, FASOM was constructed to address these limitations by linking a simple intertemporal model of the forest sector with a version of the ASM in a dynamic framework, allowing some portion of the land base in each sector to be shifted to the alternative use. Land could transfer between sectors based on its marginal profitability in all alternative forest and agricultural uses over the time horizon of the model. Management investment decisions in both sectors, including harvest timing in forestry, were made endogenous, so they too would be based on the expected profitability of an additional dollar spent on expanding future output (both timber and carbon, if valued monetarily).

The basic structure of the forest sector was based on the family of models developed to support the timber assessment component of the U.S. Forest Service's decennial Forest and Rangeland Renewable Resources Planning Act (RPA) assessment process[3]: TAMM (Timber Assessment Market Model) (Adams and Haynes, 1980, 1996; Haynes, 2003), NAPAP (North American Pulp and Paper model) (Ince, 1994; Zhang, Buongiorno, and Ince, 1993, 1996), ATLAS (Aggregate Timberland Assessment System) (Mills and Kincaid, 1992), and AREACHANGE (Alig et al., 2003, 2010a; Alig, Kline, and Lichtenstein, 2004; Alig and Plantinga, 2007). Timber inventory data and estimates of current and future timber yields were taken in large part from the ATLAS inputs used for the 2000 RPA Timber Assessment (Haynes, 2003) (these data have since been updated with information from the 2005 RPA Update assessment, as described later in this report). The AREACHANGE models provide timberland area and forest type allocations to the ATLAS model. TAMM and NAPAP are "myopic" market projection models (they project ahead one period at a time) of the solid wood and fiber products sectors in the United States and Canada. In ATLAS, harvested lands are regenerated (grown) according to exogenous assumptions about the intensity of management and associated yield volume changes. The timberland base is adjusted for gains and losses projected over time by the AREACHANGE models, including afforestation of the area moving from agriculture into forestry. Product demand relations were extracted directly from the latest versions of TAMM and NAPAP, as were product supply relations for the solid wood products and all product conversion coefficients for both solid wood and fiber commodities. Trade between the United States and Canada in all major classes of wood products is endogenous and subject to the full array of potential trade barriers and exchange rates. Timber supply also uses nearly the full set of management intensity options available in ATLAS (e.g., for the South, seven planted pine

---

[3] Adams and Haynes (2007) give a complete description of the full modeling system.

management intensity classes directly from ATLAS), and the selection of management intensity is endogenous.

In addition, detailed GHG accounting for CO_2 and major non-CO_2 GHGs was added to FASOM to create the model denoted as FASOMGHG, as described in the following paragraphs. The forest carbon accounting component of FASOMGHG is largely derived from the U.S. Forest Service's Forestry Carbon (FORCARB) modeling system, which is an empirical model of forest carbon budgets simulated across regions, forest types, land classes, forest age classes, ownership groups, and carbon pools. The U.S. Forest Service uses FORCARB, in conjunction with its economic forest-sector models (e.g., TAMM, NAPAP, ATLAS, AREACHANGE), to estimate the total amount of carbon stored in U.S. forests over time as part of the Forest Service's ongoing assessment of forest resources in general (i.e., pursuant to the RPA) and forest carbon sequestration potential in particular (Joyce, 1995; Joyce and Birdsey, 2000). Basing the model's forest carbon accounting structure on FORCARB ensures that forest carbon estimates from FASOMGHG can be compared with ongoing efforts by the Forest Service to estimate and project national forest carbon sequestration.[4] It also enables FASOMGHG to be updated over time as the FORCARB system evolves to incorporate the latest science.

After the inclusion of forest carbon accounting and some limited coverage of soil carbon changes associated with land use change, work began to widen the coverage of agricultural GHG sources and management possibilities for mitigating GHG. Schneider (2000) and McCarl and Schneider (2001) expanded ASM to account for numerous categories of GHGs and to include a detailed set of agricultural-related GHG management possibilities. That work expanded ASM to include changes in tillage, land use exchange between pasture and crops, afforestation, nitrogen fertilization alternatives, enteric fermentation, manure management, renewable fuel offsets, fossil fuel use reduction, and changes in rice cultivation. The resulting model was labeled ASMGHG.

Given the dynamic modeling and forest carbon sequestration coverage included in FASOM and the agricultural coverage in ASMGHG, it was decided to merge the agricultural alternatives into the FASOM structure. This change was manifest in the first version of FASOMGHG that was built in the context of Lee (2002). In that work, the agricultural model was expanded to have all the GHG management alternatives in ASMGHG with the additional coverage of dynamics. More recently, model modifications have been made to enhance

---

[4] Note that FASOMGHG forest carbon accounting currently reflects sequestration on private timberland. Because public forest acreage is held constant and public timber supply is exogenous, the model has assumed no change in carbon storage across scenarios. We anticipate revisiting this assumption in future modeling of ozone impacts, though, because the effects of ozone on growth rates of public forests would be expected to affect carbon sequestration on those lands.

FASOM's ability to provide detailed analyses of the agricultural and environmental impacts of bioenergy production (both liquid transportation fuels and bioelectricity) from forest and agricultural feedstocks.

In the following subsections, we provide an overview of the scope of FASOMGHG in terms of the commodities included and commodity flows between primary and secondary (processed) products, inputs used in production, U.S. regional disaggregation, land categories and allocation, market modeling, treatment of international trade, GHG accounts tracked, and other environmental impacts calculated.

Commodities

FASOMGHG includes several major groupings of agricultural and forest commodities, depending on the sector and whether they are primary commodities, processed, used for bioenergy, or mixed for livestock feed. These commodity groups are

- raw crop, livestock, forestry, and renewable fuel feedstock primary commodities grown on the land;

- processed, secondary commodities made from the raw crop, livestock, and wood products;

- bioenergy products made from renewable fuel feedstocks; and

- blended feeds for livestock consumption.

Agricultural commodities are frequently substitutable in demand. For example, sorghum is a close substitute for corn on a calorie-for-calorie basis in many uses, and beet sugar is essentially a perfect substitute for sugar derived from sugarcane. In addition, a number of feed grains are substitutes in terms of livestock feeding. Similarly, many forestry products are substitutes for one another, such as sawtimber or pulpwood derived from alternative hardwood and softwood species groups. In addition, bioenergy feedstocks derived from individual agricultural and forestry commodities are substitutes for one another (e.g., ethanol can be produced using either crop residues or logging residues, among other potential feedstocks). Thus, the mix of commodities that will be produced in a given model run depends on interactions between many related markets.

2.1.1 Primary Commodities

Primary commodity production is derived from allocation decisions that reflect the set of production possibilities for field crops, livestock, and biofuels. The allocation decisions are based on optimizing across the budgets associated with each production possibility, given prices for outputs and inputs. Budgets are based on using inputs to produce a given level of outputs.

In the model, primary commodities can be used directly or converted to secondary products via processing activities with associated costs (e.g., soybean crushing to meal and oil, livestock to meat and dairy). Primary commodities can go to livestock use, feed mixing, processing, domestic consumption, or exports. A mixture of primary commodities and processed products is supplied to meet national-level demands in each market. Table 2-1 summarizes the primary commodities currently included within FASOMGHG and their units. There are 40 primary crop products (including multiple subcategories of crops, such as grapefruit, oranges, and tomatoes), 25 primary livestock products, 12 categories of forest and agricultural residues, and 32 categories of public and private domestic and imported logs.

Table 2-1.   Primary Commodities

Commodities	Units
Crop Products	
Barley	Bushels
Canola	Hundredweight (cwt)
Corn	Bushels
Cotton	480 lb bales
Grapefruit, fresh (67 lb box)	1,000 boxes (CA, AZ)
Grapefruit, fresh (80 lb box)	1,000 boxes (TX)
Grapefruit, fresh (85 lb box)	1,000 boxes (FL)
Grapefruit, processing (67 lb box)	1,000 boxes (CA, AZ)
Grapefruit, processing (80 lb box)	1,000 boxes (TX)
Grapefruit, processing (85 lb box)	1,000 boxes (FL)
Hay	U.S. tons
Hybrid poplar	U.S. tons
Miscanthus	U.S. tons
Oats	Bushels
Orange, fresh (75 lb box)	1,000 boxes (CA, AZ)
Orange, fresh (85 lb box)	1,000 boxes (TX)
Orange, fresh (90 lb box)	1,000 boxes (FL)

(continued)

Table 2-1. Primary Commodities (continued)

Commodities	Units
Orange, processing (75 lb box)	1,000 boxes (CA, AZ)
Orange, processing (85 lb box)	1,000 boxes (TX)
Orange, processing (90 lb box)	1,000 boxes (FL)
Potatoes	cwt
Rice	cwt
Rye	Bushels
Silage	U.S. tons
Sorghum, energy	Dry metric tons
Sorghum, grain	cwt
Sorghum, sweet	U.S. tons
Sorghum, sweet (ratooned)	U.S. tons
Soybeans	Bushels
Sugar beets	U.S. tons
Sugarcane	U.S. tons
Switchgrass	U.S. tons
Tomatoes, fresh	cwt
Tomatoes, processing	U.S. tons
Wheat, durum	Bushels
Wheat, hard red spring	Bushels
Wheat, hard red winter	Bushels
Wheat, soft red winter	Bushels
Wheat, soft white	Bushels
Willow	U.S. tons
Livestock Products	
Beef cows, culled	100 lb (liveweight)
Beef slaughter, feedlot	100 lb (liveweight)
Beef slaughter, nonfed	100 lb (liveweight)
Broilers	100 lb (liveweight)
Calves, dairy	100 lb (liveweight)
Calves for slaughter	100 lb (liveweight)
Calves, heifer	100 lb (liveweight)
Calves, steer	100 lb (liveweight)
Calves, stocked	100 lb (liveweight)

(continued)

Table 2-1. Primary Commodities (continued)

Commodities	Units
Calves, stocked heifer	100 lb (liveweight)
Calves, stocked steer	100 lb (liveweight)
Dairy cows, culled	100 lb (liveweight)
Eggs	Dozens at farm level
Ewes, culled	100 lb (liveweight)
Hogs for slaughter	100 lb (liveweight)
Horses and mules	Number of head
Lamb slaughter	100 lb (liveweight)
Milk	100 lb
Pigs, feeder	100 lb (liveweight)
Sows, culled	100 lb (liveweight)
Turkeys	100 lb (liveweight)
Wool, raw	Pounds
Yearlings, stocked	100 lb (liveweight)
Yearlings, stocked heifer	100 lb (liveweight)
Yearlings, stocked steer	100 lb (liveweight)
Forest and Agricultural Residues	
Barley residues	U.S. tons
Biomanure, beef	U.S. tons
Biomanure, dairy	U.S. tons
Corn residues	U.S. tons
Logging residues, hardwood	U.S. tons
Logging residues, softwood	U.S. tons
Milling residues, hardwood	U.S. tons
Milling residues, softwood	U.S. tons
Oats residues	U.S. tons
Rice residues	U.S. tons
Sorghum residues	U.S. tons
Wheat residues	U.S. tons

(continued)

Table 2-1.   Primary Commodities (continued)

Commodities	Units
Logs From Timber Harvest	
Fuel log, hardwood privately produced	1,000 cu ft in the woods
Fuel log, hardwood publicly produced	1,000 cu ft in the woods
Fuel log, imported hardwood	1,000 cu ft in the woods
Fuel log, imported softwood	1,000 cu ft in the woods
Fuel log, softwood privately produced	1,000 cu ft in the woods
Fuel log, softwood publicly produced	1,000 cu ft in the woods
Pulp log, hardwood privately produced	1,000 cu ft in the woods
Pulp log, hardwood publicly produced	1,000 cu ft in the woods
Pulp log, softwood privately produced	1,000 cu ft in the woods
Pulp log, softwood publicly produced	1,000 cu ft in the woods
Pulp log, imported softwood	1,000 cu ft in the woods
Pulp log, imported hardwood	1,000 cu ft in the woods
Saw log, hardwood privately produced	1,000 cu ft in the woods
Saw log, hardwood publicly produced	1,000 cu ft in the woods
Saw log, imported hardwood	1,000 cu ft in the woods
Saw log, imported softwood	1,000 cu ft in the woods
Saw log, softwood privately produced	1,000 cu ft in the woods
Saw log, softwood publicly produced	1,000 cu ft in the woods
Fuel log, imported softwood	1,000 cu ft delivered to the mill
Fuel log, imported hardwood	1,000 cu ft delivered to the mill
Fuel log, hardwood privately produced	1,000 cu ft delivered to the mill
Fuel log, hardwood publicly produced	1,000 cu ft delivered to the mill
Fuel log, softwood privately produced	1,000 cu ft delivered to the mill
Fuel log, softwood publicly produced	1,000 cu ft delivered to the mill
Pulp log, imported hardwood	1,000 cu ft delivered to the mill
Pulp log, imported softwood	1,000 cu ft delivered to the mill
Pulp log, hardwood publicly produced	1,000 cu ft delivered to the mill
Pulp log, softwood publicly produced	1,000 cu ft delivered to the mill
Saw log, imported hardwood	1,000 cu ft delivered to the mill
Saw log, imported softwood	1,000 cu ft delivered to the mill
Saw log, hardwood publicly produced	1,000 cu ft delivered to the mill
Saw log, softwood publicly produced	1,000 cu ft delivered to the mill

2.1.2 Secondary Commodities

As shown in Table 2-2, FASOMGHG contains a set of processing activities that make secondary commodities using primary commodities and other inputs (included as a processing cost). Secondary commodities are generally included in the model either to represent substitution or to depict demand for components of products. For example, processing possibilities for soybeans are included depicting soybeans being crushed into soybean meal and soybean oil because these secondary commodities frequently flow into different markets. Similar possibilities exist in the forest sector. For instance, paper can be made from pulp logs or from logging residues. Thus, the model reflects a large degree of demand substitution. It includes 27 crop products, 17 livestock products, 10 processing byproducts, and 40 forestry products as secondary commodities.

Table 2-2. Secondary (Processed) Commodities

Secondary Products	Units
Crop Products	
Baked goods, sweetened	1,000 lb
Beverages, sweetened	1,000 gal
Canned goods, sweetened	1,000 gal
Canola meal	U.S. tons
Canola oil	100 gal
Confectionaries, sweetened	1,000 lb
Corn starch	1,000 lb
Corn oil	100 gal
Corn oil, nonfood, from dried distillers grains extraction	100 gal
Corn syrup	1,000 gal
Distillers grains, corn	1,000 lb
Distillers grains, corn fractionation	1,000 lb
Distillers grains, export	1,000 lb
Distillers grains, noncorn	1,000 lb
Dextrose	1,000 lb
Gluten meal	1,000 lb
Gluten feed	1,000 lb
Grapefruit juice	1,000 gal at single-strength equivalent
High-fructose corn syrup	1,000 gal
Orange juice	1,000 gal at 42 brix

(continued)

2-10

Table 2-2. Secondary (Processed) Commodities

Secondary Products	Units
Potatoes, chipped	100 lb
Potatoes, dried	100 lb
Potatoes, frozen	100 lb
Soybean meal	U.S. tons
Soybean meal equivalent, produced using feedstocks other than soybeans	U.S. tons
Soybean oil	1,000 lb
Sugar, refined	U.S. tons
Livestock Products	
American cheese	Pounds
Beef, grain-fed	100 lb (carcass weight)
Beef, grass-fed (nonfed)	100 lb (carcass weight)
Butter	Pounds
Chicken	100 lb on ready-to-cook basis
Cottage cheese	Pounds
Cream	Pounds
Evaporated condensed milk	Pounds
Fluid milk, low-fat	Pounds
Fluid milk, skim	Pounds
Fluid milk, whole	100 lb
Ice cream	Pounds
Nonfat dry milk	Pounds
Other cheese	Pounds
Pork	100 lb after dressing
Turkey	100 lb on ready-to-cook basis
Wool, clean	Pounds
Processing Byproducts	
Lard from swine slaughter	U.S. tons
Lignin produced from nonwood cellulosic ethanol processes	U.S. tons
Poultry fat from chicken and turkey slaughter	Pounds
Sugarcane bagasse	U.S. tons
Sweet sorghum pulp	U.S. tons

(continued)

Table 2-2.   Secondary (Processed) Commodities

Secondary Products	Units
Tallow, edible, from beef cattle slaughter	Pounds
Tallow, nonedible, from beef cattle slaughter	Pounds
Yellow grease (waste cooking oil)	Pounds
Wood Products	
Agrifiber, long fiber	U.S. tons
Agrifiber, short fiber	U.S. tons
Chemi-thermomechanical market pulp	Million metric tons
Coated free sheet	Million metric tons
Coated roundwood	Million metric tons
Construction paper and board	Million metric tons
Corrugated medium	Million metric tons
Dissolving pulp	Million metric tons
Hardwood kraft market pulp	Million metric tons
Hardwood lumber	Million board feet, lumber tally
Hardwood miscellaneous products	Million cubic feet
Hardwood plywood	Million square feet, 3/8″
Hardwood pulp	Million cubic meters
Hardwood pulp, moved to agricultural component of model for use in cellulosic ethanol production	U.S. tons
Hardwood residues	Million cubic meters
Hardwood used in non-OSB reconstituted panel	Million square feet, 3/8″
High-grade deinking	Million metric tons
Kraft packaging	Million metric tons
Linerboard	Million metric tons
Mixed wastepaper	Million metric tons
Newsprint	Million metric tons
Old corrugated paper	Million metric tons
Old newspapers	Million metric tons
Oriented strand board (OSB)	Million square feet, 3/8″
Pulp substitutes	Million metric tons
Recycled board	Million metric tons
Recycled market pulp	Million metric tons
Softwood kraft market pulp	Million metric tons

(continued)

Table 2-2. Secondary (Processed) Commodities

Secondary Products	Units
Solid blended board	Million metric tons
Specialty packaging	Million metric tons
Softwood lumber	Million board feet, lumber tally
Softwood miscellaneous products	Million cubic feet
Softwood plywood	Million square feet, 3/8″
Softwood pulp	Million cubic meters
Softwood pulp, moved to agricultural component of model for use in cellulosic ethanol production	U.S. tons
Softwood residues	Million cubic meters
Softwood used in non-OSB reconstituted panel	Million square feet, 3/8″
Tissue and sanitary	Million metric tons
Uncoated free sheet	Million metric tons
Uncoated roundwood	Million metric tons

Primary agricultural and forestry products are converted into processed products using processing budgets. These budgets are generally reflective of a somewhat simplified view of the resources used in processing, where the primary factors in the budgets are the use of primary commodities as inputs, the yield of secondary products, and processing costs to convert primary products into processed products. Processing costs for agricultural products are usually assumed to equal the observed price differential between the value of the outputs and the value of the inputs based on U.S. Department of Agriculture (USDA) Agricultural Statistics.[5] On the forestry side, the nonwood input supply curve provides the cost of processing wood.

The processing budgets for wood products are regionalized for all forest products with different data in the nine domestic forest production regions and the Canadian regions. Agricultural processing is regionalized for renewable fuels production, soybean crushing, wet milling, and bioelectricity generation. Processing budgets for other agricultural products are defined at a national level.

---

[5]U.S. Department of Agriculture, National Agricultural Statistics Service. Various years. USDA Agricultural Statistics (1990–2002). Available at http://www.nass.usda.gov/Publications/Ag_Statistics/.

2.1.3 Bioenergy Products

Another category of processed product that can be evaluated in FASOMGHG using a subset of primary and secondary commodities is bioenergy. In addition to the category totals shown in Table 2-3, the model tracks the quantity of each bioenergy product produced using each individual feedstock. The bioenergy sector is an important component of the FASOMGHG specification that has received a great deal of enhancement since the last major model update. Given recent policy interest and promulgation of rules greatly expanding renewable energy production and consumption, as well as the sizable potential role for bioenergy in GHG mitigation, we have been engaged in a major effort to update this component of the model in recent years. These changes have included updates to data and parameters, as well as incorporation of additional feedstocks.

Table 2-3. Bioenergy Products

Bioenergy Products	Units
Crop ethanol	1,000 gal
Cellulosic ethanol	1,000 gal
Biodiesel	1,000 gal
Bioenergy inputs to electricity production	Trillion British thermal units (Btus)

2.1.4 Blended Livestock Feeds

In addition to using the primary and secondary commodities directly as livestock feed, FASOMGHG allows for blending of livestock feeds from a number of alternative formulas. Table 2-4 summarizes the categories of blended livestock feeds that can be used to meet livestock feed demand. These blends are defined to meet nutritional requirements of the individual livestock types, but each blend can be made using a variety of mixtures of primary and secondary commodities to deliver the appropriate nutrient levels. These alternative mixtures are defined by feed and feed blending alternative and vary by market region. The actual mixtures that will be used in the market equilibrium will depend on relative prices and availability, as well as nutrient requirements. The resultant feeds are supplied for consumption by each livestock type included in the model.

Table 2-4.   Blended Livestock Feeds

Feed Item	Units
Protein feed for stockers	100 lb (cwt)
Blend of grains for cattle	100 lb (cwt)
Protein feed for cattle	100 lb (cwt)
Blend of grains for cow calf operations	100 lb (cwt)
Protein feed for cow calf operations	100 lb (cwt)
Blend of grains for pig finishing	100 lb (cwt)
Protein feed for pig finishing	100 lb (cwt)
Blend of grains for farrowing operations	100 lb (cwt)
Protein feed for farrowing operations	100 lb (cwt)
Blend of grains for feeder pigs	100 lb (cwt)
Protein feed for feeder pigs	100 lb (cwt)
Blend of grains for dairy operations	100 lb (cwt)
Blend of grains for broilers	100 lb (cwt)
Protein feed for broilers	100 lb (cwt)
Blend of grains for turkeys	100 lb (cwt)
Protein feed for turkeys	100 lb (cwt)
Blend of grains for eggs	100 lb (cwt)
Protein feed for eggs	100 lb (cwt)
Blend of grains for sheep	100 lb (cwt)
Protein feed for sheep	100 lb (cwt)

Inputs to Production

The production component of FASOMGHG includes agricultural crop and livestock operations, as well as forest industry (FI) and nonindustrial private forests (NIPF) forestry operations. FASOMGHG contains an agricultural production model for each of the primary commodities identified previously. Production of traditional agricultural crops, bioenergy crops, livestock, and forestry results in competition for suitable land. In addition to land, FASOMGHG depicts the factor supply of other resources (such as water, labor, and other agricultural inputs) in agriculture, as well as nonwood inputs in the forest sector.

In agricultural production, water and labor availability are specified on a regional basis. Supply curves for both items have a fixed-price component and an upward-sloping component, representing rising marginal costs of higher supply quantities. For water, the fixed price is available to a maximum quantity of federally provided agricultural water, whereas pumped water has an upward-sloping supply curve and is subject to maximum availability. Many other inputs (e.g., fossil fuels, capital) are assumed to be infinitely available at a fixed price (i.e., the agricultural sector is a price taker in these markets).

On the forestry side, nonwood inputs are available on an upward-sloping basis and include hauling, harvesting, and product processing costs. Other forest inputs are assumed to be infinitely available at a fixed price.

Budgets are included for all crops in the model based on data drawn from a variety of USDA and agricultural extension sources. Table 2-5 summarizes major categories of inputs included within the crop budgets that are defined and tracked in terms of quantities, typically because those quantities provide information on key energy, natural resource, GHG emissions, and other environmental impacts under a policy scenario (not all inputs are included in all crop budgets). The remainder of budget items are defined only in terms of dollars and largely aggregated for the purposes of the model. For each traditional crop, production budgets are differentiated by region, tillage choice (three choices: conventional tillage, conservation tillage, or no-till), and irrigated or dryland. The differentiation included results in thousands of cropping production possibilities (budgets) representing agricultural production in each 5-year period. Energy crop production possibilities are similar, except that irrigation is not an available option in the current FASOMGHG production possibilities; all energy crops are assumed to be produced under nonirrigated conditions and do not compete for irrigation water.

Table 2-6 summarizes the inputs included in FASOMGHG livestock production budgets in terms of quantities (not all inputs are included in all livestock budgets). A number of categories track manure management systems because they are a key source of emissions for livestock. As for crops, the remainder of the inputs identified in available livestock budgets are included only in dollar terms and aggregated for model purposes. For livestock production, budgets are included that are defined by region, animal type, enteric fermentation management alternative, manure management alternative, and feeding alternative. Hundreds of livestock production possibilities (budgets) represent agricultural production in each 5-year period.

Table 2-5. Major Categories Included in Crop Budgets in Quantities

Carbon—fertilizer production	Gasoline	Nitrogen
Carbon—fuel use	Herbicide	Nitrous oxide—fertilizer
Carbon—grain drying	Insecticide	Nitrous oxide—histosol
Carbon—irrigation water pumping	Irrigation water	Nitrous oxide—leaching
Carbon—pesticide production	Labor	Nitrous oxide—residue burning
Crop residue	Land	Nitrous oxide—volatilization
Crop yield	Lime and gypsum	Phosphorus
Diesel fuel	Methane—residue burning	Potassium
Electricity	Methane—rice cultivation	
Fungicide	Natural gas	

Table 2-6. Major Categories Included in Livestock Budgets in Quantities

Barley	Liquid volatile solids volume	Oats
Biomanure	Livestock head	Pasture
Blended feed requirements	Livestock product output	Silage
Corn	Managed manure fraction	Soybean meal
Hay	Methane—enteric fermentation	Volatile solids in manure
Head in liquid systems	Methane—manure	Wheat
Labor	Nitrous oxide—manure	

Supply curves for agricultural products are generated implicitly within the system as the outcome of competitive market forces and market adjustments. This method is in contrast to supply curves that are estimated from observed, historical data. The approach is useful here in part because FASOMGHG is often used to simulate conditions that fall well outside the range of historical observation (such as large-scale tree-planting programs or implementation of mandatory GHG mitigation policies).

The forest production component of FASOMGHG depicts the use of existing private timberland, as well as the reforestation decision on harvested land. The forest sector relies on a series of forest growth and yield values to grow the forest inventory over time and to convert harvested area into forest products. In addition, forest carbon sequestration is calculated over time based on the inventory characteristics. Timberland is differentiated by region, the age cohort of trees,[6] ownership class, forest type, site condition, management regime, and suitability of the land for agricultural use. Decisions pertaining to timber management investment are endogenous. Actions on the inventory are depicted in a framework that allows timberland owners to institute management activities that alter the inventory consistent with maximizing the net present value of the returns from the activities. The key decision for existing timber stands involves selecting the harvest age. Lands that are harvested and subsequently reforested or lands that are converted from agriculture to forestry (afforested) introduce decisions involving the choice of forest type, management regime, and future harvest age.

U.S. Regional Disaggregation

FASOMGHG includes all states in the conterminous United States, broken into 63 subregions for agricultural production and 11 market regions (see Table 2-7) (forestry production is disaggregated into the 11 market regions, but not into the 63 subregions). These regions are graphically displayed in Figure 2-1. The 11 market regions provide a consolidation of regional definitions that would otherwise differ if the forest and agricultural sectors were treated separately. Forestry production is included in 9 of the market regions (all but Great Plains and Southwest), whereas agricultural production is included in 10 of the market regions (all but Pacific Northwest—West side). The Great Plains and Southwest regions are kept separate because they reflect important differences in agricultural characteristics. Likewise, there are important differences in the two Pacific Northwest regions (PNWW, PNWE) for forestry production, and the PNWE region is considered a significant producer of agricultural commodities tracked in the model, whereas PNWW is not. Thus, the two model regions that make up the Pacific Northwest are tracked separately. Each of the production regions is uniquely mapped to one of the 11 larger market regions. The majority of production regions are defined at the state level. However, for selected major production areas with significant differences in production conditions within states, the states are broken into subregions.

---

[6] Timberlands are grouped in 21 5-year cohorts, 0 to 4 years, 5 to 9, etc., up to 100+ years. Harvesting is assumed to occur at the midyear of the cohort.

Table 2-7. Definition of FASOMGHG Production Regions and Market Regions

Key	Market Region	Production Region (States/Subregions)
NE	Northeast	Connecticut, Delaware, Maine, Maryland, Massachusetts, New Hampshire, New Jersey, New York, Pennsylvania, Rhode Island, Vermont, West Virginia
LS	Lake States	Michigan, Minnesota, Wisconsin
CB	Corn Belt	All regions in Illinois, Indiana, Iowa, Missouri, Ohio (IllinoisN, IllinoisS, IndianaN, IndianaS, IowaW, IowaCent, IowaNE, IowaS, OhioNW, OhioS, OhioNE)
GP	Great Plains (agriculture only)	Kansas, Nebraska, North Dakota, South Dakota
SE	Southeast	Virginia, North Carolina, South Carolina, Georgia, Florida
SC	South Central	Alabama, Arkansas, Kentucky, Louisiana, Mississippi, Tennessee, Eastern Texas
SW	Southwest (agriculture only)	Oklahoma, all of Texas except the eastern portion included in the SC region (Texas High Plains, Texas Rolling Plains, Texas Central Blacklands, Texas Edwards Plateau, Texas Coastal Bend, Texas South, Texas Trans Pecos)
RM	Rocky Mountains	Arizona, Colorado, Idaho, Montana, Nevada, New Mexico, Utah, Wyoming
PSW	Pacific Southwest	All regions in California (CaliforniaN, CaliforniaS)
PNWE	Pacific Northwest—East side	Oregon and Washington, east of the Cascade mountain range
PNWW	Pacific Northwest—West side (forestry only)	Oregon and Washington, west of the Cascade mountain range

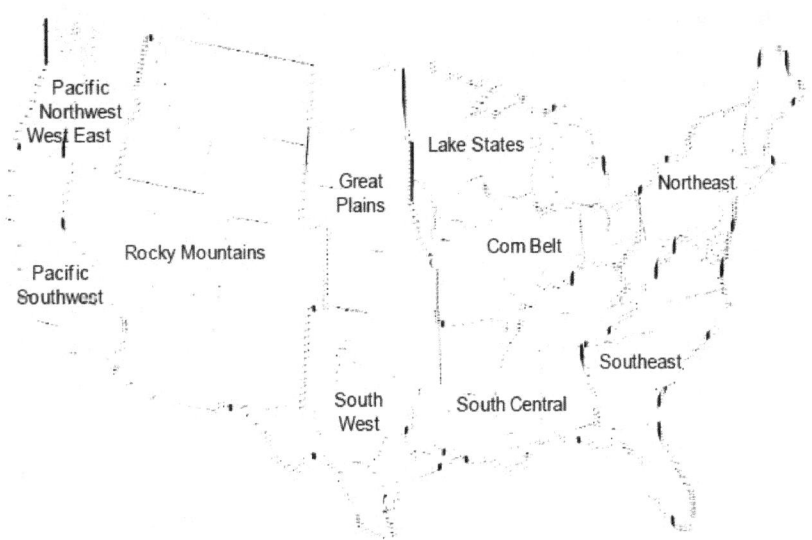

Figure 2-1. Map of the FASOMGHG Regions

When running the model, one can choose whether to keep the 63 regions or collapse to 11 regions to reduce run time. It is also possible to model agriculture explicitly in all 63 regions for an initial time period to provide maximum regional detail for the near to intermediate term and then collapse to 11 regions at a specified future time period for model size control purposes.

The full FASOMGHG can also be run at the more aggregated regional definition shown in Table 2-8, although the aggregated version of the model is more typically used for model development and testing. In addition, the wood products production and GHG accounting calculations employ an even more aggregated set of U.S. regions, following the regional definition in the North American Pulp and Paper (NAPAP) model (Zhang et al., 1993, 1996; Ince, 1994). This specification combines the Midwest and Northeast regions into a North region and does not include the Plains region because there are no forests tracked in that region.

Table 2-8. Aggregated U.S. Regions

Region	FASOMGHG Market Regions Included
Midwest	CB, LS
Northeast	NE
Plains	GP, SW
PNW_West_side	PNWW
Southern_US	SE, SC
Western_US	PNWE, RM, and PSW

Note: CB = Corn Belt; GP = Great Plains; LS = Lake States; NE = Northeast; PNWE = Pacific Northwest—East side; PNWW = Pacific Northwest—West side; PSW = Pacific Southwest; RM = Rocky Mountains; SC = South Central; SE = Southeast; SW = Southwest.

Land Use Categories

Underlying the commodity production described previously and the associated environmental impacts is the decision by landowners on how much, where, and when to allocate land across the two sectors. The inclusion of endogenous land allocation across sectors sets FASOMGHG apart from the majority of other forest and agricultural sector models of the United States. The conceptual foundation for land allocation is described in Section 2.5.4.

FASOMGHG includes all cropland, pastureland, rangeland, and private timberland[7] throughout the conterminous United States. The model tracks area used for production and area

---

[7] As noted above, although public timberland is not explicitly modeled because the focus of the model is on private decision-maker responses to changing incentives, FASOMGHG includes an exogenous timber supply from public forestlands.

idled (if any) within each land category. In addition, the model accounts for the movement of forest and agricultural lands into developed uses. We recently updated our land use categorization system to represent a more comprehensive range of categories. This process included expanding our coverage of pasturelands to explicitly represent multiple forms of public and private grazing lands (each with different animal unit grazing potential per unit of land). The FASOMGHG land base was developed based on land classifications from multiple sources, with the USDA Economic Research Service Major Land Use (MLU) database (USDA Economic Research Services [ERS], 2012) and the Natural Resources Inventory (NRI) published by the USDA-Natural Resource Conservation Service serving as our primary data sources.

These databases rely on different sampling methods and define land use categories in separate ways that each have advantages and disadvantages. To maintain consistency with other FASOMGHG input data, we rely on the ERS depiction of cropped acres to define our cropland base. However, the ERS lacks a clear distinction between grassland pasture and rangeland, while the NRI defines these as separate land categories, a distinction that we also wish to maintain given differences in ownership and productivity. Therefore, we make use of both datasets and attempt to avoid overlap between different land use categories as outlined in the following bullets. This "hybrid" NRI-MLU land categorization system is unique, and we feel that it provides FASOMGHG with a more realistic representation of public and private grazing lands, as well as regional land transition possibilities between alternative uses.

Land categories included in the model are specified as follows:

- **Cropland** is land suitable for crop production that is being used to produce either traditional crops (e.g., corn, soybeans) or dedicated energy crops (e.g., switchgrass). This category includes only cropland from which one or more crops included in FASOMGHG were harvested.[8] Cropland used for livestock grazing before or after crops were harvested is included within this category as long as crops are harvested from the land. Data used to define cropland area are directly from the ERS-MLU (USDA ERS, 2012).

- **Cropland pasture** is managed land suitable for crop production (i.e., relatively high productivity) that is being used as pasture. The ERS-MLU database defines this area as "used only for pasture or grazing that could have been used for crops without additional improvement. Also included were acres of crops hogged or grazed but not harvested prior to grazing" (USDA ERS, 2012, Glossary). Not requiring additional improvement to be suitable for crop production is a key distinction between cropland pasture and other forms of grassland pasture or rangeland. This land is assumed to be

---

[8] Note that FASOMGHG does not include every cropping activity conducted in the United States. For instance, tobacco, vineyards, and most fruits and vegetables are not included in the model.

more freely transferable with cropland than other grassland types. State totals for cropland pasture used in the model are drawn directly from the ERS-MLU Web site.

- **Pasture** is defined in an attempt to maintain a consistent definition with the NRI classification of grassland pasture but to eliminate overlap with ERS cropland or cropland pasture as defined above. For each region, we compute the initial stock of "pasture" algebraically as the maximum of (1) $(Cropland_{NRI} + Grassland\ Pasture_{NRI}) - (Cropland_{ERS} + Cropland\ Pasture_{ERS})$ or (2) zero. This procedure is necessary to avoid double counting of pasturelands between the NRI and ERS data.

- **Private grazed forest** is calculated based on woodland areas of farms reported in the Agricultural Census to be used for grazing (woodland pasture).[9] Woodland pasture is defined as "all woodland used for pasture or grazing during the census year. Woodland or forestland pastured under a per-head grazing permit was not counted as land in farms and, therefore, was not included in woodland pastured" (USDA ERS, 2012, Glossary). These lands are not included in the private timberland areas defined in FASOMGHG, and there are no forest products harvested from these lands in the model. The area in this category is fixed over time and is not allowed to transfer into forestland or other alternative uses.

- **Public grazed forest** is computed as the difference between the ERS-MLU total forest pasture stock and the private portion given by the Agricultural Census as described above.

- **Private rangeland** is defined in FASOMGHG using a combination of NRI and ERS-MLU data. Rangeland is typically unimproved land where a significant portion of the natural vegetation is native grasses and shrubs. The NRI database defines rangeland as "land on which the climax or potential plant cover is composed principally of native grasses, grass-like plants, forbs or shrubs suitable for grazing and browsing, and introduced forage species that are managed like rangeland. This would include areas where introduced hardy and persistent grasses, such as crested wheatgrass, are planted and practices, such as deferred grazing, burning, chaining, and rotational grazing, are used with little or no chemicals or fertilizer being applied. Grassland, savannas, many wetlands, some deserts, and tundra are considered to be rangeland. Certain low forb and shrub communities, such as mesquite, chaparral, mountain shrub, and pinyon-juniper, are also included as rangeland" (USDA Natural Resource Conservation Service [NRCS], 2003, p. A-6). Thus, rangeland generally has low forage productivity and is unsuitable for cultivation, and it is assumed that rangeland cannot be used for crop production or forestland. To calculate rangeland acres while avoiding double counting, we first use 2003 NRI data to provide a base definition for the rangeland class. States with no reported rangeland acres in the NRI database (USDA NRCS, 2003) are defined to have no rangeland area in FASOMGHG to be consistent with the NRI definition and to limit overlap between the NRI classification of rangeland and the ERS-MLU classification of "grassland pasture and range." Then,

---

[9] Data are available at
http://www.agcensus.usda.gov/Publications/2002/Volume_1,_Chapter_2_US_State_Level/st99_2_008_008.pdf.

to determine the state totals of private rangeland, we use USDA ERS (2012) data defining regional totals of privately held grazing land by type. These regional proportions are multiplied by corresponding state-level totals to define the private rangeland stock by state. For example, the private rangeland stock in Wyoming is calculated by multiplying the total ERS estimate for Wyoming by the proportion of private to total rangeland for the "Mountain" region in which Wyoming is located. In solving for the private rangeland area used in FASOMGHG, it is important to maintain the relationship between all grazing lands for consistency. The ERS defines all privately owned grazing lands to be equal to the sum of cropland pasture, grazed forest, and grassland pasture and range and reports a total of approximately 488 million acres. Following all of our adjustments to develop a consistent land use definition based on both NRI and ERS-MLU data, the total private grazing land base in the baseline is approximately 484 million acres.

- **Public rangeland** is calculated using the proportions described above under private rangeland and totals about 182 million acres. The total includes federal, state, and local sources.

- **Forestland** in FASOMGHG refers to private timberland, with a number of subcategories (e.g., different levels of productivity, management practices, age classes) tracked (see Section 2.5.2 for additional details). The model also reports the number of acres of private forestland existing at the starting point of the model that remains in standing forests (i.e., have not yet been harvested), the number of acres harvested, the number of harvested acres that have been reforested, and the area converted from other land uses (afforested). Public forestland area is not explicitly tracked because it is assumed to remain constant over time. Regional timberland stocks, as well as timber demand, inventory, and additional forestry sector information, are drawn from the 2005 RPA Timber Assessment (Adams and Haynes, 2007).

- **Developed** (urban) land is assumed to increase over time at an exogenous rate for each region based on projected changes in population and economic growth. It is assumed that the land value for use in development is sufficiently high that the movement of forest and agricultural land into developed land will not vary between the policy cases analyzed. All private land uses (except Conservation Research Program [CRP] land and grazed forest) are able to convert to developed land, decreasing the total land base available for forestry and agriculture over time. Land transfer rates vary by land use type over time and are consistent with the national land base assessment by Alig et al. (2010b).

- **CRP land** is specified as land that is voluntarily taken out of crop production and enrolled in the USDA's CRP. Land in the CRP is generally marginal cropland retired from production and converted to vegetative cover (such as grass, trees, or woody vegetation) to conserve soil, improve water quality, enhance wildlife habitat, or produce other environmental benefits. State- and county-level land areas enrolled in the CRP are obtained from the USDA Farm Service Agency (FSA; 2009).

Figure 2-2 shows the baseline land allocation in FASOMGHG at the national level across each of the land categories defined above. Land is allowed to move between categories over time subject to restrictions based on productivity and land suitability. The conversion costs of moving between land categories are set at the present value of the difference in the land rental rates between the alternative uses based on the assumed equilibration of land markets (see Sections 2.5.1 through 2.5.4 for additional detail on each land use category and its potential conversion to alternative land uses).

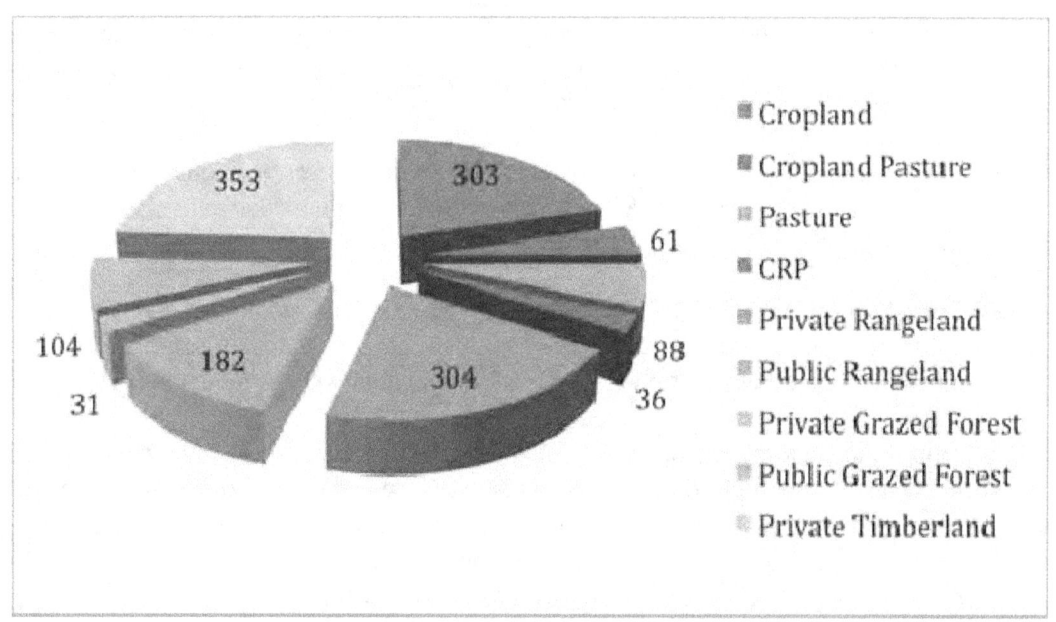

Figure 2-2. Baseline FASOMGHG U.S. Land Base by Land Use Category (million acres)

2.4.1 Agricultural Land

As described previously, cropland is land that is suitable for crop production and can potentially be used in the production of any of the crops included in FASOMGHG for the particular production region being considered. Land in the cropland category is the most productive land available for producing primary agricultural commodities, although cropland is more productive in some regions than in others. Therefore, crop yields vary across regions based on historical data. The total area of baseline cropland is based on ERS-MLU data as described above, with baseline land in production of individual crops based on USDA National Agricultural Statistics Service (NASS) historical data on county-level harvested acreage by crop. Cropland enrolled in the CRP is included under the CRP land category, and cropland used as pasture is implicitly included in the pastureland category in FASOMGHG (i.e., both of these categories of cropland are included in other categories rather than being reported under cropland). The average annual areas of cropland with failed crops[10] are not included in the reported FASOMGHG cropland and are not explicitly tracked in FASOMGHG. Cropland can potentially be converted to cropland pasture or private forestland. In addition to tracking aggregate cropland area, cropland is tracked by crop tillage system and irrigated/dryland status, as well as the duration of time the land has been in such a system.[11] This approach allows for tracking of sequestered soil carbon and the transition to a new soil carbon equilibrium after a change in tillage. Also, there are differences in crop yields between irrigated and dryland systems, as well as differences in input use, GHG emissions, and other environmental impacts. Different tillage systems also have differences in input usage and environmental impacts in FASOMGHG.

CRP land is cropland that has been enrolled in the Conservation Reserve Program, a USDA program that provides payments to encourage activities with conservation and environmental benefits. The land that farmers choose to enroll in the program is typically marginal cropland that they have agreed to retire from production for a contracted period. The area of CRP land in FASOMGHG in the baseline is based on 2007 data on CRP enrollment by state available from the USDA FSA (2009). Because landowners can choose to remove their land from the CRP program when their contract expires (or before expiration, subject to a financial penalty), FASOMGHG also tracks the area of CRP land with expiring contracts in each year. As CRP contracts expire, landowners will move land back into agricultural production if the returns to agricultural production exceed the returns associated with maintaining land in the CRP. However, based on the 2008 Farm Bill, which specifies a maximum of 32 million acres in the CRP, and indications that USDA plans to provide sufficient funding to maintain that maximum level of 32 million acres in the CRP, FASOMGHG model runs generally place a floor of 32 million acres in CRP land in future years.

---

[10] USDA data for planted area exceed the harvested area because there will inevitably be some fraction of planted cropland area that is not harvested due to crop failure associated with poor weather, extreme events, or other conditions. In that case, the cost of harvesting may exceed the value of the crop. Thus, farmers will choose not to harvest those areas.

[11] Crop tillage systems in FASOMGHG include conventional tillage, conservation tillage, and no-till. Conservation tillage and no-till reduce the exposure of carbon in the soil to oxidation and allow larger soil aggregates to form. These practices also leave crop residues on the soil, thereby potentially increasing carbon inputs. Tillage changes from more intensive conventional tillage practices, such as moldboard plowing, to conservation or zero tillage practices will generally increase levels of soil carbon over time. In addition, emission reductions may result because less-intensive tillage typically involves less direct fossil fuel use for tractors. However, there are also alterations in chemical usage (possibly increases in pesticide usage and alterations in rate of fertilization), which can potentially increase emissions associated with increased manufacture and usage. FASOMGHG has the ability to track these indirectly induced GHG effects associated with changes in tillage.

Cropland pasture, pasture, and private and public grazed forest are all suitable for livestock grazing (i.e., land that provides sufficient forage to support the needs of grazing livestock within a region), but cropland pasture tends to be more productive. Because it has sufficient quality to be used in crop production, cropland pasture can potentially be converted to crop production within the model. It can also be converted to forestland. Pasture, which is considered less productive, can be converted to forestland but not cropland. Private and public grazed forest refers to land that has varying amounts of tree cover but can also be used as pasture. Forage production on these lands tends to be relatively low, however. Neither private nor public grazed forest can be converted to any other uses. As mentioned above, FASOMGHG assumes that no timber is produced from private grazed forest.

Rangeland in FASOMGHG includes both public and private rangeland. Rangeland differs from pastureland primarily in that it is assumed to be generally unimproved land where a significant portion of the land cover is native grasses and shrubs. The productivity of rangeland varies considerably across regions of the United States. Therefore, the area of rangeland required per animal for a given species can be very different across regions. Overall, rangeland provides lower forage production per acre than pastureland and is considered unsuitable for cultivation. In addition, much of the rangeland in the United States is publicly owned. Thus, it is assumed that rangeland cannot be used for crop production or forestland.

The area of pastureland or rangeland required per animal is calculated in FASOMGHG for each combination of livestock type and pasture or rangeland category available in each region. These values are based on forage requirements for each livestock species and estimated forage productivity per acre for each category of pasture in FASOMGHG, defined on a regional basis.[12] The area of pastureland used in livestock production is limited to the pastureland inventory by time period and region. It is possible to have idle pastureland in FASOMGHG, and idle pastureland area and associated soil carbon sequestration are tracked in the model. In particular, changes in livestock populations will affect pasture and rangeland used for animal production and could increase or decrease idle land in the model. Changes in animal populations over time and impacts of policies affecting livestock markets, including use of each of the pasture and rangeland categories by each type of livestock, are tracked within FASOMGHG.

---

[12] The calculation of acres of pasture required by a given type of livestock in a particular region is implicitly based on estimates of animal unit months (AUMs) available for each category of pastureland in that region.

2.4.2 Forestland

Timberland refers to productive forestlands able to grow at least 20 cubic feet of growing stock per acre per year and that are not reserved for uses other than timber production (e.g., wilderness use). Lands under forest cover that do not produce at least 20 cubic feet per acre per year, called unproductive forestland, and timberland that is reserved for other uses are not considered part of the U.S. timber base (Haynes et al., 2007) and are therefore not tracked by the model.

In FASOMGHG, endogenous land use modeling is done only for privately held parcels, not publicly owned or publicly managed timberlands. The reason is that management of public lands is largely dictated by government decisions on management, harvesting, and other issues that account for multiple public uses of these lands rather than responses to market conditions. However, an exogenous quantity of timber harvested on U.S. public lands is accounted for within the model. Projected regional public harvest levels are drawn from the assumptions used in the baseline case of the U.S. Forest Service's 2005 RPA Timber Assessment (Haynes et al., 2007). Timber inventory levels for public timberlands are simulated based on these harvest levels.

Private timberland is tracked by its quality and its transferability between forestry and agricultural use. FASOMGHG includes three different site classes to reflect differences in forestland productivity (these site groups were defined according to ATLAS inputs [Haynes et al., 2007]), where yields vary substantially between groups[13]:

- HIGH—high site productivity group (sites that produce >85 cubic feet of live growing stock per acre per year)

- MEDIUM—medium site productivity group (sites that produce between 50 and 85 cubic feet of live growing stock per acre per year)

- LOW—low site productivity group (sites that produce between 20 and 50 cubic feet of live growing stock per acre per year)

FASOMGHG also tracks land ownership, including two private forest owner groups: forest industry (FI) and nonindustrial private forests (NIPF). The traditional definitions are used for these ownership groups: industrial timberland owners possess processing capacity for the

---

[13] Changes in ozone concentrations affect the specific forest growth rates for each region/species/management intensity/productivity class but are assumed not to result in movements between productivity classes. The primary use of the productivity classes in FASOMGHG is to aid in defining potential land use between forestland and other land uses (e.g., only high productivity forestland can be converted to cropland).

timber, and NIPF owners do not. As a result, the NIPF group includes lands owned by timber investment management organizations (TIMOs) and real estate investment trusts (REITs).

In addition, FASOMGHG tracks land in terms of the type of timber management practiced, forest type (identified by dominant species), and stand age. As shown in Table 2-9, across all regions there are 18 management intensity classes defined based on whether thinning, partial cutting, passive management, or other management methods are used. Note that some management intensity classes are defined only for a subset of regions (as identified by the region codes in parentheses) based on regional data and definitions. There are also 25 forest types, which vary by region (e.g., Douglas-fir and other species types in the West and planted pine, natural pine, and various hardwood types in the South). Stand age is explicitly accounted for in 5-year cohorts, ranging from 0 to 4 years up to 100+ years.

Table 2-9. Forest Management Intensity Classes (regions of application in parentheses)

MIC Code	Description
AFFOR	Afforestation of bottomland hardwood (SE, SC)
AFFOR_CB	Afforestation of hardwood and softwood forest types (CB)
LO	Natural regeneration (or afforestation) with low management
NAT_REGEN	Natural regeneration with low management (PNWW)
NAT_REGEN_PART_CUT_HI	Partial cutting with high level of management (PNWW)
NAT_REGEN_PART_CUT_LO	Partial cutting with medium level of management (PNWW)
NAT_REGEN_PART_CUT_MED	Partial cutting with low level of management (PNWW)
NAT_REGEN_THIN	Natural regeneration with a commercial thin (PNWW)
PART_CUT_HI	Partial cutting with medium level of management (SE, SC)
PART_CUT_HI+	Partial cutting with high level of management (SE, SC)
PART_CUT_LO	Partial cutting with low level of management (SE, SC)
PASSIVE	Passive management (minimal amount of management)
PLANT	Plant with no intermediate treatments (PNWW)
PLANT_THIN	Plant with medium level of management (PNWW)
PLANT+	Plant with high level of management (PNWW)
PLNT_HI	Planted pine with high level of management (SE, SC)
PLNT_HI_THIN	Planted pine with commercial thin and high level of management (SE, SC)
PLNT_LO_THIN	Planted pine with commercial thin and no intermediate treatments (SE, SC)
PLNT_MED	Planted pine with medium level of management (SE, SC)

(continued)

Table 2-9. Forest Management Intensity Classes (regions of application in parentheses) (continued)

MIC Code	Description
PLNT_MED_THIN	Planted pine with commercial thin and medium level of management (SE, SC)
RESERVED	Reserved from harvest
SHORT_ROTSWDS	Short rotation softwoods with high level of management (SE, SC)
TRAD_PLNT_PINE	Planted pine with no intermediate treatments (SE, SC)

Note: CB = Corn Belt; PNWW = Pacific Northwest—West side; SC = South Central; SE = Southeast.

2.4.3 Developed Land

FASOMGHG also accounts for the movement of agricultural and forestland into developed uses. The economic returns to developed land uses typically exceed the returns available to agricultural or forestry land uses. Thus, FASOMGHG assumes an exogenous rate of land conversion into developed uses by region for each of the agricultural and forestland categories included in the model (with the exception of private and public grazed forest pasture and CRP lands) based on projections of future U.S. population and income, with endogenous competition between agriculture and forestry for the remaining land base available for these uses over time. It is assumed that developed land does not convert back to other uses.

2.4.4 Land Allocation

In FASOMGHG, the initial land endowment is fixed. However, because land can move between forests and agriculture, agricultural production faces, in effect, an endogenous excess land supply "equation" from forestry. Forestry production, in turn, effectively faces an endogenous excess land supply "equation" from agriculture.

The conceptual foundation for land allocation is described in the following paragraphs. In terms of transferability between agriculture and forestry, FASOMGHG includes five land suitability classes:

- FORONLY—includes timberland acres that cannot be converted to agricultural uses

- FORCROP—includes acres that begin in timberland but can potentially be converted to cropland

- FORPAST—includes acres that begin in timberland but can potentially be converted to pastureland

- CROPFOR—includes acres that begin in cropland but can potentially be converted to timberland

- PASTFOR—includes acres that begin in pasture but can potentially be converted to timberland

Land can flow between the agricultural and forestry sectors or vice versa in the FORCROP, FORPAST, CROPFOR, and PASTFOR land suitability categories. Movements between forestry and cropland are permitted only within the high forest site productivity class. Changes in land allocation involving pastureland occur within the medium-quality forest site productivity class. In addition, land movements in forestry are allowed only in the NIPF owner category, reflecting an assumption (and lengthy historical observation) that land held by the FI ownership group will not be converted from timberland to agriculture.

As mentioned previously, the decision to move land between uses depends on the net present value of returns to alternative uses, including the costs of land conversion. Land transfers from forestry to agriculture take place only upon timber harvest and require an investment to clear stumps from the land, level the land, and otherwise prepare it for planting agricultural crops. Agricultural land can move to other uses during any of the 5-year model periods, but when afforested it begins in the youngest age cohort of timberland.

In addition to the endogenous land allocation decision, land moves out of agricultural and forestry uses into developed uses (e.g., shopping centers, housing, and other developed and infrastructural uses) at an exogenous rate. Rates at which forest and agricultural land are converted to developed uses in FASOMGHG are based on land use modeling for a national land base assessment by the U.S. Forest Service and cooperators. Thus, although land can move between forest, cropland, and pasture, the total land area devoted to agricultural and forestry production is trending downward over time as more land is shifted to developed uses.

Another potential source of land is CRP land moving back into production. There are, however, environmental benefits associated with land in CRP and indications that USDA plans to retain 32 million acres of land in CRP. Because of this assumption that no less than 32 million acres can be allocated to CRP land, there are only about 5 million acres of land in FASOMGHG in the 2000 base period that can move from CRP to cropland over time.

Market Modeling

FASOMGHG uses commodity supply and demand curves for the U.S. market that are calibrated to historic price and production data with constant differentials between regional and national prices for some crops. In addition, the model includes supply and demand data for major commodities traded on world markets, such as corn, wheat, soybeans, rice, and sorghum (see Section 2.7 for additional discussion of international trade modeling and foreign regions included). Transportation costs clearly influence equilibrium exports, and FASOMGHG includes data on transportation costs to all regions included within the model and between foreign regions for those commodities where trade is explicitly modeled.

The model solution requires that all markets are in equilibrium (i.e., the quantity supplied is equal to the quantity demanded in every market modeled at the set of market prices in the model solution). The demand and supply curves included within the model that need to be in equilibrium in each 5-year period include

- regional product supply;

- national raw product demand;

- regional or national processed commodity demand;

- regional or national supply of processed commodities;

- regional or national (depending on commodity) export demand;

- regional or national (depending on commodity) import supply;

- regional feed supply and demand;

- regional direct livestock demand;

- interregional transport perfectly elastic supply;

- international transport perfectly elastic supply; and

- country-specific excess demand and supply of rice, sorghum, corn, soybeans, and the five individual types of wheat modeled.

In the case of forestry products, commodities are typically produced regionally and are then transported to meet a national demand at a fixed regional transport cost. Harvests from public forestlands are included in the model but are treated as exogenously determined by the government. For agricultural products, processed commodities such as soybean meal, gluten feed, starch, and all livestock feeds are manufactured and used on the 11-market region basis but are supplied into a single national domestic market to meet export demand.

International Trade

FASOMGHG accounts for international trade in both forestry and agricultural products, with the commodities included in the trade component and their treatment varying based on the importance of trade to the U.S. market and available data.

2.6.1 Forestry

For the forest sector, trade of forest products with Canada and trade of softwood lumber with the rest of the world are endogenous. These are the largest (by volume or weight) U.S. forest products trade flows. All other product movements are exogenous and, in the baseline case, follow projections derived from the Forest Service's 2005 RPA Timber Assessment Update (Haynes et al., 2007).

Product movements from Canadian producing regions to the United States are endogenous and subject to appropriate transport costs, exchange rates, and tariffs. Supplies of logs in Canada derive primarily from public lands ("Crown" lands) governed by individual provinces, with small volumes from private lands. Harvests from these lands vary over time based on provincial policies, extraction and delivery costs, and market prices for logs. These supplies are represented by a set of (log price sensitive) delivered log supply equations for both sawlogs and pulpwood in each Canadian region.

Softwood lumber imports into the United States from non-Canadian sources are based on a linear import supply function drawn from the 2005 RPA Timber Assessment Update (Haynes et al., 2007), which shifts over time to correspond to the base scenario in the Update.

2.6.2 Agriculture

Three types of agricultural commodity trade arrangements are represented. Agricultural primary and secondary commodities may be portrayed

- with trade occurring in explicit international markets using a Takayama and Judge (1971) style, spatial equilibrium submodel that portrays country/region-level excess demand on behalf of a set of foreign countries/regions, excess supply on behalf of a set of foreign countries/regions, and interregional trade between the foreign countries/regions themselves and with the United States;

- with the United States facing a single excess supply or excess demand relationship on behalf of the rest of world (ROW); or

- without being subject to international trade.

FASOMGHG has explicit trade functions between the United States and 29 distinct foreign trading partners for agricultural commodities with detailed trade data available. For the remaining commodities traded internationally, excess supply and demand functions are specified to capture net trade flows with the rest of the world as one composite trade region. Demand levels are parameterized based on the USDA Static World Policy Simulation Model (SWOPSIM) database and USDA annual statistics.

International regions are generally defined in a more simple way than domestic regions, with individual region-level supply and demand curves specified only for the commodities with the largest trade volumes, such as corn, wheat, soybeans, sorghum, and rice. In addition, only certain regions are defined for exporters and importers of a given commodity. For many other commodities (e.g., cotton, oats, barley, beef, pork, poultry), trade is modeled as total excess import supply and export demand functions for the ROW facing the United States rather than individual region supply and demand. In these cases, there are single curves representing the import supply and export demand facing the United States. In addition, there are many commodities without any explicit opportunities for international trade, such as hay, silage, energy crops, livestock, and many processed commodities. Generally, trade is not explicitly modeled for commodities where international trade volumes for the United States are small or the commodity is not actively traded.

When commodities are subject to explicit spatial interregional trade with spatial equilibrium submodels, then trading is portrayed among the 29 individual countries/foreign regions currently included in FASOMGHG. In those countries/foreign regions that are major importers or exporters of an explicitly traded commodity, explicit supply and demand functions are defined. Table 2-10 presents the commodities that are traded and the countries/regions that supply and demand them in the model. Note that when a country supplies certain commodities, it can export them either to another explicitly defined country/foreign region or to the United States. Similarly, demand in a country/region can be met from imports from other countries or from the United States.

Table 2-10. Explicitly Traded Commodities and Countries/Regions Trading with the United States

FASOMGHG Commodity	Exporting Countries	Importing Countries
Canola	Canada	NA
Canola oil	Canada	NA
Canola meal	Canada	NA
Corn	Argentina, Brazil, China, USSR, W-Africa	Canada, Caribbean, E-Mexico, Indonesia, Japan, N-Africa, NC-Euro, Philippines, SE-Asia, S-Korea, Taiwan, W-Asia
Rice	E-Medit, India, Myanmar, N-Africa, Pakistan, Thailand, Vietnam	Bangladesh, Brazil, Caribbean, China, Indonesia, Japan, N-Korea, NC-Euro, Philippines, S-Africa, SE-Asia, Taiwan, USSR, W-Africa, WS-America
Sorghum	Argentina, Australia, China	E-Mexico, Japan, NC-Euro, S-Korea, Taiwan
Soybeans	Argentina, Brazil, Canada, Caribbean, USSR	China, E-Europe, E-Mexico, Indonesia, Japan, N-Africa, NC-Euro, SE-Asia, S-Korea, Taiwan, W-Africa, W-Asia
Wheat, durum	Canada	Brazil, Indonesia, Japan, N-Africa, Philippines, SE-Asia, S-Korea, Taiwan, USSR
Wheat, hard red spring	Australia, Canada	Brazil, Caribbean, China, Indonesia, Japan, N-Africa, Philippines, SE-Asia, S-Korea, Taiwan, USSR, W-Africa, W-Asia
Wheat, hard red sinter	Argentina, Australia, Canada	Brazil, China, E-Mexico, Indonesia, Japan, N-Africa, Philippines, SE-Asia, S-Korea, Taiwan, USSR, W-Africa, W-Asia
Wheat, soft red winter	Argentina, Australia, Canada	Brazil, China, E-Mexico, Indonesia, Japan, N-Africa, Philippines, SE-Asia, S-Korea, Taiwan, USSR, W-Africa, W-Asia
Wheat, soft white	Australia, Canada, NC-Euro	Brazil, China, E-Mexico, Indonesia, Japan, N-Africa, Philippines, SE-Asia, S-Korea, Taiwan, USSR, W-Africa, W-Asia

For commodities where trade is important to the U.S. market, but data on trade flows with individual countries/foreign regions are more limited, U.S. trade is modeled at an aggregate level with the ROW. When U.S. trade is included in the model with only ROW excess import supply and export demand functions, then the curves represent the sum of ROW exports and imports that are faced at the national U.S. market level. The commodities currently included in the model in this way are listed in Table 2-11, identifying whether they are included in the import supply or export demand functions.

Table 2-11. Commodities with Only ROW Export or Import Possibilities

FASOMGHG Commodity	Imported into the United States	Exported from the United States
Canola	Y	N
Canola oil	Y	N
Canola meal	Y	N
Cotton	N	Y
Distillers grains	N	Y
Oats	N	N
Barley	Y	Y
Sugarcane	N	N
Potatoes	Y	Y
Tomatoes, fresh	Y	Y
Tomatoes, processed	N	N
Oranges, fresh (75 lb box)	Y	Y
Grapefruit, fresh (85 lb box)	Y	Y
Eggs	Y	Y
Orange juice	Y	Y
Grapefruit juice	Y	Y
Soybean meal	N	Y
Soybean oil	N	Y
High-fructose corn syrup	N	Y
Confection	Y	N
Gluten feed	N	Y
Frozen potatoes	Y	Y
Dried potatoes	Y	Y
Chipped potatoes	N	Y
Refined sugar	Y	Y
Fed beef	N	Y

(continued)

Table 2-11. Commodities with Only ROW Export or Import Possibilities (continued)

FASOMGHG Commodity	Imported into the United States	Exported from the United States
Nonfed beef	Y	N
Feedlot beef slaughter	Y	N
Stocked calf	Y	N
Stocked steer calf	Y	N
Pork	Y	Y
Chicken	N	Y
Turkey	N	Y
Wool, clean	Y	Y
Evaporated condensed milk	Y	Y
Nonfat dry milk	Y	Y
Butter	Y	Y
American cheese	Y	Y
Other cheese	Y	Y

Commodities without explicit trade are generally specified as such because either the trade numbers are small or the commodity is not traded. These include the commodities listed in Table 2-12, as well as all of the blended feeds.

Table 2-12. Commodities without International Trade Possibilities Modeled

Baking	Feeder pigs	Oranges, processing (75 lb box)
Beverages	Fluid milk	Oranges, processing (85 lb box)
Biodiesel	Grapefruit, fresh (67 lb box)	Oranges, processing (90 lb box)
Broilers	Grapefruit, fresh (80 lb box)	Refined sugar
Calf slaughter	Grapefruit, processing (67 lb box)	Silage
Canning	Grapefruit, processing (80 lb box)	Skim milk
Corn oil	Grapefruit, processing (85 lb box)	Steer calves
Corn starch	Hay	Stocked heifer calves
Corn syrup	Heifer calves	Stocked heifer yearlings
Cottage cheese	Hogs for slaughter	Stocked steer yearlings
Cream	Horses and mules	Stocked yearlings
Cull beef cows	Hybrid poplar	Sugar beet
Cull dairy cows	Ice cream	Switchgrass

(continued)

Table 2-12. Commodities without International Trade Possibilities Modeled (continued)

Cull ewes	Lamb slaughter	Tbtus
Cull sow	Milk	Turkeys
Dairy calves	Nonfed slaughter	Willow
Dextrose	Oranges, fresh (85 lb box)	Wool
Ethanol	Oranges, fresh (90 lb box)	

Note: FASOMGHG does not explicitly include ethanol trade, but in applications for biofuels analyses, we have assumed that exogenous levels of mandated ethanol volumes would be provided by imports based on information from other models.

GHG Accounts

FASOMGHG quantifies the stocks of GHGs emitted from and sequestered by agriculture and forestry, as well as the carbon stock on lands in the model that are converted to nonagricultural, nonforest developed usage. In addition, the model tracks GHG emission reductions in other sectors caused by mitigation actions in the forest and agricultural sectors.

The GHGs tracked by the model include CO_2, CH_4, and N_2O. Given the multi-GHG impact of the agricultural and forestry sectors, there are multidimensional trade-offs between model variables and net GHG emissions. To consider these trade-offs, all GHGs are converted to carbon or carbon dioxide equivalent basis using 100-year global warming potential (GWP) values for application of GHG incentives.

GWPs compare the abilities of different GHGs to trap heat in the atmosphere. They are based on the radiative forcing (heat-absorbing ability) and decay rate of each gas relative to that of CO_2. The GWP allows one to convert emissions of various GHGs into a common measure, which allows for aggregating the radiative impacts of various GHGs into a single measure denominated in CO_2 or C equivalents. Extensive discussion of GWPs can be found in the documents of the Intergovernmental Panel on Climate Change (IPCC). The IPCC updated its estimates of GWPs for key GHGs in 2001 (IPCC, 2001) and again in 2007 (IPCC, 2007), but these estimates are still under debate. As a result, the FASOMGHG model continues to use the 1996 GWPs for the GHGs covered by the model:

- $CO_2 = 1$

- $CH_4 = 21$

- $N_2O = 310$

When CO_2 equivalent results are converted to a C equivalent basis, a transformation is done based on the molecular weight of C in the CO_2. The CO_2 equivalent quantities of gas are divided by 3.667 to compute the carbon equivalent quantities.

A list of all categories included in the model's GHG accounting appears in Table 2-13, totaling 57 categories. Brief summaries of the major categories are presented in the following subsections.

Table 2-13. Categories of GHG Sources and Sinks in FASOMGHG

Forest_SoilSequest	Carbon in forest soil
Forest_LitterUnder	Carbon in litter and understory of forests that remain forests
Forest_ContinueTree	Carbon in trees of forests that remain forests
Forest_AfforestSoilSequest	Carbon in forest soil of afforested forests
Forest_AfforestLitterUnder	Carbon in litter and understory of afforested forests
Forest_AfforestTree	Carbon in trees of afforested forests
Forest_USpvtProduct	Carbon from U.S. private forests consumed producing forest products
Forest_USpubProduct	Carbon from U.S. public forests consumed producing forest products
Forest_CANProduct	Carbon in U.S. consumed but Canadian produced forest products
Forest_USExport	Carbon in U.S. produced but exported forest products
Forest_USImport	Carbon in U.S. consumed but imported from non-Canadian source
Forest_USFuelWood	Carbon in U.S. consumed fuelwood
Forest_USFuelResidue	Carbon in U.S. residue that is burned
Forest_USresidProduct	Carbon from U.S. residues consumed producing forest products
Forest_CANresidProduct	Carbon from Canadian residues consumed producing forest products
Carbon_For_Fuel	Carbon emissions from forest use of fossil fuel
Dev_Land_from_Ag	Carbon on land after it moves from agriculture into developed use
Dev_Land_from_Forest	Carbon on land after it moves from forest into developed use
AgSoil_CropSequest_Initial	Carbon in cropped agricultural soil with initial tillage
AgSoil_CropSequest_TillChange	Carbon in cropped agricultural soil with change in tillage
AgSoil_PastureSequest	Carbon in pastureland
Carbon_AgFuel	Carbon emissions from agricultural use of fossil fuels
Carbon_Dryg	Carbon emissions from grain drying
Carbon_Fert	Carbon emissions from fertilizer production

(continued)

Table 2-13. Categories of GHG Sources and Sinks in FASOMGHG (continued)

Carbon_Pest	Carbon emissions from pesticide production
Carbon_Irrg	Carbon emissions from water pumping
Carbon_Ethl_Offset	Carbon emission offset by conventional ethanol production
Carbon_Ethl_Haul	Carbon emissions in hauling for conventional ethanol production
Carbon_Ethl_Process	Carbon emissions in processing of conventional ethanol production
Carbon_CEth_Offset	Carbon emission offset by cellulosic ethanol production
Carbon_CEth_Haul	Carbon emissions in hauling for cellulosic ethanol production
Carbon_CEth_Process	Carbon emissions in processing of cellulosic ethanol production
Carbon_BioElec_Offset	Carbon emission offset from bioelectricity production
Carbon_BioElec_Haul	Carbon emissions in hauling for bioelectricity production
Carbon_BioElec_Process	Carbon emissions in processing of for bioelectricity production
Carbon_Biodiesel_Offset	Carbon emission offset from biodiesel production
Carbon_Biodiesel_Process	Carbon emissions in processing of biodiesel production
Methane_Liquidmanagement	Methane from emission savings from improved manure technologies
Methane_EntericFerment	Methane from enteric fermentation
Methane_Manure	Methane from manure management
Methane_RiceCult	Methane from rice cultivation
Methane_AgResid_Burn	Methane from agricultural residue burning
Methane_BioElec	Net change in methane emissions from bioelectricity relative to coal-fired
Methane_Biodiesel	Net change in methane emissions from biodiesel production relative to diesel
Methane_Ethl	Net change in methane emissions from ethanol production relative to gasoline
Methane_CEth	Net change in methane emissions from cellulosic ethanol production relative to gasoline
NitrousOxide_Manure	Livestock manure practices under managed soil categories under AgSoilMgmt
NitrousOxide_BioElec	Net change in nitrous oxide emissions from bioelectricity relative to coal-fired
NitrousOxide_Biodiesel	Net change in nitrous oxide emissions from biodiesel production relative to diesel
NitrousOxide_Ethl	Net change in nitrous oxide emissions from noncellulosic ethanol processing relative to gasoline
NitrousOxide_CEth	Net change in nitrous oxide emissions from cellulosic ethanol processing relative to gasoline
NitrousOxide_Fert	Nitrous oxide emissions from nitrogen inputs including nitrogen fertilizer application practices, crop residue retention, and symbiotic nitrogen fixation under managed soil categories under AgSoilMgmt
NitrousOxide_Pasture	Nitrous oxide emissions from pasture

(continued)

Table 2-13. Categories of GHG Sources and Sinks in FASOMGHG (continued)

NitrousOxide_Histosol	Emissions from temperate histosol area
NitrousOxide_Volat	Indirect soils volatilization
NitrousOxide_Leach	Indirect soils leaching runoff
NitrousOxide_AgResid_Burn	Agricultural residue burning

2.7.1 Forest GHG Accounts

As identified in Table 2-13, forest GHG accounting includes carbon sequestered, carbon emitted, and fossil fuel–related carbon emissions avoided. Sequestration accounting encompasses carbon in standing (live and dead) trees, forest soils, the forest understory vegetation, forest floor including litter and large woody debris, and wood products both in use and in landfills. The sequestration accounting involves both increases and reductions in stocks, with changes in specific accounts to reflect land movement into forest use through afforestation, net growth of forests not of afforestation origin, and placement of products in long-lasting uses or landfills.[14] Reductions arise when land is migrated to agriculture or development and products decay in their current uses.

Forest-related emissions accounting includes GHGs emitted when fossil fuels are used in forest production. Forest-related GHG accounting calculates the estimated amount of fossil fuels (and associated GHG emissions) that are saved when wood products are combusted in place of fossil fuels, particularly when milling residues are burned to provide energy (generally for use at the mill). In addition, woody biomass may be used as a bioenergy feedstock.

Forest carbon accounts also include the carbon content of products imported into, or exported out of, the United States. In particular, there is explicit accounting for products

- processed in and coming from Canada,

- imported from other countries, and

- exported to other countries.

---

[14] In the case of wood product accounts, note that these accounts have increases in C sequestration when more products are made, but the forest carbon accounts are simultaneously reduced to account for C reduced by harvesting.

These categories may or may not be included in an incentive scheme for GHG mitigation, because they will generally be accounted for elsewhere. Nonetheless, the accounts are included in the model in case they are needed for policy analysis.

2.7.2 Agricultural GHG Accounts

On the agricultural side, the categories tracked in the model are also listed in Table 2-13. Agricultural emissions arise from crop and livestock production, principally from

- fossil fuel use,

- nitrogen fertilization usage,

- other nitrogen inputs to crop production,

- agricultural residue burning,

- rice production,

- enteric fermentation, and

- manure management.

In addition, changes in carbon sequestration are tracked within the model. Agricultural sequestration involves the amount of carbon sequestered in agricultural soils, due principally to choice of tillage, and irrigation along with changes to crop mix choice. Sequestration is also considered in terms of grasslands versus cropland or mixed usage, where cropland can be moved to pasture use or vice versa. The sequestration accounting can yield either positive or negative quantities, depending on the direction of change in tillage between the three available options (conventional, conservation, or zero tillage) and irrigation choices, along with pasture land (grassland)/cropland conversions and movements between agriculture and forestry. With movements from forestry to agriculture, gains in the agricultural soil carbon account are typically more than offset by losses in the forest soil carbon account (e.g., forest soils typically store more carbon per acre than soils in agricultural uses). When moving from agricultural land uses to forestland, on the other hand, there are typically net increases in soil carbon sequestration.

As with forest products, certain agricultural commodities can also be used as bioenergy feedstocks.

2.7.3 Bioenergy GHG Accounts

Selected agricultural and forestry commodities can be used as feedstocks for biofuel production processes in FASOMGHG, possibly affecting fossil fuel usage and associated GHG emissions after accounting for emissions during hauling and processing of bioenergy feedstocks. Four major forms of bioenergy production are included:

- Biodiesel: usage of canola oil, corn oil, lard, poultry fat, soybean oil, tallow, or yellow grease in the production of biodiesel, which replaces petroleum-based diesel fuel

- Bioelectricity: usage of bagasse, crop residues, energy sorghum, hybrid poplar, lignin, manure, miscanthus, sweet sorghum pulp, switchgrass, willow, wood chips, logging residues, or milling residues as inputs to electric generating power plants in place of coal (through either cofiring or dedicated biomass plants)

- Cellulosic ethanol: usage of bagasse, crop residues, energy sorghum, hybrid poplar, miscanthus, sweet sorghum pulp, switchgrass, willow, wood chips, logging residues, or milling residues to produce cellulosic ethanol, which replaces gasoline

- Starch or sugar-based ethanol: usage of barley, corn, oats, rice, sorghum, sugar, sweet sorghum, or wheat for conversion to ethanol and replacement of gasoline

In all of these cases, the GHG reduction provided by bioenergy production is equal to the GHGs emitted from burning and producing the fossil fuel replaced less the GHG emissions of producing, transporting, and processing the bioenergy feedstock.

2.7.4 Developed Land GHG

FASOMGHG incorporates exogenous data that specify the rate of conversion of agriculture and forestry lands to nonagricultural and nonforestry developed uses. Simplified accounting is employed to estimate the carbon sequestered on these lands.

Other Environmental Impacts

FASOMGHG considers a number of environmental indicators above and beyond the GHG accounts. The main components are nitrogen and phosphorus application and runoff, soil erosion, irrigation water usage, and a number of descriptions of total resource use and activity within the agricultural and forestry sectors (e.g., total land use, total pasture use, manure load, livestock numbers, total afforestation).

SECTION 3
METHODS USED TO DEVELOP ESTIMATES OF OZONE EFFECTS ON CROP AND FOREST PRODUCTIVITY

Incorporating the impacts of different ambient ozone concentration levels into FASOMGHG requires determining crop yield and forest productivity impacts associated with changes in concentrations. Productivity impacts are required for each crop/region and forest type/region combination included within the model. In this section, we describe our methods for calculating relative yield losses (RYLs) and relative yield gains (RYGs) of crops and tree species under alternative ambient ozone concentration levels.

These data are essential for our market analysis because crop and forest yields play an important role in determining the economic returns to agricultural and forest production activities. Thus, they affect landowner decisions regarding land use, crop mix, forest rotation lengths, production practices, and others. Alterations in ambient ozone concentration levels will therefore change the supply curves of U.S. agricultural and forest commodities, resulting in new market equilibriums. Because both the changes in ozone concentrations and the distribution of ozone-sensitive crops and tree species vary spatially, there may be substantial differences in the net impacts across regions. There may also be distributional impacts as commodity production shifts between regions in response to changes in relative productivity.

Ambient Ozone Concentration Data

There are several alternative metrics used for assessing ozone concentrations (see Lehrer et al. [2007] for more information). For this assessment, we are using the W126 metric, which is a weighted sum of all ozone concentrations observed from 8 a.m. to 8 p.m. More specifically, we are using W126 ozone concentration surfaces generated using enhanced Voronoi Neighbor Averaging (eVNA). W126 concentration surfaces based on meeting the current ozone standard[15] were provided by EPA to serve as the baseline for this analysis. In addition, EPA provided W126 ozone concentration surfaces for current conditions, as well as for 15 ppm-hr, 11 ppm-hr, and 7 ppm-hr W126 standards. According to information provided by EPA, the eVNA W126 ozone surface is built from monitor data fused with Community Multiscale Air Quality (CMAQ) model-based gradient interpolations. The spatial resolution of the ozone surface in ArcGIS Shapefile format is 12 km.

---

[15]The current primary and secondary ozone standards are 0.075 parts per million (ppm) (or 75 parts per billion) based on the annual fourth-highest daily maximum 8-hr concentration, averaged over 3 years. For the purposes of calculating impacts on crop yields and forest growth rates, we used the W126 equivalent of the current standard.

County-level values were extracted from the eVNA W126 ozone surface using ArcGIS. Only the ozone concentrations for the cropland and forestland portions of the W126 ozone surface are used to derive the county-level average crop and forest W126 ozone levels, respectively. These weighting adjustments were made to better reflect the ozone concentration that would affect the specific portions of each county containing forested land or cropland, rather than basing county-level exposure on the ozone concentration across the whole county. Data from the 2006 National Land Cover Database (NLCD) are used to extract the cropland and forestland portions from the ozone surface.

Calculation of Relative Yield Loss

The median W126 ozone concentration response (CR) functions for crops and tree seedlings in the 2007 EPA technical report (Lehrer et al., 2007) are used to calculate the RYLs for crops and tree species under each ambient ozone concentration scenario used in this analysis.

Table 3-1 presents the α and β parameters being used in the W126 ozone CR function for different crops and tree species. The W126 ozone CR function is as follows: $RYL = 1 - e^{-(W126/\alpha)^\beta}$.

Table 3-1. Parameter Values Used for Crops and Tree Species

	α	β
Crops		
Corn	98.3	2.973
Sorghum	205.9	1.963
Soybean	110.0	1.367
Winter wheat	53.7	2.391
Potato	99.5	1.242
Cotton	94.4	1.572
Tree Species		
Ponderosa	159.63	1.1900
Red alder	179.06	1.2377
Black cherry	38.92	0.9921
Tulip poplar	51.38	2.0889
Sugar maple	36.35	5.7785
Eastern white	63.23	1.6582
Red maple	318.12	1.3756
Douglas fir	106.83	5.9631
Quaking aspen	109.81	1.2198
Virginia pine	1,714.64	1.0000

3.1.1 Relative Yield Loss for Crops

Specifically, for crops, we first calculate the FASOMGHG subregion RYLs for crops that have W126 ozone CR functions using the subregion-level, cropland-based ozone concentration values under each scenario. The FASOMGHG subregion-level ozone concentration values are initially calculated for all crops as the simple averages of the county-level ozone concentration values. For crops that do not have W126 ozone CR functions, we assign them W126 ozone CR functions based on the crop proxy mapping shown in Table 3-2. This crop mapping was based on the authors' judgment and previous experience.[16] In addition, for oranges, rice, and tomatoes, which have ozone CR functions that are not W126-based (they are defined based on alternative measures of ozone levels), we directly used the median RYG values under the 13 ppm-hr ozone level reported in Table G-7 of Lehrer et al. (2007). More details on RYG are presented in further subsections.

Table 3-2.   Mapping of Ozone Impacts on Crops to FASOMGHG Crops

Crops Used for Estimating Ozone Impacts	FASOMGHG Crops
W126 Crops	
Corn	Corn
Cotton	Cotton
Potatoes	Potatoes
Winter wheat	Soft white wheat, hard red winter wheat, soft red winter wheat, durum wheat, hard red spring wheat, oats, barley, rye, sugar beet, grazing wheat, and improved pasture
Sorghum	Sorghum, silage, hay, sugarcane, switchgrass, miscanthus, energy sorghum, and sweet sorghum
Soybeans	Soybeans and canola
Aspen (tree)	Hybrid poplar, willow (FASOMGHG places short-rotation woody biomass production in the crop sector rather than in the forest sector)
Non-W126 Crops	
Oranges	Orange fresh/processed, grapefruit fresh/processed
Rice	Rice
Tomatoes	Tomato fresh/processed

Moreover, for crops that have county-level production data and W126 ozone CR functions (including functions based on proxy crops), we updated the RYLs with production-weighted W126 values. The 2007 USDA Census of Agriculture (Ag Census) county-level

---

[16] Also, note that FASOMGHG defines short-rotation woody trees such as hybrid poplar and willow as crops. Ozone impacts on short-rotation woody trees were based on ozone RYLs for aspen.

production data are used to derive the weighted FASOMGHG subregion RYLs, following Formula (3.1).

$$wRYL_{ik} = Ozone\ CR\ Function_k(\frac{\sum_j Prod_{ijk}*W126_{ij}}{\sum_j Prod_{ijk}}),\qquad (3.1)$$

where i denotes FASOMGHG subregion, j indicates county, and k represents crop. Ozone CR Function$_k$ refers to the ozone concentration response function for crop k. Prod$_{ijk}$ represents the county-level production level of crop k, and W126$_{ij}$ represents the cropland-based ozone value for county j in subregion i. Finally, wRYL$_{ik}$ stands for the weighted FASOMGHG subregion RYL for crop k. RYLs are calculated for each ozone concentration level being considered.

3.1.2 Relative Yield Loss for Trees

The ozone CR functions for tree seedlings were used to calculate RYLs for FASOMGHG trees over their whole life span. To derive the FASOMGHG region-level RYLs for trees under each ozone concentration scenario, we used FASOMGHG region ozone values and the mapping in Table 3-3.

Table 3-3. **Mapping of Ozone Impacts on Forests to FASOMGHG Forest Types**

Tree Species Used for Estimating Ozone Impacts	FASOMGHG Forest Type	FASOMGHG Region(s)
Black cherry, tulip poplar	Upland hardwood	SC, SE
Douglas fir	Douglas fir	PNWW
Eastern white pine	Softwood	CB, LS
Ponderosa pine	Softwood	PNWE, PNWW, PSW, RM
Quaking aspen	Hardwood	RM
Quaking aspen, black cherry, red maple, sugar maple, tulip poplar	Hardwood	CB, LS, NE
Red alder	Hardwood	PNWE, PNWW, PSW
Red maple	Bottomland hardwood	SC, SE
Virginia pine	Natural pine, oak-pine, planted pine	SC
Virginia pine, eastern white pine	Natural pine, oak-pine, planted pine	SE
Virginia pine, eastern white pine	Softwood	NE

Note: CB = Corn Belt; LS = Lake States; NE = Northeast; PNWE = Pacific Northwest—East side; PNWW = Pacific Northwest—West side; PSW = Pacific Southwest; RM = Rocky Mountains; SC = South Central; SE = Southeast.

Specifically, the FASOMGHG region-level RYLs are first calculated for each tree species listed in first column of Table 3-3. Then, a simple average of RYLs for each tree species mapped to a FASOMGHG forest type in a given region is calculated. The mapping of tree species to FASOMGHG forest types is based on Elbert L. Little, Jr.'s *Atlas of United States Trees* (1971, 1976, 1977, 1978). Note that crop RYLs are generated at the FASOMGHG subregion level, whereas forest RYLs are calculated at the FASOMGHG region level, consistent with the greatest level of regional disaggregation available for these sectors within FASOMGHG.

Calculation of Relative Yield Gain

As described by Lehrer et al. (2007), the RYL is the relative yield loss compared with the baseline yield under a "clean air" environment. For implementation within FASOMGHG, we calculate the RYG for crops and trees from moving between ambient ozone concentrations (i.e., RYG is calculated as a change in RYL when moving between scenarios).

Thus, to obtain the RYG for crops and trees under alternative ozone concentrations, we need the RYLs under each scenario. For example, to derive RYG under the current standard 75 ppb scenario relative to current conditions "currcond," we use Formula (3.2):

$$RYG_{75ppb} = \frac{1-RYL_{75ppb}}{1-RYL_{currcond}} - 1 = \frac{RYL_{currcond}-RYL_{75ppb}}{1-RYL_{currcond}} \tag{3.2}$$

The FASOMGHG subregion-level crop RYGs and the FASOMGHG region-level tree RYGs for changes associated with moving from one scenario to another were calculated for additional comparisons in the same way.

Conducting Model Scenarios in FASOMGHG

The current crop/forest budgets included in FASOMGHG are assumed to reflect input/output relationships under current ambient ozone concentrations because these budgets are based on historical data. To model the effects of changing ozone concentrations on the agricultural and forest sectors, the following five scenarios were constructed and run through the model:

1. "current" scenario, where no RYGs of crops and trees are considered (assumed to be consistent with current ambient ozone concentration levels);

2. 75 ppb scenario, where crop and forest yields are assumed to increase by the percentages calculated in RYG_{75ppb}, calculated relative to the current scenario;

3. 15 ppm-hr scenario, using $RYG_{15ppm-hr}$, calculated relative to both current and 75 ppb scenarios

4. 11 ppm-hr scenario, using $RYG_{11ppm-hr}$, calculated relative to both current and 75 ppb scenarios; and

5. 7 ppm-hr scenario, using $RYG_{7ppm-hr}$, calculated relative to both current and 75 ppb scenarios.

Our primary comparisons in this report are between the current standard and the 15 ppm-hr, 11 ppm-hr, and 7 ppm-hr cases. The results of those comparisons are included in Sections 5 and 6 of this report. However, we also calculated comparisons between current conditions and the current standard (75 ppb), 15 ppm-hr, 11 ppm-hr, and 7 ppm-hr scenarios that are reported in Appendixes A, B, and C.

The time scope of the FASOMGHG model scenarios used for these analyses is 2000–2050, solved in 5-year time steps.[17] The crop and tree RYGs are introduced into the model starting in 2010, and they remain constant at those percentage changes relative to baseline for the rest of the modeling period.

Figure 3-1 presents the modeling process of simulating ozone scenarios using FASOMGHG. The changes in crop and tree yield growth potentially lead to new market equilibriums for agricultural and forestry commodities, as well as land use changes between agricultural and forestry uses and consequent GHG emissions and sequestration changes.

By comparing the market equilibriums under different scenarios, we can calculate the welfare, land use, and GHG impacts of alternative ozone standards on the U.S. agricultural and forest sector, including changes in consumer and producer welfare, land use allocation, and GHG mitigation potential over time.

---

[17] Because of terminal period effects, the model is run out to the 2050 time period, but only results through the 2040 model time period (representative of 2040–2044) are used in our analyses.

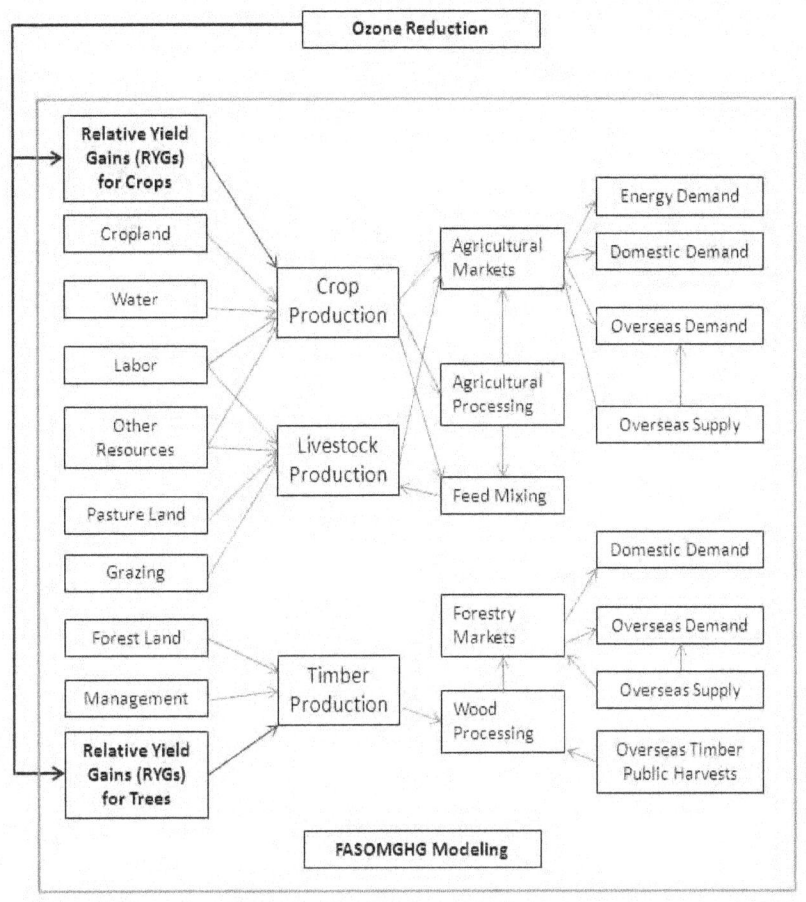

Figure 3-1.   FASOMGHG Modeling Flowchart

SECTION 4
DATA INPUTS

In this section, we summarize the input data used in the FASOMGHG scenarios specified for this assessment. Following the methods described in Section 3, we calculated W126 ozone concentration levels by region and crop. Effects on crop yields and forest productivity were calculated for each FASOMGHG region. We present the values used as model inputs in tabular and map format, with a primary focus here on comparison of the more stringent scenarios to the current standard.

Ambient Ozone Concentration Data

The county-level forested and cropland W126 ozone values were aggregated at regional and subregional levels, respectively. Table 4-1 presents the forestland W126 ozone values for each of the five scenarios. Comparing the 75 ppb current standard scenario with the current conditions scenario, one can see that the Pacific Southwest, South Central, Southeast, and Rocky Mountains regions would have the largest ground-level ozone reductions, if attained. The Corn Belt, Southwest, and Northeast regions would also experience significant ozone reductions. The Great Plains, Lake States, and Pacific Northwest regions are the regions least affected by attainment of current ozone standards.

Table 4-1. Forestland W126 Ozone Values under Alternative Scenarios

FASOMGHG Region	currcond	75 ppb	15 ppm-hr	11 ppm-hr	7 ppm-hr
CB	11.78	3.41	3.41	2.46	1.61
GP[a]	8.99	3.10	3.05	2.97	2.52
LS	6.32	1.15	1.15	1.14	1.12
NE	8.50	0.69	0.69	0.55	0.42
PNWE	5.56	1.92	1.86	1.80	1.76
PNWW	3.74	1.83	1.81	1.80	1.79
PSW	17.15	1.53	1.40	1.26	1.13
RM	13.31	3.72	3.09	2.28	1.96
SC	11.84	2.49	2.49	2.07	1.53
SE	13.05	2.35	2.35	2.13	1.41
SW[a]	10.05	1.83	1.82	1.79	1.78

[a] GP and SW are modeled as agriculture-only regions in FASOMGHG.

Note: CB = Corn Belt; GP = Great Plains; LS = Lake States; NE = Northeast; PNWE = Pacific Northwest—East side; PNWW = Pacific Northwest—West side; PSW = Pacific Southwest; RM = Rocky Mountains; SC = South Central; SE = Southeast; SW = Southwest.

Comparing the most stringent 7 ppm-hr standard scenario with the 75 ppb scenario, the Corn Belt and Rocky Mountains regions are the areas that will have the most notable further ozone reductions.

Table 4-2 displays the agricultural W126 ozone values at the subregion level. Similar to the forest ozone values, the Pacific Southwest, South Central, Southeast, and Rocky Mountains regions experience the greatest agricultural ozone reductions under the 75 ppb scenario compared with current conditions. The Corn Belt, Southwest, and Northwest regions also see noteworthy agricultural ozone reductions.

Table 4-2. Cropland W126 Ozone Values under Modeled Scenarios

FASOMGHG Region	FASOMGHG Subregion	Current Conditions	75 ppb	15 ppm-hr	11 ppm-hr	7 ppm-hr
Corn Belt (CB)	Illinois, Northern	7.80	3.14	3.14	2.44	1.73
	Illinois, Southern	11.16	3.60	3.60	2.53	1.54
	Indiana, Northern	10.35	3.60	3.60	2.65	1.72
	Indiana, Southern	13.01	4.27	4.27	2.96	1.75
	Iowa, Central	5.68	1.09	1.09	1.07	1.04
	Iowa, Northeast	6.24	0.73	0.73	0.73	0.73
	Iowa, Southern	6.53	1.57	1.57	1.35	1.14
	Iowa, Western	5.71	1.80	1.80	1.78	1.66
	Missouri	11.41	3.41	3.41	2.47	1.64
	Ohio, Northeast	12.74	3.19	3.19	2.30	1.46
	Ohio, Northwest	12.03	3.90	3.90	2.90	1.93
	Ohio, Southern	13.75	3.42	3.42	2.39	1.45
Great Plains (GP)	Kansas	10.90	2.35	2.15	1.87	1.72
	Nebraska	8.79	2.86	2.62	2.31	2.00
	North Dakota	4.46	1.96	1.96	1.96	1.70
	South Dakota	5.64	2.13	2.11	2.09	1.81
Lake States (LS)	Michigan	9.51	2.35	2.35	2.29	2.22
	Minnesota	4.95	1.11	1.11	1.11	1.04
	Wisconsin	7.01	1.28	1.28	1.27	1.25

(continued)

Table 4-2. Cropland W126 Ozone Values under Modeled Scenarios (continued)

FASOMGHG Region	FASOMGHG Subregion	Current Conditions	75 ppb	15 ppm-hr	11 ppm-hr	7 ppm-hr
Northeast (NE)	Connecticut	11.74	0.24	0.24	0.24	0.24
	Delaware	17.23	0.38	0.38	0.38	0.36
	Maine	3.71	0.39	0.39	0.39	0.39
	Maryland	17.36	0.76	0.76	0.70	0.56
	Massachusetts	9.90	0.27	0.27	0.27	0.27
	New Hampshire	5.71	0.31	0.31	0.31	0.31
	New Jersey	16.76	0.30	0.30	0.30	0.30
	New York	8.34	0.16	0.16	0.16	0.16
	Pennsylvania	11.97	0.53	0.53	0.46	0.40
	Rhode Island	11.61	0.40	0.40	0.40	0.40
	Vermont	5.52	0.28	0.28	0.28	0.28
	West Virginia	10.86	2.25	2.25	1.60	0.95
Pacific Northwest East (PNWE)	Oregon	6.39	2.25	2.16	2.05	2.00
	Washington	4.95	1.99	1.93	1.86	1.83
Pacific Southwest (PSW)	California, Northern	21.91	1.50	1.31	1.11	0.93
	California, Southern	20.06	3.08	2.72	2.31	2.01
Rocky Mountains (RM)	Arizona	13.75	5.52	4.27	2.68	2.20
	Colorado	16.34	4.75	3.77	2.50	1.95
	Idaho	12.54	3.16	2.72	2.17	1.94
	Montana	6.59	2.32	2.31	2.29	1.92
	Nevada	15.55	2.77	2.47	2.10	1.95
	New Mexico	12.54	3.48	2.78	1.88	1.64
	Utah	18.04	4.88	3.90	2.59	2.20
	Wyoming	14.13	4.22	3.76	3.20	2.62
South Central (SC)	Alabama	13.22	3.07	3.07	2.81	1.86
	Arkansas	12.43	2.41	2.41	2.14	1.89
	Kentucky	13.77	3.91	3.91	2.71	1.61
	Louisiana	9.60	1.20	1.20	1.20	1.19
	Mississippi	11.32	1.56	1.56	1.52	1.44

(continued)

4-3

Table 4-2. Cropland W126 Ozone Values under Modeled Scenarios (continued)

FASOMGHG Region	FASOMGHG Subregion	Current Conditions	75 ppb	15 ppm-hr	11 ppm-hr	7 ppm-hr
	Tennessee	15.53	3.76	3.76	2.66	1.62
	Texas, East	9.92	1.59	1.59	1.59	1.59
Southeast (SE)	Florida	9.24	2.93	2.93	2.78	2.08
	Georgia	12.86	2.68	2.68	2.47	1.64
	North Carolina	14.28	2.00	2.00	1.84	1.20
	South Carolina	12.88	1.71	1.71	1.59	1.07
	Virginia	12.57	2.09	2.09	1.86	1.22
Southwest (SW)	Oklahoma	11.50	1.87	1.82	1.76	1.74
	Texas, Central Blacklands	9.31	1.96	1.96	1.96	1.96
	Texas, Coastal Bend	7.28	1.73	1.73	1.73	1.73
	Texas, Edwards Plateau	9.14	1.93	1.85	1.75	1.73
	Texas, High Plains	11.76	2.60	2.18	1.65	1.52
	Texas, Rolling Plains	10.71	2.06	1.99	1.89	1.87
	Texas, South	4.38	1.41	1.41	1.40	1.40
	Texas, Trans Pecos	11.78	3.59	3.20	2.70	2.57

Figure 4-1 presents the incremental ozone reductions under alternative ozone standards with respect to the current 75 ppb standard. As the standard is tightened from the 15 ppm-hr to 7 ppm-hr scenario, the greatest ozone reductions are observed in the southern half of the Rocky Mountains region. The southern Corn Belt and northern areas of the South Central regions, along with selected places in the Southeast and Pacific Southwest near urban areas, also see substantial ozone reductions. These ozone reductions would affect the production of crops and timber that are susceptible to ground-level ozone in these regions.

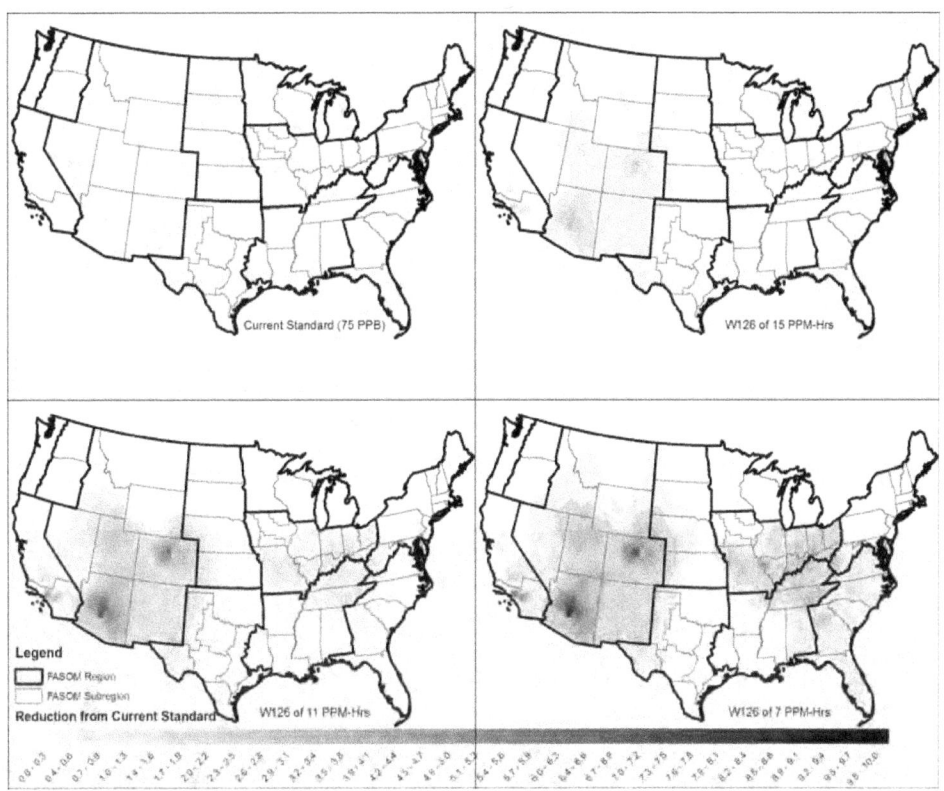

Figure 4-1.   Ozone Reductions with Respect to 75 ppb under Alternative Scenarios

Changes in Crop and Forest Yields with Respect to 75 ppb Scenario

Figures 4-2 through 4-7 display major crops' RYGs under alternative ozone standards scenarios at the FASOMGHG subregion level, with respect to the current 75 ppb standard. These are the values that were directly incorporated into FASOMGHG to define the scenarios modeled. See Appendix A for maps of county-level yield changes for major crops. Figures 4-8 and 4-9 display changes in forest RYGs. As discussed previously, the Rocky Mountains, Corn Belt, and parts of the southern regions of the United States (e.g., within the Pacific Southwest, South Central, and Southeast regions) would experience the most significant further ozone reductions. Hence, one would expect to see the most sizable increases in RYGs for crops and trees grown in those regions. This finding is consistent with our calculations.

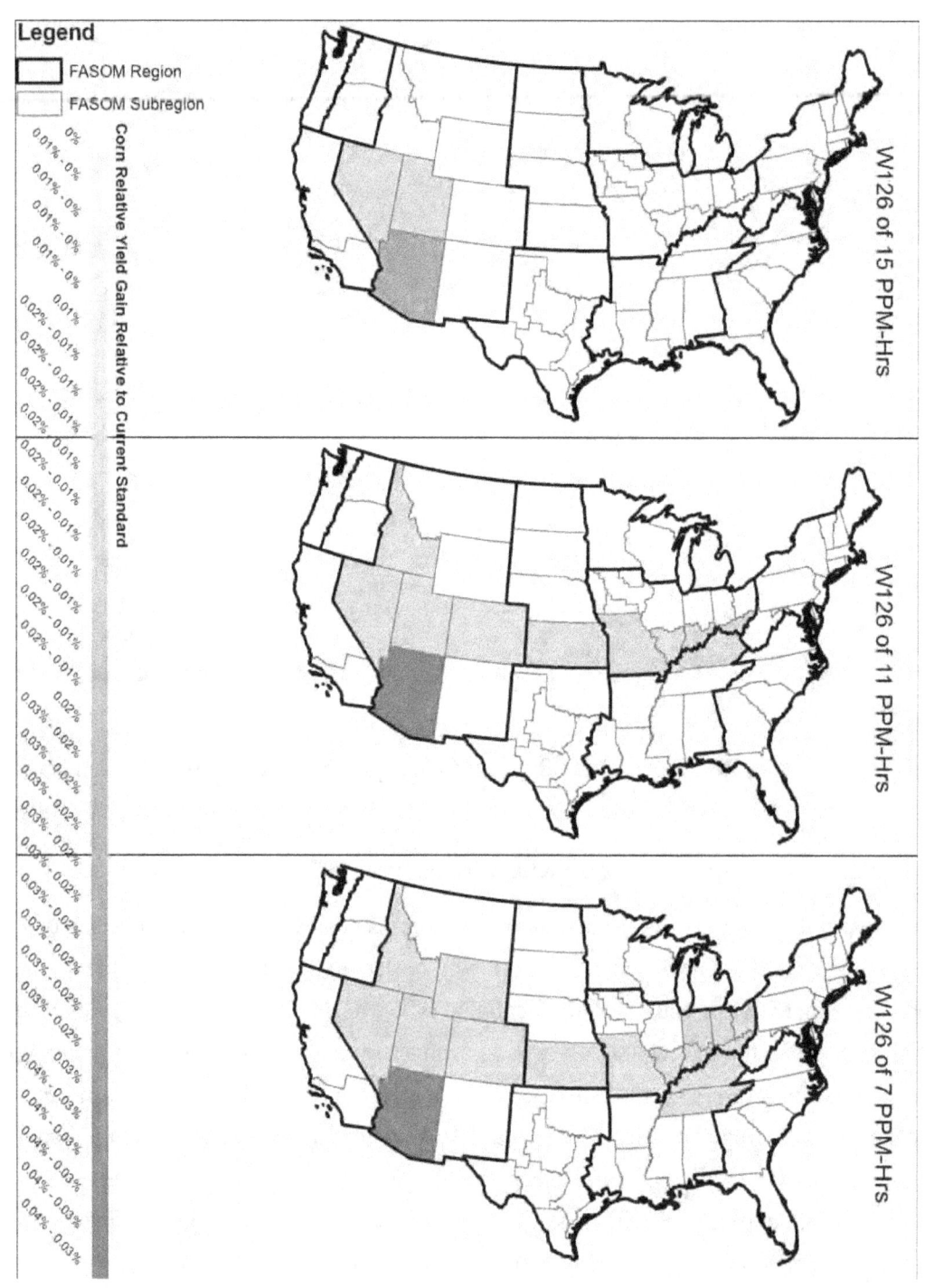

Figure 4-2.   Percentage Changes in Corn RYGs with Respect to the 75 ppb Scenario

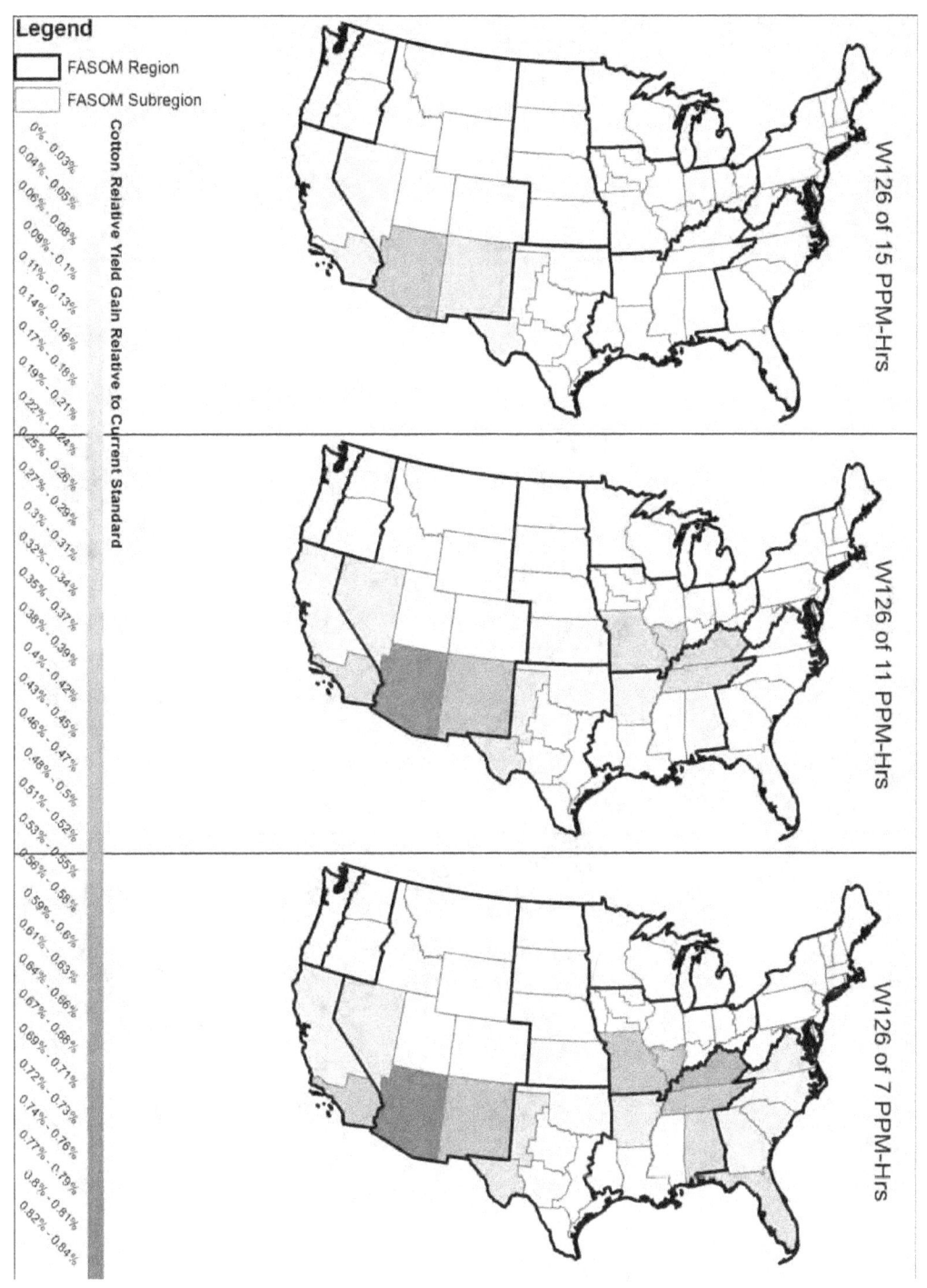

Figure 4-3. **Percentage Changes in Cotton RYGs with Respect to the 75 ppb Scenario**

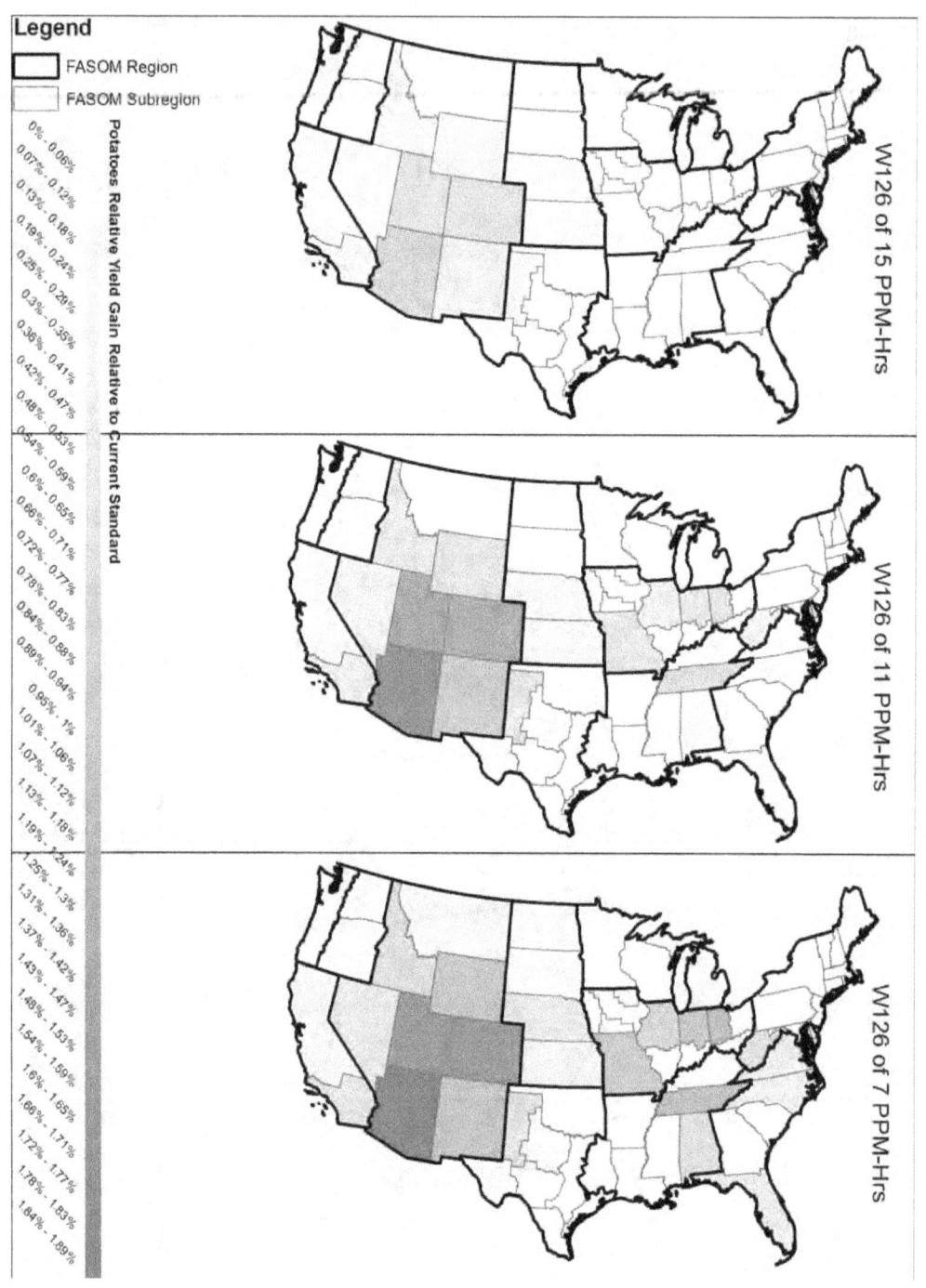

Figure 4-4. **Percentage Changes in Potato RYGs with Respect to the 75 ppb Scenario**

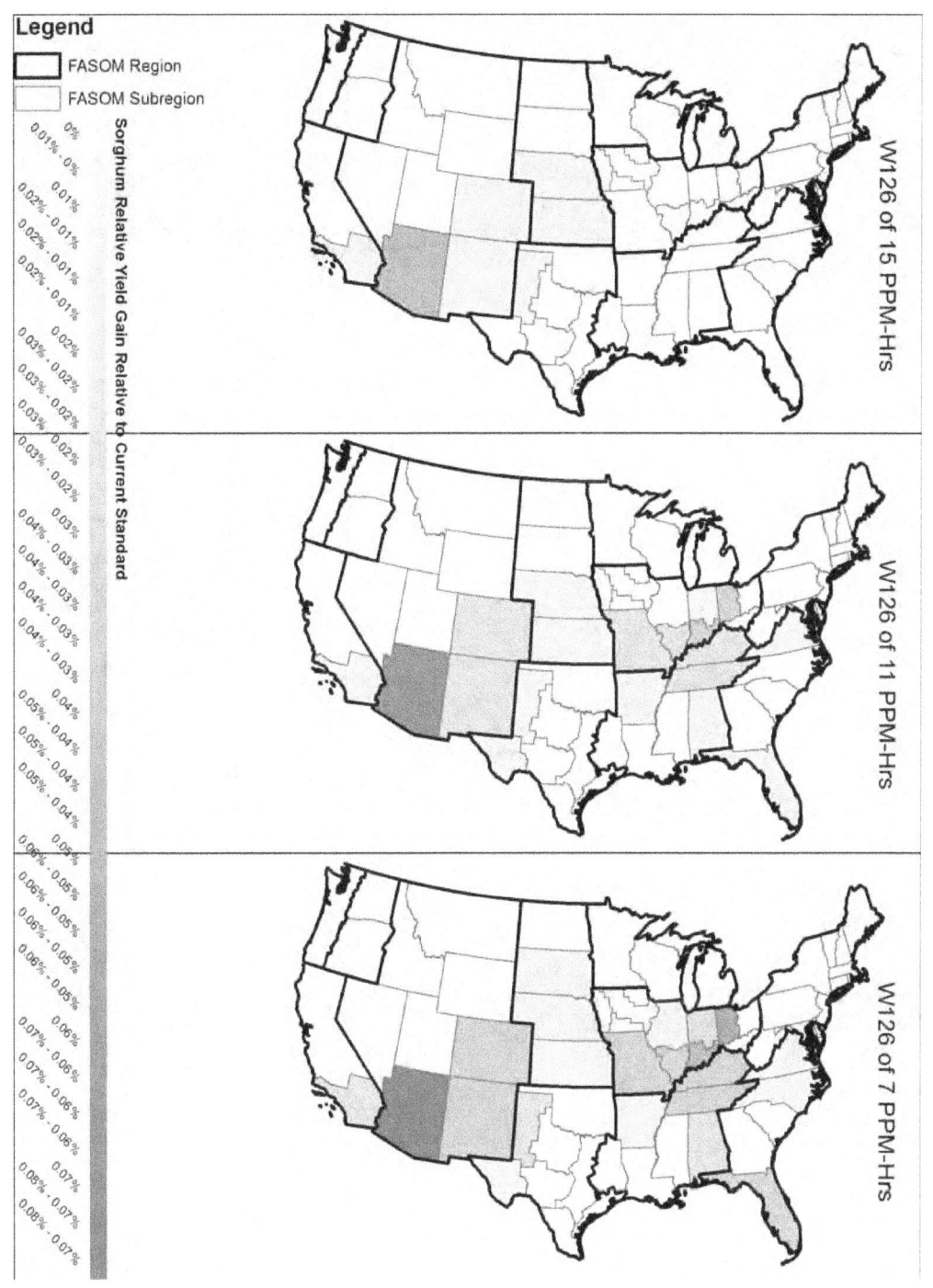

Figure 4-5.   Percentage Changes in Sorghum RYGs with Respect to the 75 ppb Scenario

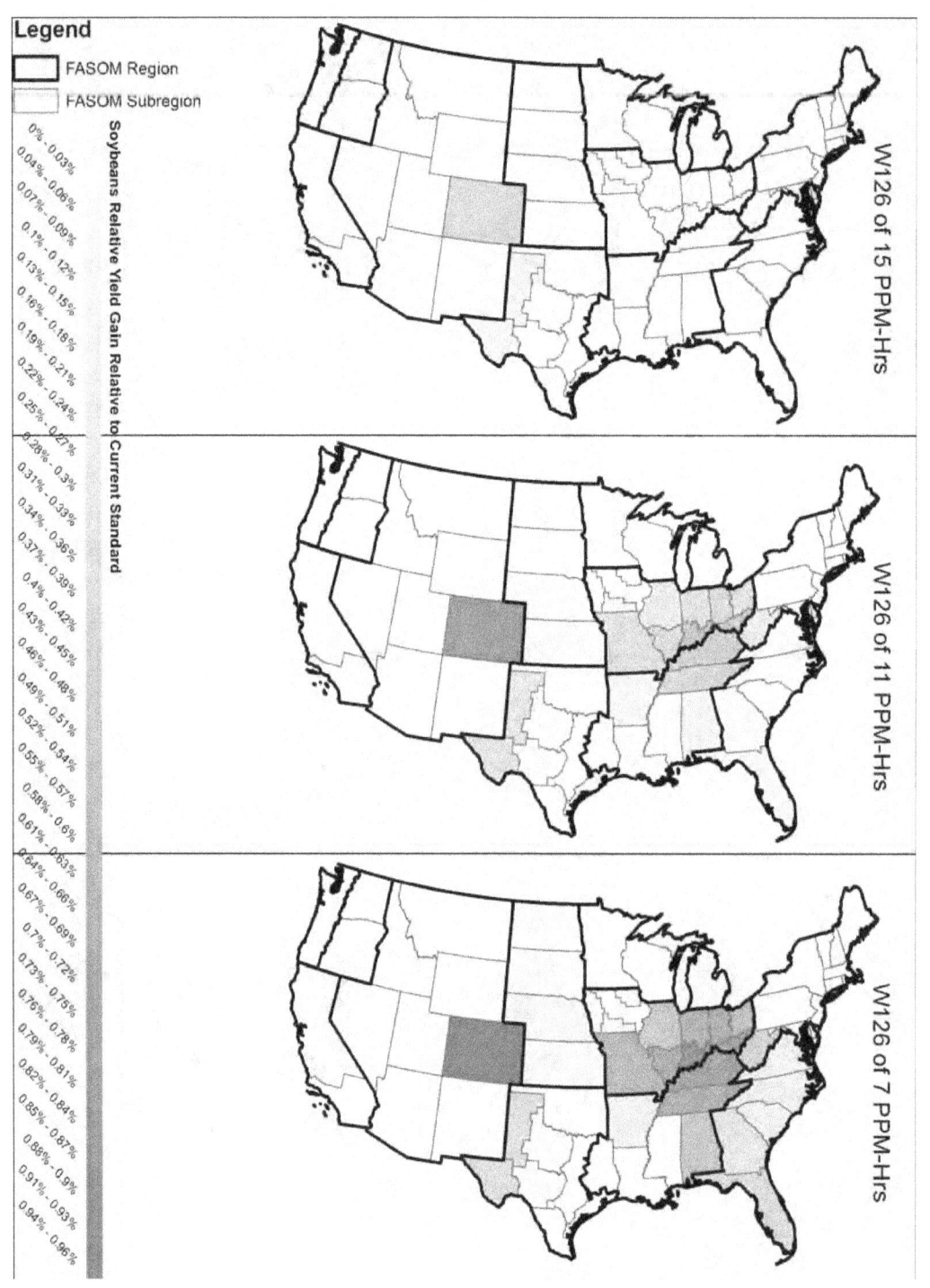

Figure 4-6.   Percentage Changes in Soybean RYGs with Respect to the 75 ppb Scenario

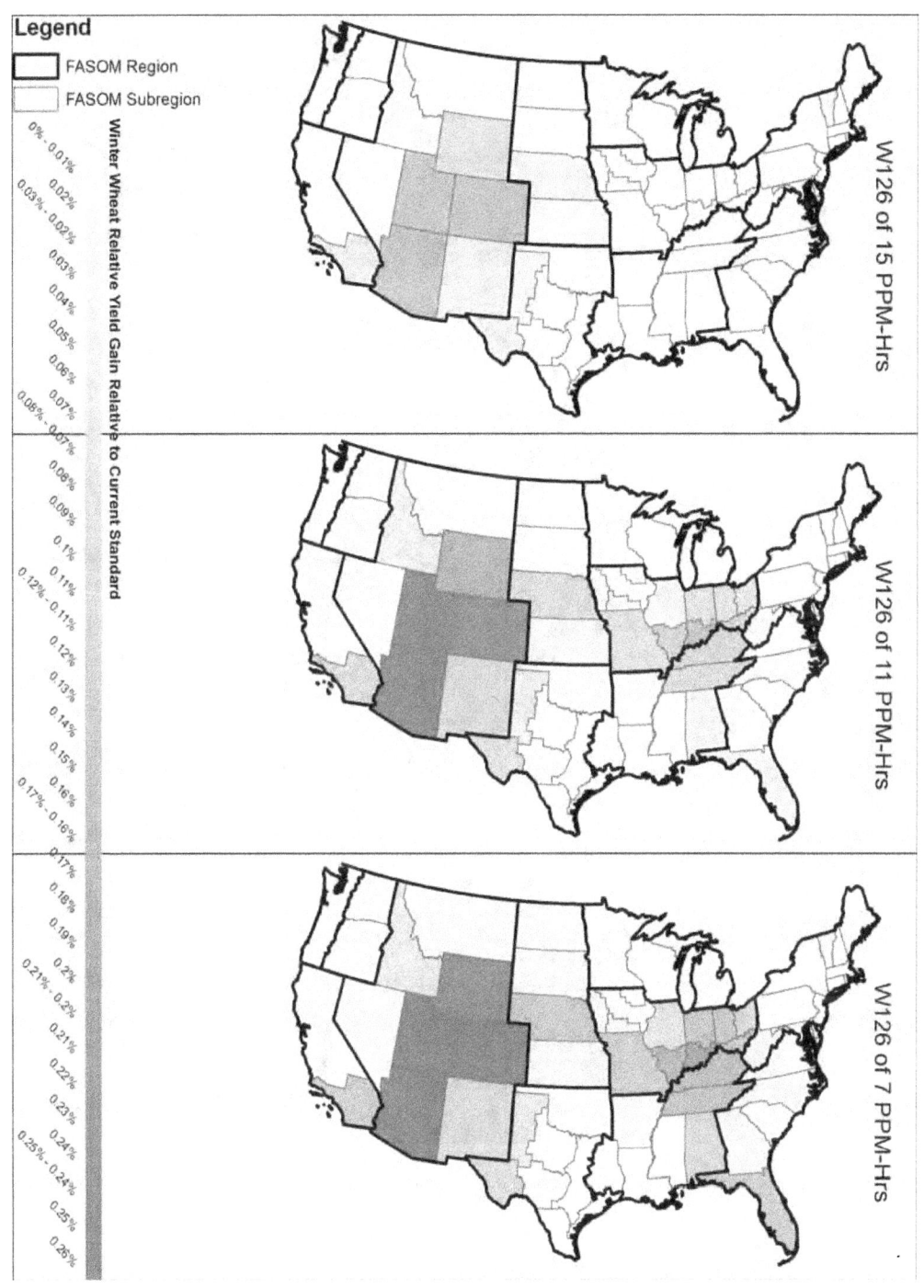

Figure 4-7. **Percentage Changes in Winter Wheat RYGs with Respect to the 75 ppb Scenario**

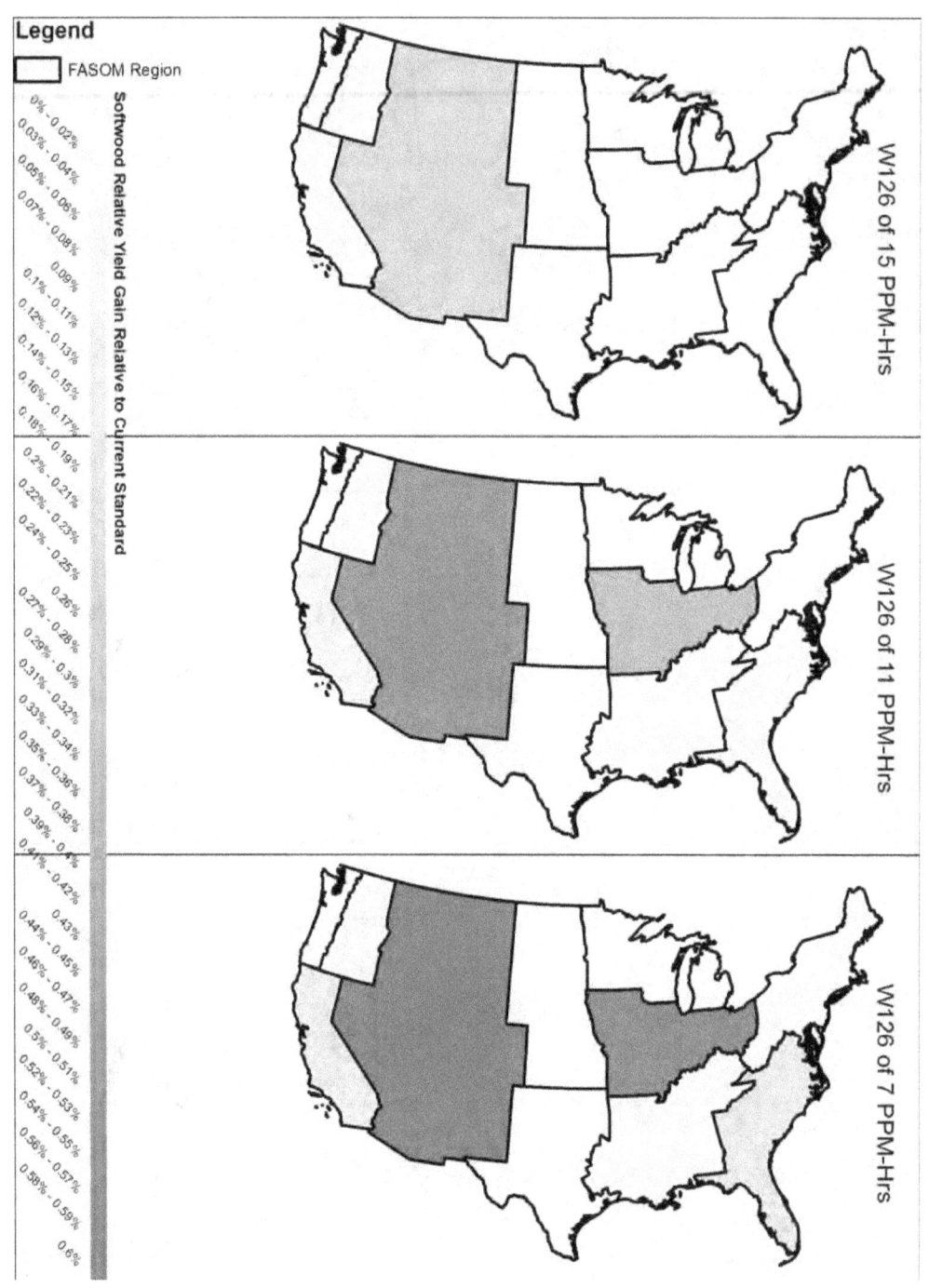

Figure 4-8.   Percentage Changes in Softwood RYGs with Respect to the 75 ppb Scenario

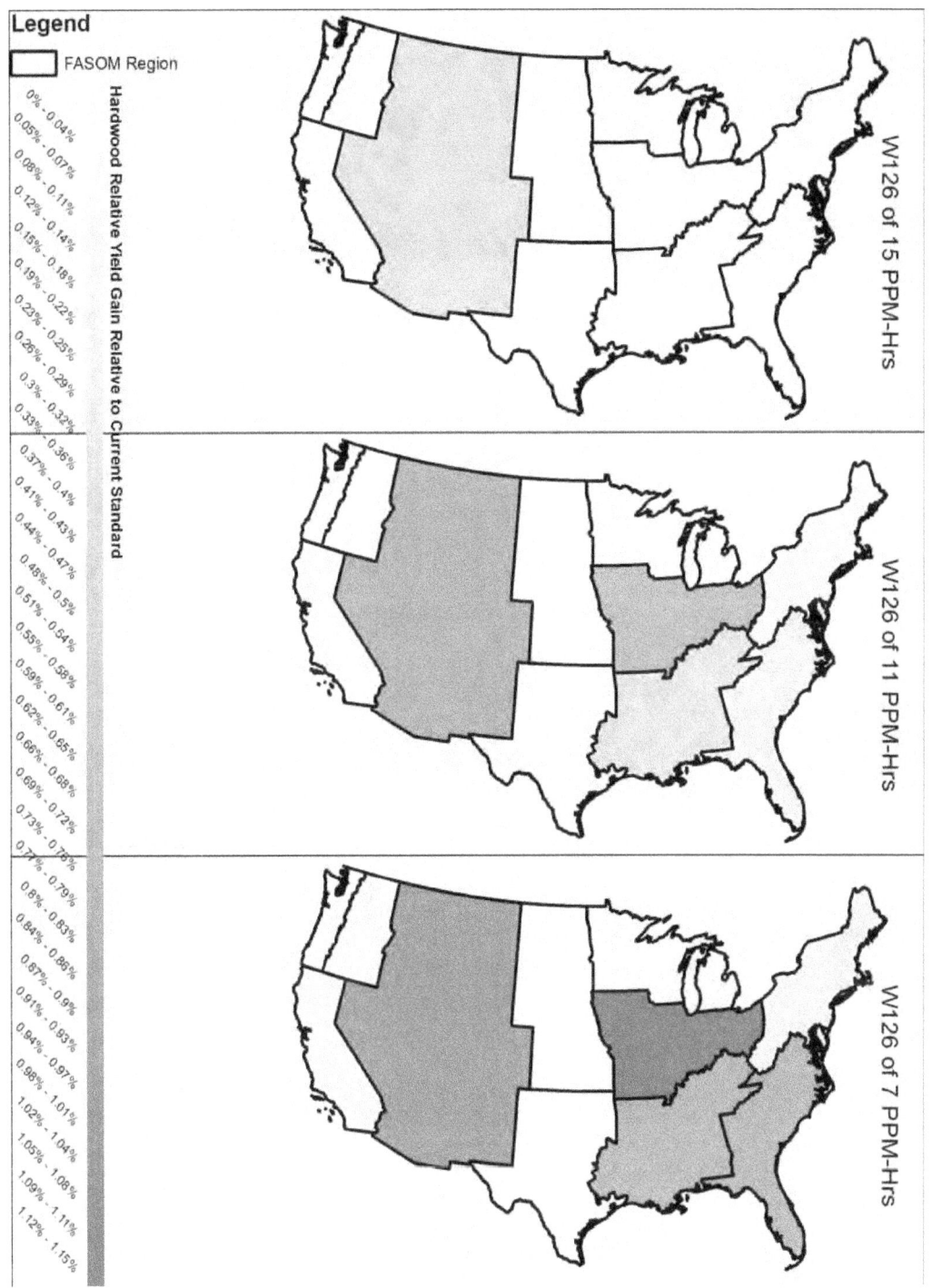

Figure 4-9.   Percentage Changes in Hardwood RYGs with Respect to the 75 ppb Scenario

For more information about RYL and RYG data from which the percentage changes in RYGs for crops and forests were derived, see Appendix A.

SECTION 5
MODEL RESULTS

FASOMGHG was used to estimate the projected effects of alternative ozone concentration standards on the U.S. agricultural and forestry sectors. As introduced earlier, the comparisons considered for this report focus on the differences between a scenario assuming compliance with the existing 2008 standards (75 ppb) and scenarios in which three more stringent ozone standards are met. Those three scenarios are 15 ppm-hr, 11 ppm-hr, and 7 ppm-hr W126 standards. Our analysis included changes to production, prices, forest inventory, land use, welfare, and GHG mitigation potential associated with achieving each of the more stringent standards.

Agricultural Sector

Ozone negatively affects growth in many plants, leading to lower crop yields. The reductions in ozone concentrations that would be achieved under a given standard vary across regions. In addition, some crops are more sensitive to ozone than others, so the percentage changes in yield will vary by crop and region. However, reducing ambient ozone concentrations would generally increase agricultural yields and total production. Our analysis began by determining the extent to which current yield losses caused by ozone could be reversed by reducing ozone levels. Increased crop yields lead to a greater available supply of most agricultural crops, which in turn tends to reduce market prices. There is also an overall tendency toward acreage shifting away from ozone-sensitive crops. In general, impacts in the agricultural sector are relatively limited, especially when compared with the forestry sector. By and large, more stringent standards led to increased incremental impact, but the additional impact in moving to increasingly stringent ozone standards was relatively small.[18]

5.0.1 *Production and Prices*

Changes in U.S. agricultural production and prices were measured using Fisher indices (sees Tables 5-1 and 5-2).[19] Both primary and secondary commodity production levels are projected to increase by 2040 as a result of heightened productivity. Agricultural production changes were generally relatively small across products, rarely exceeding an increase of 0.50% with respect to the current standard and often changing by 0.01% or less.

---

[18] As shown in the appendixes, there are large impacts associated with reductions in ozone consistent with moving from current conditions to meeting any of the four standards examined. The differences between standards are much smaller, though.

[19] The Fisher price index is known as the "ideal" price index. It is calculated as the geometric mean of an index of current prices and an index of past prices.

Table 5-1. Agricultural Production Fisher Indices (Current conditions =100)

Sector	Policy	2010	2020	2030	2040
		\multicolumn{4}{Primary Commodities}			
Crops	75 ppb	101.01	100.77	100.97	100.87
	15 ppm-hr	101.01	100.77	100.95	100.87
	11 ppm-hr	101.05	100.79	100.96	100.88
	7 ppm-hr	101.08	100.83	100.96	100.90
Livestock	75 ppb	100.13	100.03	100.36	100.88
	15 ppm-hr	100.13	100.03	100.36	100.90
	11 ppm-hr	100.14	100.03	100.36	100.92
	7 ppm-hr	100.14	100.03	100.36	100.92
Farm products[a]	75 ppb	100.58	100.42	100.43	100.40
	15 ppm-hr	100.58	100.42	100.65	100.89
	11 ppm-hr	100.61	100.43	100.65	100.90
	7 ppm-hr	100.62	100.45	100.65	100.91
		\multicolumn{4}{Secondary Commodities}			
Processed	75 ppb	100.17	100.10	100.08	100.28
	15 ppm-hr	100.16	100.10	100.08	100.28
	11 ppm-hr	100.19	100.10	100.08	100.28
	7 ppm-hr	100.20	100.12	100.08	100.28
Meats	75 ppb	100.03	100.01	100.43	100.40
	15 ppm-hr	100.05	100.01	100.43	100.41
	11 ppm-hr	100.05	100.01	100.43	100.41
	7 ppm-hr	100.05	100.01	100.43	100.41
Mixed feeds	75 ppb	100.06	100.01	100.64	100.53
	15 ppm-hr	100.07	100.01	100.64	100.54
	11 ppm-hr	100.08	100.01	100.64	100.54
	7 ppm-hr	100.07	100.01	100.64	100.54

[a] Farm Products is the composite of Crops and Livestock.

Table 5-2. Agricultural Price Fisher Indices (Current Conditions = 100)

Sector	Policy	2010	2020	2030	2040
			Primary Commodities		
Crops	75 ppb	97.51	98.35	99.05	98.73
	15 ppm-hr	97.49	98.36	99.05	98.71
	11 ppm-hr	97.49	98.33	99.02	98.72
	7 ppm-hr	97.46	98.31	98.97	98.66
Livestock	75 ppb	99.78	99.31	100.26	98.90
	15 ppm-hr	99.25	99.07	100.31	99.31
	11 ppm-hr	98.97	98.97	99.86	98.84
	7 ppm-hr	99.33	99.50	100.34	99.85
Farm products	75 ppb	97.51	98.35	99.75	99.51
	15 ppm-hr	97.49	98.36	99.04	98.70
	11 ppm-hr	97.49	98.33	99.02	98.72
	7 ppm-hr	97.46	98.31	98.97	98.65
			Secondary Commodities		
Processed	75 ppb	97.60	98.34	98.69	98.22
	15 ppm-hr	97.58	98.34	98.69	98.22
	11 ppm-hr	97.58	98.29	98.65	98.21
	7 ppm-hr	97.56	98.29	98.61	98.16
Meats	75 ppb	99.86	100.00	99.75	99.51
	15 ppm-hr	99.86	100.00	99.75	99.51
	11 ppm-hr	99.86	100.00	99.73	99.53
	7 ppm-hr	99.94	100.00	99.71	99.52
Mixed feeds	75 ppb	97.48	98.98	99.11	98.91
	15 ppm-hr	98.53	96.29	98.55	99.48
	11 ppm-hr	97.52	99.48	100.65	98.81
	7 ppm-hr	100.05	97.40	99.80	99.32

Increased production led to a general decline in market prices because the equilibrium price adjusts to higher levels of supply. This result is consistent with expectations because higher productivity leads to greater supply, which tends to decrease market prices. Changes in price were generally more pronounced than changes in production, with the largest decreases in the

Farm Products, Livestock, and Mixed Feeds categories. Agricultural prices tend to decline by a greater percentage than production increases because the demand for most agricultural commodities is inelastic.[20] However, almost all declines in price were less than 1.0% of prices at the current standards, and most were less than 0.5%. Often, the change in price from the current standard to 15 ppm-hr was minimal, at less than 0.01%, with the larger changes occurring only at the 11 ppm-hr and 7 ppm-hr levels.

5.0.2 Crop Acreage

Crop acreage was projected to decline with the introduction of the ozone standards because additional productivity per acre reduces the demand for crop acreage. In aggregate, farmers will be able to meet the demand for agricultural commodities using less land under scenarios with lower ozone concentrations. Consistent with these expectations, the total cropped area is slightly smaller for each model year in the alternative standard cases. However, land allocation also depends on relative returns across various uses and is influenced by forest harvest timing. Changes in land allocation between the agricultural and forestry sectors are discussed later in this section.

Table 5-3 provides projections of acreage in each of the major U.S. crops, as well as composites of all remaining crops and total cropland. The absolute change relative to the current standard is presented for each alternative standard. Larger changes occurred in soybean and sorghum acreage, whereas only minor changes occurred in all other crops, leading to an overall slight decrease in crop acreage across all crops. This shift occurred largely because of differential crop sensitivity to ozone concentrations. Note that the sum of the crop-specific changes will not necessarily equal the total changes shown in Table 5-3 because some double-cropping is reflected in the model (e.g., soybeans and winter barley).

---

[20] Demand elasticities are measures of the responsiveness of the quantity demanded to a change in price. Commodities with inelastic demands are those where consumers change the quantity of a good they purchase by a smaller percentage than the change in market price. Many food products fall into this category because they are relatively low-priced necessities.

Table 5-3. Major Crop Acreage, Million Acres

Crop	Policy	2010	2020	2030	2040
Corn	75 ppb	92.9	87.1	79.7	70.8
	Change with Respect to Current Standard				
	15 ppm-hr	0.00	0.00	−0.01	0.00
	11 ppm-hr	−0.01	0.03	−0.04	−0.01
	7 ppm-hr	−0.01	0.00	−0.02	−0.01
Soybeans	75 ppb	73.3	71.9	71.7	69.9
	Change with Respect to Current Standard				
	15 ppm-hr	0.01	0.00	−0.02	−0.01
	11 ppm-hr	0.02	−0.01	−0.19	−0.14
	7 ppm-hr	0.06	−0.04	−0.09	−0.10
Hay	75 ppb	44.0	42.0	41.0	39.2
	Change with Respect to Current Standard				
	15 ppm-hr	−0.01	−0.01	−0.01	0.00
	11 ppm-hr	0.01	−0.10	0.00	−0.01
	7 ppm-hr	−0.01	−0.02	−0.03	0.00
Hard red winter wheat	75 ppb	20.6	19.5	15.8	13.1
	Change with Respect to Current Standard				
	15 ppm-hr	0.04	−0.02	−0.01	−0.01
	11 ppm-hr	0.06	−0.07	−0.07	0.02
	7 ppm-hr	0.02	−0.01	−0.02	0.01
Cotton	75 ppb	14.6	15.4	15.9	15.7
	Change with Respect to Current Standard				
	15 ppm-hr	−0.01	−0.01	0.00	0.00
	11 ppm-hr	−0.01	−0.03	−0.02	0.01
	7 ppm-hr	−0.01	−0.02	−0.01	0.01
Hard red spring wheat	75 ppb	13.7	12.5	12.2	11.4
	Change with Respect to Current Standard				
	15 ppm-hr	0.00	0.00	−0.01	0.00
	11 ppm-hr	−0.01	−0.02	0.01	0.00
	7 ppm-hr	0.00	0.00	0.01	0.00

(continued)

Table 5-3. Major Crop Acreage, Million Acres (continued)

Crop	Policy	2010	2020	2030	2040
Sorghum	75 ppb	10.7	10.7	11.3	11.6
	Change with Respect to Current Standard				
	15 ppm-hr	0.00	0.57	0.33	−0.87
	11 ppm-hr	−0.01	0.57	0.37	−0.88
	7 ppm-hr	0.00	0.55	0.35	−0.88
Switchgrass	75 ppb	0.0	13.4	11.2	10.3
	Change with Respect to Current Standard				
	15 ppm-hr	0.00	0.00	0.00	0.00
	11 ppm-hr	0.00	0.00	0.00	0.01
	7 ppm-hr	0.00	0.00	−0.01	0.00
All others[a]	75 ppb	41.7	40.7	43.3	44.0
	Change with Respect to Current Standard				
	15 ppm-hr	−0.01	0.01	−0.02	−0.02
	11 ppm-hr	0.00	0.00	−0.04	−0.01
	7 ppm-hr	−0.03	0.00	−0.05	−0.02
Total	75 ppb	311.5	313.8	302.5	285.1
	Change with Respect to Current Standard				
	15 ppm-hr	0.01	−0.03	−0.07	−0.03
	11 ppm-hr	0.04	−0.19	−0.31	−0.14
	7 ppm-hr	0.03	−0.12	−0.19	−0.11

[a] Canola, durum wheat, fresh grapefruit, fresh orange, fresh tomato, grazing wheat, hybrid poplar, oats, potato, processed grapefruit, processed orange, processed tomato, rice, rye, silage, soft red winter wheat, soft white wheat, spring barley, sugar beet, sugarcane, sweet sorghum, winter barley.

Forestry Sector

As with agricultural crops, ozone diminishes growth in most tree species, and our analysis began by estimating how much of this diminished growth would be reversed under the more stringent ozone standards. Impacts are significantly higher in the forestry sector, especially in hardwood species and species more prevalent in the southern regions. Impacts are more significant for southern regions because of the higher baseline ozone concentrations in the South Central and Southeast regions. Higher initial concentrations resulted in higher reductions to meet the alternative standards and, thus, higher impacts to tree growth. This relationship also contributes to the larger changes in the forestry sector as a whole.

5.1.1 Production and Prices

Reducing ozone concentrations led to increased forest growth, which was reflected in increased production in FASOMGHG. Some of the most substantial ozone standard impacts occurred in saw log and pulp log harvest quantities and prices. Compared with the current standard, alternative standard cases had consistently higher production except for softwood pulp logs, where production increased only marginally and at times fell below the baseline estimates, especially in 2040. The most significant impacts occurred in hardwood saw logs, where harvests were projected to be more than 1% higher than under the current standard level by 2040. There are some cases where production of pulp logs and saw logs moved in opposite directions, with two primary explanations. The first is that as saw log production expands, the price for saw logs drops, allowing processors to substitute saw logs for instances in which pulp logs are traditionally used. The second is that even when the primary log size being harvested is pulp logs, saw logs will generally also be present because of natural variation in tree growth rates (and vice versa for harvest of saw logs). With higher growth rates, there would tend to be more saw logs in stands harvested primarily for pulp logs over time.

The largest changes in production occurred at the 7 ppm-hr level. Changes from the current standard to 15 ppm-hr were fairly small, whereas changes from 15 ppm-hr to 11 ppm-hr and from 11 ppm-hr to 7 ppm-hr are generally of the same magnitude and range from 0.3% to 2.5% of the levels under current standards, depending on the product and year. Table 5-4 presents these changes by major product.

The impact of policy intervention on timber market prices was more substantial than the change in production in terms of percentage changes compared with the current standard. Although increases in production of forest products did not exceed 2.5% compared with the current standard, changes in price were as large as 8.7%. As with agricultural products, many forest products have relatively inelastic demand so prices tend to change by a larger percentage than quantities. Table 5-5 lists absolute changes with respect to the current standard, whereas Table 5-6 lists the percentage change in forest product prices for each year, alternative standard, and forest product.

Table 5-4. Forest Products Production, Million Cubic Feet

Product	Policy	2010	2020	2030	2040
Hardwood saw logs	75 ppb	3,697	3,472	3,776	4,843
	Change with Respect to Current Standard				
	15 ppm-hr	1	2	10	1
	11 ppm-hr	27	17	37	-4
	7 ppm-hr	61	10	35	-13
Hardwood pulp logs	75 ppb	2,592	2,277	2,601	2,529
	Change with Respect to Current Standard				
	15 ppm-hr	1	1	−12	14
	11 ppm-hr	29	17	−29	20
	7 ppm-hr	65	35	−25	31
Softwood saw logs	75 ppb	4,666	5,186	5,614	6,696
	Change with Respect to Current Standard				
	15 ppm-hr	0	0	1	8
	11 ppm-hr	0	28	9	44
	7 ppm-hr	13	27	11	48
Softwood pulp logs	75 ppb	3,469	3,923	4,324	4,326
Change with Respect to Current Standard					
	15 ppm-hr	1	3	0	−12
	11 ppm-hr	2	12	5	−25
	7 ppm-hr	6	16	6	−31

Table 5-5. Forest Product Prices, U.S. Dollars per Cubic Foot

Product	Policy	2010	2020	2030	2040
Hardwood saw logs	75 ppb	0.69	0.65	0.39	0.19
	Change with Respect to Current Standard				
	15 ppm-hr	0.00	0.00	0.00	0.00
	11 ppm-hr	−0.01	−0.01	−0.01	−0.01
	7 ppm-hr	−0.01	−0.02	−0.02	−0.01

(continued)

Table 5-5. Forest Product Prices, U.S. Dollars per Cubic Foot (continued)

Product	Policy	2010	2020	2030	2040
Hardwood pulp logs	75 ppb	0.24	0.44	0.22	0.12
	Change with Respect to Current Standard				
	15 ppm-hr	0.00	0.00	0.00	0.00
	11 ppm-hr	0.00	0.00	−0.01	0.00
	7 ppm-hr	0.00	−0.01	−0.02	−0.01
Softwood saw logs	75 ppb	2.31	1.91	1.60	1.31
	Change with Respect to Current Standard				
	15 ppm-hr	0.00	−0.01	−0.01	−0.01
	11 ppm-hr	−0.01	−0.02	−0.02	−0.02
	7 ppm-hr	−0.01	−0.03	−0.03	−0.03
Softwood pulp logs	75 ppb	1.42	1.12	1.34	0.94
	Change with Respect to Current Standard				
	15 ppm-hr	0.00	0.00	0.00	0.00
	11 ppm-hr	−0.01	0.00	0.00	0.00
	7 ppm-hr	−0.01	0.00	−0.01	−0.02

Table 5-6. Forest Product Prices and Percentage Change, U.S. Dollars per Cubic Foot

Product	Policy	2010	2020	2030	2040
Hardwood saw logs	75 ppb	0.69	0.65	0.39	0.19
	Percentage Change with Respect to Current Standard				
	15 ppm-hr	−0.28	0.13	−0.16	0.94
	11 ppm-hr	−0.79	0.13	−2.52	−1.51
	7 ppm-hr	−1.59	−2.60	−8.72	−7.12
Hardwood pulp logs	75 ppb	0.24	0.44	0.22	0.12
	Percentage Change with Respect to Current Standard				
	15 ppm-hr	0.00	−0.15	−0.08	−0.08
	11 ppm-hr	−0.87	−1.95	−2.06	−2.64
	7 ppm-hr	−2.10	−3.52	−4.92	−6.23

(continued)

Table 5-6. **Forest Product Prices and Percentage Change, U.S. Dollars per Cubic Foot (continued)**

Product	Policy	2010	2020	2030	2040
Softwood saw logs	75 ppb	2.31	1.91	1.60	1.31
	Percentage Change with Respect to Current Standard				
	15 ppm-hr	−0.09	−0.33	−0.44	−0.69
	11 ppm-hr	−0.26	−1.24	−1.32	−1.40
	7 ppm-hr	−0.46	−1.54	−1.91	−2.28
Softwood pulp logs	75 ppb	1.42	1.12	1.34	0.94
	Percentage Change with Respect to Current Standard				
	15 ppm-hr	−0.14	0.12	0.15	0.18
	11 ppm-hr	−0.43	0.13	−0.19	−0.51
	7 ppm-hr	−1.03	−0.42	−0.82	−2.17

5.1.2 Forest Acres Harvested

Harvested acres are projected to decline as a result of higher productivity in the policy cases. The most significant reductions occurred in species found in the southern regions where the largest ozone reductions would occur: natural pine and upland hardwoods. The difference between the hardwood harvested acres in the current standard case and in the alternative standards widens from 2010 to 2040, increasing to a difference of more than 4%. The impact to total acres of softwood harvested was smaller, with differences remaining at less than 1% of the current standard levels. Impacts to harvesting are larger for more stringent ozone standards, with the largest shifts occurring between the 11 ppm-hr and 7 ppm-hr standards. Table 5-7 presents the model results for forest acres harvested.

Table 5-7. **Forest Acres Harvested, Thousand Acres**

Product	Policy	2010	2020	2030	2040
Total hardwood	75 ppb	14,421	10,177	10,187	10,410
	Change with Respect to Current Standard				
	15 ppm-hr	−12	4	−20	53
	11 ppm-hr	101	−30	−167	−136
	7 ppm-hr	274	−130	−257	−424

(continued)

Table 5-7. Forest Acres Harvested, Thousand Acres (continued)

Product	Policy	2010	2020	2030	2040
Total softwood	75 ppb	17,335	15,205	14,740	17,826
		Change with Respect to Current Standard			
	15 ppm-hr	8	13	−10	−27
	11 ppm-hr	6	60	−18	1
	7 ppm-hr	23	134	−81	62

5.1.3 Forest Inventory

Under FASOMGHG definitions, existing inventory includes only trees that have been standing since the initial model year of 2000. All trees planted since then, including both reforestation and afforestation, are included in new inventory. The model projected significant increases in existing inventory for both hardwood and softwood species, although the increase was greater in hardwood species. As with the crops, this difference is largely explained by differential sensitivity to ozone between species. Hardwood species show a much higher sensitivity to ozone levels and are thus modeled to respond more dramatically to reductions in ozone concentration. The gap between the current standard and alternative standards widened over time as inventory continued to accumulate. Because there is greater existing inventory in each of the alternative standards, there is less demand for reforestation and afforestation to meet future demand for forest products and therefore less new inventory than under the current standard.

Some relatively large differences between ozone standards occurred in the forest inventory projections. For example, existing hardwood inventory was projected to be 4.0% higher under the 11 ppm-hr case than the 7 ppm-hr case by 2040. New hardwood inventory is similarly sensitive, with the model projecting a 2.6% increase for this same comparison. For new and existing inventory of both hardwoods and softwoods, the largest impacts occurred at the 11 ppm-hr standard. This type of nonlinear response can occur because of differences in the relative impacts on alternative forest and agricultural products that lead to land reallocation. Table 5-8 presents the model results for forest inventory.

Table 5-8. Existing and New Forest Inventory, Million Cubic Feet

Product	Policy	2010	2020	2030	2040
Existing hardwood	75 ppb	302,813	345,013	400,552	459,892
	Percentage Change with Respect to Current Standard				
	15 ppm-hr	12	–6	18	79
	11 ppm-hr	1,745	6,337	15,223	29,155
	7 ppm-hr	698	2,357	5,530	10,454
Existing softwood	75 ppb	190,790	173,039	160,470	161,991
	Percentage Change with Respect to Current Standard				
	15 ppm-hr	46	62	113	223
	11 ppm-hr	360	1,134	2,355	3,965
	7 ppm-hr	182	561	867	1,858
New hardwood	75 ppb	1,968	10,008	20,162	33,159
	Percentage Change with Respect to Current Standard				
	15 ppm-hr	0	–2	–6	0
	11 ppm-hr	5	59	208	1,317
	7 ppm-hr	1	15	76	446
New softwood	75 ppb	9,135	64,727	109,869	118,954
	Percentage Change with Respect to Current Standard				
	15 ppm-hr	–2	–17	–59	–51
	11 ppm-hr	12	–131	–1,596	–2,209
	7 ppm-hr	0	–98	–560	–1,016

Cross-Sectoral Policy Impacts

One of the advantages of a model such as FASOMGHG for analysis of impacts on major land-using activities is the ability to account for shifts in land use. Differentiated impacts on productivity across products will lead to changes in market prices and in the relative profitability of alternative land uses. In response, landowners will change their allocation of land across different productive activities, which will contribute to market impacts. In addition, these changes in land use have implications for GHG emissions and other environmental impacts. In this section, we discuss changes in land use, net GHG emissions, and producer and consumer welfare across the agricultural and forest sectors.

5.2.1 Land Use

FASOMGHG projected changes in eight land use categories: existing forest, reforestation, afforestation, cropland, pasture, cropland pasture,[21] and lands enrolled in the Conservation Research Program (CRP).[22] The largest impacts were projected in afforestation. The general projected pattern within these categories was a decline in total forest- and cropland, coupled with small increases in both traditional pasture and cropland pasture. The areas of total forest- and cropland declined as productivity and inventory increased because of decreased ozone concentrations, implying that less land would be required to meet market demands. There was almost no change in acreage retained in CRP.

The incremental impact of more stringent ozone standards was apparent in most of these cases, especially reforestation, afforestation, and cropland pasture. Each of the standard levels, the current and all alternatives, followed the same pattern of continuous decline in afforestation over the projection years, though these declines were more pronounced at more stringent standard levels. Generally, the additional impact of moving from the 11 ppm-hr standard to 7 ppm-hr was higher than shifts between other standard levels. Table 5-9 presents the model results by major land use type.

Table 5-9. Land Use by Major Category, Thousand Acres

Product	Policy	2010	2020	2030	2040
Existing forest	75 ppb	341,843	328,746	314,445	307,835
	Change with Respect to Current Standard				
	15 ppm-hr	−5	−4	45	−2
	11 ppm-hr	−45	−66	73	39
	7 ppm-hr	−58	−189	88	58
Reforested	75 ppb	70,879	112,945	139,607	158,639
	Change with Respect to Current Standard				
	15 ppm-hr	−4	12	63	39
	11 ppm-hr	73	−7	−144	−283
	7 ppm-hr	250	−203	−536	−927

(continued)

---

[21] Cropland pasture is managed land suitable for crop production (i.e., relatively high productivity) that is being used as pasture.

[22] Rangeland estimates are also included, but rangeland is held fixed in FASOMGHG by assumption because it cannot be allocated to any other use.

Table 5-9. Land Use by Major Category, Thousand Acres (continued)

Product	Policy	2010	2020	2030	2040
Afforested	75 ppb	12,656	9,748	4,566	4,461
	Change with Respect to Current Standard				
	15 ppm-hr	−5	−4	−19	−19
	11 ppm-hr	−11	−10	−55	−55
	7 ppm-hr	−14	−104	−78	−78
Cropland	75 ppb	311,714	313,784	304,833	293,396
	Change with Respect to Current Standard				
	15 ppm-hr	14	−27	−62	−17
	11 ppm-hr	25	−117	−157	−115
	7 ppm-hr	44	−195	−274	−114
Pasture	75 ppb	84,429	85,113	86,810	86,075
	Change with Respect to Current Standard				
	15 ppm-hr	0	0	4	0
	11 ppm-hr	34	0	21	32
	7 ppm-hr	34	−27	−12	−4
Cropland pasture	75 ppb	47,585	47,992	58,179	66,114
	Change with Respect to Current Standard				
	15 ppm-hr	−12	28	11	16
	11 ppm-hr	−4	192	72	54
	7 ppm-hr	−11	420	207	69
Rangeland	75 ppb	302,210	301,104	300,049	299,039
	Change with Respect to Current Standard				
	15 ppm-hr	0	0	0	0
	11 ppm-hr	0	0	0	0
	7 ppm-hr	0	0	0	0
CRP	75 ppb	36,534	34,480	34,480	34,480
	Change with Respect to Current Standard				
	15 ppm-hr	3	3	3	3
	11 ppm-hr	−9	−9	−9	−9
	7 ppm-hr	−9	−9	−9	−9

5.2.2 Welfare

Welfare impacts resulting from the implementation of alternative standard levels followed the same pattern between the agriculture and forestry sectors, although it was more pronounced in forestry. Consumer surplus typically increased in both cases as higher productivity under reduced ozone conditions tended to increase total production and reduce market prices. Because demand for most forestry and agricultural commodities is inelastic, there are more instances in which producer surplus declines. In some year/ozone concentration combinations, the effect of falling prices on producer profits more than outweighs the effects of higher production levels.

Percentage changes in agricultural sector consumer and producer surplus between the current standard and the alternative standards were relatively small in many cases, with the largest percentage change being a 0.4% decline in producer surplus in the 2035 model period. However, the agricultural sector is a very large market, and even small percentage changes in welfare can result in annualized values of tens or even hundreds of millions of dollars. Table 5-10 provides consumer and producer surplus for the agricultural sectors under the current standard, along with the change in surplus for each alternative standard. There is considerable variability in the magnitude of consumer and producer impacts from year to year, which is not surprising given the dynamic nature of the model and numerous adjustments taking place over time in response to changes in net returns associated with alternative land uses.

Table 5-10. Consumer and Producer Surplus in Agriculture, Million 2010 U.S. Dollars

Product	Policy	2010	2015	2020	2025	2030	2035	2040
Consumer surplus	75 ppb	1,918,082	1,940,673	1,968,142	1,995,346	2,023,022	2,050,791	2,076,018
	Change with Respect to Current Standard							
	15 ppm-hr	15	−2	1	6	−7	10	3
	11 ppm-hr	19	24	13	51	42	20	13
	7 ppm-hr	−31	46	36	104	90	26	46
Producer surplus	75 ppb	725,364	831,565	815,072	863,165	878,986	836,692	863,308
	Change with Respect to Current Standard							
	15 ppm-hr	612	−1,255	980	−961	90	41	697
	11 ppm-hr	1,474	−2,197	1,013	230	232	−3,413	2,189
	7 ppm-hr	269	−1,873	1,780	423	264	−1,052	2,991

The impacts of the scenarios with more stringent ozone standards were larger in the forestry sector, with bigger increases in consumer surplus and greater declines in producer surplus. Table 5-11 presents the model results of the welfare analysis in the forestry sector.

Table 5-11. Consumer and Producer Surplus in Forestry, Million 2010 U.S. Dollars

Product	Policy	2010	2015	2020	2025	2030	2035	2040
Consumer surplus	75 ppb	721,339	793,234	809,271	826,375	875,620	894,705	934,882
	Change with Respect to Current Standard							
	15 ppm-hr	7	31	118	105	2	36	597
	11 ppm-hr	44	48	360	202	688	56	712
	7 ppm-hr	86	187	694	224	734	91	779
Producer surplus	75 ppb	93,322	121,476	153,997	146,275	145,913	146,115	133,132
	Change with Respect to Current Standard							
	15 ppm-hr	−11	−7	−141	−161	15	−46	−839
	11 ppm-hr	−41	20	−503	−178	−880	55	−858
	7 ppm-hr	−136	−48	−892	−37	−786	156	−766

Because of the complex dynamics of the agriculture and forestry sectors and variability in welfare impacts over time, it is often helpful to summarize the impacts in terms of annualized values. Table 5-12 summarizes the annualized impacts of alternative ozone standards on consumer and producer surplus in the agricultural and forestry sectors for 2010–2044.[23] The impacts of alternative standards on consumer surplus are positive for each of the tighter standards for both agricultural and forestry sectors, with the benefits increasing with more stringent requirements. For producer surplus, on the other hand, annualized impacts are negative for the 15 ppm-hr and 11 ppm-hr scenarios, but a large positive value for the 7 ppm-hr case. For the forestry sector, consumer surplus changes are positive for each scenario and increasing with stringency, while changes in producer surplus are negative for all cases, becoming more negative as stringency is increased. Overall, total surplus across both sectors decreases in the 15 ppm-hr and 11 ppm-hr scenarios but increases substantially in the 7 ppm-hr case.

---

[23] Each model period in FASOMGHG is representative of the 5-year period starting with that year, so results reported for 2040 are representative of 2040–2044. Thus, we use values through 2044 in the annualization calculations.

Table 5-12. Annualized Changes in Consumer and Producer Surplus in Agriculture and Forestry, 2010–2044, Million 2010 U.S. Dollars (4% Discount Rate)

Product	Policy	Agriculture	Forestry	Total
Consumer surplus	75 ppb	NA	NA	NA
	Change with Respect to Current Standard			
	15 ppm-hr	4.5	88.1	92.5
	11 ppm-hr	25.4	236.9	262.3
	7 ppm-hr	36.7	344.0	380.7
Producer surplus	75 ppb	NA	NA	NA
	Change with Respect to Current Standard			
	15 ppm-hr	−4.7	−112.2	−116.9
	11 ppm-hr	−4.6	−264.4	−269.0
	7 ppm-hr	194.4	−318.4	−124.0
Total surplus	75 ppb	NA	NA	NA
	15 ppm-hr	−0.2	−24.2	−24.4
	11 ppm-hr	20.8	−27.5	−6.7
	7 ppm-hr	231.1	25.6	256.7

5.2.3 Greenhouse Mitigation Potential

The capacity for both the agricultural and forest sectors to sequester carbon is enhanced in each of the alternative standard cases, with increasing magnitude as policy stringency is increased. Although FASOMGHG projects fewer acres of forestland and total cropland, the accelerated storage of carbon in trees and forestland and cropland soils outweighs any decline from reductions in covered area. Carbon storage in both sectors is consistently higher in the alternative standard cases, with the gap widening over time (see Figure 5-1 for change in forest carbon stock). By 2040, the agricultural sector sequestered 0.1% more carbon under the alternative standard cases and the forestry sector up to 2% more, resulting in gains of more than 1,600 million metric tons CO_2 equivalent (MMtCO$_2$e). Table 5-13 presents carbon sequestration projections under the current standard and changes under each alternative standard.[24] Note that negative values in the row for the current standard indicate sequestration or carbon storage. Negative values in the change rows indicate that the alternative standard stores more carbon than the current standard (and vice versa for positive changes).

---

[24] These are total stocks of net GHG emissions over time, not annual emissions. If the total stock of GHG is becoming more negative over time, more net sequestration is taking place than emissions. If the total stock of GHG is becoming less negative or positive over time, emissions are greater than the increase in sequestration.

Notice that for the agricultural sector, the overall stock of net GHG would decrease over time in the baseline because cropping activities involve fertilizer and chemical usage, fossil fuels, running machinery, livestock emissions from enteric fermentation and manure management, and so forth—all these GHG emissions are being released each year, while soil carbon sequestration moves toward equilibrium within 25 years of a change in tillage. As soil carbon reaches equilibrium, little additional sequestration is taking place each year but annual emissions from other sources continue. Thus, over time, the annual emissions tend to outweigh the increase in carbon stocked in agricultural soils, and net stock of GHG tends to become less negative and eventually positive relative to the starting point.[25]

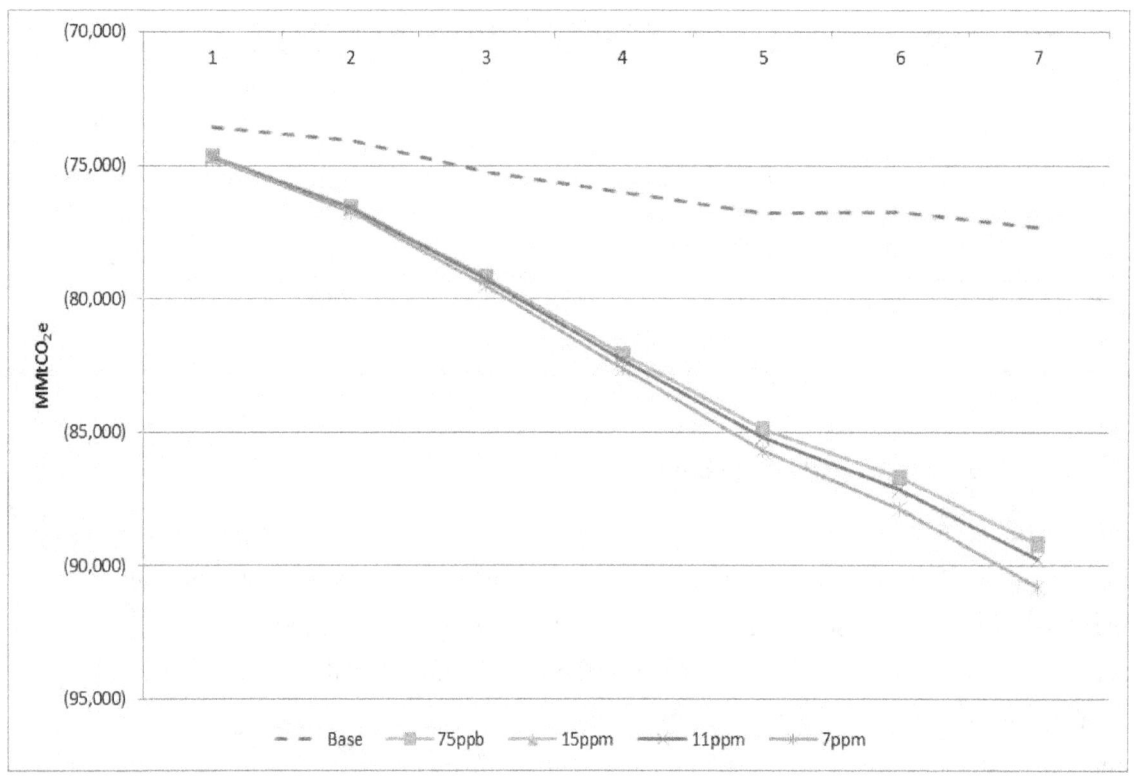

Figure 5-1. Carbon Storage in Forestry Sector, MMtCO$_2$e

---

[25] This change is consistent with the fact that U.S. agriculture is a net source of emissions on an annual basis. The value of the total GHG stock associated with agriculture is starting at a negative value because of the FASOMGHG convention of accounting for total carbon sequestration present in agricultural soils in the first year of the model run. A large stock of carbon is sequestered, but it does not increase by much over time.

Table 5-13. Carbon Storage, MMtCO₂e

Product	Policy	2010	2020	2030	2040
Agriculture	75 ppb	−18,748	−15,363	−12,002	−8,469
	Change with Respect to Current Standard				
	15 ppm-hr	0	−1	−1	−4
	11 ppm-hr	−2	−5	−6	−10
	7 ppm-hr	−3	−4	−6	−9
Forestry	75 ppb	−74,679	−79,171	−84,863	−89,184
	Change with Respect to Current Standard				
	15 ppm-hr	−1	0	−16	−13
	11 ppm-hr	−19	−103	−312	−593
	7 ppm-hr	−50	−305	−832	−1,602

Changes in forestry sector carbon sequestration are largely driven by changes in forest management, which include the increases in tree yield in the lower ozone environments. The increased sequestration in this category outweighs losses in sequestration in the other major forestry categories: afforestation and forest soil. Table 5-13 presents the detailed changes in forestry carbon sequestration.

Table 5-13. Forestry Carbon Sequestration, MMtCO₂e

Product	Policy	2010	2020	2030	2040
Afforestation, trees	75 ppb	−696	−1,516	−800	−1,054
	Change with Respect to Current Standard				
	15 ppm-hr	0	0	1	2
	11 ppm-hr	0	0	6	7
	7 ppm-hr	0	22	6	7
Afforestation, soils	75 ppb	−691	−538	−373	−362
	Change with Respect to Current Standard				
	15 ppm-hr	0	0	2	2
	11 ppm-hr	1	1	5	5
	7 ppm-hr	1	6	8	8

(continued)

Table 5-13. Forestry Carbon Sequestration, MMtCO₂e (continued)

Product	Policy	2010	2020	2030	2040
Forest management	75 ppb	−41,337	−43,022	−47,266	−48,770
		Change with Respect to Current Standard			
	15 ppm-hr	0	0	−12	−14
	11 ppm-hr	−17	−103	−305	−589
	7 ppm-hr	−44	−335	−825	−1,596
Forest soils	75 ppb	−28,194	−27,566	−27,011	−26,774
		Change with Respect to Current Standard			
	15 ppm-hr	0	0	−5	1
	11 ppm-hr	4	7	−10	−6
	7 ppm-hr	5	17	−9	−5

Summary

Impacts to both sectors generally mirror one another, although they are more prominent in the forestry sector. Not only are tree species more responsive to changes in ozone, but the largest reductions to meet the alternative standards will occur in regions with large forestry sectors: South Central, Southeast, and Rocky Mountains. Reductions in agricultural regions are comparatively moderate. Productivity of both crops and forests is projected to increase at each of the alternative standard levels. This increase in supply resulted in decreased prices for forest products and agricultural commodities, which benefits consumer welfare while reducing producer welfare. Unless there are significant changes to wood products markets in particular, producers will be forced to sell at reduced prices to absorb the increased supply. Nonetheless, gains to consumers become increasingly large with more stringent ozone standards and, unlike the 15 ppm-hr and 11 ppm-hr scenarios, we observe agricultural producer surplus gains in the 7 ppm-hr scenario. Gains to agricultural producers in the most stringent case are associated with a decline in forestry returns that results in a net shift in land use toward agriculture. Gains to consumers and agricultural producers under the 7 ppm-hr scenario are large enough to more than offset producer surplus losses in the forestry sector, yielding an estimated annualized net benefit of $256.7 million.

Increased productivity is also projected to affect land use both within and between the agricultural and forest sectors. Within sectors, acreage is projected to shift from crops and tree species that are more sensitive to ozone to those that are less sensitive because productivity in the former will be more substantially affected by reductions in ozone concentrations. For ozone-sensitive crops and species, producers are projected to require less land to produce at the same or higher levels. Forest acreage in particular is projected to decline sharply, driven by declines in both reforestation and afforestation.

Despite reductions in crop and forest area, carbon sequestration is expected to increase over time, led almost entirely by increased forest sequestration. Although there is less reforestation and afforestation and lower sequestration in new inventory, the change is a result of existing inventories becoming so much larger as trees grow faster. Lower sequestration in new inventory is outweighed by increased inventory in standing forests, represented in the model as a change in forest management.

Increased stringency in the ozone standard generally produces larger impacts on all of the model outputs. However, the additional impact of moving from the current standard to 15 ppm-hr or 11 ppm-hr to 7 ppm-hr was sometimes marginal compared with changes occurring between 15 ppm-hr and 11 ppm-hr. In particular, the impacts to the forestry sector, most notably in forest inventories and the forest sector welfare analysis, tended to increase at a decreasing rate after meeting the 11 ppm-hr standard.

The model results are subject to several limitations: First, the ozone concentration response functions applied to crops and trees were using "median" parameters in Lehrer et al. (2007)—the RYLs and RYGs calculated are thus "median" ones; second, the use of crop proxy mapping and the forest-type mapping due to incomplete data specified in Section 4 adds to the uncertainty of these model results; third, the potential changes in tree species mixes within forest types due to ground ozone-level changes were not considered; and last, the international trade component in FASOMGHG that assumes USDA-based future projections under current conditions may present another uncertainty for the model results, especially when soybeans and wheat are among the major crop commodities for U.S. exports and have relatively large responses to changed ozone environments.

SECTION 6
COUNTY-LEVEL AGRICULTURAL WELFARE

In addition to information on the estimated net benefits at the national level, it is important to consider the regional distribution of benefits. In this study, we focused on approximating the distribution of changes in agricultural producer surplus for major crops at the county level. The decision to focus on agricultural producer surplus from crop production was based on data availability, heterogeneity of impacts across crops and regions, and available resources for conducting analyses. Impacts on consumer surplus are expected to be positive in all regions, whereas impacts on forest producer surplus are generally expected to be negative, although regions with the largest increases in productivity may experience net gains.

Calculating County-Level Agricultural Welfare

As introduced earlier, the agricultural component of FASOMGHG consists of 63 production regions (subregions) and 10 market regions (regions) simulating the U.S. agricultural sector. To gain an understanding of the scenario effects at the county level, we employ a downscaling calculation procedure to further disaggregate the 63-subregion simulation results.

The data on county-level crop production from the 2007 USDA Census of Agriculture (Ag Census) are used to generate the county-level cropping patterns that reflect the production differences between counties. The allocation of production across counties within a FASOMGHG subregion is held constant under alternative scenarios. Specifically, for a select crop, the county-level production percentage shares of that crop in a FASOMGHG subregion under new market equilibriums are assumed to equal the percentage shares of the 2007 Ag Census county-level production with respect to their FASOMGHG subregion's 2007 Ag Census production. In mathematical terms, the county-level agricultural welfare calculation involves Formulas (6.1) through (6.3)—from sector to region to county, they are as follows:

$$Ag\ Select\ Crop\ Welfare_{Subregion} =$$

$$\frac{\sum_{select\ crops} Crop\ Production\ (crop) * Price\ (crop)}{\sum_{crop\ \&\ livestock\ commodities} Commodity\ Production\ (commodity) * Price\ (commodity)} *$$

$$Ag\ Producer\ Welfare_{Subregion}$$

(6.1)

where the select crops' portion in agricultural producer welfare at FASOMGHG subregion level is extracted. Note that the subregion-level ratio $\frac{Select\ Crops'\ Production * Price}{Crop\ \&\ Livestock\ Production * Price}$ would change under different ozone environments, because new equilibriums of production and price levels could result under alternative ozone scenarios. Figure 6-1 illustrates how the select crops' portion is defined over agricultural commodities produced in FASOMGHG subregions.

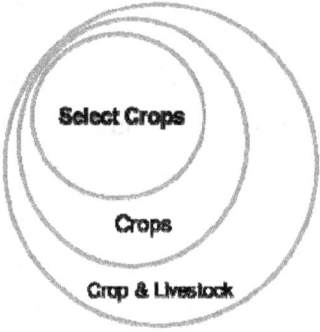

Figure 6-1.   Relationship between Select Crops Used for Analysis and Total Agricultural Production

$$\overline{\hspace{7cm}} \hspace{3cm} (6.2)$$

where the county-level agricultural welfare is thus obtained. Note that the county-level ratio $\overline{\hspace{4.5cm}}$ may also vary under different scenarios, because both the production levels and prices could change under new ozone environments. Figure 6-2 depicts how a county's portion is defined over the counties within a FASOMGHG subregion.

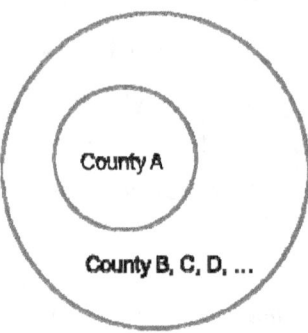

Figure 6-2.   County Share of Total FASOMGHG Subregion Production Is Used to Calculate County Gross Revenue and Production

$$\overline{\hspace{7cm}} \hspace{3cm} (6.3)$$

where the county-level production estimates for select crops are derived based on both the 2007 Ag Census data and the FASOMGHG simulation results. The calculation of the county share of the total value across all counties within a FASOMGHG subregion as depicted in Figure 6-2 also applies here.

In the actual calculation procedure, Formula (6.3) would be carried out first, followed by Formulas (6.2) and (6.1).

Notice that the agricultural producer welfare at FASOMGHG subregion level in Formula (6.1)—from which the select crops' portion is extracted—is defined as the area above the supply curve(s) of inputs and endowments involved in agricultural commodities production and below the equilibrium price(s) of the commodities, as shown in Figure 6-3.

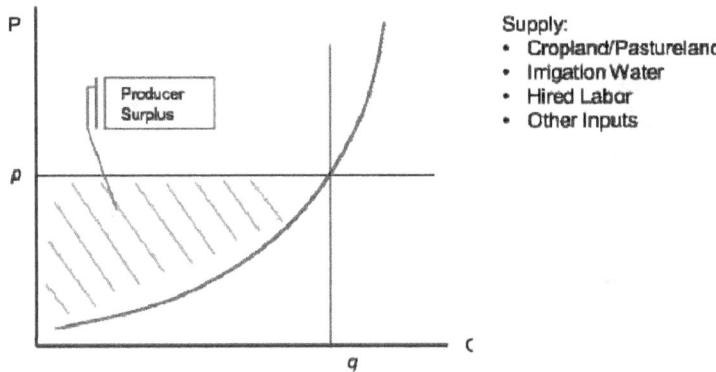

Figure 6-3.   Area of Producer Surplus

In addition, a mapping of USDA crops to FASOMGHG crops is involved in Formula (6.3), as presented in Table 6-1.

The following FASOMGHG crops are not included in the county-level agricultural welfare calculation because they lack 2007 Ag Census county-level production data: silage, potato, tomato (fresh and processed), orange (fresh and processed; 75, 90, and 85 lb boxes), grapefruit (fresh and processed; 67, 85, and 80 lb boxes), sweet sorghum, hybrid poplar, willow, switchgrass, and crop residues.

Table 6-1. Mapping of USDA Crops to FASOMGHG Crops

FASOMGHG Crop	USDA 2007 Census Crop
Canola	Canola
Corn	Corn
Durum wheat	Durum wheat
Hard red spring wheat	Spring wheat
Hard red winter wheat	Winter wheat
Soft red winter wheat	Winter wheat
Soft white wheat	Wheat excluding spring, winter, and durum wheat
Hay	Hay
Oats	Oats
Rice	Rice
Sorghum	Sorghum
Soybeans	Soybeans
Spring barley	Barley
Winter barley	Barley
Sugar beet	Sugar beet
Sugarcane	Sugarcane

Changes in County-Level Agricultural Welfare: Alternative Scenarios versus Current Standard (75 ppb)

When comparing the W126 ozone values under alternative standards and the current standard 75 ppb scenario (as presented in Section 4), one can notice the following:

1. Under 15 ppm-hr, slight further reductions of ground ozone levels occurred in the southern Rocky Mountains region.

2. Under 11 ppm-hr, larger further reductions of ground ozone levels occurred in the Rocky Mountains and Pacific Southwest (southern California) regions; noticeable further reductions occurred in the southern Corn Belt and the northern South Central regions.

3. Under the most stringent 7 ppm-hr standard, significant further ground ozone reductions occurred in the mid and southern Rocky Mountains regions and the Pacific Southwest region. Noticeable reductions also occurred in the southern Corn Belt and Great Plains regions. The South Central and Southeast regions experienced slight ozone reductions.

Taking those three observations into consideration, one can expect that the production of wheat crops, which have relatively larger RYGs under reduced ozone environments, would expand in one of its major production regions—the Rocky Mountains region, which experiences significant ground ozone reductions in its southern half The county-level welfare increases in the Rocky Mountains region, shown in Figure 6-4, correspond to this wheat production expansion.

The effects of this Rocky Mountains wheat expansion has implications for other wheat-producing regions—such as the Lake States region, where wheat production decreases. Compared with the Rocky Mountains region, producing wheat in the Lake States region becomes less efficient in terms of enhancing producer and consumer welfare at the national level.

The Lake States region would also see production changes for other crops—because the wheat production contraction implies more room for other alternatives—in particular, the highly profitable ones. Soybean production in the Lake States region thus expanded, and this expansion induces regional shifts of soybean production at the national level—the Great Plains and the Corn Belt regions experience soybean production decreases. Moreover, the ripple effects on the Great Plains region include larger corn production as well, as part of its soybean production shifts to the Lake States region.

The consequences of the regional production shifts, reflected in county-level welfare, are the increases in part of the Lake States and Rocky Mountains regions, as well as the decreases in some of the Corn Belt and Great Plains counties—for earlier periods in 2010 under the 15 ppm-hr scenario.

Under the 11 ppm-hr and 7 ppm-hr scenarios in which the southern Corn Belt and northern South Central regions also see noticeable ground ozone reductions, the regional shifts that were principally propelled by the Rocky Mountains wheat changes under 15 ppm-hr would now have to accommodate new changes from corn and soybean production in the Corn Belt region—soybean is another crop that has large RYGs under reduced-ozone environments. In addition, as Rocky Mountains ozone reductions get even greater, cotton production and revenues in the South would be also influenced—the increased cotton supply has led to a price decrease, and the cotton revenues in the South Central region have thus decreased.

Integrating the major changes induced by Corn Belt soybean and Rocky Mountains cotton and wheat production stated previously, one would then see welfare increases in the majority of southern Corn Belt counties and welfare decreases along the Mississippi River region in the South Central region.

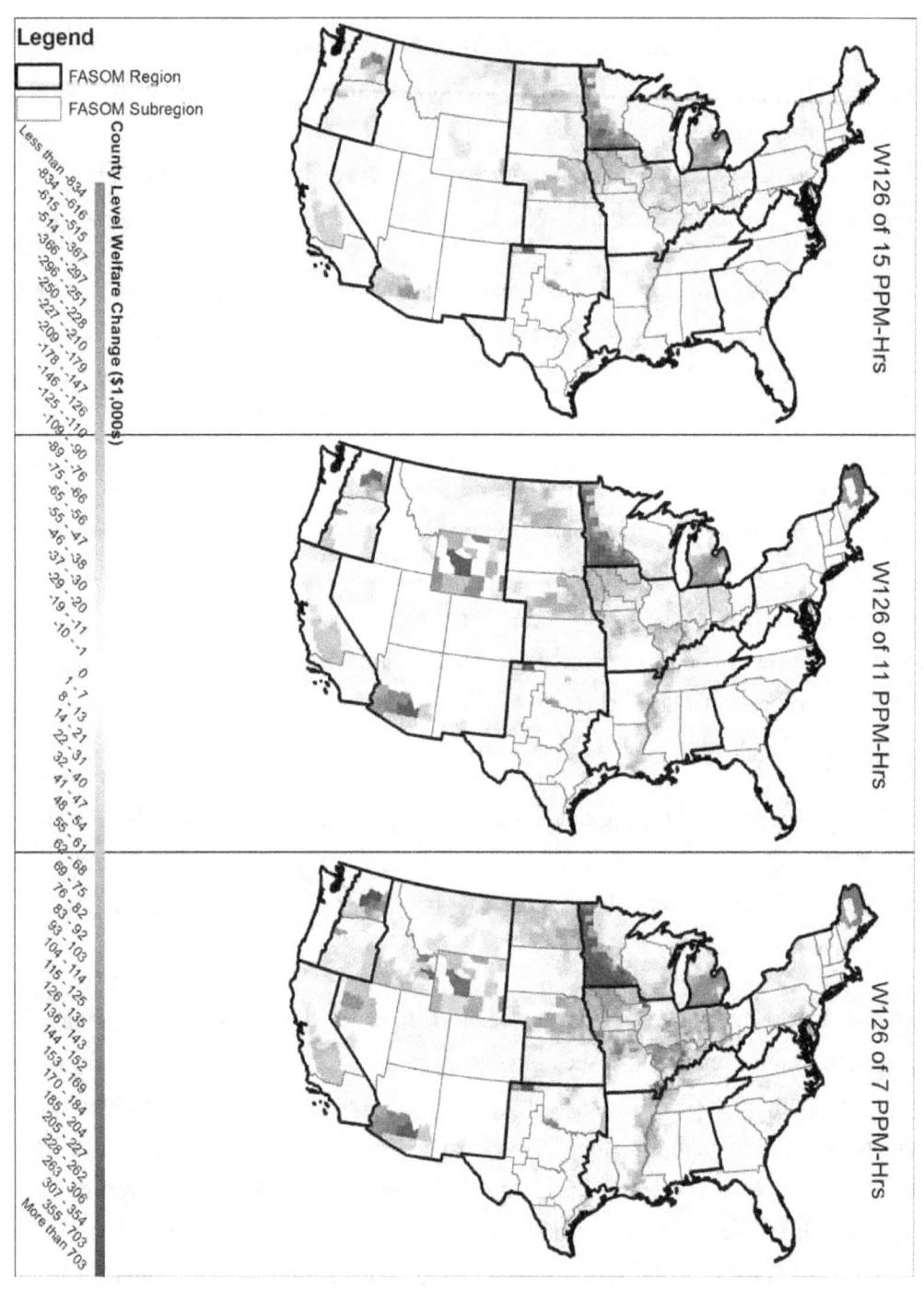

Figure 6-4. **Changes in Crop Producer Welfare under Alternative Standards with Respect to 75 ppb, 2010, Thousand 2010 U.S. Dollars**

When comparing the 11 ppm-hr and 7 ppm-hr scenarios, one can notice that the 7 ppm-hr scenario effects on county-level crop producer welfare are essentially an intensification of the 11 ppm-hr scenario effects.

Figure 6-5 shows that by 2020, the Rocky Mountains effects are largely contained in the region—the ripple effects on other regions, including Lake States, Corn Belt, and Great Plains, are quite limited. This is because, by 2020, as the Renewable Fuel Standards (RFS2) become fully phased in, the production capacity of corn in the Corn Belt and Lake States regions becomes much more utilized—as does the production capacity of soybeans in these regions. The further ground-level ozone reduction under the 15 ppm-hr scenario thus did not lead to major nationwide changes in 2020.

Nonetheless, under the 11 ppm-hr and 7 ppm-hr scenarios in which the Corn Belt and South Central regions experience noticeable further ozone reductions, the regional distributional effects of the ozone standards start to become visible. In both cases, soybean production expanded in the Corn Belt and South Central regions, contributing to county-level welfare increases in these regions.

The ripple effects of Corn Belt and South Central soybean production expansion also impact the Great Plains region, where less soybean production occurred and more production of other grain crops took place—in particular, barley and wheat. In turn, the Great Plains changes induced decreases of barley and wheat production in the Rocky Mountains region, the southern area of which incurred even greater cotton production. The greater cotton production is reflected in county-level welfare; the northern Rocky Mountains counties see welfare decreases, whereas the southern Rocky Mountains counties see welfare increases.

The 7 ppm-hr scenario effects are generally intensified relative to the 11 ppm-hr scenario effects again for 2020, in terms of county-level welfare changes. However, under this case, the further enhanced soybean yields in the Corn Belt region have led to reduced corn production in that region, despite the strong effects of RFS2 on corn production. This change leads to welfare decreases in corn-producing counties, as shown in Figure 6-5.

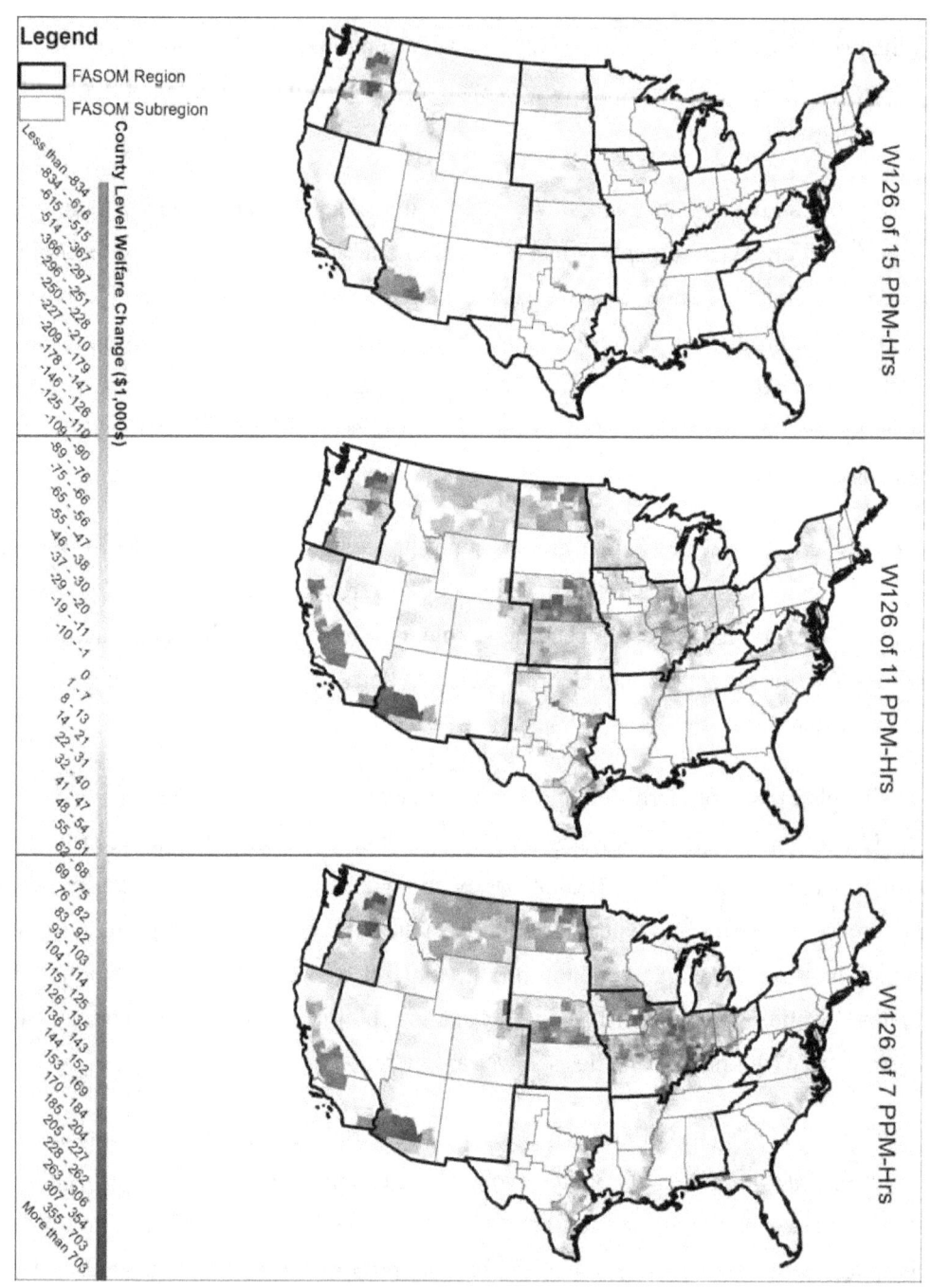

Figure 6-5. Changes in Crop Producer Welfare under Alternative Standards with Respect to 75 ppb, 2020, Thousand 2010 U.S. Dollars

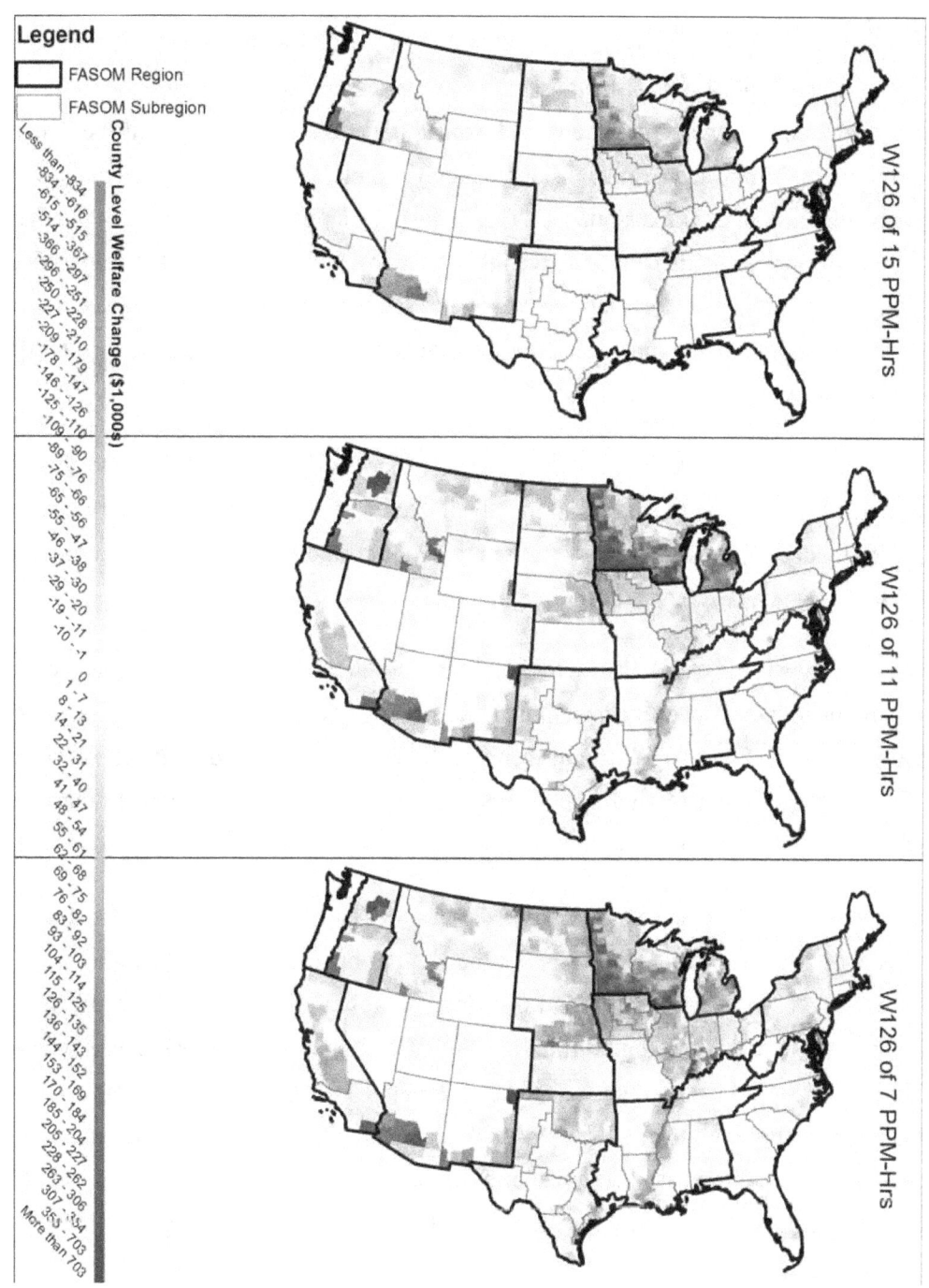

Figure 6-6.   Changes in Crop Producer Welfare under Alternative Standards with Respect to 75 ppb, 2030, Thousand 2010 U.S. Dollars

As shown in Figure 6-6, by 2030, in addition to the RFS2 effects and the ozone standard-induced RYGs, the implications of land use changes also occurred for crop production—Section 5 detailed that cropland pasture would expand under alternative standard scenarios compared with the 75 ppb scenario. Consider, as the yield increase effects accumulate over time for the trees, that less land would be needed for the forestry industry, assuming that the market situation remains generally unchanged. Given that crop production also benefits from reduced-ozone environments and, thus, less land would be needed for cropping, these reduced demands for land would result in more land available for grazing and would in turn induce greater demand for feed crops as livestock herds increase.

The South Central region is among the areas experiencing large RYGs for tree growth, and consequently, it sees more land becoming available for grazing use. The pasture land increase also induced greater corn production in this region, and the ripple effects reached out to the Lake States region, reducing corn production there.

Other things being equal, the Lake States corn production decrease would imply welfare decreases in this region. However, the opposite occurred for the Lake States counties, because by 2030, the increases in overall agricultural welfare—due to livestock production expansion as presented in Section 5—outweigh the decreases in crop production; hence, welfare increases still occurred for the Lake States region.

When the Corn Belt and South Central soybean RYG effects, along with the Rocky Mountains cotton RYG effects, come in under the 11 ppm-hr scenario, greater soybean production would occur in the South Central region. Because the land use change effects discussed previously would become further intensified under the 11 ppm-hr scenario, corn production in the South Central region would also expand. These increases, however, lead to reduction in production of one of the most valuable crops in this region: cotton. The cotton price decreases due to overall increased cotton supply further decreased the cotton revenues in this region, and this cotton effect outweighs the corn and soybean effects, leading to welfare decreases along the Mississippi River region.

Under the 7 ppm-hr scenario, soybean production in the southern Corn Belt region increases further, contributing to the welfare increases in the region's southern area and welfare decreases in the region's northern area.

By 2040 (see Figure 6-7), the effects of land use changes start to appear in the Great Plains region, where more land is used for grazing. This also raises the demand for feed crops—corn production expands in this region and the adjacent Lake States region. The South Central region correspondingly reduced its corn production—the land use effects have already been applied in the South Central region in 2030. Notice that the corn production increases in the Great Plains and Lake States regions led to decreases in production of other crops, especially wheat. These wheat production reductions led to net welfare decreases in counties in Great Plains and Lake States regions.

Under the more stringent 11 ppm-hr and 7 ppm-hr scenarios, the effects of land use changes on the Great Plains region would further increase, resulting in more corn production in the Great Plains region, meanwhile reducing the soybean and wheat production in this region. The ripple effects of these Great Plains changes result in less corn production in the southern regions and the Lake States regions for this time and in a shift of more soybean production to the Corn Belt region. Hence, under the more stringent 11 ppm-hr and 7 ppm-hr scenarios, there would be considerable welfare increases in the southern Corn Belt counties and welfare decreases in the Mississippi River region. The effects of heightened soybean RYGs were thus intensified in the southern Corn Belt region and outweighed by the land use effects in the South Central region.

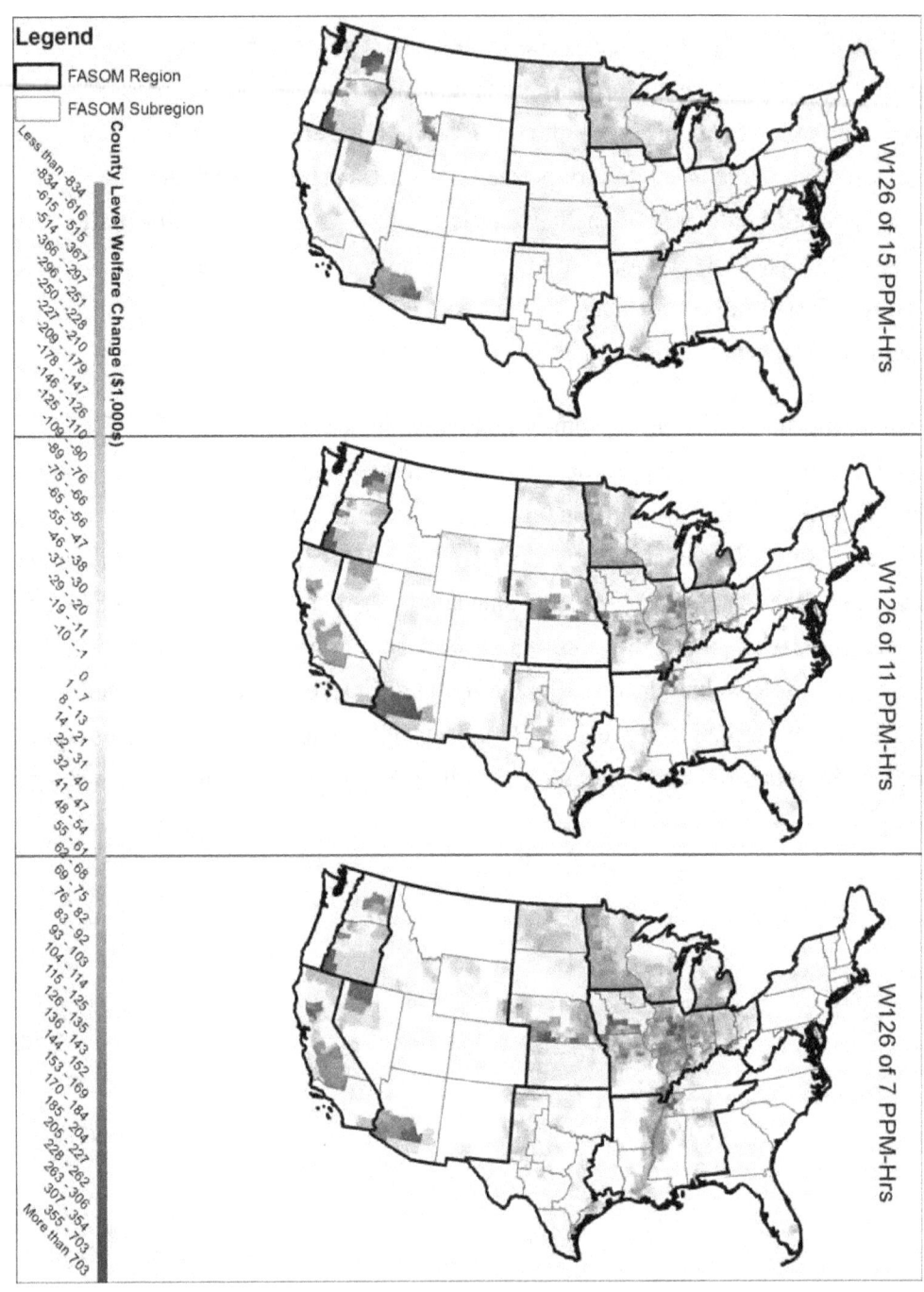

Figure 6-7. Changes in Crop Producer Welfare under Alternative Standards with Respect to 75 ppb, 2040, Thousand 2010 U.S. Dollars

Summary

Table 6-2 lists the driving factors causing crop producer welfare increases and decreases across the counties discussed previously, by scenario and period modeled.

Table 6-2.   Driving Factors Behind County-Level Welfare Changes

Scenario\Period	2010	2020	2030	2040
15 ppm-hr vs. 75 ppb	▪ Cotton and wheat RYGs in RM	▪ Cotton and wheat RYGs in RM ▪ RFS2 policy effects	▪ Cotton and wheat RYGs in RM ▪ Land use effects in SC	▪ Cotton and wheat RYGs in RM ▪ Land use effects in GP
11 ppm-hr vs. 7 ppb	▪ Cotton and wheat RYGs in RM ▪ Soybean and cotton RYGs across southern regions	▪ Cotton and wheat RYGs in RM ▪ Soybean and cotton RYGs across southern regions ▪ RFS2 policy effects	▪ Cotton and wheat RYGs in RM ▪ Soybean and cotton RYGs across southern regions ▪ Land use effects in SC	▪ Cotton and wheat RYGs in RM ▪ Soybean and cotton RYGs across southern regions ▪ Land use effects in GP
7 ppm-hr vs. 75 ppb	▪ Intensified cotton and wheat RYGs in RM ▪ Intensified soybean and cotton RYGs in the South	▪ Intensified cotton and wheat RYGs in RM ▪ Intensified soybean and cotton RYGs in the South ▪ RFS2 policy effects	▪ Intensified cotton and wheat RYGs in RM ▪ Intensified soybean and cotton RYGs in the South ▪ Land use effects in SC	▪ Intensified cotton and wheat RYGs in RM ▪ Intensified soybean and cotton RYGs in the South ▪ Land use effects in GP

To summarize, the crop producers' welfare in the southern Corn Belt region and the southern Rocky Mountains region would generally experience increases in most policy cases and across periods, whereas the counties along the Mississippi River region would experience welfare decreases as ozone standards get more stringent. The counties in the Great Plains and the Lake States regions, however, would experience alternate increases and decreases across periods.

SECTION 7
REFERENCES

Adams, D.M. and R.W. Haynes. 1980. *The 1980 Softwood Timber Assessment Market Model: Structure, Projections, and Policy Simulations*. Forest Science Monograph 22, 64 p.

Adams, D.M. and R.W. Haynes. 1996. *The 1993 Timber Assessment Market Model: Structure, Projections, and Policy Simulations*. General Technical Report PNW-GTR-368. Portland, OR: U.S. Department of Agriculture, Forest Service, Pacific Northwest Research Station, 58 p.

Adams, D.M. and R.W. Haynes (Eds.). 2007. Resource and Market Projections for Forest Policy Development: Twenty-Five Years of Experience with the US RPA Timber Assessment. In *Managing Forest Ecosystems Series #14*. Springer Publishers, Dordrecht, The Netherlands. 615 p.

Adams, D., R. Alig, B.A. McCarl, and B.C. Murray. 2005. FASOMGHG Conceptual Structure and Specification: Documentation. Available at: http://agecon2.tamu.edu/people/faculty/mccarl-bruce/papers/1212FASOMGHG_doc.pdf.

Adams, R.M., D.M. Adams, J.M. Callaway, C.C. Chang, and B.A. McCarl. 1993. Sequestering Carbon on Agricultural Land: Social Cost and Impacts on Timber Markets. *Contemporary Policy Issues* 11:76-87.

Adams, R.M., J.M. Callaway, and B.A. McCarl. 1986. Pollution, Agriculture and Social Welfare: The Case of Acid Deposition. *Canadian Journal of Agricultural Economics* 34:3-19.

Adams, R.M., J.D. Glyer, B.A. McCarl, and D.J. Dudek. 1988. The Implications of Global Change for Western Agriculture. *Western Journal of Agricultural Economics* 13(December):348-356.

Adams, R.M., C. Rosenzweig, R.M. Peart, J.T. Ritchie, B.A. McCarl, J.D. Glyer, R.B. Curry, J.W. Jones, K.J. Boote, and L.H. Allen, Jr. 1990. Global Change and U.S. Agriculture. *Nature* 345:219-224.

Adams, R.M., B.A. McCarl, K. Segerson, C. Rosenzweig, K.J. Bryant, B.L. Dixon, R. Connor, R.E. Evenson, and D. Ojima. 1999. The Economic Effects of Climate Change on U.S. Agriculture. In *The Economics of Climate Change*, R. Mendelsohn and J. Neumann, Eds., pp. 19-54. New York: Cambridge University Press.

Adams, R.M., C.C. Chen, B.A. McCarl, and D.E. Schimmelpfenning. 2001. Climate Variability and Climate Change. In *Advances in the Economics of Environmental Resources*, Vol. 3, D. Hall and R. Howarth, Eds., pp. 115-148. London: JAI Press.

Adams, R.M., S.A. Hamilton, and B.A. McCarl. 1984. The Economic Effects of Ozone on Agriculture. Research Monograph EPA/600-3-84-90. Corvallis, OR: USEPA, Office of Research and Development.

Alig, R.J. and A. Plantinga. 2007. Methods for Projecting Areas of Private Timberland and Forest Cover Types. In *Resource and Market Projections for Forest Policy Development: Twenty-Five Years of Experience with the U.S. RPA Timber Assessment*, D.M. Adams and R.W. Haynes, Eds. Dordrecht, The Netherlands: Springer.

Alig, R.J., A. Plantinga, S. Ahn, and J. Kline. 2003. *Land Use Changes Involving Forestry for the United States: 1952 to 1997, with Projections to 2050*. U.S. Forest Service General Technical Report 587, Pacific Northwest Research Station. Portland, OR: U.S. Department of Agriculture, Forest Service, Pacific Northwest Research Station, 92 p.

Alig, R.J., A. Plantinga, D. Haim, and M. Todd. 2010a. *Area Changes in U.S. Forests and other Major Land Uses, 1982-2002, with Projections to 2062*. U.S. Forest Service General Technical Report PNW-GTR-815, Pacific Northwest Research Station. Portland, OR: U.S. Department of Agriculture, Forest Service, Pacific Northwest Research Station, 98 p.

Alig, R.J., G. Latta, D.M. Adams, and B.A. McCarl. 2010b. Mitigating Greenhouse Gases: The Importance of Land Base Interactions Between Forests, Agriculture, and Residential Development in the Face of Changes in Bioenergy and Carbon Prices. *Forest Policy and Economics* 12(1):67-75.

Alig, R.J., J.D. Kline, and M. Lichtenstein. 2004. Urbanization on the U.S. Landscape: Looking Ahead in the 21st Century. *Landscape and Urban Planning* 69(2-3):219-234.

Baumes, H. 1978. A Partial Equilibrium Sector Model of U.S. Agriculture Open to Trade: A Domestic Agricultural and Agricultural Trade Policy Analysis. PhD dissertation. West Lafayette, IN: Purdue University.

Beach, R.H., D.M. Adams, R.J. Alig, J.S. Baker, G.S. Latta, B.A. McCarl, B.C. Murray, S.K. Rose, and E.M. White. 2010. Model Documentation for the Forest and Agricultural Sector Optimization Model with Greenhouse Gases (FASOMGHG). Available at: http://www.cof.orst.edu/cof/fr/research/tamm/FASOM_Documentation.htm.

Beach, R.H. and B.A. McCarl. 2010. *Impacts of the Energy Independence and Security Act on U.S. Agriculture and Forestry: FASOM Results and Model Description*. Prepared for the U.S. Environmental Protection Agency, Office of Transportation and Air Quality.

Burton, R.O. 1982. Reduced Herbicide Availability: An Analysis of the Economic Impacts on U.S. Agriculture. PhD dissertation. West Lafayette, IN: Purdue University.

Burton, R.O. and M.A. Martin. 1987. Restrictions on Herbicide Use: An Analysis of Economic Impacts on U.S. Agriculture. *North Central Journal of Agricultural Economics* 9:181-194.

Chang, C.C., J.D. Atwood, K. Alt, and B.A. McCarl. 1994. Economic Impacts of Erosion Management Measures in Coastal Drainage Basins. *Journal of Soil and Water Conservation* 49(6):606-611.

Chattin, B.L. 1982. By-Product Utilization from Biomass Conversion to Ethanol. PhD dissertation. West Lafayette, IN: Purdue University.

Hamilton, S.A. 1985. The Economic Effects of Ozone on U.S. Agriculture: A Sector Modeling Approach. PhD dissertation. Corvallis, OR: Oregon State University.

Haynes, R.W. (Technical coordinator). 2003. *An Analysis of the Timber Situation in the United States: 1952 to 2050.* General Technical Report PNW-GTR-560. Portland, OR: U.S. Department of Agriculture, Forest Service, Pacific Northwest Research Station, 254 p.

Haynes, R.W., D.M. Adams, R.J. Alig, P.J. Ince, J.R. Mills, and X. Zhou. 2007. *The 2005 RPA Timber Assessment Update.* General Technical Report PNW-GTR-699. Portland, OR: U.S. Department of Agriculture, Forest Service, Pacific Northwest Research Station, 212 p.

Hickenbotham, T.L. 1987. Vegetable Oil as a Diesel Fuel Alternative: An Investigation of Selected Impacts on U.S. Agricultural Sector. PhD dissertation. St. Paul, MN: University of Minnesota.

Ince, P.J. 1994. *Recycling and Long-Range Timber Outlook.* General Technical Report RM-242. Ft. Collins, CO: U.S. Department of Agriculture, Forest Service, Rocky Mountain Forest and Range Experiment Station, 23 p.

Intergovernmental Panel on Climate Change (IPCC). 2001. *Climate Change 2001: The Scientific Basis. Contribution of Working Group I to the Third Assessment Report of the Intergovernmental Panel on Climate Change* [Houghton, J.T.,Y. Ding, D.J. Griggs, M. Noguer, P.J. van der Linden, X. Dai, K. Maskell, and C.A. Johnson (eds.)]. Cambridge, United Kingdom, and New York, NY: Cambridge University Press, 881pp.

IPCC. 2007. *Climate Change 2007: The Physical Science Basis. Contribution of Working Group I to the Fourth Assessment Report of the Intergovernmental Panel on Climate Change* [Solomon, S., D. Qin, M. Manning, Z. Chen, M. Marquis, K.B. Averyt, M. Tignor, and H.L. Miller (eds.)]. Cambridge, United Kingdom, and New York, NY: Cambridge University Press, 996 pp.

Joyce, L.A. (Ed.). 1995. *Productivity of America's Forests and Climate Change.* General Technical Report RM-271. Fort Collins, CO: U.S. Forest Service, Rocky Mountain Forest and Range Experiment Station.

Joyce, L.A., and R.A. Birdsey (Eds.). 2000. *The Impact of Climate Change on America's Forests.* RMRS-GTR-59. Fort Collins, CO: U.S. Forest Service, Rocky Mountain Research Station.

Lee, H.-C. 2002. The Dynamic Role for Carbon Sequestration by the U.S. Agricultural and Forest Sectors in Greenhouse Gas Emission Mitigation. PhD dissertation. College Station, TX: Department of Agricultural Economics, Texas A&M University.

Lehrer, J.A., M. Bacou, B. Blankespoor, D. McCubbin, J. Sacks, C.R. Taylor, and D.A. Weinstein. 2007. *Technical Report on Ozone Exposure, Risk, and Impact Assessments for Vegetation*. EPA 452/R-07-002.

Little, E.L., Jr. 1971. *Atlas of United States Trees, Volume 1, Conifers and Important Hardwoods*. U.S. Department of Agriculture Miscellaneous Publication 1146, 9 p., 200 maps.

Little, E.L., Jr. 1976. *Atlas of United States Trees, Volume 3, Minor Western Hardwoods*. U.S. Department of Agriculture Miscellaneous Publication 1314, 13 p., 290 maps.

Little, E.L., Jr. 1977. *Atlas of United States Trees, Volume 4, Minor Eastern Hardwoods*. U.S. Department of Agriculture Miscellaneous Publication 1342, 17 p., 230 maps.

Little, E.L., Jr. 1978. *Atlas of United States Trees, Volume 5, Florida*. U.S. Department of Agriculture Miscellaneous Publication 1361, 262 maps.

McCarl, B.A. 1999. *Economic Assessments under National Climate Change Assessment*. Presented at Meeting of National Climate Change Assessment Group, Washington, DC.

McCarl, B.A. and U.A. Schneider. 2001. The Cost of Greenhouse Gas Mitigation in US Agriculture and Forestry. *Science* 294(December):2481-2482.

Mills, J. and J. Kincaid. 1992. *The Aggregate Timberland Assessment System—ATLAS: A Comprehensive Timber Projection Model*. General Technical Report PNW-281. Portland, OR: U.S. Department of Agriculture, Forest Service, Pacific Northwest Research Station, 160 p.

Reilly, J., F. Tubiello, B. McCarl, and J. Melillo. 2000. Climate Change and Agriculture in the United States. In *Climate Change Impacts on the United States: The Potential Consequences of Climate Variability and Change*, pp. 379-403. Report for the U.S. Global Change Research Program. New York: Cambridge University Press.

Reilly, J.M., F. Tubiello, B.A. McCarl, D.G. Abler, R. Darwin, K. Fuglie, S.E. Hollinger, R.C. Izaurralde, S. Jagtap, J.W. Jones, L.O. Mearns, D.S. Ojima, E.A. Paul, K. Paustian, S.J. Riha, N.J. Rosenberg, and C. Rosenzweig. 2002. U.S. Agriculture and Climate Change: New Results. *Climatic Change* 57:43-69.

Schneider, U.A. 2000. Agricultural Sector Analysis on Greenhouse Gas Emission Mitigation in the U.S. PhD dissertation. College Station, TX: Department of Agricultural Economics, Texas A&M University.

Takayama, T. and G.G. Judge. 1971. *Spatial and Temporal Price Allocation Models*. Amsterdam: North-Holland Publishing Co., 210 p.

Tyner, W., M. Abdallah, C. Bottum, O. Doering, B.A. McCarl, W.L. Miller, B. Liljedahl, R. Peart, C. Richey, S. Barber, and V. Lechtenberg. 1979. *The Potential of Producing Energy from Agriculture*. Report to the Office of Technology Assessment. West Lafayette, IN: Purdue University School of Agriculture.

U.S. Department of Agriculture, Economic Research Service (USDA ERS). 2012. *Major Land Uses*. Dataset available at: http://www.ers.usda.gov/Data/MajorLandUses/. Last updated January 27, 2012.

U.S. Department of Agriculture, Farm Service Agency (USDA FSA). 2009. *CRP Contract Summary and Statistics*. Available at: http://www.fsa.usda.gov/FSA/webapp?area=home&subject=copr&topic=rns-css.

U.S. Department of Agriculture, National Agricultural Statistics Service (USDA NASS). Various years. *USDA Agricultural Statistics (1990–2002)*. Available at: http://www.nass.usda.gov/Publications/Ag_Statistics/.

U.S. Department of Agriculture, Natural Resource Conservation Service (USDA NRCS). 2003. *Annual NRI—Land Use*. Available at: http://www.nrcs.usda.gov/technical/NRI.

Zhang, D., J. Buongiorno, and P. Ince. 1993. *PELPS III: A Microcomputer Price Endogenous Linear Programming System for Economic Modeling: Version 1.0*. Research Paper FPL-526. Madison, WI: USDA, Forest Service, Forest Products Laboratory, 43 p.

Zhang, D., J. Buongiorno, and P. Ince. 1996. A Recursive Linear Programming Analysis of the Future of the Pulp and Paper Industry in the United States: Changes in Supplies and Demands, and the Effects of Recycling. *Annals of Operations Research* 68:109-139.

APPENDIX A
DATA INPUTS: FOCUSING ON COMPARISON OF ALTERNATIVE OZONE STANDARDS WITH CURRENT CONDITIONS

Our primary focus in the analyses conducted for this report involved comparisons between impacts on the forestry and agricultural sectors under three ozone standards that are more stringent than the current standard. Those inputs and results are presented in Sections 4–6 of this report. In the appendixes, we supplement the information with additional tables, figures, and maps that compare all four scenarios where ozone standards are achieved and current conditions. Because current ozone concentrations are well above the current standards in many U.S. regions, meeting any of the four ozone standards examined would result in a large reduction in ambient ozone concentrations. The reductions in concentrations associated with moving from current conditions to any of the ozone standards are considerably larger than the incremental reductions from moving between the standards examined.

In Appendix A, we present the data inputs used for assessing the impacts of achieving each of the four ozone standards considered relative to current conditions. This appendix includes a map of the change in ozone concentrations relative to current conditions, along with maps, tables, and figures showing the relative yield losses (RYLs) and relative yield gains (RYGs) by major crop for each FASOMGHG subregion. We also include maps breaking out county-level changes in yields, These maps primarily serve to highlight the counties that grow a given crop and are affected by changes in ozone concentrations.

A.1   Ambient Ozone Concentration

Figure A-1 presents the changes in W126 ozone values for alternative ozone standards with respect to current conditions. The regional patterns of ozone reductions are virtually the same across the 75 ppb, 15 ppm-hr, 11 ppm-hr, and 7 ppm-hr scenarios, although the magnitude of regional reductions increases as the standard is tightened. For a closer examination of changes arising from the alternative ozone standards (15 ppm-hr, 11 ppm-hr, and 7 ppm-hr) with respect to the current 75 ppb standard, see Section 4.

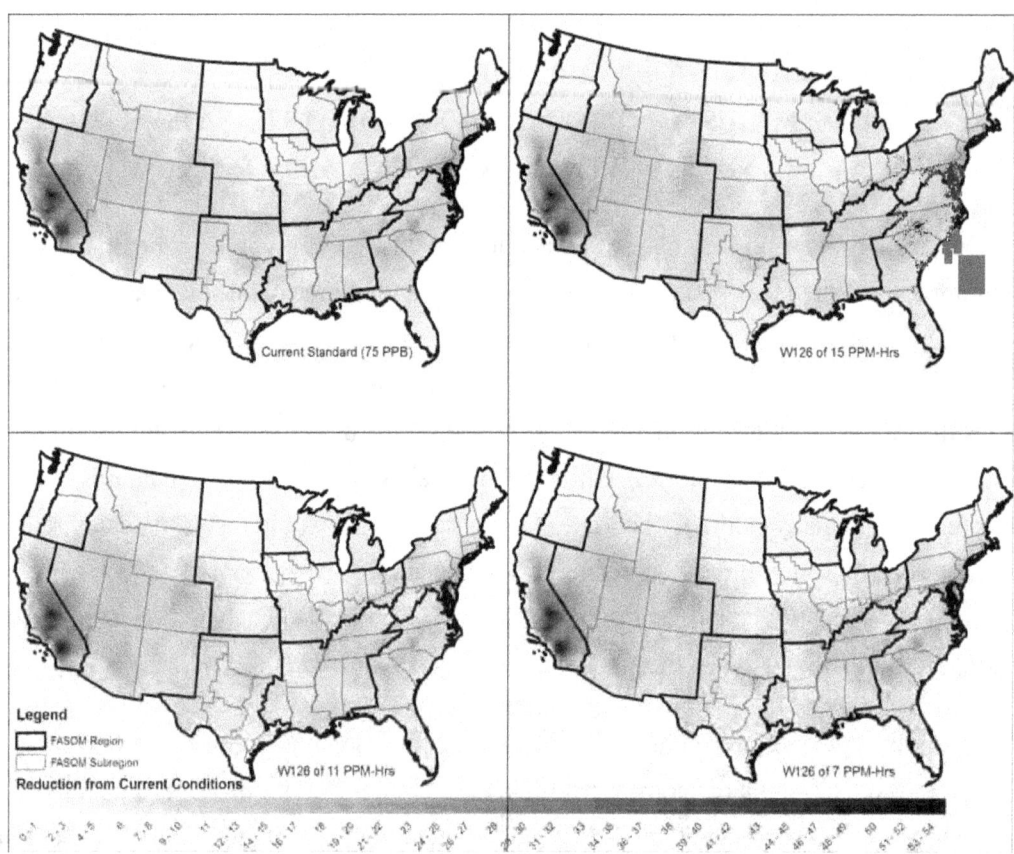

Figure A-1. Ozone Reduction with Respect to Current Conditions under Alternative Scenarios

Table A-1 summarizes the information in Figure A-1. It presents the grouping of FASOMGHG regions into three qualitative categories based on the relative level ozone reductions under the current standard versus current conditions. The table also lists the major crops grown in each FASOMGHG agricultural region. The underscored crops are the species that are more susceptible to ground-level ozone effects than other crops, as their RYGs will demonstrate.

Table A-1 also suggests that the agricultural production in the Pacific Southwest, Rocky Mountains, South Central, and Southeast regions are likely to experience the largest changes under ozone control because they would have significant ozone reductions, and the most "sensitive" crop species are present in these regions. Sizable effects would also be expected in the Corn Belt region, given the level of ozone reductions and the exceptionally large crop production in that region.

Table A-1. General Ozone Reduction Levels with Respect to Current Conditions by FASOMGHG Regions

Ozone Reduction			FASOMGHG Region	Major Crops[a]
High			PSW	<u>Cotton</u>, Rice, <u>Wheat</u>
High			SE	<u>Cotton</u>, Corn, <u>Soybeans</u>, <u>Wheat</u>
High			SC	<u>Cotton</u>, Corn, Rice, <u>Soybeans</u>, <u>Wheat</u>
High			RM	Barley, <u>Cotton</u>, <u>Wheat</u>
	Medium		CB	Corn, <u>Soybeans</u>
	Medium		SW	<u>Cotton</u>, Sorghum, <u>Wheat</u>
	Medium		NE	Corn, <u>Soybeans</u>, <u>Wheat</u>
		Low	GP	Corn, Sorghum, <u>Soybeans</u>, <u>Wheat</u>
		Low	LS	Corn, <u>Soybeans</u>, <u>Wheat</u>
		Low	PNWE	Barley, <u>Wheat</u>

[a] Underscored crops are the species that are more susceptible to ground-level ozone effects than other crops.

Note: CB = Corn Belt; GP = Great Plains; LS = Lake States; NE = Northeast; PNWE = Pacific Northwest—East side; PSW = Pacific Southwest; RM = Rocky Mountains; SC = South Central; SE = Southeast; SW = Southwest.

A.2   Summarized Relative Yield Losses and Gains for Crops and Forest Types

This subsection presents data on RYLs and RYGs for crops and trees. See Section 3 for a discussion of these measures and the equations used to calculate them.

Figures A-2 through A-7 show the subregion-level RYL estimates for major crops under the current ambient ozone concentrations. These RYLs were calculated using the cropland ozone surfaces. Table 3-3 in Section 3 presented the mapping of proxy crops to FASOMGHG crops, and Table 2-7 in Section 2 provided definitions of FASOMGHG regions. Based on the relationships between ozone concentrations and yield reductions applied (see Section 3.2), there are currently substantial yield reductions due to ground-level ozone.

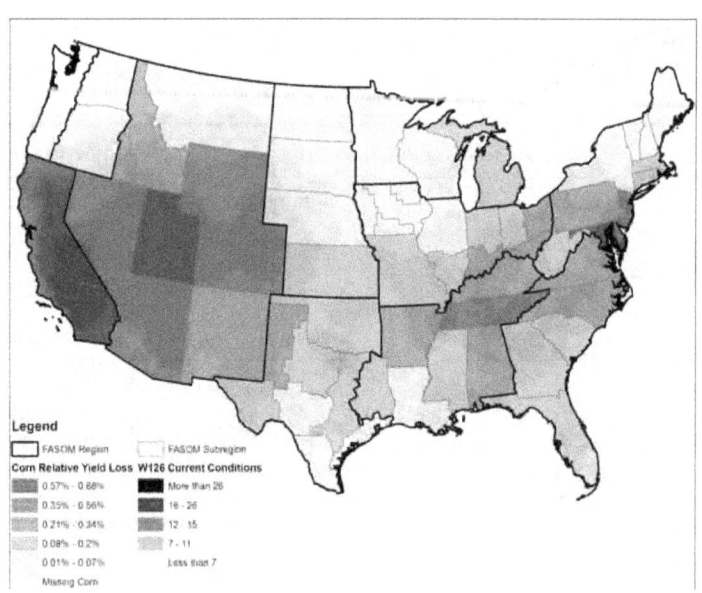

Figure A-2. Map of Corn RYLs under Current Conditions

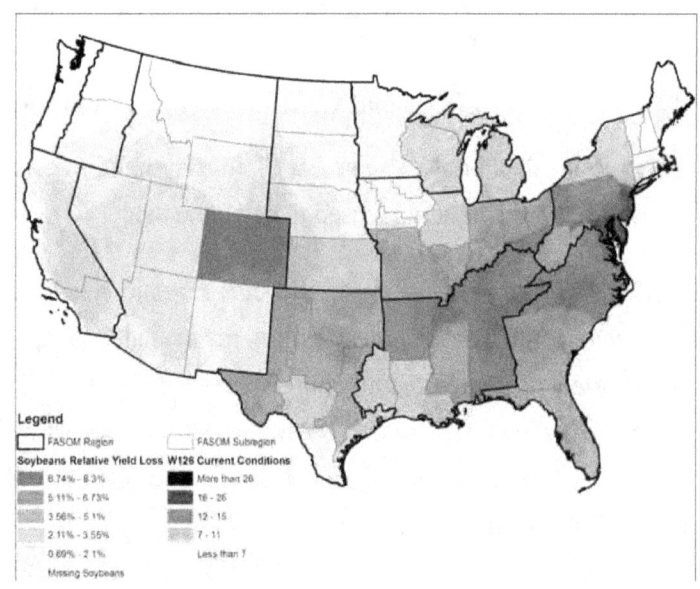

Figure A-3. Map of Soybean RYLs under Current Conditions

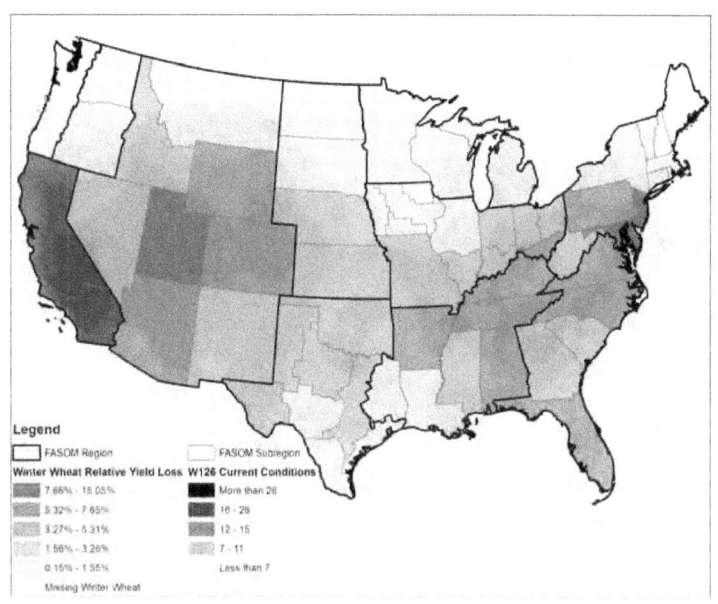

Figure A-4. Map of Wheat RYLs under Current Conditions

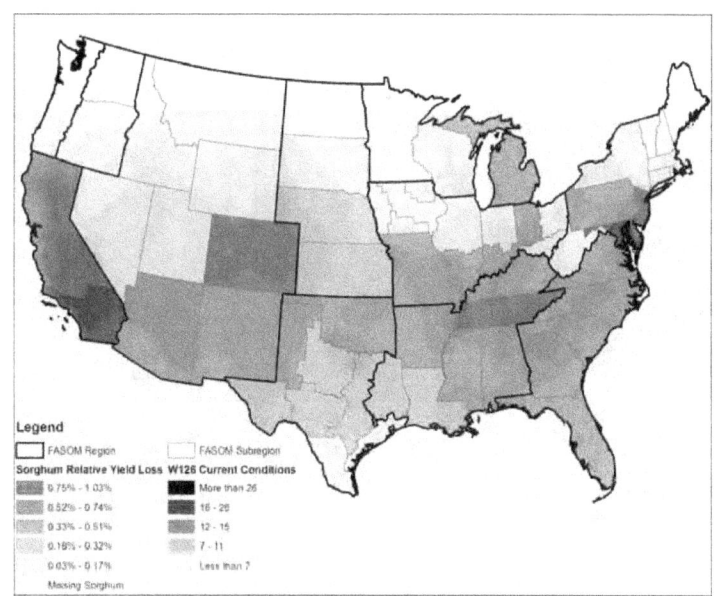

Figure A-5. Map of Sorghum RYLs under Current Conditions

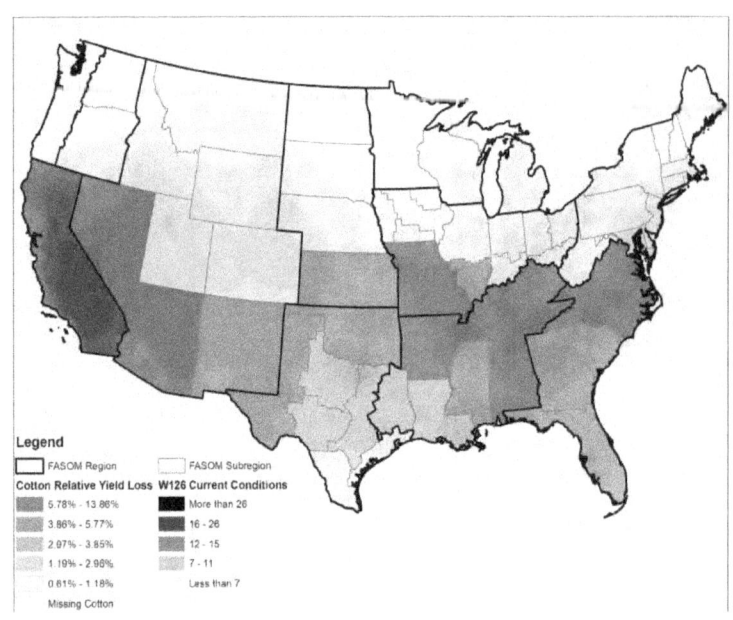

Figure A-6. Map of Cotton RYLs under Current Conditions

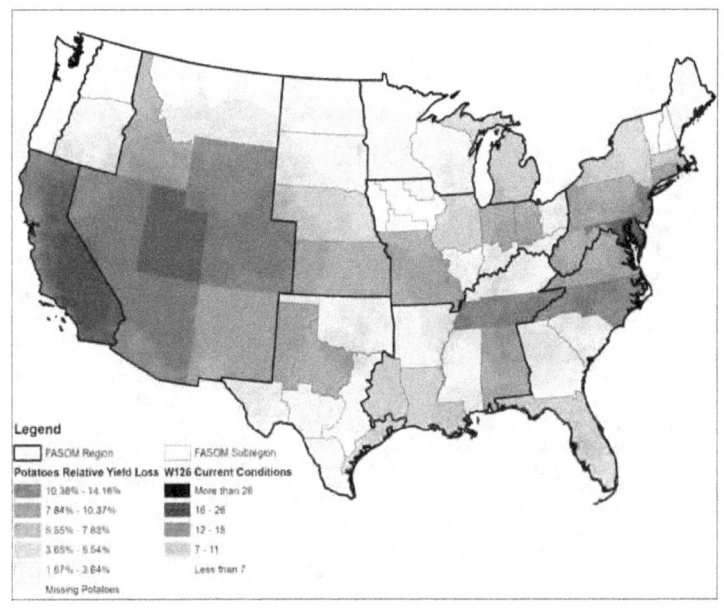

Figure A-7. Map of Potato RYLs under Current Conditions

As noted earlier, to implement the examination of scenarios with alternative ozone impacts, we use the RYL differences to calculate RYGs. Among the major crops, winter wheat and soybeans are more sensitive to ambient ozone concentration levels than corn and sorghum—as indicated in Figures A-2 through A-7—which implies that they would benefit more from ozone control in terms of RYGs. Correspondingly, FASOMGHG crops mapped to winter wheat or soybeans as proxy crops would have larger RYGs than other crops in general.

For RYLs under alternative ozone levels, one can expect that, in general, the yield losses would become much smaller compared with the current ozone standard, because of the substantial ozone reductions associated with meeting that standard (as presented in Figure A-1). This is shown to be the case in the primary analyses presented in the main body of this report.

The magnitude of RYGs essentially depends on two factors: (1) the sensitivity of the (proxy) crop to its ambient ozone concentration level, and (2) the difference between the ozone levels being compared. In this appendix, we focus on comparing current conditions and the "alternative" ozone levels defined by the standards considered. For FASOMGHG subregions such as Minnesota and North Dakota, the RYG estimates are virtually zero because the room for air quality improvement in these subregions is limited.

Figures A-8 through A-13 show the RYGs for major crops under alternative scenarios relative to current conditions. RYGs for California are generally much larger than in other regions, which largely reflects the significant room for improvement in ozone concentrations in the Pacific Southwest region. Some major crops, including corn and sorghum, are estimated to incur less positive effects from the improved ozone environments because of their relatively moderate sensitivities to ambient ozone concentration levels. Note that in many cases, subregions that show no change in yield for a given crop have no production of that crop in that subregion in FASOMGHG. For instance, soybeans are relatively sensitive to ozone concentrations and there are large reductions in ozone in California, but there are no impacts on soybean yields in that region because no soybeans are produced in California in FASOMGHG.

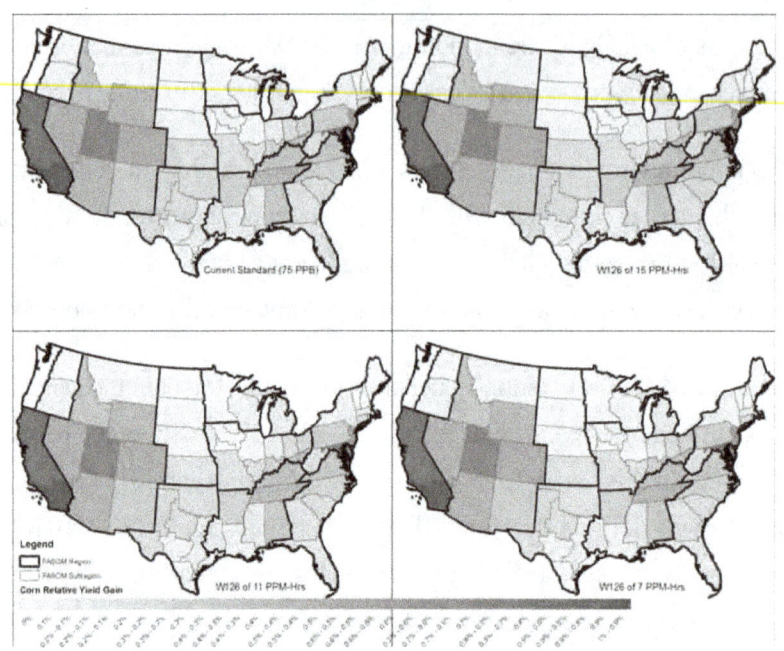

Figure A-8.   Map of Corn RYGs under Alternative Scenarios

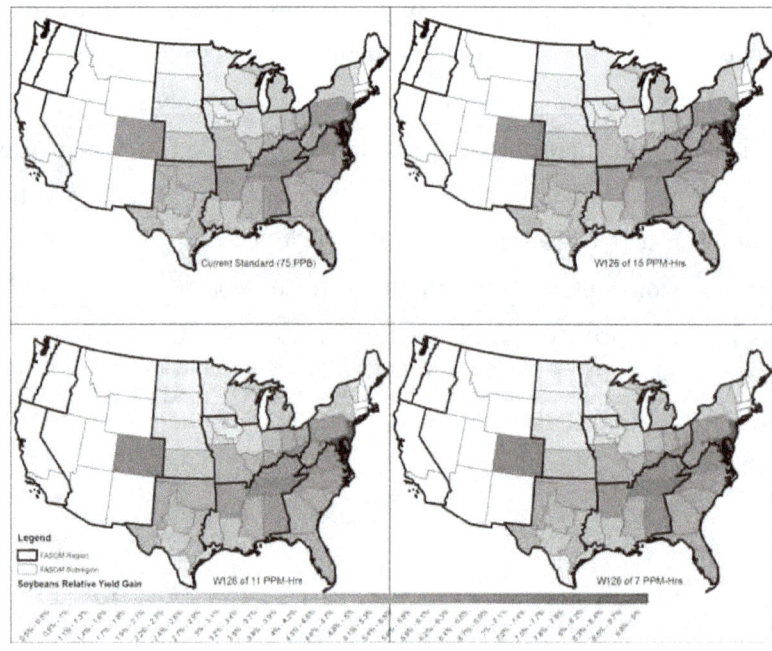

Figure A-9.   Map of Soybean RYGs under Alternative Scenarios

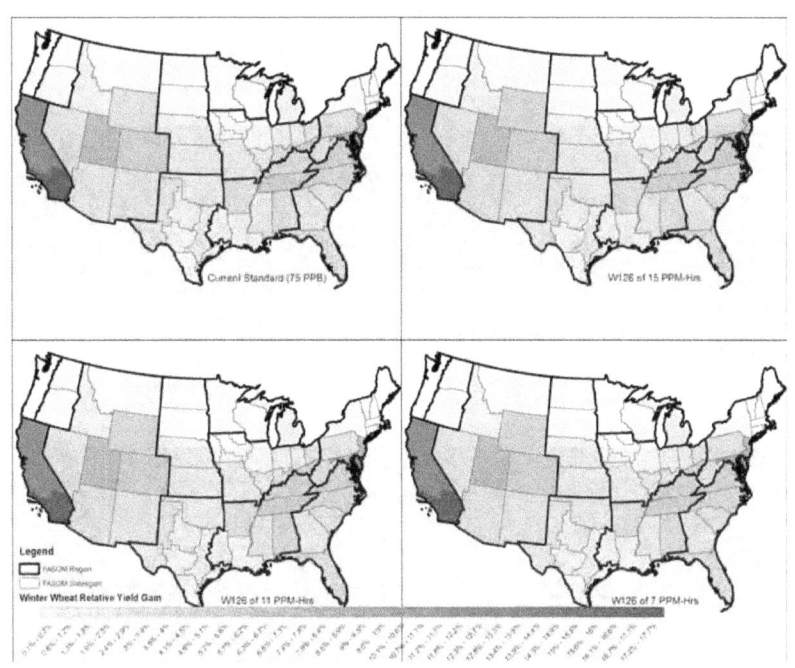

Figure A-10. Map of Wheat RYGs under Alternative Scenarios

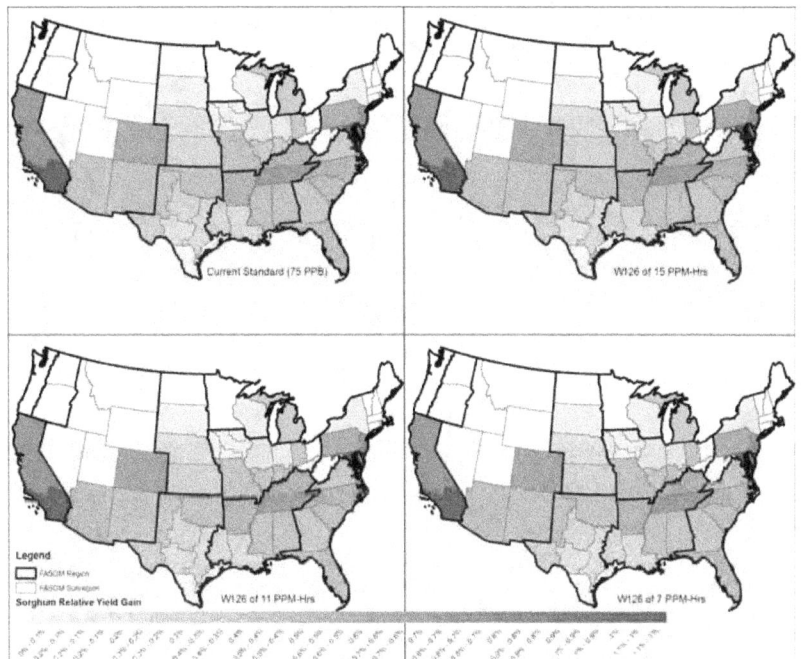

Figure A-11. Map of Sorghum RYGs under Alternative Scenarios

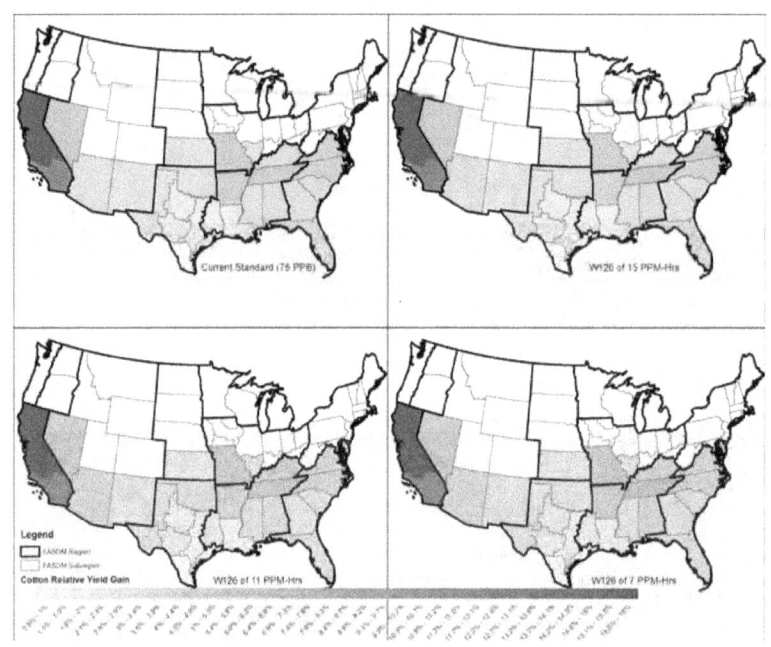

Figure A-12. Map of Cotton RYGs under Alternative Scenarios

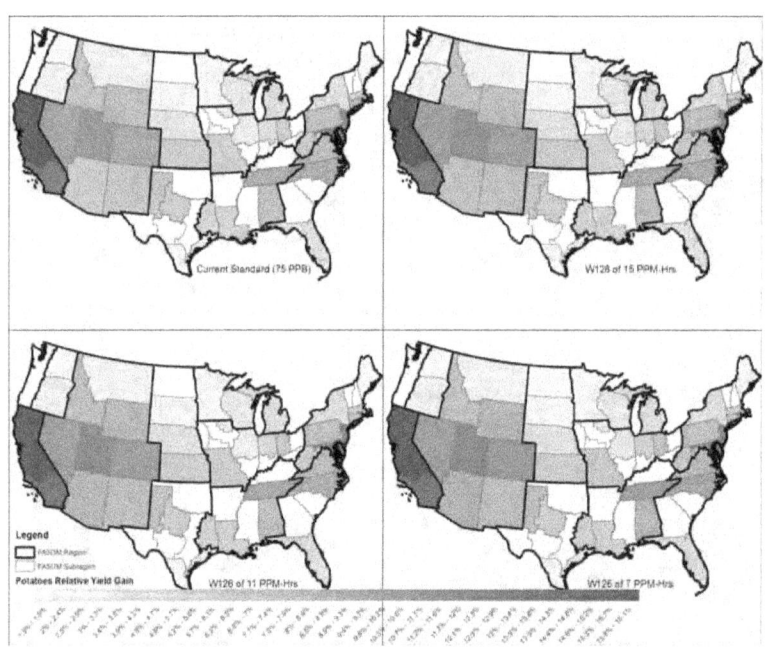

Figure A-13. Map of Potato RYGs under Alternative Scenarios

The region-specific RYLs for softwood and hardwood forest types from which the RYGs were derived are presented in Figures A-14 and A-15, respectively. Black cherry in the South is the most sensitive of the tree species examined. FASOMGHG assumes that there is no substantial forest products production in the Great Plains or Southwest regions, so forests in those regions are not modeled.

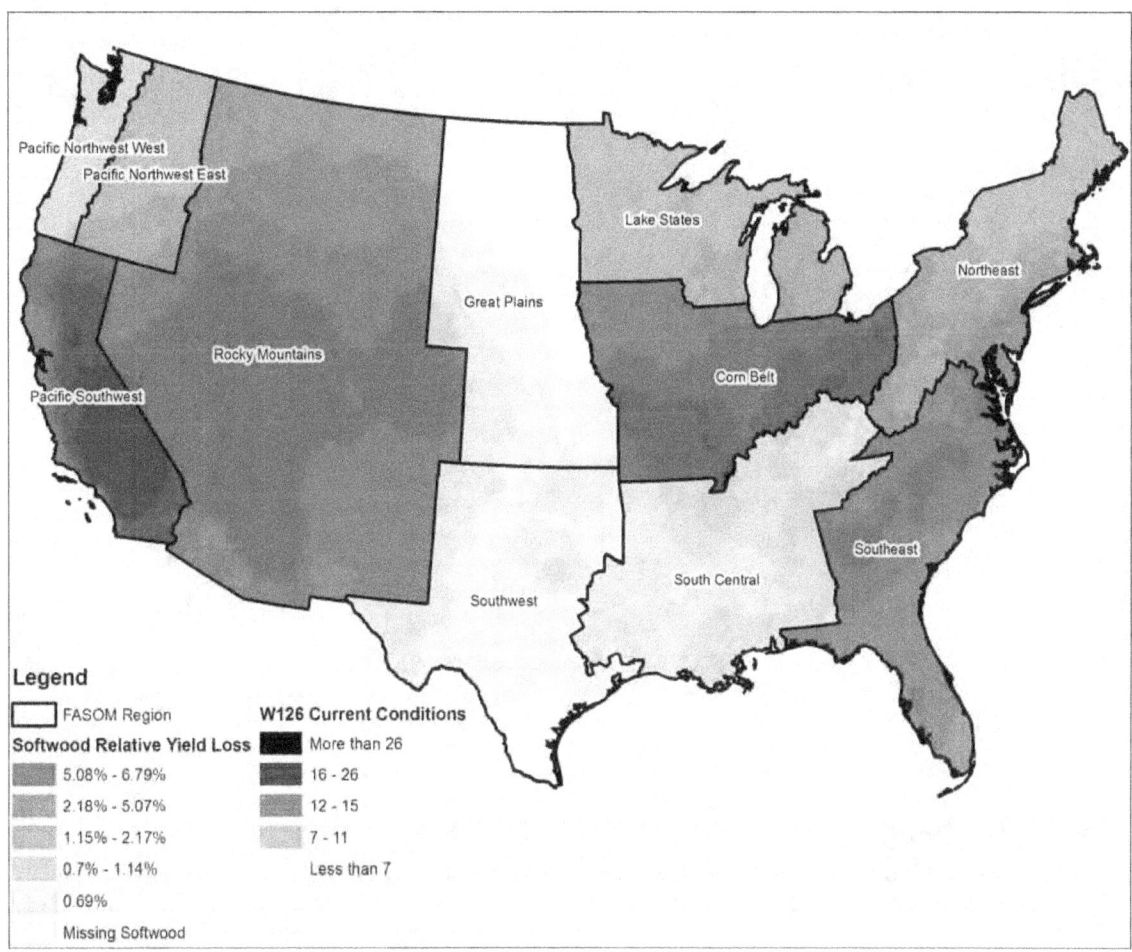

Figure A-14. Map of Softwood RYLs under Current Conditions

Note: RYL displayed in the Pacific Northwest—West region is for softwoods excluding Douglas fir. Douglas fir is
estimated to have no RYL associated with current ozone conditions.

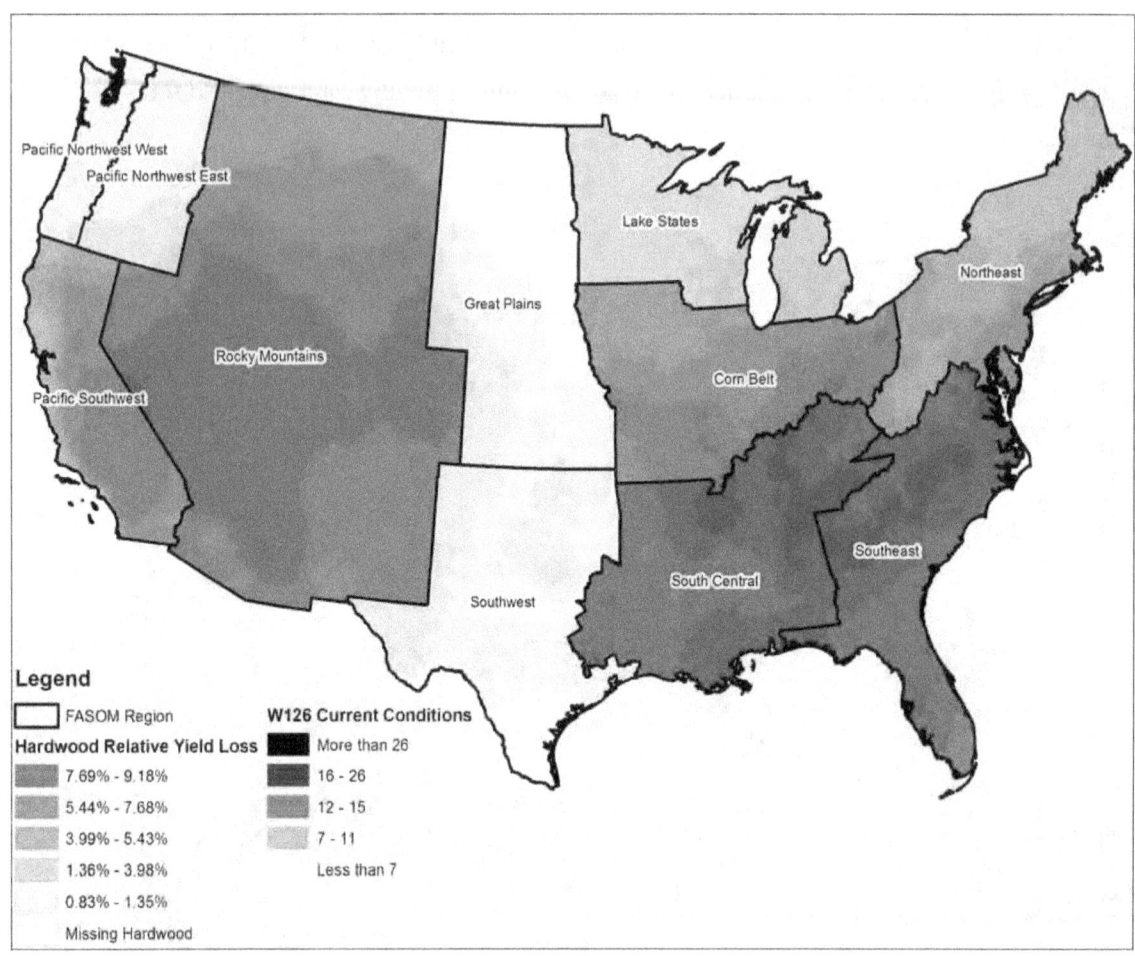

Figure A-15. Map of Hardwood RYLs under Current Conditions

Note: RYL displayed in the Southeast and South Central regions is an average of upland and bottomland hardwood
   RYL (15.5% and 1.08%, respectively, in South Central and 17.13% and 1.23%, respectively, in Southeast). These
   forest types are aggregated within FASOMGHG in other regions, consistent with the level of detail available from
   U.S. Forest Service data.

Figures A-16 and A-17 present the derived RYG estimates for FASOMGHG forest types
relative to current conditions. The upland hardwood forests in the South Central and Southeast
regions have the largest RYGs among the various forest types. In addition, softwood and
hardwood forests in the Pacific Southwest region would incur relatively larger yield increases
than other forest types.

As for the agricultural sector, the differences between RYGs under alternative scenarios
are generally small. See the results presented in Sections 5 and 6 of the main body of this report
for more information comparing impacts across scenarios with reductions in ozone
concentrations.

Figure A-16. Map of Softwood RYGs under Alternative Scenarios

Figure A-17. Map of Hardwood RYGs under Alternative Scenarios

A.3 Detailed Data on RYLs and RYGs for Crops and Forest Types

Tables A-2 through A-13 display the RYLs and RYGs for major crops. The bar charts in Figures A-18 through A-29 accompany the tables and present additional visuals for the RYLs and RYGs.

Table A-2. Corn RYLs under Alternative Scenarios (% change)

FASOMGHG Subregion	Current Conditions	75 ppb	15 ppm-hr	11 ppm-hr	7 ppm-hr
Alabama	0.31	0.00	0.00	0.00	0.00
Arizona	0.41	0.03	0.01	0.00	0.00
Arkansas	0.25	0.00	0.00	0.00	0.00
CaliforniaN	0.74	0.00	0.00	0.00	0.00
CaliforniaS	0.88	0.00	0.00	0.00	0.00
Colorado	0.45	0.01	0.01	0.00	0.00
Connecticut	0.17	0.00	0.00	0.00	0.00
Delaware	0.56	0.00	0.00	0.00	0.00
Florida	0.14	0.00	0.00	0.00	0.00
Georgia	0.16	0.00	0.00	0.00	0.00
Idaho	0.30	0.00	0.00	0.00	0.00
IllinoisN	0.05	0.00	0.00	0.00	0.00
IllinoisS	0.15	0.01	0.01	0.00	0.00
IndianaN	0.12	0.01	0.01	0.00	0.00
IndianaS	0.23	0.01	0.01	0.00	0.00
IowaW	0.02	0.00	0.00	0.00	0.00
IowaCent	0.02	0.00	0.00	0.00	0.00
IowaNE	0.03	0.00	0.00	0.00	0.00
IowaS	0.03	0.00	0.00	0.00	0.00
Kansas	0.17	0.00	0.00	0.00	0.00
Kentucky	0.32	0.01	0.01	0.00	0.00
Louisiana	0.07	0.00	0.00	0.00	0.00
Maine	0.01	0.00	0.00	0.00	0.00
Maryland	0.68	0.00	0.00	0.00	0.00
Massachusetts	0.10	0.00	0.00	0.00	0.00
Michigan	0.10	0.00	0.00	0.00	0.00
Minnesota	0.02	0.00	0.00	0.00	0.00
Mississippi	0.14	0.00	0.00	0.00	0.00
Missouri	0.17	0.01	0.01	0.00	0.00
Montana	0.05	0.00	0.00	0.00	0.00
Nebraska	0.06	0.00	0.00	0.00	0.00

(continued)

Table A-2. Corn RYLs under Alternative Scenarios (% change) (continued)

FASOMGHG Subregion	Current Conditions	75 ppb	15 ppm-hr	11 ppm-hr	7 ppm-hr
Nevada	0.42	0.00	0.00	0.00	0.00
NewHampshire	0.02	0.00	0.00	0.00	0.00
NewJersey	0.48	0.00	0.00	0.00	0.00
NewMexico	0.26	0.00	0.00	0.00	0.00
NewYork	0.07	0.00	0.00	0.00	0.00
NorthCarolina	0.25	0.00	0.00	0.00	0.00
NorthDakota	0.01	0.00	0.00	0.00	0.00
OhioNW	0.20	0.01	0.01	0.00	0.00
OhioS	0.34	0.01	0.01	0.00	0.00
OhioNE	0.23	0.00	0.00	0.00	0.00
Oklahoma	0.20	0.00	0.00	0.00	0.00
Oregon	0.04	0.00	0.00	0.00	0.00
Pennsylvania	0.26	0.00	0.00	0.00	0.00
RhodeIsland	0.17	0.00	0.00	0.00	0.00
SouthCarolina	0.18	0.00	0.00	0.00	0.00
SouthDakota	0.02	0.00	0.00	0.00	0.00
Tennessee	0.39	0.01	0.01	0.00	0.00
TxHiPlains	0.22	0.00	0.00	0.00	0.00
TxRolingPl	0.18	0.00	0.00	0.00	0.00
TxCntBlack	0.10	0.00	0.00	0.00	0.00
TxEast	0.11	0.00	0.00	0.00	0.00
TxEdplat	0.07	0.00	0.00	0.00	0.00
TxCoastBe	0.04	0.00	0.00	0.00	0.00
TxSouth	0.01	0.00	0.00	0.00	0.00
TxTranspec	0.18	0.01	0.00	0.00	0.00
Utah	0.67	0.01	0.01	0.00	0.00
Vermont	0.02	0.00	0.00	0.00	0.00
Virginia	0.30	75 ppb	15 ppm-hr	11 ppm-hr	7 ppm-hr
Washington	0.01	0.00	0.00	0.00	0.00
WestVirginia	0.16	0.00	0.00	0.00	0.00
Wisconsin	0.04	0.00	0.00	0.00	0.00
Wyoming	0.39	0.01	0.01	0.00	0.00

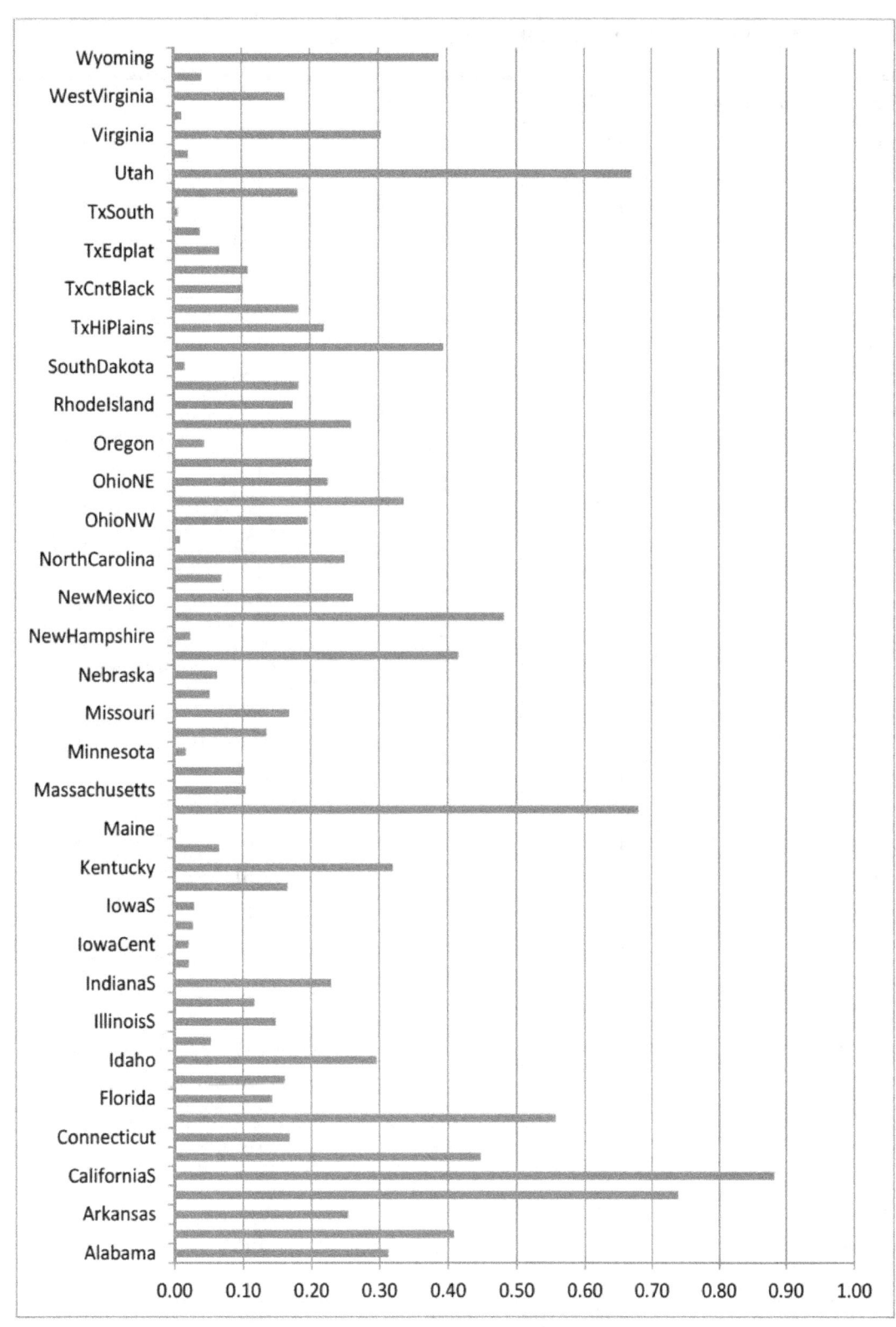

Figure A-18. Corn RYLs under Current Conditions (% change)

Table A-3. Corn RYGs under Alternative Scenarios (% change)

FASOMGHG Subregion	75 ppb	15 ppm-hr	11 ppm-hr	7 ppm-hr
Alabama	0.31	0.31	0.31	0.31
Arizona	0.38	0.40	0.41	0.41
Arkansas	0.25	0.25	0.25	0.25
CaliforniaN	0.74	0.74	0.74	0.74
CaliforniaS	0.89	0.89	0.89	0.89
Colorado	0.44	0.44	0.45	0.45
Connecticut	0.17	0.17	0.17	0.17
Delaware	0.56	0.56	0.56	0.56
Florida	0.14	0.14	0.14	0.14
Georgia	0.16	0.16	0.16	0.16
Idaho	0.29	0.29	0.30	0.30
IllinoisN	0.05	0.05	0.05	0.05
IllinoisS	0.14	0.14	0.15	0.15
IndianaN	0.11	0.11	0.11	0.12
IndianaS	0.22	0.22	0.23	0.23
IowaW	0.02	0.02	0.02	0.02
IowaCent	0.02	0.02	0.02	0.02
IowaNE	0.03	0.03	0.03	0.03
IowaS	0.03	0.03	0.03	0.03
Kansas	0.16	0.16	0.17	0.17
Kentucky	0.31	0.31	0.32	0.32
Louisiana	0.07	0.07	0.07	0.07
Maine	0.01	0.01	0.01	0.01
Maryland	0.69	0.69	0.69	0.69
Massachusetts	0.10	0.10	0.10	0.10
Michigan	0.10	0.10	0.10	0.10
Minnesota	0.02	0.02	0.02	0.02
Mississippi	0.14	0.14	0.14	0.14
Missouri	0.16	0.16	0.17	0.17
Montana	0.05	0.05	0.05	0.05
Nebraska	0.06	0.06	0.06	0.06
Nevada	0.41	0.42	0.42	0.42
NewHampshire	0.02	0.02	0.02	0.02
NewJersey	0.48	0.48	0.48	0.48
NewMexico	0.26	0.26	0.26	0.26
NewYork	0.07	0.07	0.07	0.07
NorthCarolina	0.25	0.25	0.25	0.25

(continued)

Table A-3. Corn RYGs under Alternative Scenarios (% change) (continued)

FASOMGHG Subregion	75 ppb	15 ppm-hr	11 ppm-hr	7 ppm-hr
NorthDakota	0.01	0.01	0.01	0.01
OhioNW	0.19	0.19	0.19	0.20
OhioS	0.33	0.33	0.34	0.34
OhioNE	0.22	0.22	0.22	0.23
Oklahoma	0.20	0.20	0.20	0.20
Oregon	0.04	0.04	0.04	0.04
Pennsylvania	0.26	0.26	0.26	0.26
RhodeIsland	0.17	0.17	0.17	0.17
SouthCarolina	0.18	0.18	0.18	0.18
SouthDakota	0.02	0.02	0.02	0.02
Tennessee	0.39	0.39	0.39	0.40
TxHiPlains	0.22	0.22	0.22	0.22
TxRolingPl	0.18	0.18	0.18	0.18
TxCntBlack	0.10	0.10	0.10	0.10
TxEast	0.11	0.11	0.11	0.11
TxEdplat	0.07	0.07	0.07	0.07
TxCoastBe	0.04	0.04	0.04	0.04
TxSouth	0.01	0.01	0.01	0.01
TxTranspec	0.18	0.18	0.18	0.18
Utah	0.66	0.67	0.67	0.67
Vermont	0.02	0.02	0.02	0.02
Virginia	0.30	0.30	0.30	0.30
Washington	0.01	0.01	0.01	0.01
WestVirginia	0.16	0.16	0.16	0.16
Wisconsin	0.04	0.04	0.04	0.04
Wyoming	0.38	0.38	0.38	0.39

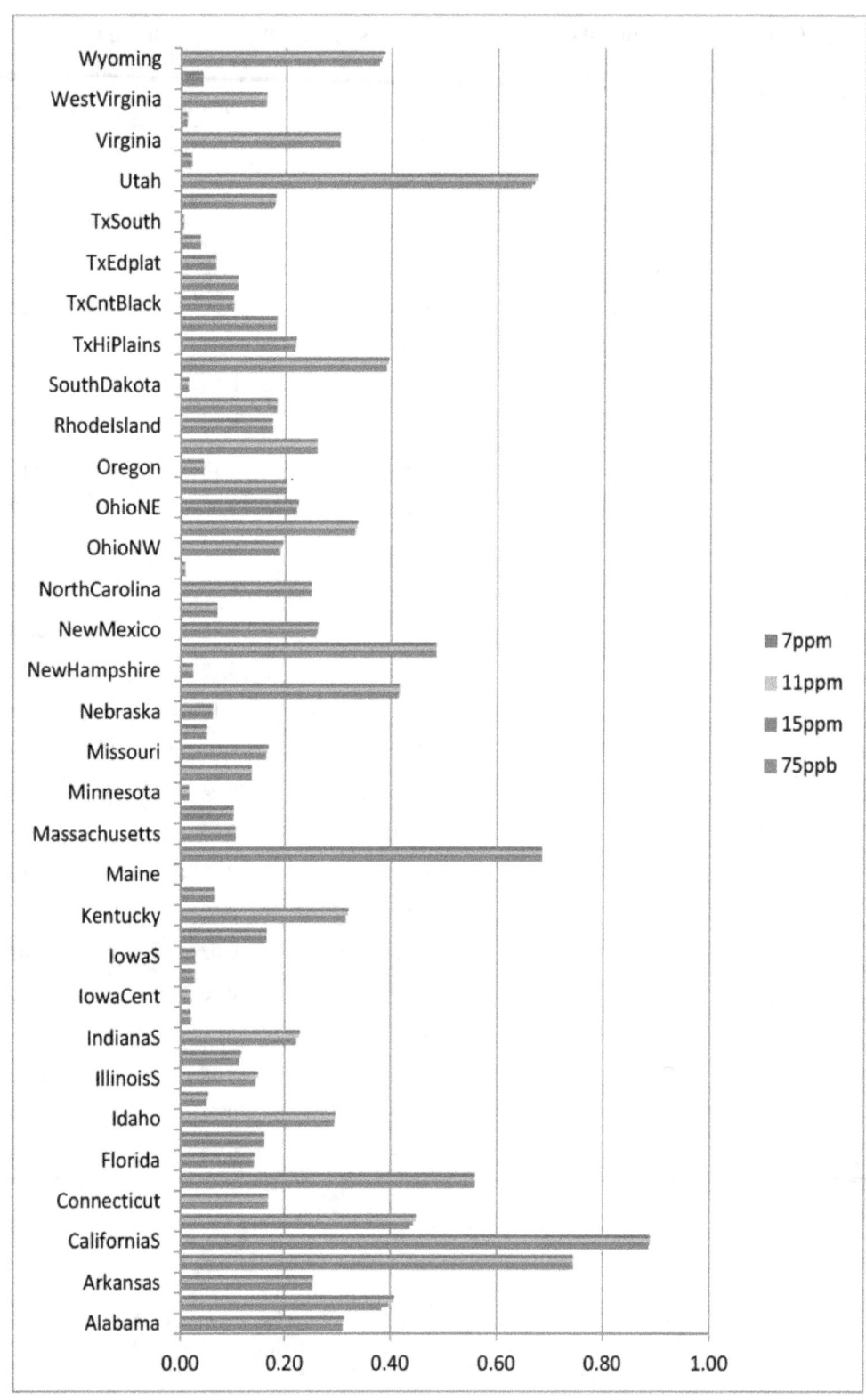

Figure A-19. Corn RYGs under Alternative Scenarios (% change)

Table A-4. Soybean RYLs under Alternative Scenarios (% change)

FASOMGHG Subregion	Current Conditions	75 ppb	15 ppm-hr	11 ppm-hr	7 ppm-hr
Alabama	6.15	0.88	0.88	0.75	0.42
Arkansas	5.68	0.60	0.60	0.51	0.44
Colorado	6.73	1.39	1.05	0.65	0.45
Delaware	7.65	0.04	0.04	0.04	0.04
Florida	4.84	0.75	0.75	0.68	0.40
Georgia	4.57	0.53	0.53	0.48	0.28
IllinoisN	2.69	0.79	0.79	0.56	0.34
IllinoisS	4.28	0.94	0.94	0.58	0.29
IndianaN	3.91	0.94	0.94	0.62	0.34
IndianaS	5.10	1.13	1.13	0.69	0.34
IowaW	1.72	0.36	0.36	0.35	0.32
IowaCent	1.70	0.18	0.18	0.18	0.17
IowaNE	1.93	0.10	0.10	0.10	0.10
IowaS	2.03	0.28	0.28	0.24	0.20
Kansas	3.41	0.43	0.42	0.38	0.35
Kentucky	5.95	1.05	1.05	0.64	0.32
Louisiana	3.45	0.19	0.19	0.19	0.19
Maine	0.69	0.03	0.03	0.03	0.03
Maryland	8.30	0.09	0.09	0.08	0.06
Michigan	3.55	0.57	0.57	0.54	0.51
Minnesota	1.49	0.19	0.19	0.19	0.18
Mississippi	4.40	0.29	0.29	0.28	0.27
Missouri	4.37	0.91	0.91	0.59	0.33
Nebraska	2.10	0.46	0.44	0.41	0.36
NewJersey	7.76	0.03	0.03	0.03	0.03
NewYork	2.96	0.01	0.01	0.01	0.01
NorthCarolina	5.35	0.37	0.37	0.34	0.20
NorthDakota	1.14	0.30	0.30	0.30	0.25
OhioNW	4.70	1.02	1.02	0.68	0.39
OhioS	6.20	0.99	0.99	0.61	0.31
OhioNE	5.00	0.80	0.80	0.52	0.29
Oklahoma	4.52	0.31	0.31	0.30	0.30
Pennsylvania	5.38	0.06	0.06	0.06	0.05
SouthCarolina	4.63	0.32	0.32	0.29	0.18
SouthDakota	1.51	0.37	0.37	0.37	0.31
Tennessee	6.60	0.95	0.95	0.60	0.32

(continued)

Table A-4. Soybean RYLs under Alternative Scenarios (% change) (continued)

FASOMGHG Subregion	Current Conditions	75 ppb	15 ppm-hr	11 ppm-hr	7 ppm-hr
TxHiPlains	4.72	0.60	0.47	0.33	0.30
TxRolingPl	4.06	0.43	0.41	0.39	0.38
TxCntBlack	3.94	0.35	0.35	0.35	0.35
TxEast	3.33	0.19	0.19	0.19	0.19
TxEdplat	3.28	0.40	0.37	0.35	0.34
TxCoastBe	2.57	0.38	0.38	0.38	0.38
TxSouth	0.69	0.20	0.20	0.20	0.20
TxTranspec	4.61	0.92	0.79	0.63	0.59
Vermont	1.81	0.03	0.03	0.03	0.03
Virginia	6.04	0.44	0.44	0.40	0.23
WestVirginia	4.98	0.58	0.58	0.38	0.18
Wisconsin	2.39	0.27	0.27	0.26	0.25

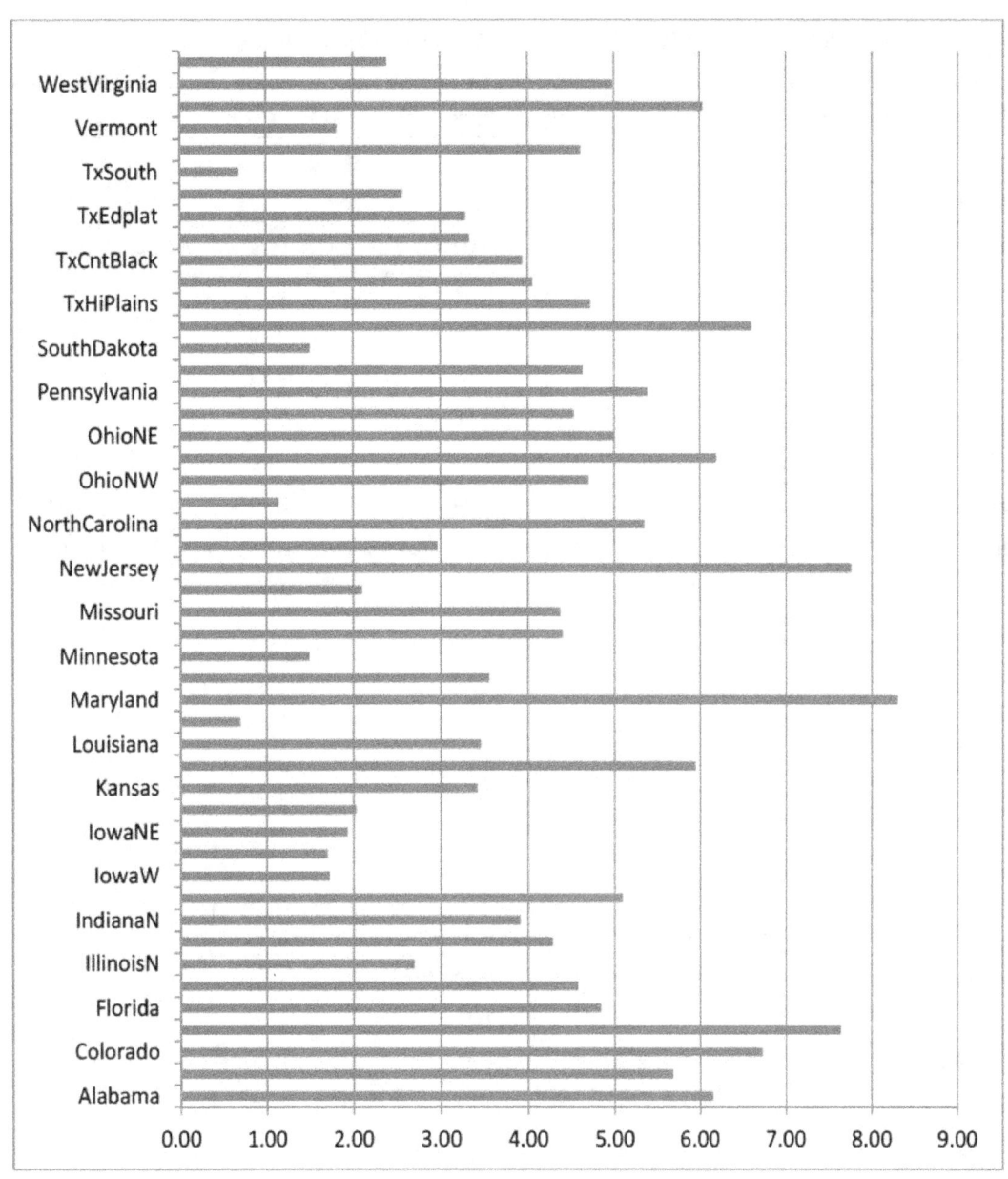

Figure A-20. Soybean RYLs under Current Conditions (% change)

Table A-5. Soybean RYGs under Alternative Scenarios (% change)

FASOMGHG Subregion	75 ppb	15 ppm-hr	11 ppm-hr	7 ppm-hr
Alabama	5.61	5.61	5.75	6.11
Arkansas	5.38	5.38	5.48	5.56
Colorado	5.72	6.09	6.52	6.73
Delaware	8.23	8.23	8.23	8.24
Florida	4.29	4.29	4.36	4.66
Georgia	4.24	4.24	4.29	4.49
IllinoisN	1.96	1.96	2.20	2.42
IllinoisS	3.49	3.49	3.86	4.16
IndianaN	3.09	3.09	3.42	3.71
IndianaS	4.17	4.17	4.64	5.01
IowaW	1.39	1.39	1.39	1.43
IowaCent	1.55	1.55	1.55	1.56
IowaNE	1.86	1.86	1.86	1.86
IowaS	1.78	1.78	1.83	1.87
Kansas	3.09	3.10	3.14	3.17
Kentucky	5.21	5.21	5.64	5.99
Louisiana	3.38	3.38	3.38	3.38
Maine	0.66	0.66	0.66	0.66
Maryland	8.95	8.95	8.96	8.98
Michigan	3.09	3.09	3.12	3.15
Minnesota	1.32	1.32	1.32	1.34
Mississippi	4.30	4.30	4.31	4.31
Missouri	3.62	3.62	3.96	4.22
Nebraska	1.68	1.70	1.73	1.78
NewJersey	8.37	8.37	8.37	8.37
NewYork	3.04	3.04	3.04	3.04
NorthCarolina	5.26	5.26	5.30	5.45
NorthDakota	0.85	0.85	0.85	0.90
OhioNW	3.86	3.86	4.21	4.52
OhioS	5.55	5.55	5.96	6.28
OhioNE	4.42	4.42	4.71	4.96
Oklahoma	4.41	4.41	4.42	4.42
Pennsylvania	5.62	5.62	5.63	5.64

(continued)

Table A-5. Soybean RYGs under Alternative Scenarios (% change) (continued)

FASOMGHG Subregion	75 ppb	15 ppm-hr	11 ppm-hr	7 ppm-hr
SouthCarolina	4.51	4.51	4.54	4.67
SouthDakota	1.15	1.15	1.16	1.21
Tennessee	6.05	6.05	6.43	6.73
TxHiPlains	4.33	4.46	4.61	4.64
TxRolingPl	3.77	3.80	3.83	3.83
TxCntBlack	3.73	3.73	3.73	3.73
TxEast	3.25	3.25	3.25	3.25
TxEdplat	2.98	3.00	3.03	3.04
TxCoastBe	2.25	2.25	2.25	2.25
TxSouth	0.49	0.49	0.49	0.49
TxTranspec	3.86	4.00	4.17	4.21
Vermont	1.82	1.82	1.82	1.82
Virginia	5.96	5.96	6.01	6.18
WestVirginia	4.63	4.63	4.85	5.06
Wisconsin	2.18	2.18	2.18	2.19

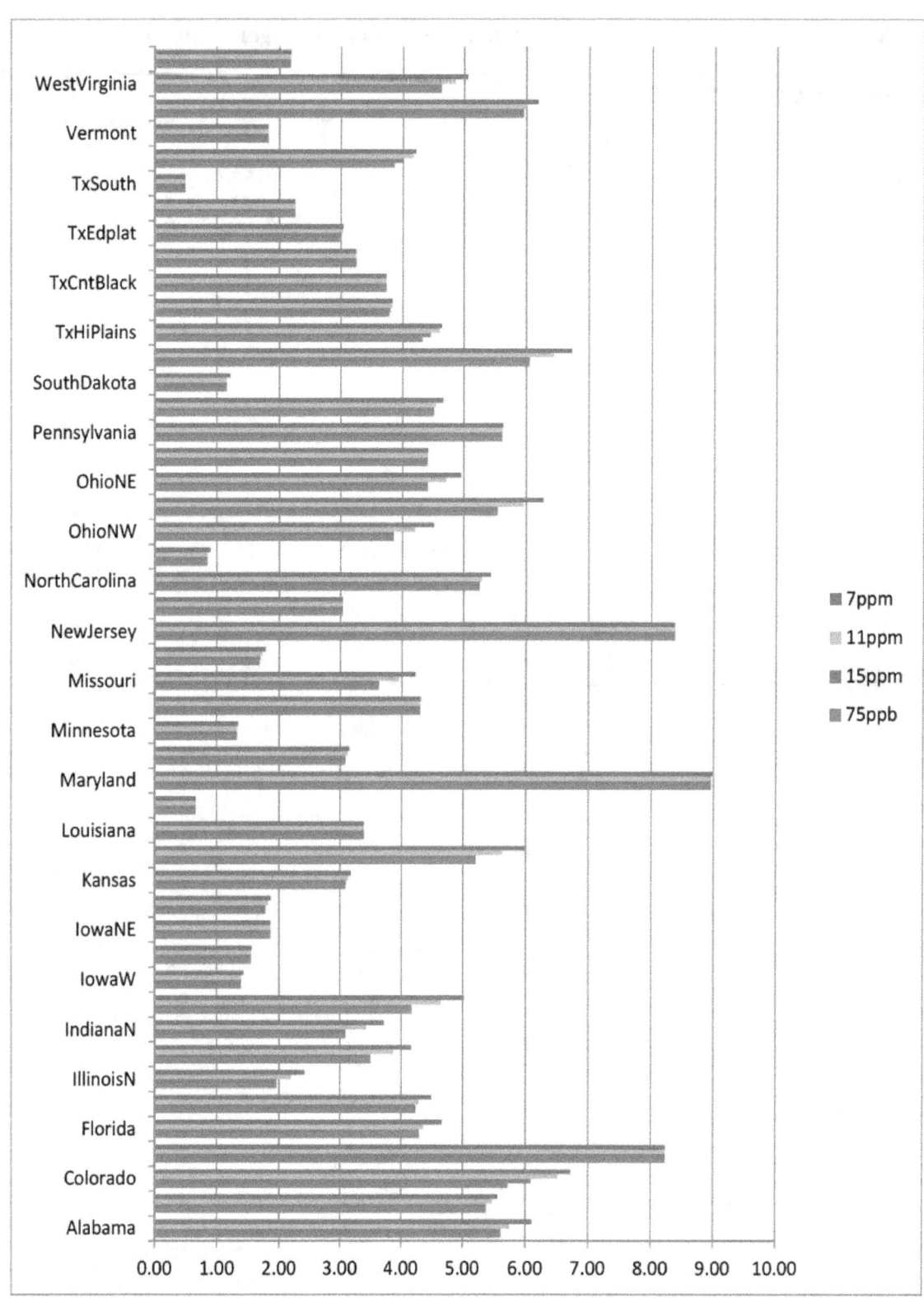

Figure A-21. Soybean RYGs under Alternative Scenarios (% change)

Table A-6. Winter Wheat RYLs under Alternative Scenarios (% change)

FASOMGHG Subregion	Current Conditions	75 ppb	15 ppm-hr	11 ppm-hr	7 ppm-hr
Alabama	3.95	0.14	0.14	0.11	0.04
Arizona	3.59	0.30	0.16	0.06	0.04
Arkansas	3.57	0.06	0.06	0.05	0.04
CaliforniaN	10.95	0.02	0.01	0.01	0.01
CaliforniaS	15.05	0.16	0.10	0.06	0.03
Colorado	5.31	0.29	0.17	0.07	0.04
Delaware	6.43	0.00	0.00	0.00	0.00
Florida	3.51	0.19	0.19	0.16	0.06
Georgia	2.68	0.05	0.05	0.05	0.02
Idaho	2.08	0.09	0.06	0.04	0.03
IllinoisN	0.97	0.11	0.11	0.06	0.03
IllinoisS	2.65	0.18	0.18	0.08	0.02
IndianaN	2.01	0.15	0.15	0.07	0.03
IndianaS	3.20	0.22	0.22	0.09	0.03
IowaW	0.46	0.05	0.05	0.05	0.04
IowaCent	0.56	0.02	0.02	0.02	0.01
IowaS	0.84	0.04	0.04	0.02	0.01
Kansas	2.44	0.07	0.05	0.04	0.03
Kentucky	4.41	0.17	0.17	0.07	0.02
Louisiana	1.49	0.01	0.01	0.01	0.01
Maryland	7.65	0.00	0.00	0.00	0.00
Michigan	1.50	0.05	0.05	0.05	0.05
Minnesota	0.26	0.01	0.01	0.01	0.01
Mississippi	2.65	0.02	0.02	0.02	0.02
Missouri	2.83	0.16	0.16	0.07	0.02
Montana	0.66	0.05	0.05	0.05	0.03
Nebraska	2.64	0.18	0.14	0.09	0.05
Nevada	2.93	0.06	0.05	0.04	0.04
NewJersey	6.55	0.00	0.00	0.00	0.00
NewMexico	3.26	0.12	0.07	0.03	0.02
NewYork	1.21	0.00	0.00	0.00	0.00
NorthCarolina	3.86	0.04	0.04	0.03	0.01
NorthDakota	0.28	0.04	0.04	0.04	0.03

(continued)

Table A-6. Winter Wheat RYLs under Alternative Scenarios (% change) (continued)

FASOMGHG Subregion	Current Conditions	75 ppb	15 ppm-hr	11 ppm-hr	7 ppm-hr
OhioNW	2.40	0.15	0.15	0.08	0.03
OhioS	4.16	0.15	0.15	0.07	0.02
OhioNE	2.91	0.11	0.11	0.05	0.02
Oklahoma	2.59	0.05	0.04	0.03	0.03
Oregon	0.41	0.05	0.05	0.04	0.04
Pennsylvania	3.69	0.00	0.00	0.00	0.00
SouthCarolina	2.90	0.02	0.02	0.02	0.01
SouthDakota	0.60	0.07	0.07	0.06	0.04
Tennessee	5.03	0.16	0.16	0.07	0.02
TxHiPlains	2.93	0.08	0.05	0.02	0.02
TxRolingPl	1.95	0.04	0.04	0.03	0.03
TxCntBlack	2.25	0.04	0.04	0.04	0.04
TxEast	1.55	0.01	0.01	0.01	0.01
TxEdplat	1.51	0.03	0.03	0.03	0.03
TxCoastBe	0.64	0.02	0.02	0.02	0.02
TxSouth	0.15	0.01	0.01	0.01	0.01
TxTranspec	2.81	0.13	0.09	0.05	0.04
Utah	7.26	0.30	0.16	0.05	0.04
Vermont	0.53	0.00	0.00	0.00	0.00
Virginia	4.56	0.04	0.04	0.03	0.01
Washington	0.36	0.04	0.03	0.03	0.03
WestVirginia	2.97	0.04	0.04	0.02	0.01
Wisconsin	0.95	0.03	0.03	0.03	0.03
Wyoming	4.60	0.32	0.24	0.16	0.09

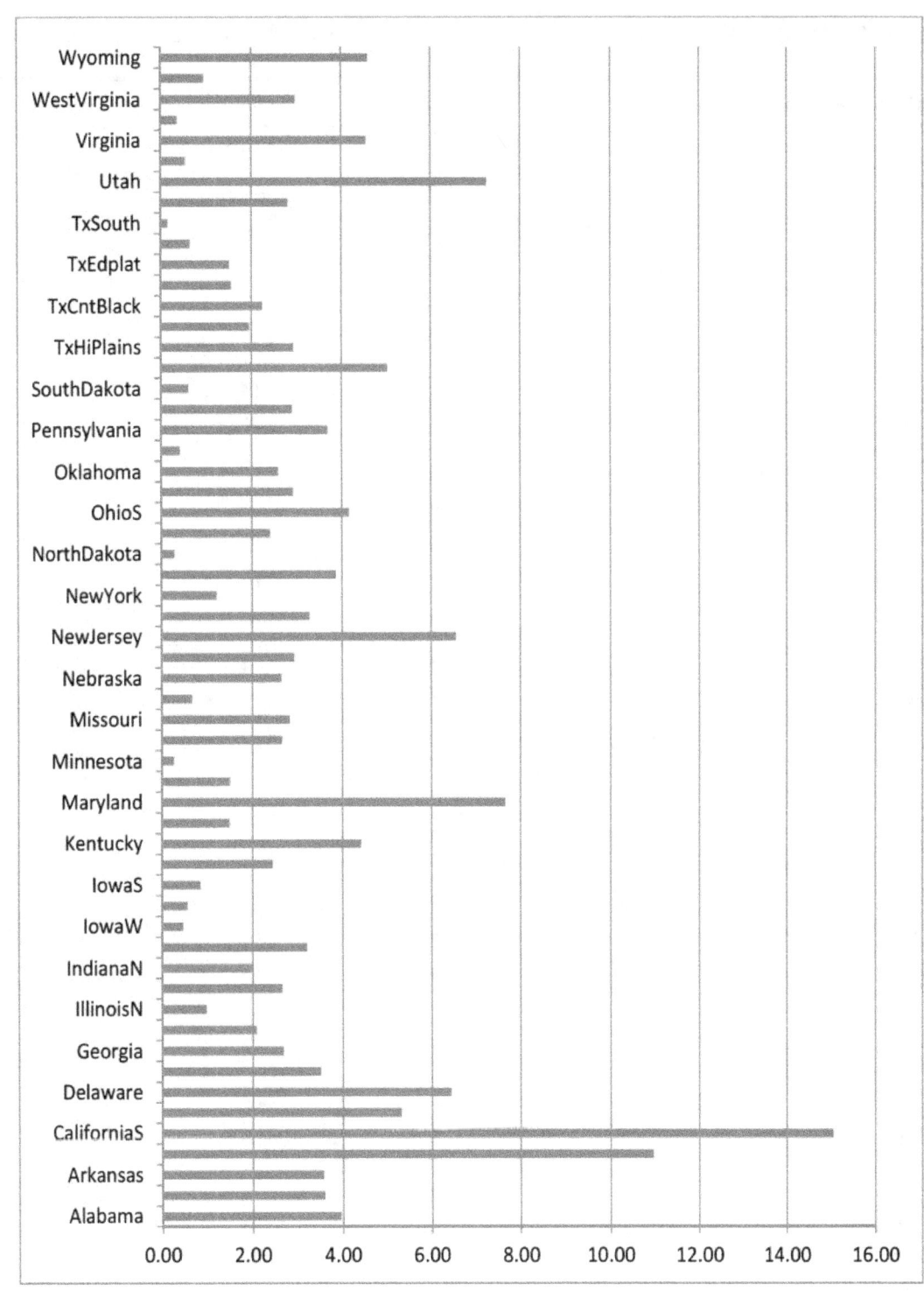

Figure A-22. Winter Wheat RYLs under Current Conditions (% change)

Table A-7. Winter Wheat RYGs under Alternative Scenarios (% change)

FASOMGHG Subregion	75 ppb	15 ppm-hr	11 ppm-hr	7 ppm-hr
Alabama	3.97	3.97	4.01	4.08
Arizona	3.41	3.55	3.66	3.68
Arkansas	3.64	3.64	3.65	3.66
CaliforniaN	12.28	12.28	12.29	12.29
CaliforniaS	17.53	17.59	17.65	17.68
Colorado	5.30	5.43	5.54	5.57
Delaware	6.88	6.88	6.88	6.88
Florida	3.44	3.44	3.47	3.57
Georgia	2.69	2.69	2.70	2.73
Idaho	2.04	2.06	2.09	2.10
IllinoisN	0.88	0.88	0.92	0.96
IllinoisS	2.53	2.53	2.64	2.70
IndianaN	1.89	1.89	1.97	2.02
IndianaS	3.08	3.08	3.22	3.28
IowaW	0.41	0.41	0.42	0.43
IowaCent	0.54	0.54	0.54	0.55
IowaS	0.81	0.81	0.83	0.84
Kansas	2.43	2.45	2.47	2.47
Kentucky	4.44	4.44	4.55	4.60
Louisiana	1.50	1.50	1.50	1.50
Maryland	8.29	8.29	8.29	8.29
Michigan	1.47	1.47	1.47	1.47
Minnesota	0.25	0.25	0.25	0.25
Mississippi	2.70	2.70	2.70	2.70
Missouri	2.75	2.75	2.84	2.88
Montana	0.62	0.62	0.62	0.63
Nebraska	2.53	2.57	2.63	2.66
Nevada	2.96	2.96	2.97	2.98
NewJersey	7.01	7.01	7.01	7.01
NewMexico	3.25	3.30	3.35	3.35
NewYork	1.23	1.23	1.23	1.23
NorthCarolina	3.98	3.98	3.99	4.01
NorthDakota	0.24	0.24	0.24	0.25

(continued)

Table A-7. Winter Wheat RYGs under Alternative Scenarios (% change) (continued)

FASOMGHG Subregion	75 ppb	15 ppm-hr	11 ppm-hr	7 ppm-hr
OhioNW	2.30	2.30	2.38	2.43
OhioS	4.19	4.19	4.28	4.32
OhioNE	2.88	2.88	2.95	2.98
Oklahoma	2.61	2.61	2.62	2.62
Oregon	0.36	0.36	0.37	0.37
Pennsylvania	3.83	3.83	3.83	3.83
SouthCarolina	2.96	2.96	2.96	2.97
SouthDakota	0.54	0.54	0.54	0.56
Tennessee	5.12	5.12	5.22	5.27
TxHiPlains	2.94	2.97	2.99	3.00
TxRolingPl	1.95	1.95	1.96	1.96
TxCntBlack	2.25	2.25	2.25	2.25
TxEast	1.56	1.56	1.56	1.56
TxEdplat	1.50	1.50	1.51	1.51
TxCoastBe	0.62	0.62	0.62	0.62
TxSouth	0.14	0.14	0.14	0.14
TxTranspec	2.75	2.80	2.84	2.85
Utah	7.51	7.65	7.77	7.79
Vermont	0.54	0.54	0.54	0.54
Virginia	4.74	4.74	4.74	4.76
Washington	0.32	0.32	0.33	0.33
WestVirginia	3.01	3.01	3.03	3.05
Wisconsin	0.93	0.93	0.93	0.93
Wyoming	4.49	4.57	4.66	4.73

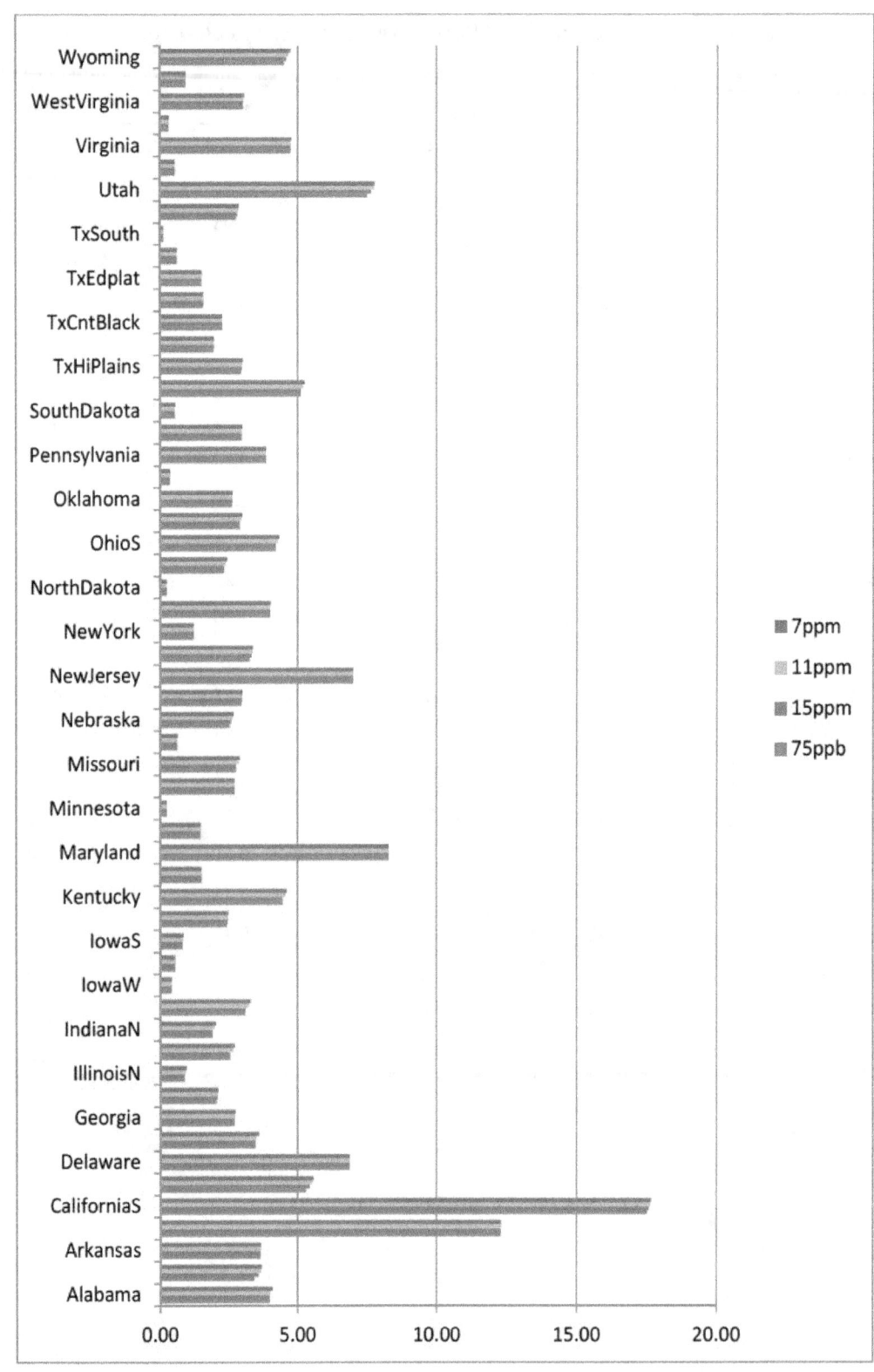

Figure A-23. Winter Wheat RYGs under Alternative Scenarios (% change)

Table A-8. Sorghum RYLs under Alternative Scenarios (% change)

FASOMGHG Subregion	Current Conditions	75 ppb	15 ppm-hr	11 ppm-hr	7 ppm-hr
Alabama	0.40	0.03	0.03	0.02	0.01
Arizona	0.47	0.08	0.04	0.02	0.01
Arkansas	0.50	0.02	0.02	0.01	0.01
CaliforniaN	0.67	0.00	0.00	0.00	0.00
CaliforniaS	1.03	0.03	0.02	0.01	0.01
Colorado	0.57	0.04	0.02	0.01	0.01
Delaware	0.74	0.00	0.00	0.00	0.00
Florida	0.45	0.04	0.04	0.04	0.02
Georgia	0.36	0.01	0.01	0.01	0.01
IllinoisN	0.17	0.02	0.02	0.01	0.01
IllinoisS	0.37	0.04	0.04	0.02	0.01
IndianaN	0.16	0.03	0.03	0.02	0.01
IndianaS	0.37	0.04	0.04	0.02	0.01
IowaW	0.11	0.01	0.01	0.01	0.01
IowaCent	0.10	0.00	0.00	0.00	0.00
IowaNE	0.10	0.00	0.00	0.00	0.00
IowaS	0.08	0.00	0.00	0.00	0.00
Kansas	0.29	0.02	0.01	0.01	0.01
Kentucky	0.51	0.04	0.04	0.02	0.01
Louisiana	0.23	0.00	0.00	0.00	0.00
Maryland	0.93	0.00	0.00	0.00	0.00
Michigan	0.35	0.02	0.02	0.02	0.02
Mississippi	0.39	0.01	0.01	0.01	0.01
Missouri	0.41	0.04	0.04	0.02	0.01
Nebraska	0.22	0.02	0.02	0.01	0.01
NewJersey	0.65	0.00	0.00	0.00	0.00
NewMexico	0.40	0.03	0.02	0.01	0.01
NewYork	0.13	0.00	0.00	0.00	0.00
NorthCarolina	0.48	0.01	0.01	0.01	0.00
OhioNW	0.39	0.06	0.06	0.03	0.01
Oklahoma	0.39	0.01	0.01	0.01	0.01
Pennsylvania	0.50	0.00	0.00	0.00	0.00
SouthCarolina	0.42	0.01	0.01	0.01	0.00

(continued)

Table A-8. Sorghum RYLs under Alternative Scenarios (% change) (continued)

FASOMGHG Subregion	Current Conditions	75 ppb	15 ppm-hr	11 ppm-hr	7 ppm-hr
SouthDakota	0.12	0.02	0.02	0.02	0.01
Tennessee	0.62	0.03	0.03	0.02	0.01
TxHiPlains	0.37	0.02	0.01	0.01	0.01
TxRolingPl	0.27	0.01	0.01	0.01	0.01
TxCntBlack	0.25	0.01	0.01	0.01	0.01
TxEast	0.26	0.01	0.01	0.01	0.01
TxEdplat	0.21	0.01	0.01	0.01	0.01
TxCoastBe	0.09	0.00	0.00	0.00	0.00
TxSouth	0.03	0.00	0.00	0.00	0.00
TxTranspec	0.32	0.02	0.02	0.01	0.01
Virginia	0.34	0.01	0.01	0.01	0.00
Wisconsin	0.13	0.01	0.01	0.01	0.01

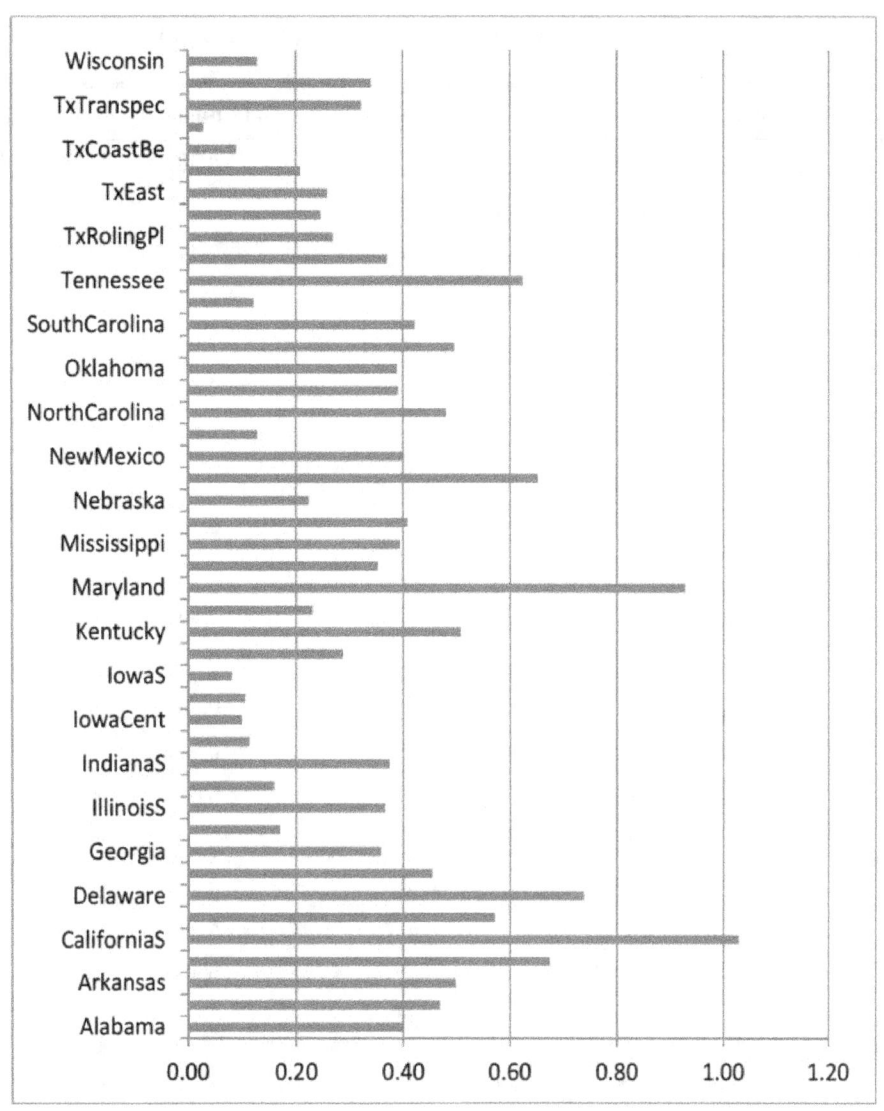

Figure A-24. Sorghum RYLs under Current Conditions (% change)

Table A-9. Sorghum RYGs under Alternative Scenarios (% change)

FASOMGHG Subregion	75 ppb	15 ppm-hr	11 ppm-hr	7 ppm-hr
Alabama	0.37	0.37	0.38	0.39
Arizona	0.39	0.43	0.45	0.46
Arkansas	0.48	0.48	0.49	0.49
CaliforniaN	0.68	0.68	0.68	0.68
CaliforniaS	1.01	1.02	1.02	1.03
Colorado	0.54	0.55	0.56	0.57
Delaware	0.74	0.74	0.74	0.74
Florida	0.41	0.41	0.42	0.44
Georgia	0.35	0.35	0.35	0.35
IllinoisN	0.15	0.15	0.15	0.16
IllinoisS	0.33	0.33	0.35	0.36
IndianaN	0.13	0.13	0.14	0.15
IndianaS	0.33	0.33	0.36	0.37
IowaW	0.10	0.10	0.10	0.10
IowaCent	0.09	0.09	0.09	0.09
IowaNE	0.10	0.10	0.10	0.10
IowaS	0.08	0.08	0.08	0.08
Kansas	0.27	0.28	0.28	0.28
Kentucky	0.47	0.47	0.49	0.50
Louisiana	0.23	0.23	0.23	0.23
Maryland	0.94	0.94	0.94	0.94
Michigan	0.34	0.34	0.34	0.34
Mississippi	0.39	0.39	0.39	0.39
Missouri	0.37	0.37	0.39	0.40
Nebraska	0.20	0.21	0.21	0.21
NewJersey	0.66	0.66	0.66	0.66
NewMexico	0.37	0.38	0.39	0.40
NewYork	0.13	0.13	0.13	0.13
NorthCarolina	0.47	0.47	0.47	0.48
OhioNW	0.33	0.33	0.36	0.38
Oklahoma	0.38	0.38	0.38	0.38
Pennsylvania	0.50	0.50	0.50	0.50
SouthCarolina	0.42	0.42	0.42	0.42

(continued)

Table A-9. Sorghum RYGs under Alternative Scenarios (% change) (continued)

FASOMGHG Subregion	75 ppb	15 ppm-hr	11 ppm-hr	7 ppm-hr
SouthDakota	0.10	0.10	0.10	0.11
Tennessee	0.59	0.59	0.61	0.62
TxHiPlains	0.35	0.36	0.36	0.37
TxRolingPl	0.26	0.26	0.26	0.26
TxCntBlack	0.24	0.24	0.24	0.24
TxEast	0.25	0.25	0.25	0.25
TxEdplat	0.20	0.20	0.20	0.20
TxCoastBe	0.08	0.08	0.08	0.08
TxSouth	0.02	0.02	0.02	0.02
TxTranspec	0.30	0.30	0.31	0.31
Virginia	0.33	0.33	0.34	0.34
Wisconsin	0.12	0.12	0.12	0.12

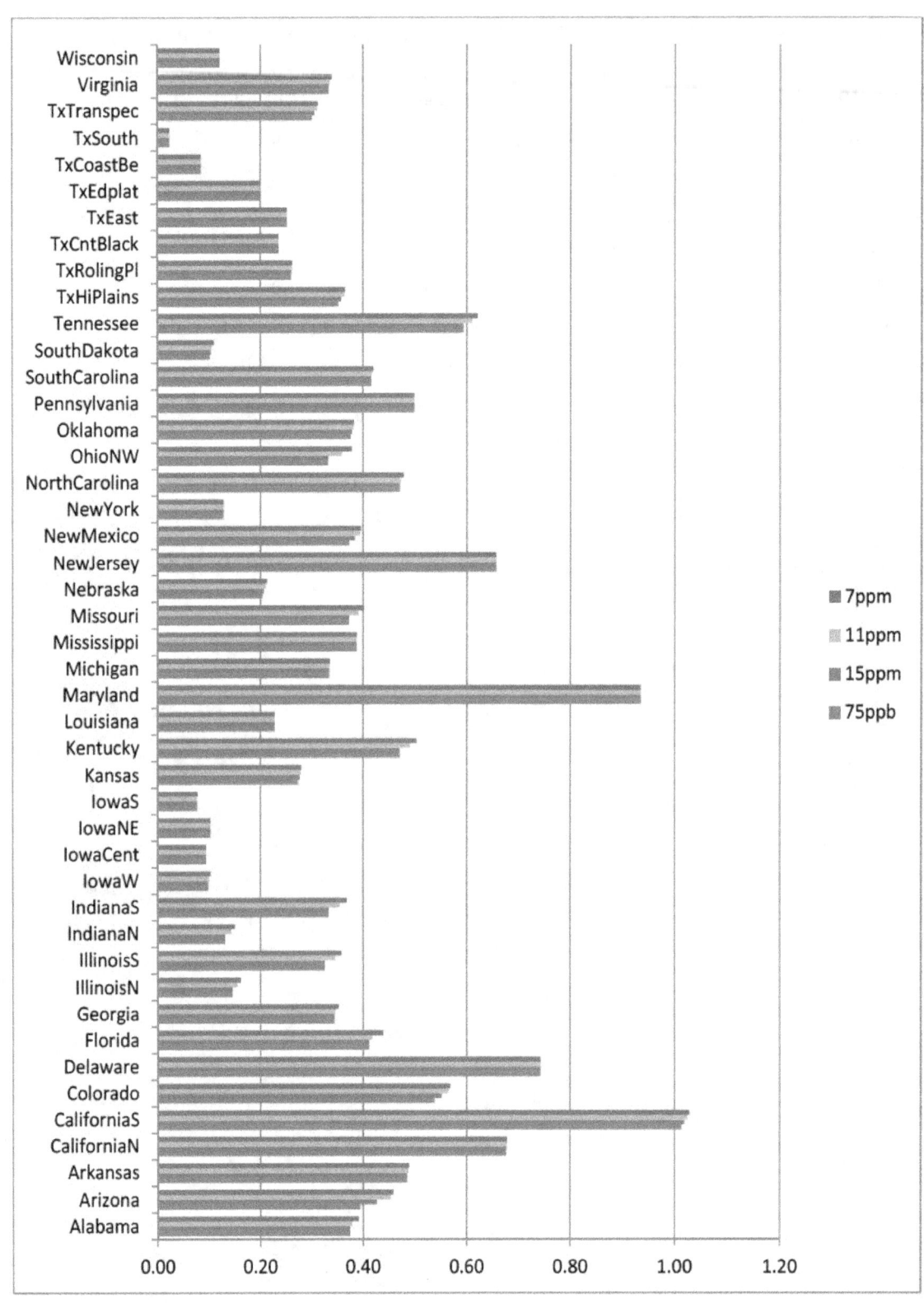

Figure A-25. Sorghum RYGs under Alternative Scenarios (% change)

Table A-10. Cotton RYLs under Alternative Scenarios (% change)

FASOMGHG Subregion	Current Conditions	75 ppb	15 ppm-hr	11 ppm-hr	7 ppm-hr
Alabama	4.51	0.51	0.51	0.44	0.23
Arizona	4.65	1.07	0.70	0.32	0.24
Arkansas	4.91	0.42	0.42	0.32	0.24
CaliforniaN	13.86	0.20	0.16	0.12	0.08
CaliforniaS	10.50	0.59	0.47	0.33	0.26
Florida	3.85	0.50	0.50	0.45	0.24
Georgia	3.42	0.25	0.25	0.22	0.12
IllinoisS	3.43	0.59	0.59	0.34	0.15
Kansas	3.37	0.20	0.18	0.16	0.15
Kentucky	4.73	0.67	0.67	0.38	0.17
Louisiana	2.31	0.07	0.07	0.07	0.07
Mississippi	3.45	0.14	0.14	0.14	0.14
Missouri	5.49	0.66	0.66	0.41	0.23
Nevada	5.70	0.39	0.32	0.25	0.22
NewMexico	3.72	0.64	0.45	0.25	0.20
NorthCarolina	4.43	0.17	0.17	0.15	0.08
Oklahoma	3.62	0.27	0.25	0.24	0.23
SouthCarolina	3.68	0.19	0.19	0.17	0.09
Tennessee	5.77	0.58	0.58	0.34	0.18
TxHiPlains	3.47	0.32	0.25	0.17	0.15
TxRolingPl	2.91	0.23	0.21	0.19	0.18
TxCntBlack	2.48	0.21	0.21	0.21	0.21
TxEast	2.96	0.25	0.25	0.25	0.25
TxEdplat	2.70	0.23	0.21	0.18	0.18
TxCoastBe	1.18	0.11	0.11	0.11	0.11
TxSouth	0.61	0.12	0.12	0.12	0.12
TxTranspec	3.76	0.67	0.57	0.46	0.43
Virginia	4.65	0.28	0.28	0.25	0.13

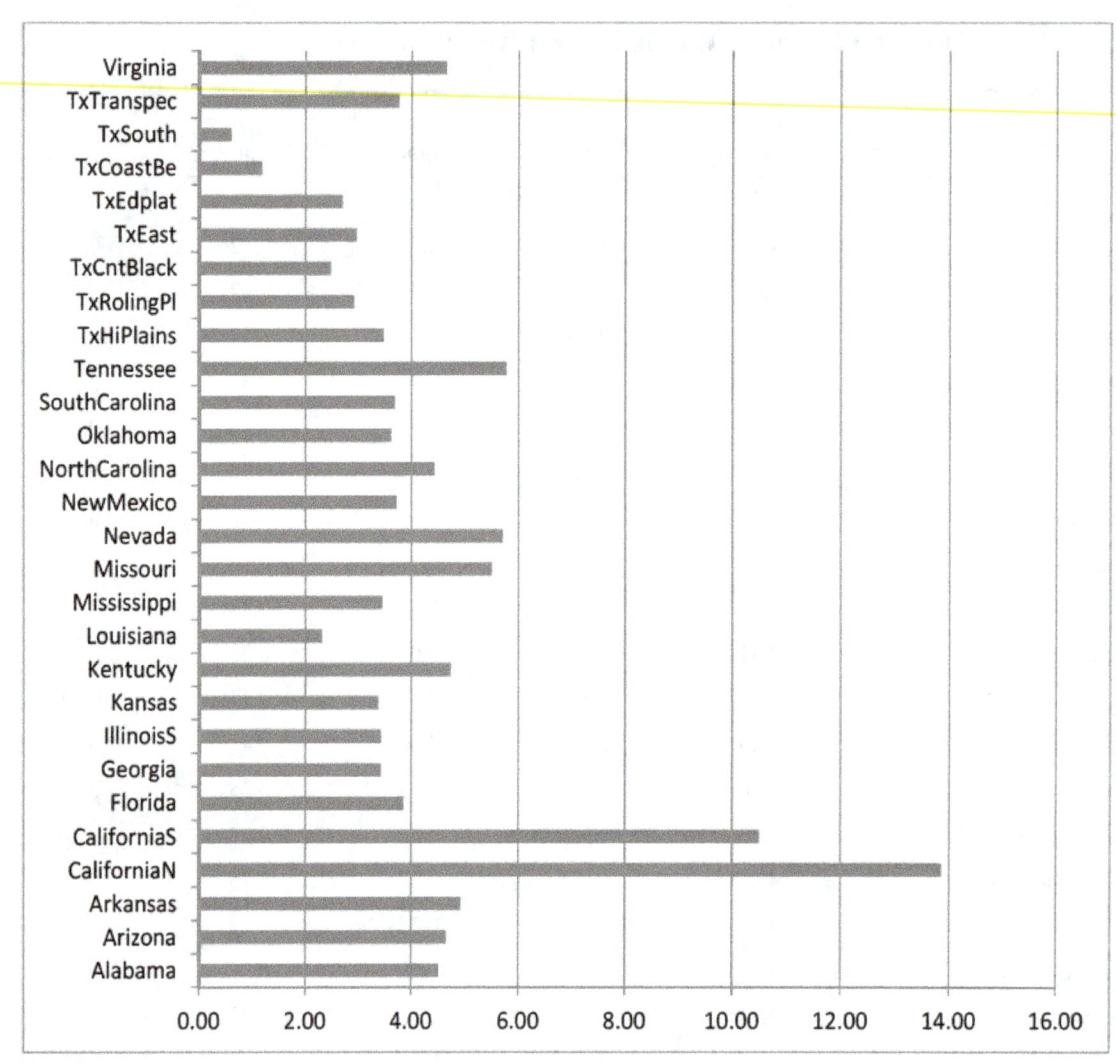

Figure A-26. Cotton RYLs under Current Conditions (% change)

Table A-11. Cotton RYGs under Alternative Scenarios (% change)

FASOMGHG Subregion	75 ppb	15 ppm-hr	11 ppm-hr	7 ppm-hr
Alabama	4.19	4.19	4.27	4.48
Arizona	3.76	4.15	4.54	4.63
Arkansas	4.73	4.73	4.83	4.92
CaliforniaN	15.87	15.91	15.96	16.00
CaliforniaS	11.06	11.20	11.35	11.44
Florida	3.49	3.49	3.54	3.76
Georgia	3.29	3.29	3.31	3.41
IllinoisS	2.94	2.94	3.20	3.39
Kansas	3.28	3.30	3.33	3.33
Kentucky	4.27	4.27	4.57	4.79
Louisiana	2.30	2.30	2.30	2.30
Mississippi	3.42	3.42	3.42	3.43
Missouri	5.11	5.11	5.38	5.57
Nevada	5.64	5.70	5.78	5.81
NewMexico	3.20	3.39	3.61	3.65
NorthCarolina	4.46	4.46	4.48	4.56
Oklahoma	3.48	3.49	3.51	3.51
SouthCarolina	3.63	3.63	3.65	3.73
Tennessee	5.50	5.50	5.76	5.93
TxHiPlains	3.26	3.34	3.42	3.44
TxRolingPl	2.76	2.78	2.81	2.81
TxCntBlack	2.32	2.32	2.32	2.32
TxEast	2.80	2.80	2.80	2.80
TxEdplat	2.54	2.56	2.59	2.59
TxCoastBe	1.08	1.08	1.08	1.08
TxSouth	0.49	0.50	0.50	0.50
TxTranspec	3.21	3.31	3.42	3.45
Virginia	4.59	4.59	4.62	4.74

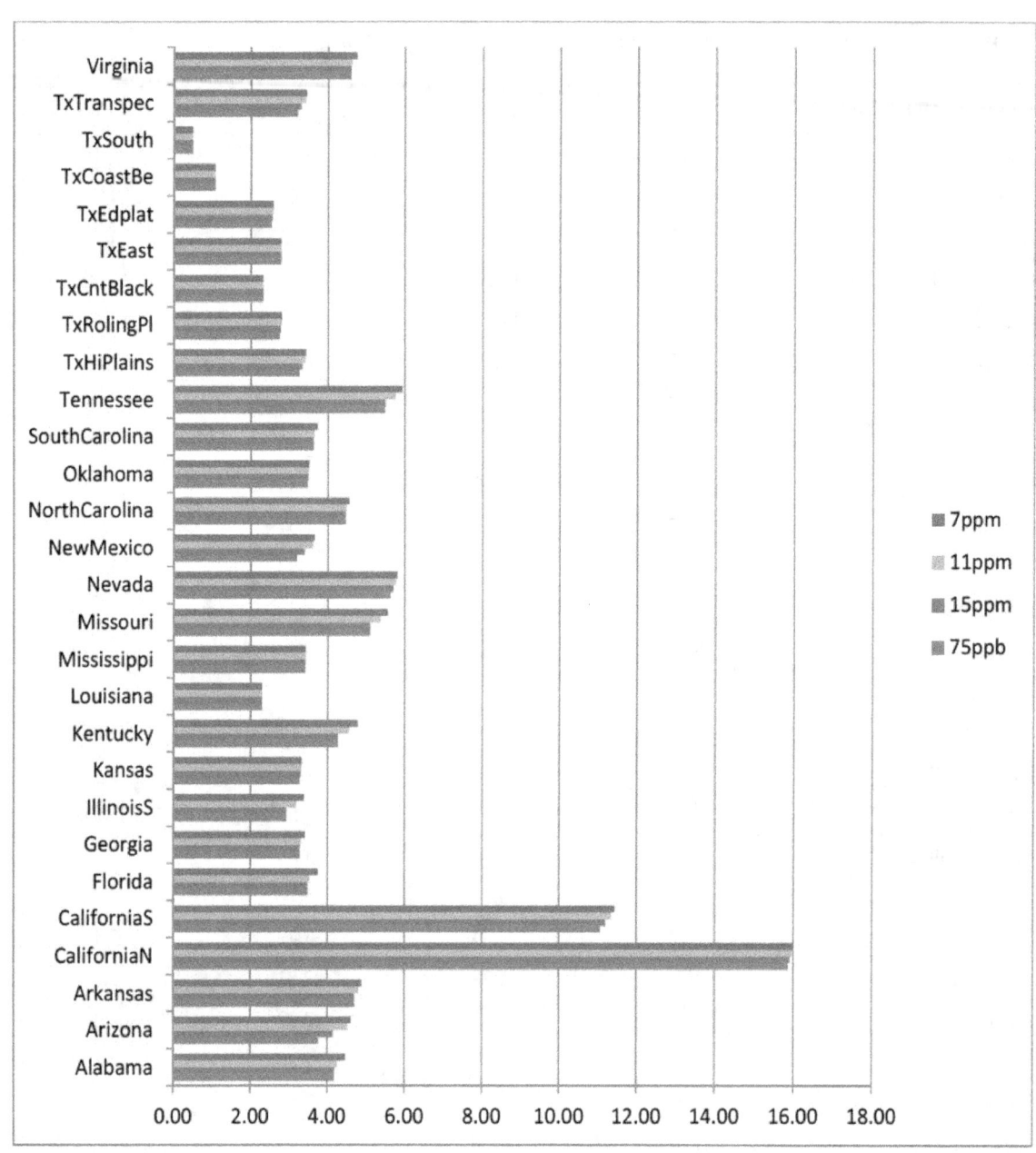

Figure A-27. Cotton RYGs under Alternative Scenarios (% change)

Table A-12. Potato RYLs under Alternative Scenarios (% change)

FASOMGHG Subregion	Current Conditions	75 ppb	15 ppm-hr	11 ppm-hr	7 ppm-hr
Alabama	7.83	1.32	1.32	1.18	0.71
Arizona	8.21	2.72	1.98	1.12	0.88
CaliforniaN	14.16	0.54	0.46	0.37	0.30
CaliforniaS	12.79	1.33	1.14	0.93	0.78
Colorado	10.07	2.26	1.70	1.02	0.75
Connecticut	6.79	0.06	0.06	0.06	0.06
Delaware	10.71	0.10	0.10	0.10	0.09
Florida	5.09	1.25	1.25	1.17	0.82
Idaho	7.35	1.37	1.14	0.86	0.75
IllinoisN	4.15	1.36	1.36	1.00	0.65
IndianaN	5.84	1.61	1.61	1.10	0.64
IowaNE	3.16	0.22	0.22	0.22	0.22
Kansas	6.22	0.95	0.85	0.72	0.64
Louisiana	5.33	0.41	0.41	0.41	0.41
Maine	1.67	0.10	0.10	0.10	0.10
Maryland	10.80	0.23	0.23	0.21	0.16
Massachusetts	5.53	0.07	0.07	0.07	0.07
Michigan	5.27	0.95	0.95	0.92	0.88
Minnesota	2.38	0.38	0.38	0.38	0.34
Missouri	6.56	1.50	1.50	1.01	0.61
Montana	3.38	0.94	0.93	0.92	0.74
Nebraska	4.79	1.21	1.09	0.93	0.78
Nevada	9.49	1.16	1.01	0.83	0.75
NewJersey	10.37	0.07	0.07	0.07	0.07
NewMexico	7.35	1.54	1.17	0.72	0.61
NewYork	4.50	0.03	0.03	0.03	0.03
NorthCarolina	8.58	0.78	0.78	0.70	0.41
NorthDakota	2.09	0.76	0.76	0.76	0.64
OhioNW	6.99	1.78	1.78	1.23	0.74
Oregon	3.25	0.90	0.86	0.80	0.78
Pennsylvania	6.95	0.15	0.15	0.13	0.10
RhodeIsland	6.70	0.11	0.11	0.11	0.11
SouthDakota	2.79	0.84	0.83	0.82	0.69

(continued)

Table A-12. Potato RYLs under Alternative Scenarios (% change) (continued)

FASOMGHG Subregion	Current Conditions	75 ppb	15 ppm-hr	11 ppm-hr	7 ppm-hr
Tennessee	9.48	1.69	1.69	1.11	0.60
TxHiPlains	6.81	1.08	0.86	0.61	0.55
TxRolingPl	6.08	0.81	0.77	0.73	0.71
TxEast	5.54	0.58	0.58	0.58	0.58
TxCoastBe	3.81	0.65	0.65	0.65	0.65
TxSouth	2.05	0.50	0.50	0.50	0.50
Utah	11.30	2.34	1.77	1.07	0.87
Virginia	7.37	0.82	0.82	0.71	0.42
Washington	2.38	0.77	0.74	0.71	0.70
WestVirginia	6.19	0.90	0.90	0.59	0.31
Wisconsin	3.64	0.45	0.45	0.44	0.44
Wyoming	8.47	1.96	1.70	1.39	1.09

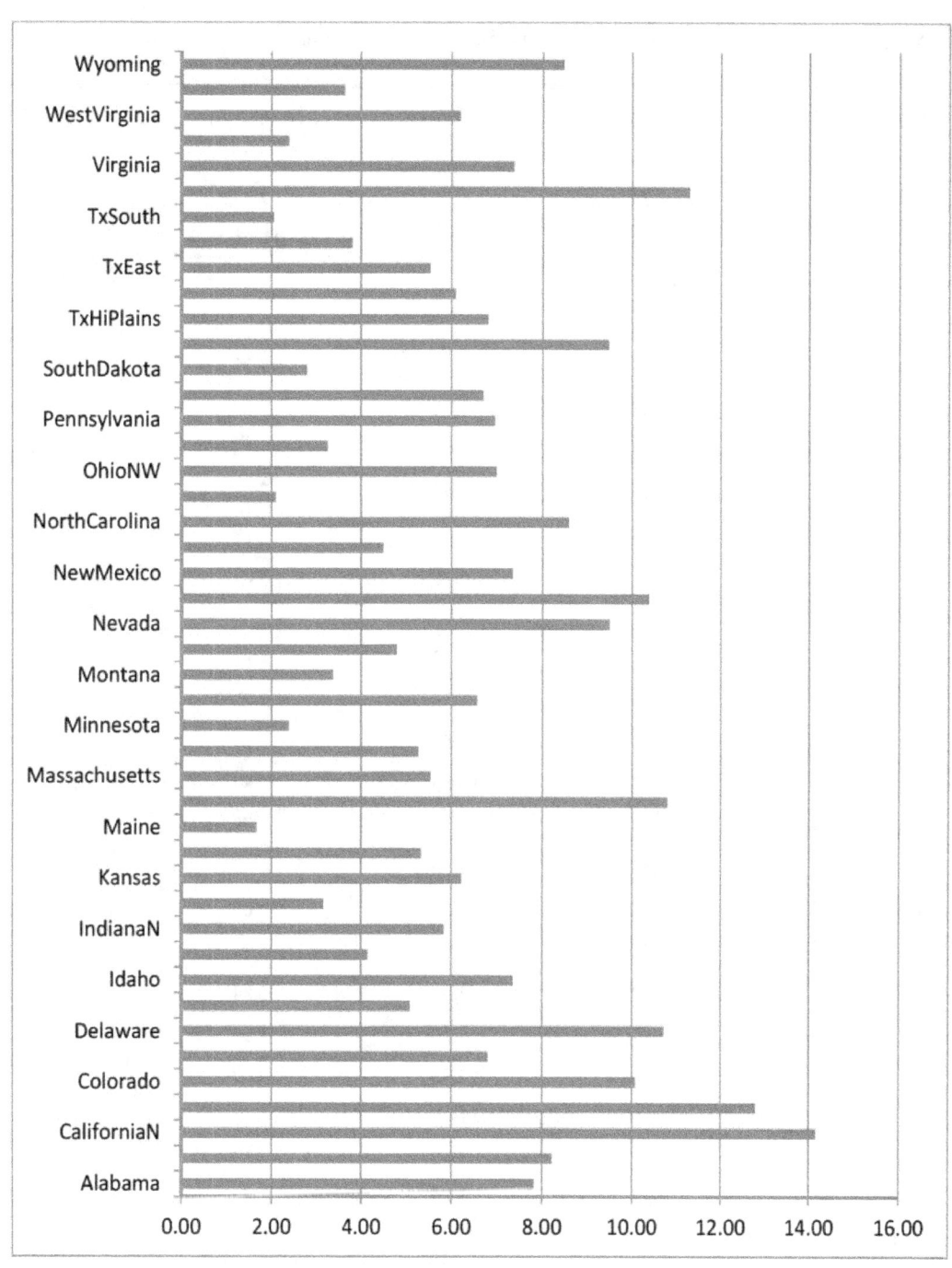

Figure A-28. Potato RYLs under Current Conditions (% change)

Table A-13. Potato RYGs under Alternative Scenarios (% change)

FASOMGHG Subregion	75 ppb	15 ppm-hr	11 ppm-hr	7 ppm-hr
Alabama	7.06	7.06	7.21	7.73
Arizona	5.98	6.78	7.72	7.98
CaliforniaN	15.86	15.96	16.06	16.14
CaliforniaS	13.14	13.35	13.59	13.76
Colorado	8.68	9.30	10.05	10.35
Connecticut	7.23	7.23	7.23	7.23
Delaware	11.88	11.88	11.89	11.89
Florida	4.05	4.05	4.13	4.50
Idaho	6.46	6.71	7.01	7.13
IllinoisN	2.91	2.91	3.29	3.65
IndianaN	4.49	4.49	5.03	5.51
IowaNE	3.03	3.03	3.03	3.03
Kansas	5.61	5.72	5.87	5.94
Louisiana	5.20	5.20	5.20	5.20
Maine	1.59	1.59	1.59	1.59
Maryland	11.85	11.85	11.87	11.93
Massachusetts	5.79	5.79	5.79	5.79
Michigan	4.56	4.56	4.59	4.63
Minnesota	2.05	2.05	2.05	2.08
Missouri	5.42	5.42	5.94	6.37
Montana	2.53	2.53	2.54	2.73
Nebraska	3.77	3.90	4.06	4.22
Nevada	9.20	9.38	9.57	9.66
NewJersey	11.49	11.49	11.49	11.49
NewMexico	6.27	6.67	7.16	7.28
NewYork	4.67	4.67	4.67	4.67
NorthCarolina	8.53	8.53	8.61	8.93
NorthDakota	1.36	1.36	1.36	1.48
OhioNW	5.61	5.61	6.19	6.72
Oregon	2.43	2.48	2.53	2.56
Pennsylvania	7.32	7.32	7.34	7.36
RhodeIsland	7.07	7.07	7.07	7.07
SouthDakota	2.01	2.01	2.02	2.16

(continued)

Table A-13. Potato RYGs under Alternative Scenarios (% change) (continued)

FASOMGHG Subregion	75 ppb	15 ppm-hr	11 ppm-hr	7 ppm-hr
Tennessee	8.60	8.60	9.25	9.80
TxHiPlains	6.15	6.37	6.64	6.71
TxRolingPl	5.62	5.66	5.70	5.72
TxEast	5.25	5.25	5.25	5.25
TxCoastBe	3.29	3.29	3.29	3.29
TxSouth	1.57	1.57	1.58	1.58
Utah	10.10	10.74	11.53	11.75
Virginia	7.07	7.07	7.19	7.50
Washington	1.65	1.67	1.71	1.72
WestVirginia	5.64	5.64	5.97	6.27
Wisconsin	3.32	3.32	3.32	3.33
Wyoming	7.12	7.40	7.74	8.07

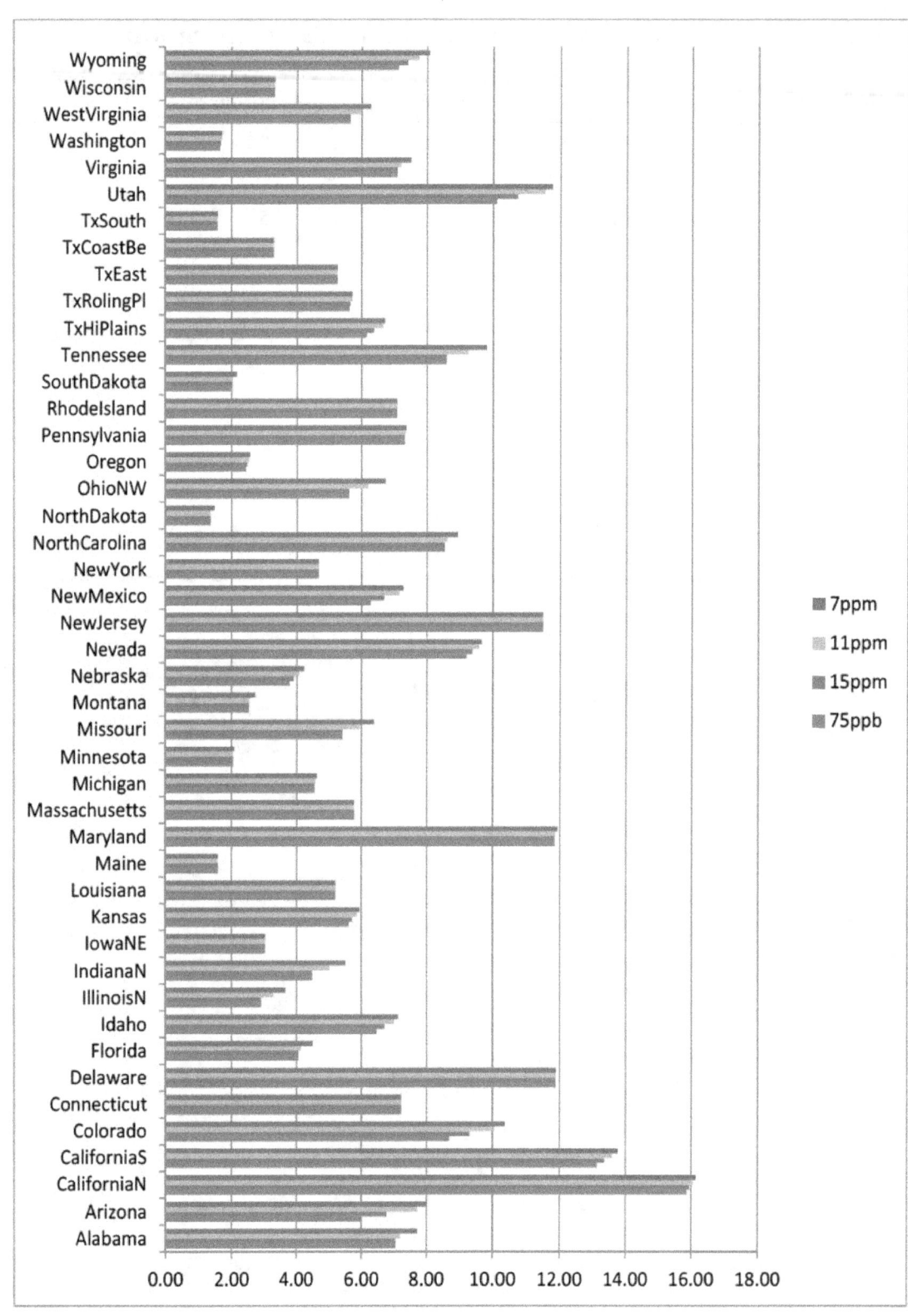

Figure A-29. Potato RYGs under Alternative Scenarios (% change)

A.4   Derived County-Level Relative Yield Gains for Crops

This subsection presents the county-level RYGs and RYG changes for major crops under alternative scenarios. The 75 ppb RYGs and RYG changes for alternative ozone standards versus the 75 ppb scenario are selected for presentation in this subsection. The crop-specific, 2007 USDA Census of Agriculture-based county mappings are applied in this subsection. Figures A-30 through A-35 present the RYGs for corn, soybeans, winter wheat, sorghum, cotton, and potatoes under the 75 ppb scenario, respectively. Figures A-36 through A-41 display the RYG changes for those crops under alternative ozone standards scenarios with respect to the 75 ppb scenario.

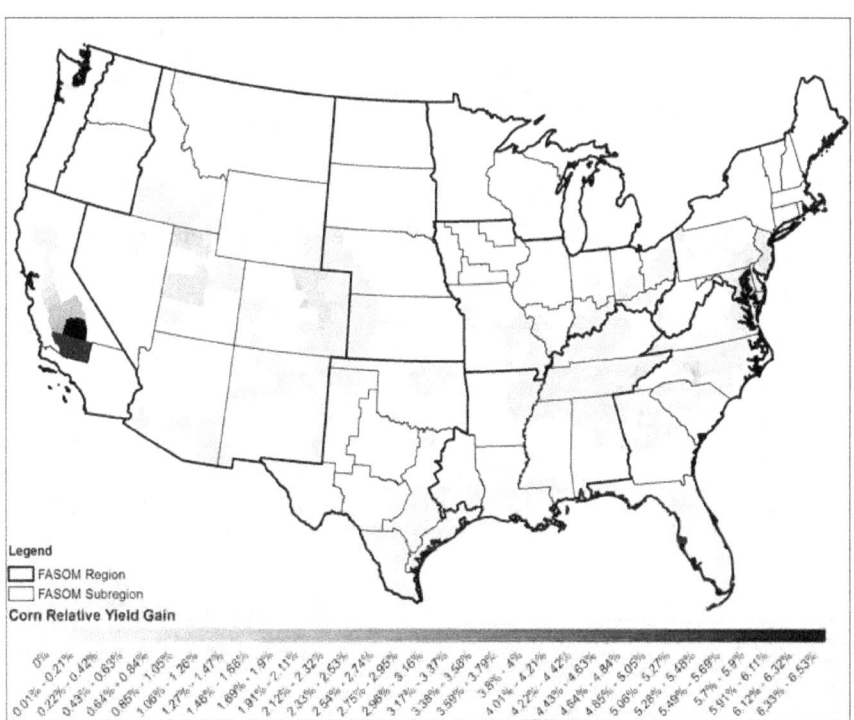

Figure A-30.   County-Level Corn RYGs under the 75 ppb Scenario

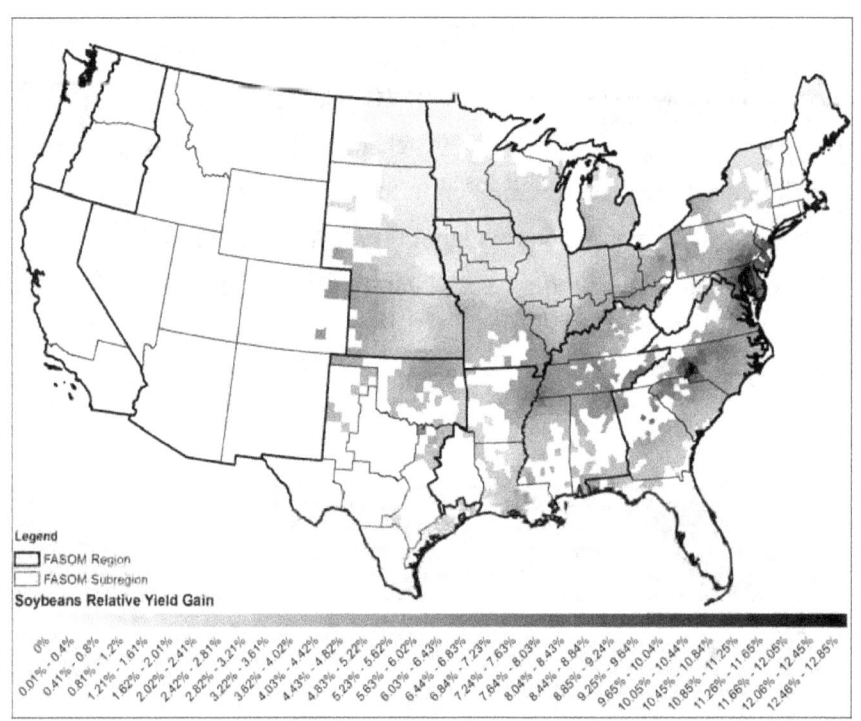

Figure A-31. County-Level Soybean RYGs under the 75 ppb Scenario

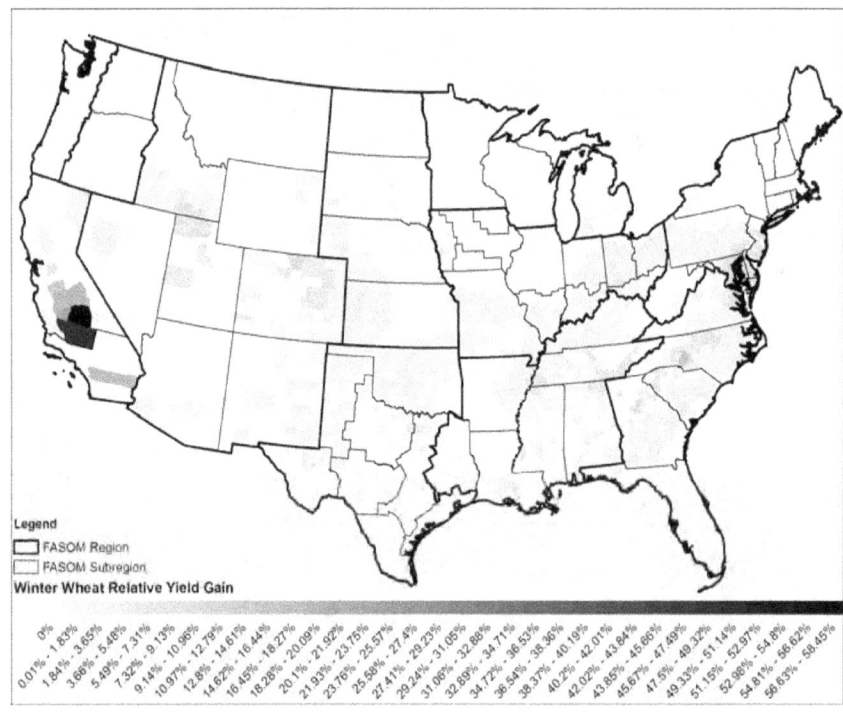

Figure A-32. County-Level Winter Wheat RYGs under the 75 ppb Scenario

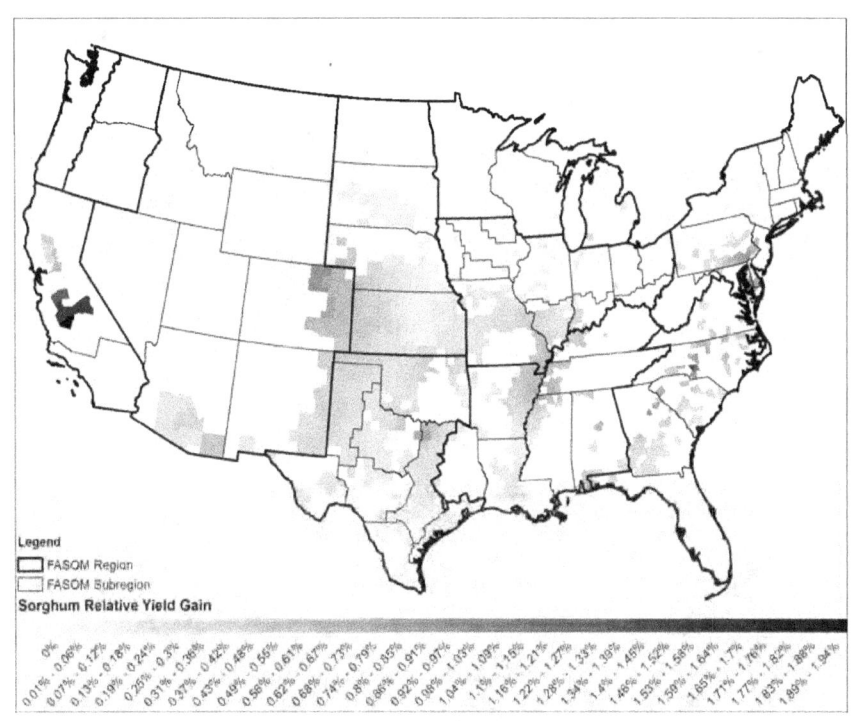

Figure A-33. County-Level Sorghum RYGs under the 75 ppb Scenario

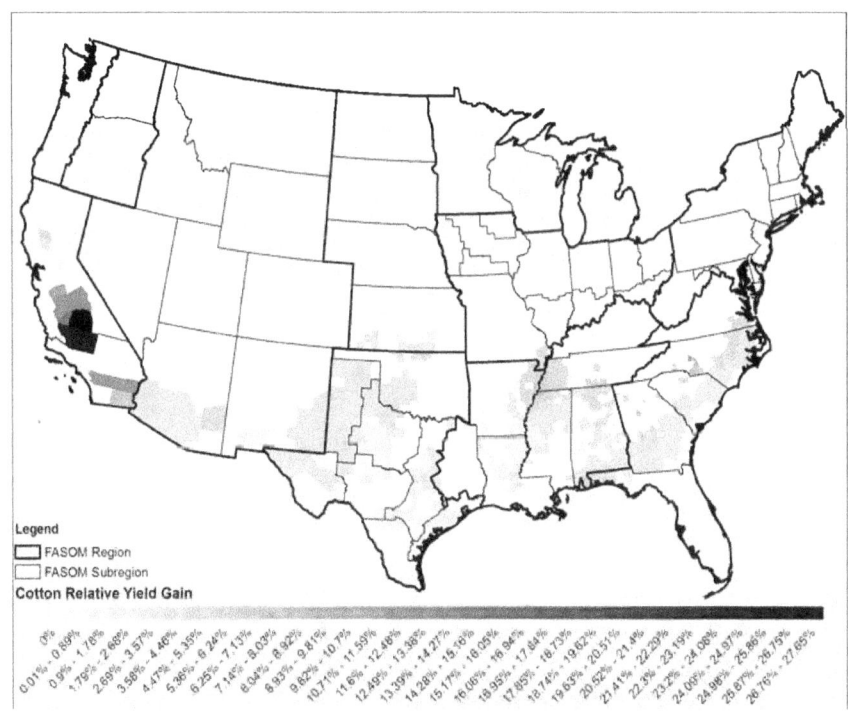

Figure A-34. County-Level Cotton RYGs under the 75 ppb Scenario

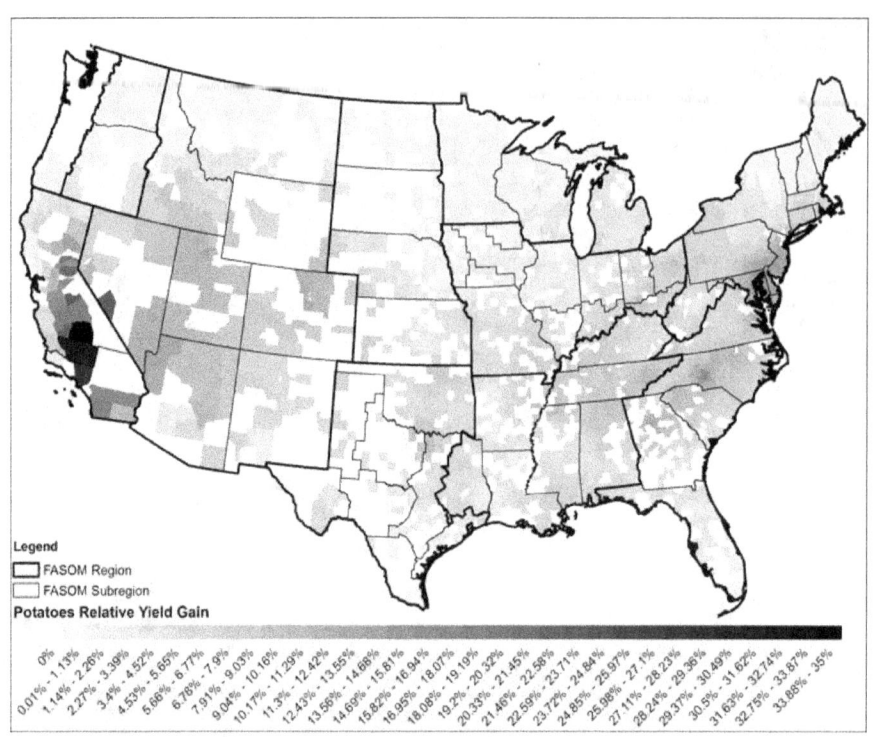

Figure A-35. County-Level Potato RYGs under the 75 ppb Scenario

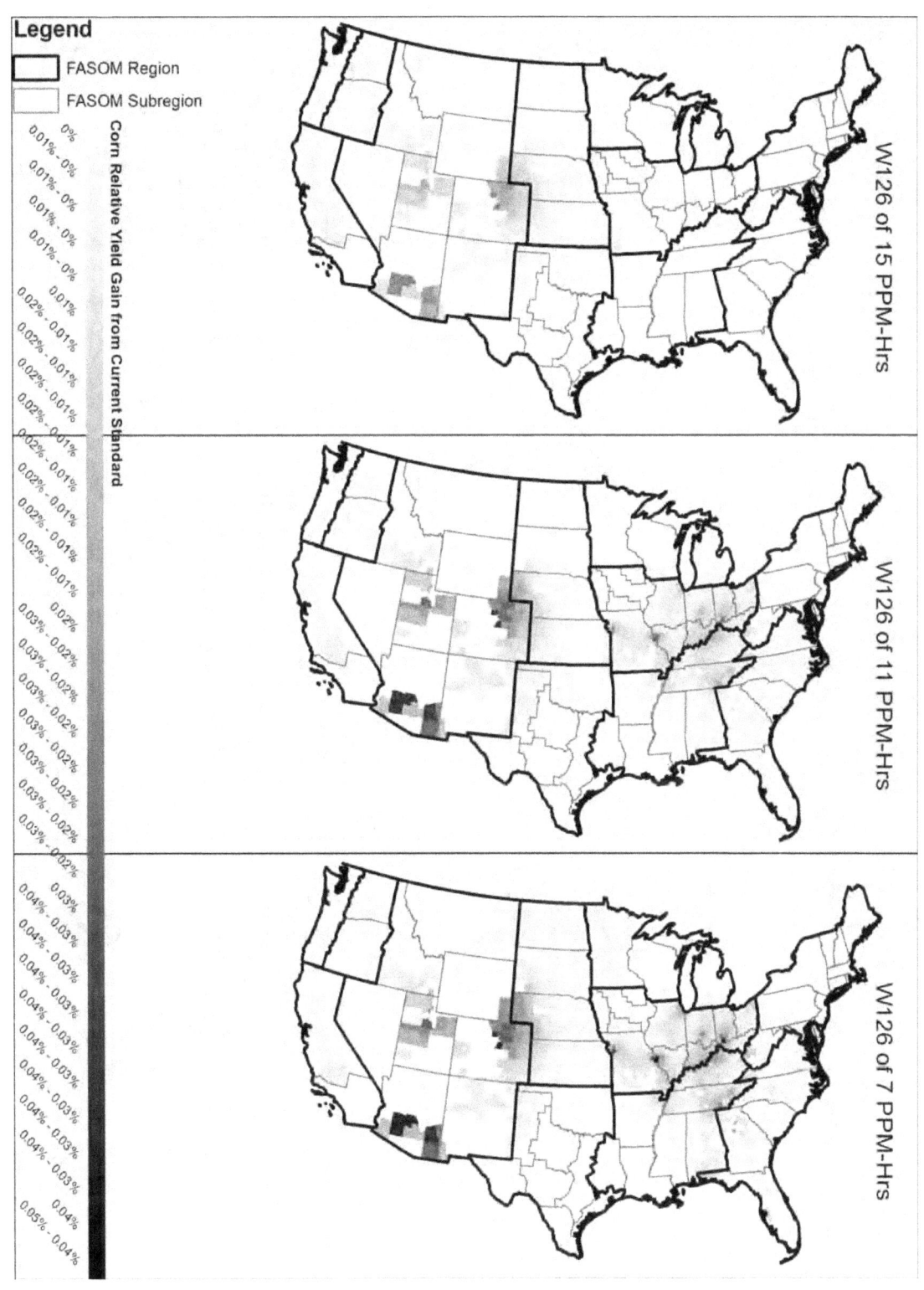

Figure A-36. County-Level Changes in Corn RYGs under Alternative Scenarios with Respect to 75 ppb

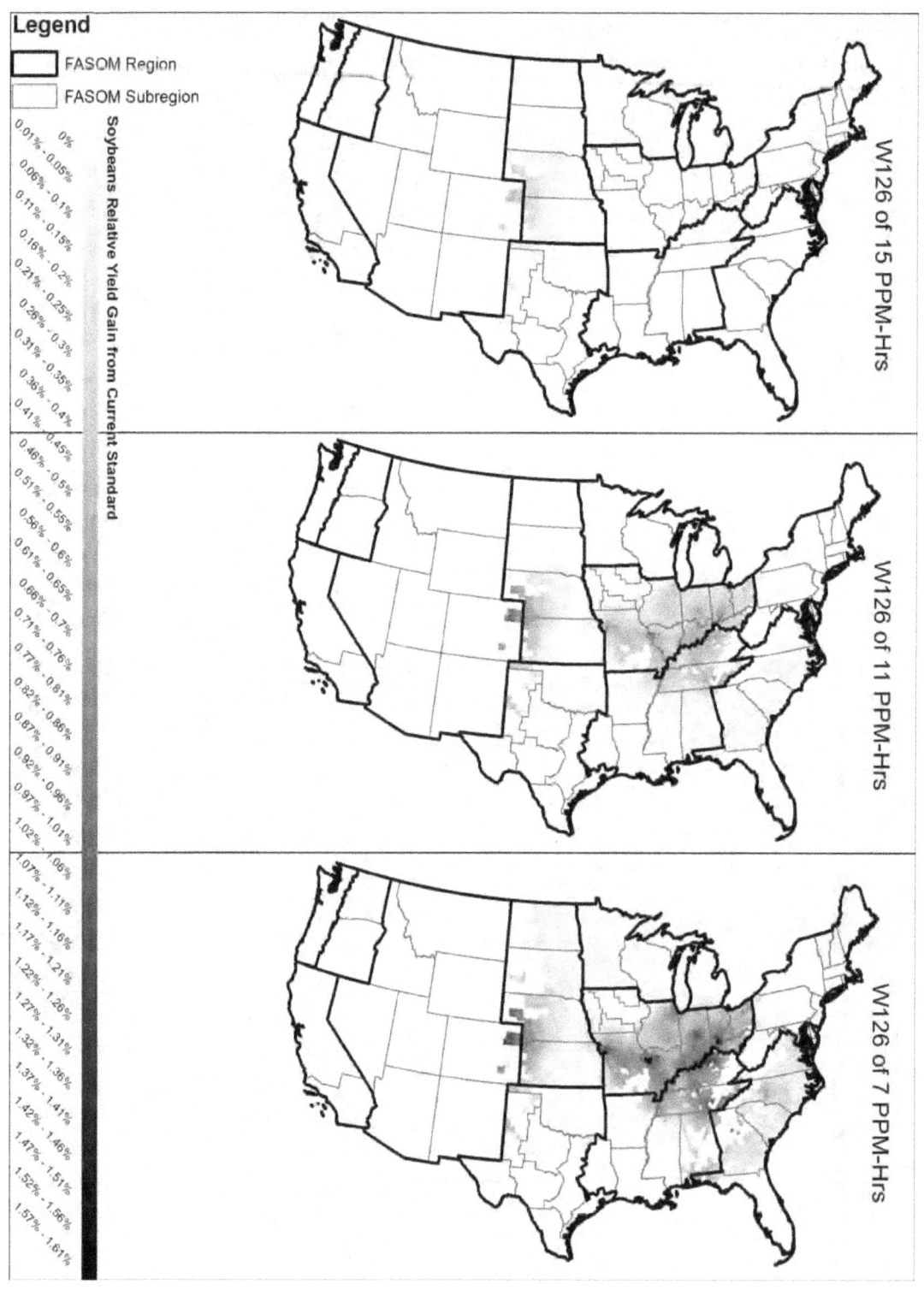

Figure A-37. County-Level Changes in Soybean RYGs under Alternative Scenarios with Respect to 75 ppb

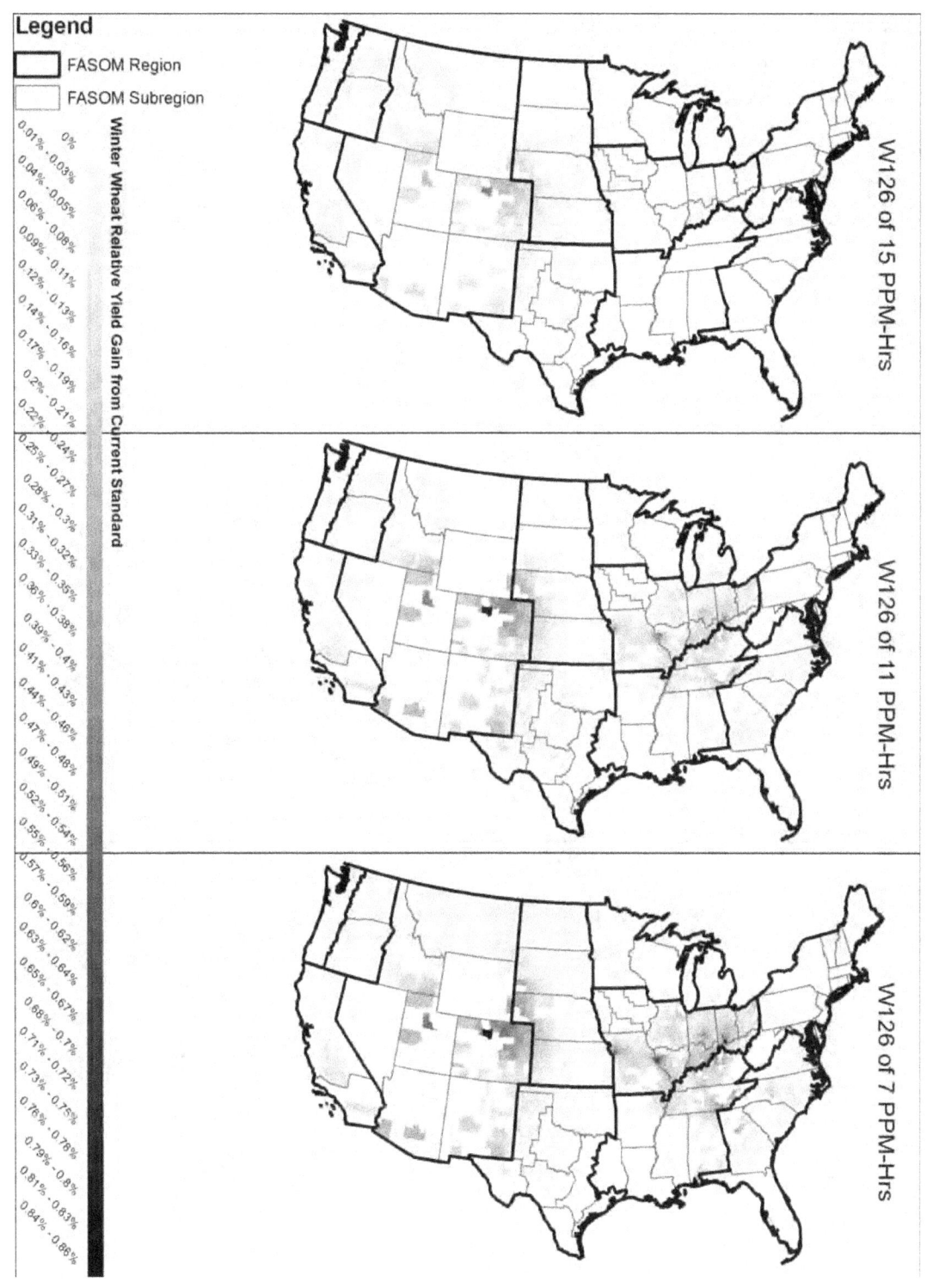

Figure A-38. County-Level Changes in Winter Wheat RYGs under Alternative Scenarios with Respect to 75 ppb

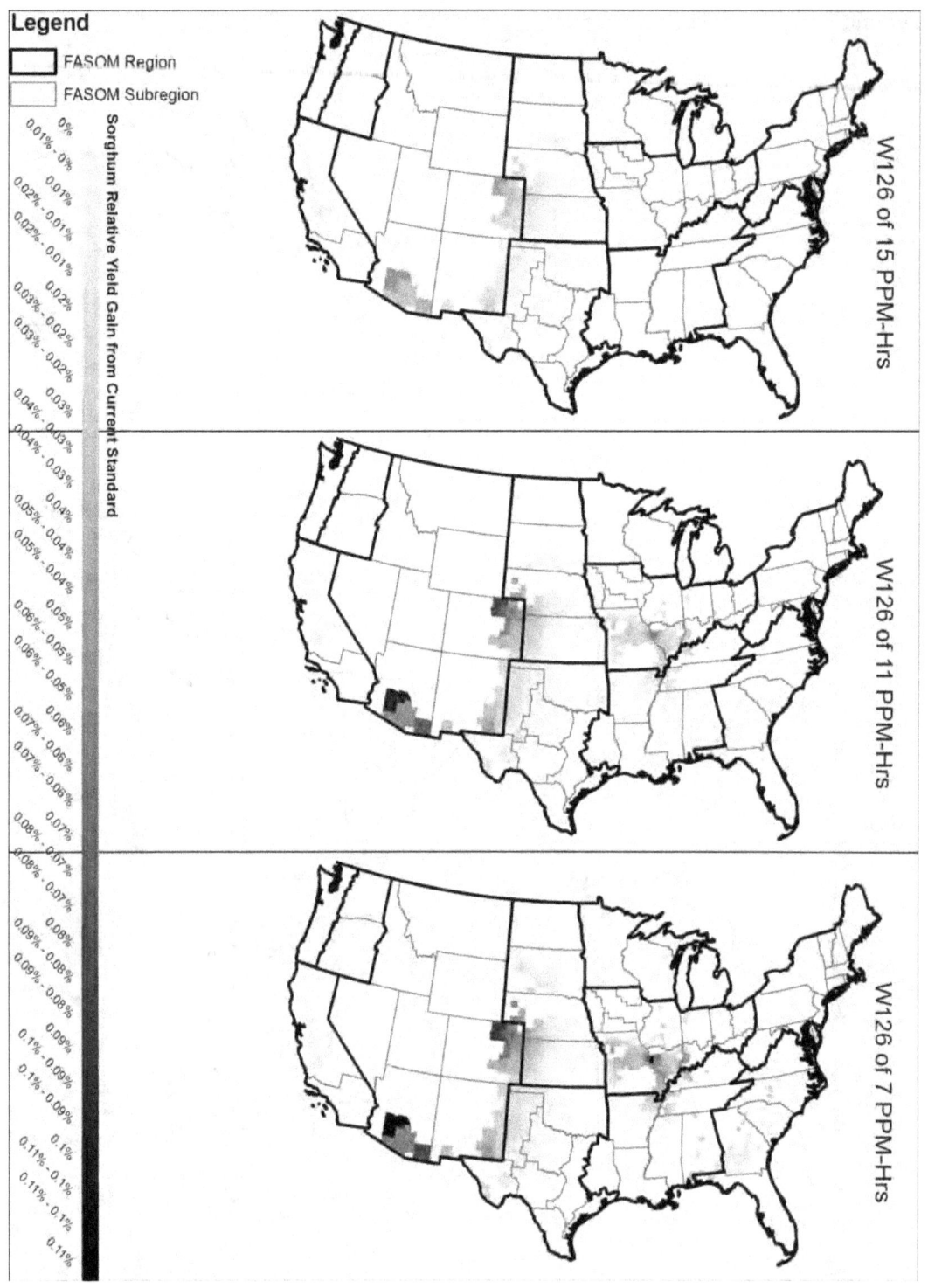

Figure A-39. County-Level Changes in Sorghum RYGs under Alternative Scenarios with Respect to 75 ppb

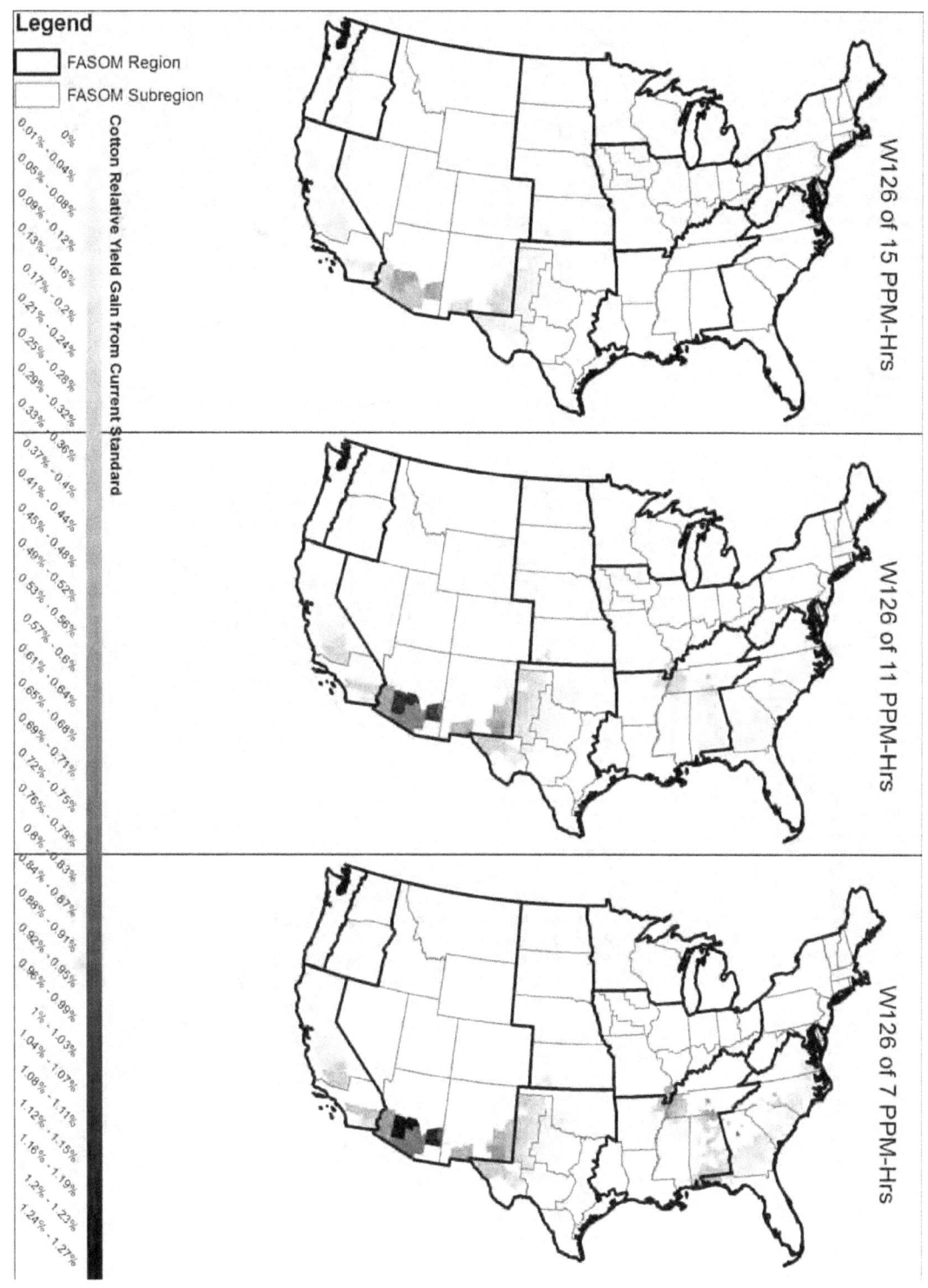

Figure A-40. County-Level Changes in Cotton RYGs under Alternative Scenarios with Respect to 75 ppb

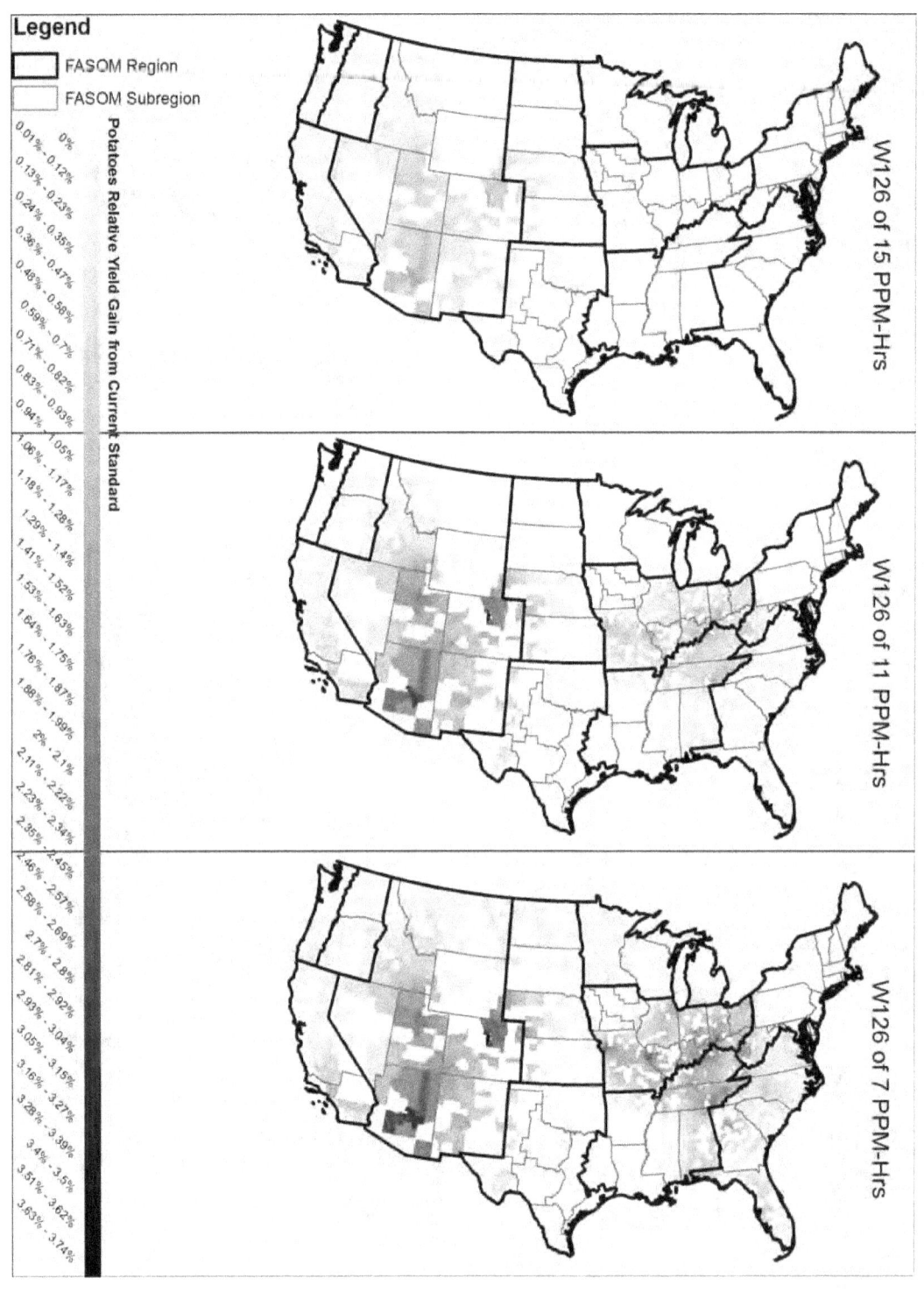

Figure A-41. County-Level Changes in Potato RYGs under Alternative Scenarios with Respect to 75 ppb

APPENDIX B
MODEL RESULTS: FOCUSING ON COMPARISON WITH CURRENT CONDITIONS

This appendix section focuses on presentation of model results for a comparison of the current standard versus current conditions. Under current conditions, we assumed that ozone concentrations continued at current levels, which exceed the current standard (75 ppb). As for the results presented in Section 5, we present our findings for changes to production, prices, forest inventory, land use, welfare, and greenhouse gas (GHG) mitigation potential. This section is designed to be a standalone analysis providing an alternative comparison point to that provided in Section 5.

B.1   Agricultural Sector

As discussed in the main body of the report, ozone negatively affects growth in many plants, leading to lower yields of agricultural crops. Thus, meeting the existing or more stringent ozone standards would increase agricultural production by reducing ozone concentrations and alleviating existing detrimental impacts. Consistent with expectations, the incremental impacts when moving from current conditions to compliance with existing or more stringent standards are found to be considerably larger than the incremental impacts of increasing stringency relative to the current standard presented in Section 5. This is simply because the reductions in ozone concentrations relative to current conditions are much bigger than those in moving between the standards considered.

B.1.1   Production and Prices

Changes in U.S. agricultural production and prices were measured using Fisher indices (see Tables B-1 and B-2).[1] Both primary and secondary commodity production levels are projected to increase by 2040 as a result of increased productivity when the current ozone standard is met. There is a greater difference in the production of primary commodities than secondary commodities. The reason for secondary commodities experiencing smaller impacts is that they use a number of other inputs in combination with primary commodities, so the percentage impact on cost of production is smaller.

---

[1] The Fisher price index is known as the "ideal" price index. It is calculated as the geometric mean of an index of current prices and an index of past prices.

Table B-1. Agricultural Production Fisher Indices (Current Conditions = 100)

Sector	Policy	2010	2020	2030	2040
		Primary Commodities			
Crops	75 ppb	101.01	100.77	100.97	100.87
Livestock	75 ppb	100.13	100.03	100.36	100.88
Farm products[a]	75 ppb	100.58	100.42	100.43	100.40
		Secondary Commodities			
Processed	75 ppb	100.17	100.10	100.08	100.28
Meats	75 ppb	100.03	100.01	100.43	100.40
Mixed feeds	75 ppb	100.06	100.01	100.64	100.53

[a] Farm products is the composite of crops and livestock.

Consistent with expectations, increased production led to a general decline in market prices because the equilibrium price adjusts to higher levels of supply.

Table B-2. Agricultural Price Fisher Indices (Current Conditions = 100)

Sector	Policy	2010	2020	2030	2040
		Primary Commodities			
Crops	75 ppb	97.51	98.35	99.05	98.73
Livestock	75 ppb	99.78	99.31	100.26	98.90
Farm products	75 ppb	97.51	98.35	99.75	99.51
		Secondary Commodities			
Processed	75 ppb	97.60	98.34	98.69	98.22
Meats	75 ppb	99.86	100.00	99.75	99.51
Mixed feeds	75 ppb	97.48	98.98	99.11	98.91

B.1.2 Crop Acreage

Total crop acreage declines with the introduction of the ozone standards because higher productivity per acre reduces the demand for cropland. In aggregate, farmers are able to meet the demand for agricultural commodities using less land under the ozone standard scenarios. Consistent with these expectations, the total cropped area is slightly smaller for each model year in the ozone standard cases than under current conditions. However, land allocation over time also depends on relative returns across various uses and is influenced by forest harvest timing.

Discussion of the changes in land allocation between the agricultural and forestry sectors follows later in this section.

Table B-3 provides baseline projections of acreage in each of the major U.S. crops, as well as composites of all remaining crops and total cropland. The absolute change relative to current conditions is presented for each policy scenario. Overall, there tended to be reallocation of land from soybeans, winter wheat, and cotton to corn, hay, spring wheat, and other minor crops. Increases in wheat for livestock grazing[2] and spring barley dominated changes in other minor crops. This shift occurred largely because of differential crop sensitivity to ozone concentrations. Soybeans, winter wheat, and cotton are all relatively sensitive to ozone concentrations, indicating that lowering ozone concentrations would significantly increase yield for these crops and decrease demand for cropland. Corn, however, is not very sensitive to ozone concentrations. Corn acreage increases are driven by increased demand for feed as livestock and meat demand increase over time, as discussed in Section B.1.1. The sum of the crop-specific changes will not necessarily equal the total changes shown in Table B-3 because some double-cropping is reflected in the model (e.g., soybeans and winter barley).

---

[2] FASOMGHG includes categories of selected small grains (e.g., wheat, oats) that are planted to provide grazing for livestock rather than being harvested. These crops would generally be planted as winter cover crops that also provide grazing for livestock.

Table B-3. Major Crop Acreage, Million Acres

Crop	Policy	2010	2020	2030	2040
Corn	Current conditions	92.7	87.2	79.0	70.8
	Change with Respect to Current Conditions				
	75 ppb	0.14	−0.02	0.63	0.00
	15 ppm-hr	0.14	−0.02	0.63	0.00
	11 ppm-hr	0.13	0.01	0.60	−0.01
	7 ppm-hr	0.14	−0.02	0.61	−0.01
Soybeans	Current conditions	73.1	72.0	72.2	69.9
	Change with Respect to Current Conditions				
	75 ppb	0.20	−0.12	−0.53	0.00
	15 ppm-hr	0.21	−0.12	−0.55	−0.01
	11 ppm-hr	0.23	−0.13	−0.72	−0.14
	7 ppm-hr	0.26	−0.16	−0.62	−0.10
Hay	Current conditions	44.2	42.6	41.4	39.2
	Change with Respect to Current Conditions				
	75 ppb	−0.22	−0.52	−0.44	0.00
	15 ppm-hr	−0.23	−0.53	−0.44	0.00
	11 ppm-hr	−0.21	−0.62	−0.43	−0.01
	7 ppm-hr	−0.23	−0.54	−0.46	0.00
Hard red winter wheat	Current conditions	20.7	19.4	15.8	13.5
	Change with Respect to Current Conditions				
	75 ppb	−0.06	0.07	0.07	−0.02
	15 ppm-hr	−0.02	0.05	0.06	−0.03
	11 ppm-hr	0.00	0.00	0.00	0.00
	7 ppm-hr	−0.04	0.36	3.68	4.60
Cotton	Current conditions	15.1	15.9	16.3	15.7
	Change with Respect to Current Conditions				
	75 ppb	−0.48	−0.44	−0.38	0.00
	15 ppm-hr	−0.48	−0.45	−0.38	0.00
	11 ppm-hr	−0.49	−0.47	−0.40	0.01
	7 ppm-hr	−0.48	−0.46	−0.39	0.01

(continued)

B-4

Table B-3. Major Crop Acreage, Million Acres (continued)

Crop	Policy	2010	2020	2030	2040
Hard red spring wheat	Current conditions	13.7	12.5	12.2	11.4
	Change with Respect to Current Conditions				
	75 ppb	0.01	0.02	−0.01	0.00
	15 ppm-hr	0.01	0.02	−0.02	0.00
	11 ppm-hr	0.00	0.00	0.00	0.00
	7 ppm-hr	0.01	0.02	0.00	0.00
Sorghum	Current conditions	10.6	11.3	11.6	10.7
	Change with Respect to Current Conditions				
	75 ppb	0.14	0.03	0.07	0.00
	15 ppm-hr	0.14	0.03	0.06	0.00
	11 ppm-hr	0.13	0.03	0.11	−0.01
	7 ppm-hr	0.14	0.01	0.09	−0.01
Switchgrass	Current conditions	0.0	13.4	11.2	10.3
	Change with Respect to Current Conditions				
	75 ppb	0.00	0.00	0.00	−0.01
	15 ppm-hr	0.00	0.00	0.01	−0.01
	11 ppm-hr	0.00	0.00	0.00	0.00
	7 ppm-hr	0.00	0.00	0.00	−0.01
All others[a]	Current conditions	42.0	40.9	43.8	44.0
	Change with Respect to Current Conditions				
	75 ppb	−0.25	−0.22	−0.45	0.00
	15 ppm-hr	−0.25	−0.21	−0.46	−0.02
	11 ppm-hr	−0.25	−0.21	−0.48	−0.01
	7 ppm-hr	−0.28	−0.22	−0.50	−0.02
Total	Current conditions	311.9	315.5	304.7	286.4
	Change with Respect to Current Conditions				
	75 ppb	−0.33	−1.69	−2.17	−1.28
	15 ppm-hr	−0.31	−1.71	−2.24	−1.31
	11 ppm-hr	−0.28	−1.88	−2.48	−1.42
	7 ppm-hr	−0.30	−1.80	−2.36	−1.39

[a] Canola, durum wheat, fresh grapefruit, fresh orange, fresh tomato, hybrid poplar, oats, potato, processed grapefruit, processed orange, processed tomato, rice, rye, silage, soft red winter wheat, soft white wheat, spring barley, sugar beet, sugarcane, sweet sorghum, wheat for livestock grazing, winter barley.

B.2   Forestry Sector

As with agricultural crops, ozone diminishes growth in most tree species, and our analysis began by estimating how much of this diminished growth would be reversed under the proposed ozone standards. Impacts are significantly higher in the forestry sector, especially in hardwood species and species more prevalent in the southern regions. Impacts are more significant for southern regions because of the higher baseline ozone concentrations in the South Central and Southeast regions than for other regions with major forestry sectors. Higher initial concentrations resulted in greater reductions to meet the proposed standards and, thus, higher impacts to tree growth. This relationship also contributes to the larger changes in the forestry sector as a whole.

B.2.1   Production and Prices

Reducing ozone concentrations leads to increased forest growth, which is reflected in increased production and lower prices in FASOMGHG. Some of the most substantial ozone standard impacts occurred in saw and pulp log harvest quantities and prices. Compared with current conditions, the current ozone standard had consistently higher production except for softwood pulp logs, where production increased only marginally and at times fell below the baseline estimates, especially in 2040. The most significant impacts occurred in hardwood saw logs, where harvests were projected to be more than 20% higher than baseline harvest by 2040 (see Figure B-1). Additional policy stringency resulted in marginal additional impacts for each of the forest products (see Sections 4, 5, and 6 for additional information on the impacts associated with moving between ozone concentration standards). Figure B-1 depicts the differences in hardwood saw log production under current conditions, compared with each of the policy cases. Table B-4 presents these changes in numerical form for all the log products. Table B-5 presents the prices of forest products per cubic foot.

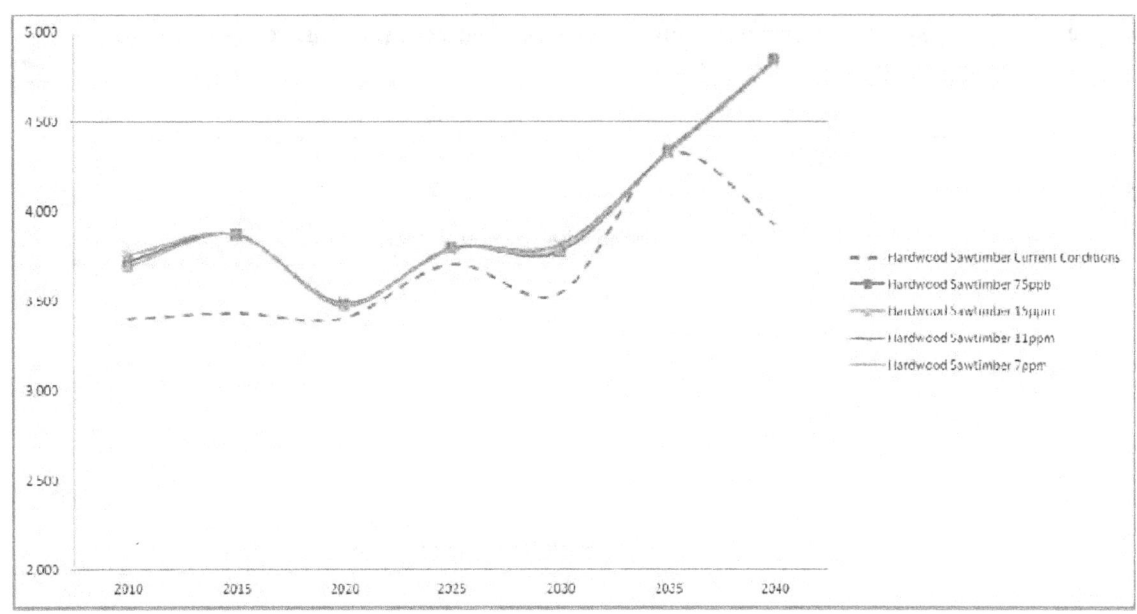

Figure B-1. Hardwood Saw Log Production, Million Cubic Feet

Table B-4. Forest Products Production, Million Cubic Feet

Product	Policy	2010	2020	2030	2040
Hardwood saw logs	Current conditions	3,401	3,405	3,541	3,929
	Change with Respect to Current Conditions				
	75 ppb	296	67	235	915
	15 ppm-hr	297	68	245	916
	11 ppm-hr	324	83	272	910
	7 ppm-hr	358	77	270	901
Hardwood pulp logs	Current conditions	2,258	2,175	2,585	2,306
	Change with Respect to Current Conditions				
	75 ppb	335	102	17	223
	15 ppm-hr	336	103	5	237
	11 ppm-hr	363	119	−13	244
	7 ppm-hr	400	137	−9	254

(continued)

Table B-4. Forest Products Production, Million Cubic Feet (continued)

Product	Policy	2010	2020	2030	2040
Softwood saw logs	Current conditions	4,521	5,049	5,458	6,223
	Change with Respect to Current Conditions				
	75 ppb	145	137	156	473
	15 ppm-hr	145	137	157	481
	11 ppm-hr	145	137	156	473
	7 ppm-hr	158	165	167	521
Softwood pulp logs	Current conditions	3,398	3,872	4,241	4,567
	Change with Respect to Current Conditions				
	75 ppb	70	51	−1,134	−1,896
	15 ppm-hr	72	54	−1,135	−1,908
	11 ppm-hr	72	64	−1,129	−1,921
	7 ppm-hr	77	68	−1,129	−1,927

The impact of reduced ozone concentrations on timber market prices was even more substantial compared with current conditions, although these differences are also linked to significant differences in projected baseline prices and the impacts of increased harvest quantities.

Table B-5. Forest Product Prices, U.S. Dollars per Cubic Foot

Product	Policy	2010	2020	2030	2040
Hardwood saw logs	Current conditions	1.07	1.19	1.11	1.05
	Change with Respect to Current Conditions				
	75 ppb	−0.38	−0.55	−0.72	−0.86
	15 ppm-hr	−0.38	−0.55	−0.72	−0.86
	11 ppm-hr	−0.38	−0.56	−0.73	−0.86
	7 ppm-hr	−0.39	−0.57	−0.74	−0.87

(continued)

Table B-5. Forest Product Prices, U.S. Dollars per Cubic Foot (continued)

Product	Policy	2010	2020	2030	2040
Hardwood pulp logs	Current conditions	0.49	0.92	0.86	0.80
	Change with Respect to Current Conditions				
	75 ppb	−0.25	−0.66	−0.42	−0.46
	15 ppm-hr	−0.25	−0.66	−0.41	−0.46
	11 ppm-hr	−0.25	−0.66	−0.41	−0.46
	7 ppm-hr	−0.25	−0.66	−0.43	−0.48
Softwood saw logs	Current conditions	2.61	2.25	2.08	1.88
	Change with Respect to Current Conditions				
	75 ppb	−0.30	−0.35	−0.48	−0.57
	15 ppm-hr	−0.31	−0.35	−0.49	−0.58
	11 ppm-hr	−0.31	−0.37	−0.50	−0.59
	7 ppm-hr	−0.32	−0.38	−0.51	−0.60
Softwood pulp logs	Current conditions	1.68	1.58	1.82	1.59
	Change with Respect to Current Conditions				
	75 ppb	−0.26	−0.46	−0.48	−0.64
	15 ppm-hr	−0.27	−0.46	−0.48	−0.64
	11 ppm-hr	−0.27	−0.46	−0.48	−0.65
	7 ppm-hr	−0.28	−0.47	−0.49	−0.66

B.2.2 Forest Acres Harvested

Harvested acres are projected to decline as a result of higher productivity in the policy cases (i.e., higher growth rates lead to increased quantities of timber per acre such that demand for forestry products can be made by harvesting fewer acres). The most significant reductions occurred in species found in the southern regions where the largest ozone reductions would occur: bottomland hardwoods, oak-pine, and planted pine. The difference between the harvested acres of hardwoods under current conditions and the policy scenarios widens significantly from 2010 to 2040, increasing for a difference of approximately 5% to more than 30%. Impacts to total harvested acres of softwoods are not as large in terms of percentage change, with differences ranging from 2% to 10% by year. Table B-6 presents the number of forest acres harvested, by thousand acres.

Table B-6.   Forest Acres Harvested, Thousand Acres

	Policy	2010	2020	2030	2040
Total hardwood	Current conditions	13,777	11,685	13,486	14,601
	Change with Respect to Current Conditions				
	75 ppb	644	−1,508	−3,300	−4,191
	15 ppm-hr	631	−1,503	−3,320	−4,138
	11 ppm-hr	745	−1,537	−3,467	−4,326
	7 ppm-hr	918	−1,638	−3,557	−4,615
Total softwood	Current conditions	16,971	16,119	15,278	18,273
	Change with Respect to Current Conditions				
	75 ppb	363	−914	−538	−447
	15 ppm-hr	371	−902	−548	−474
	11 ppm-hr	369	−854	−556	−446
	7 ppm-hr	387	−781	−620	−385

B.2.3   Forest Inventory

The impacts observed for forest inventory were similar to those presented in Section 5, but the magnitude of the difference is much larger because the change in ozone concentrations relative to current conditions is substantially larger. The model projected significant increases in existing inventory for both hardwood and softwood species, although the increase was greater in hardwood species. As with the crops, this difference is largely explained by differential sensitivity to ozone between species. Hardwood species show a much higher sensitivity to ozone levels and are thus modeled to respond more dramatically to reductions in ozone concentration. The gap between current conditions and the policy cases widened over time as inventory continued to accumulate, and by 2040, hardwood inventory in each policy case was nearly twice the inventory under current conditions. Accompanying the presence of increased existing inventory in each of the policy cases is that new inventory in the reduced ozone environment was lower than for current conditions beginning in 2025, with the gap increasing over time. Continued growth in existing inventory reduces the incentive for expansion of new inventory.

The largest incremental standard differences occurred in the forest inventory projections. In general, there were incremental differences between alternative ozone standards, with the greatest impact occurring at the 11 ppm-hr standard. Figures B-2 and B-3 depict FASOMGHG projections for existing and new inventory under current conditions compared with each of the policy cases.

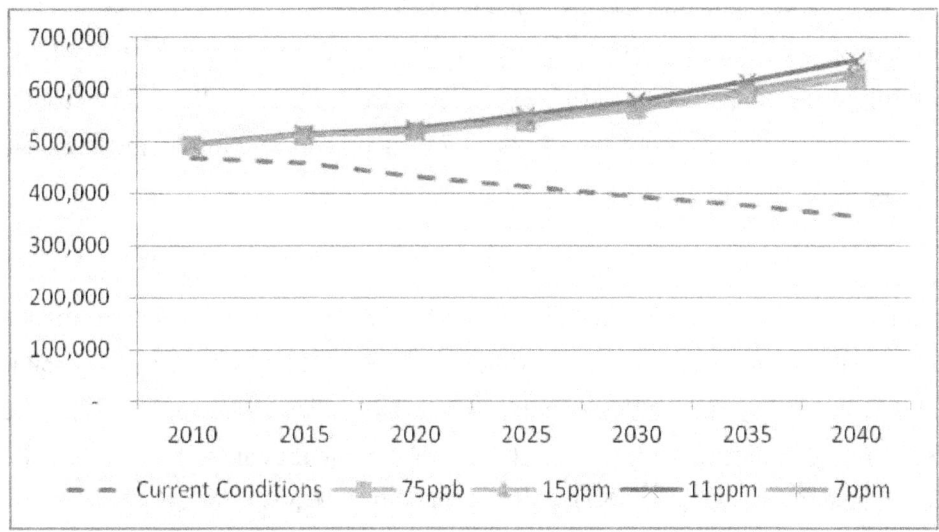

Figure B-2. Existing Forest Inventory, Million Cubic Feet

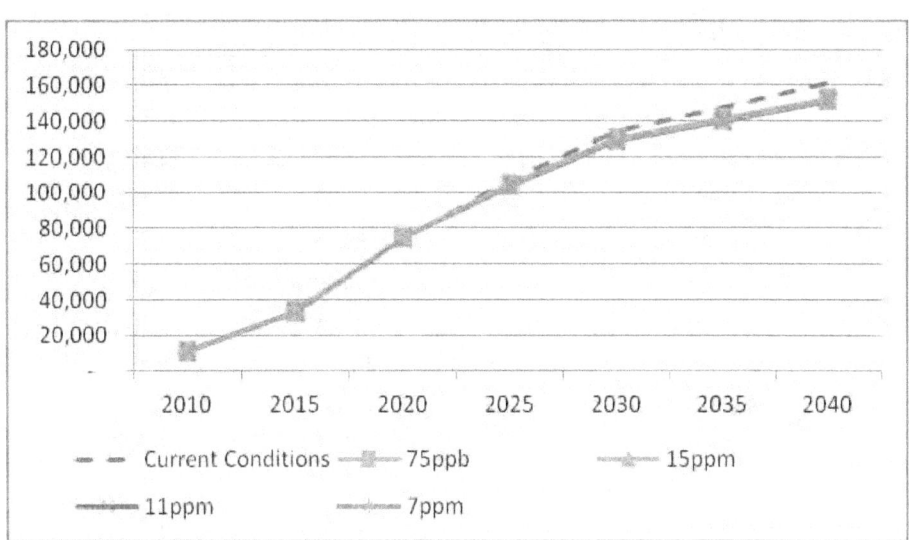

Figure B-3. New Forest Inventory, Million Cubic Feet

B.3 Cross-Sectoral Policy Impacts

B.3.1 Land Use

The general pattern observed for allocation of land use over time within these categories was a decline in forestland coupled with relatively stable cropland and small increases in both traditional pasture and cropland pasture. Forestland declines as productivity and inventory increase because of decreased ozone concentrations, implying that less forestland would be required to meet market demands. Cropland increased to accommodate increases in feed demand induced by livestock product expansion. Acreage retained in CRP also increased compared with current conditions, due to reduced pressure to convert previously agricultural lands back to agriculture because of higher crop yields. Table B-7 presents the land use by major category per thousand acres.

The incremental impact of more stringent ozone policies is apparent in most of these cases, especially reforestation, afforestation, and both types of pasture. The largest impacts were projected in afforestation. Under current conditions, afforestation was projected to decline from 2010 to 2030, followed by substantial growth to 2040. The policy cases not only decline at a faster rate but also continue to decline after 2030. This divergence causes large differences in later projections, with policy cases projecting less than 30% of the projected afforestation under current conditions.

Table B-7. Land Use by Major Category, Thousand Acres

	Policy	2010	2020	2030	2040
Existing forest	Current conditions	343,180	331,162	318,911	317,690
	Change with Respect to Current Conditions				
	75 ppb	−1,337	−2,420	−4,467	−9,854
	15 ppm-hr	−1,342	−2,424	−4,422	−9,857
	11 ppm-hr	−1,383	−2,485	−4,394	−9,815
	7 ppm-hr	−1,395	−2,609	−4,379	−9,796
Reforested	Current conditions	70,445	115,416	147,303	176,004
	Change with Respect to Current Conditions				
	75 ppb	435	−2,472	−7,693	−17,363
	15 ppm-hr	431	−2,460	−7,630	−17,324
	11 ppm-hr	509	−2,479	−7,838	−17,646
	7 ppm-hr	685	−2,674	−8,229	−18,290
Afforested	Current conditions	13,429	11,692	8,884	14,708
	Change with Respect to Current Conditions				
	75 ppb	−773	−1,947	−4,319	−10,247
	15 ppm-hr	−777	−1,951	−4,338	−10,266
	11 ppm-hr	−784	−1,957	−4,374	−10,301
	7 ppm-hr	−787	−2,051	−4,397	−10,325

(continued)

Table B-7. Land Use by Major Category, Thousand Acres (continued)

	Policy	2010	2020	2030	2040
Cropland	Current conditions	312,035	315,471	306,401	293,902
	Change with Respect to Current Conditions				
	75 ppb	−321	−1,687	−1,568	−506
	15 ppm-hr	−308	−1,714	−1,630	−523
	11 ppm-hr	−296	−1,804	−1,725	−621
	7 ppm-hr	−277	−1,881	−1,843	−621
Pasture	Current conditions	84,036	84,617	85,487	81,939
	Change with Respect to Current Conditions				
	75 ppb	394	497	1,323	4,136
	15 ppm-hr	394	497	1,327	4,136
	11 ppm-hr	428	497	1,345	4,168
	7 ppm-hr	428	470	1,311	4,132
Cropland pasture	Current conditions	46,722	45,332	54,413	60,840
	Change with Respect to Current Conditions				
	75 ppb	871	2,665	3,767	5,280
	15 ppm-hr	859	2,693	3,778	5,296
	11 ppm-hr	867	2,857	3,839	5,333
	7 ppm-hr	860	3,085	3,974	5,349
Rangeland	Current conditions	302,210	301,104	300,049	299,039
	Change with Respect to Current Conditions				
	75 ppb	0	0	0	0
	15 ppm-hr	0	0	0	0
	11 ppm-hr	0	0	0	0
	7 ppm-hr	0	0	0	0
CRP	Current conditions	36,141	33,535	33,535	33,535
	Change with Respect to Current Conditions				
	75 ppb	393	945	945	945
	15 ppm-hr	396	948	948	948
	11 ppm-hr	384	935	935	935
	7 ppm-hr	384	935	935	935

B.3.2 Welfare

Welfare impacts resulting from the implementation of policy cases followed the same pattern between the agriculture and forestry sectors, although the magnitude of the impacts was larger in forestry. Consumer surplus increased substantially in both the agricultural and forestry sectors in both cases, as higher productivity under reduced ozone conditions tends to increase total production and reduce market prices. Because demand for most forestry and agricultural commodities is inelastic, producer surplus tends to decline, with higher productivity as the effect of falling prices on profits more than outweighs the effects of higher production levels.

Differences in agriculture between current conditions and the policy cases were small percentage changes relative to the surplus values, representing changes of between 0.1% and 0.3% from 2010 to 2040. In later years, more stringent policy had a more pronounced impact on declines in producer surplus, although the level of policy intervention has little impact in other areas. Table B-8 provides consumer and producer surplus for the agricultural sectors under current conditions, along with the change to surplus for each policy case. Table B-9 provides the consumer and producer surplus in the forestry sector.

Table B-8. **Consumer and Producer Surplus in Agriculture Sector, Million 2010 U.S. Dollars**

	Policy	2010	2015	2020	2025	2030	2035	2040
Consumer surplus	Current conditions	1,916,213	1,939,184	1,967,111	1,993,473	2,022,038	2,049,773	2,074,636
	Change with Respect to Current Conditions							
	75 ppb	1,854	1,491	1,031	1,867	990	1,008	1,379
	15 ppm-hr	1,869	1,489	1,032	1,873	984	1,018	1,382
	11 ppm-hr	1,873	1,514	1,043	1,918	1,032	1,028	1,391
	7 ppm-hr	1,823	1,537	1,066	1,971	1,081	1,034	1,425
Producer surplus	Current conditions	735,135	836,572	821,723	859,440	874,133	835,532	865,536
	Change with Respect to Current Conditions							
	75 ppb	−9,771	−5,007	−6,650	3,725	4,853	1,160	−2,227
	15 ppm-hr	−9,159	−6,262	−5,671	2,764	4,943	1,201	−1,530
	11 ppm-hr	−8,297	−7,204	−5,637	3,955	5,085	−2,252	−38
	7 ppm-hr	−9,501	−6,880	−4,871	4,148	5,117	108	763

Table B-9. Consumer and Producer Surplus in Forestry Sector, Million 2010 U.S. Dollars

	Policy	2010	2015	2020	2025	2030	2035	2040
Consumer surplus	Current conditions	717,612	787,431	804,351	820,803	866,378	888,448	925,814
	Change with Respect to Current Conditions							
	75 ppb	3,728	5,803	4,920	5,572	9,242	6,257	9,068
	15 ppm-hr	3,734	5,834	5,037	5,677	9,244	6,292	9,666
	11 ppm-hr	3,772	5,852	5,279	5,774	9,930	6,313	9,780
	7 ppm-hr	3,813	5,990	5,613	5,795	9,976	6,348	9,848
Producer surplus	Current conditions	97,659	128,377	157,483	149,474	154,962	149,226	140,291
	Change with Respect to Current Conditions							
	75 ppb	−4,337	−6,900	−3,486	−3,199	−9,050	−3,111	−7,158
	15 ppm-hr	−4,347	−6,907	−3,628	−3,361	−9,034	−3,157	−7,997
	11 ppm-hr	−4,378	−6,880	−3,990	−3,377	−9,929	−3,056	−8,016
	7 ppm-hr	−4,473	−6,948	−4,378	−3,236	−9,835	−2,955	−7,924

The policy impact was more substantial in the forestry sector, especially in declines in producer surplus. Consumer surplus was higher in each ozone standard case than under current conditions by between 0.5% and 1.1%, whereas producer surplus was up to 6.4% less than under current conditions. Incremental differences in the ozone standard scenarios were also more apparent in forestry producer surplus: Additional declines occurred for each policy level until the most stringent 7 ppm-hr case, where the policy impact declined or remained the same.

B.3.3 Greenhouse Mitigation Potential

The capacity for both the agricultural and forest sectors to sequester carbon is enhanced by each of the policy cases, with increasing magnitude as policy stringency is increased. Although FASOMGHG projects fewer acres of forestland and total cropland, the accelerated storage of carbon in trees and forestland and cropland soils outweighs any decline from reductions in area. Carbon storage in both sectors is consistently higher in the policy cases, with the gap widening over time. By 2040, the agricultural sector sequestered 8% more carbon under the policy cases and the forestry sector between 15 and 17%, resulting in gains of nearly 14,000 million metric tons CO_2 equivalent (MMtCO$_2$e). Table B-10 presents the carbon sequestration projections for current conditions and each policy case. Negative values under current conditions indicate sequestration or carbon storage. Negative values in the change rows indicate that the alternative stores more carbon than current conditions. Figure B-4 depicts the forestry values.

Table B-10. Carbon Storage, MMtCO$_2$e

	Policy	2010	2020	2030	2040
Agriculture	Current conditions	−18,690	−15,246	−11,815	−7,863
	Change with Respect to Current Conditions				
	75 ppb	−58	−118	−187	−606
	15 ppm-hr	−58	−119	−188	−610
	11 ppm-hr	−60	−123	−193	−616
	7 ppm-hr	−61	−122	−193	−615
Forestry	Current conditions	−73,577	−75,239	−76,781	−77,343
	Change with Respect to Current Conditions				
	75 ppb	−1,102	−3,932	−8,082	−11,840
	15 ppm-hr	−1,103	−3,932	−8,099	−11,853
	11 ppm-hr	−1,121	−4,035	−8,394	−12,433
	7 ppm-hr	−1,152	−4,237	−8,915	−13,442

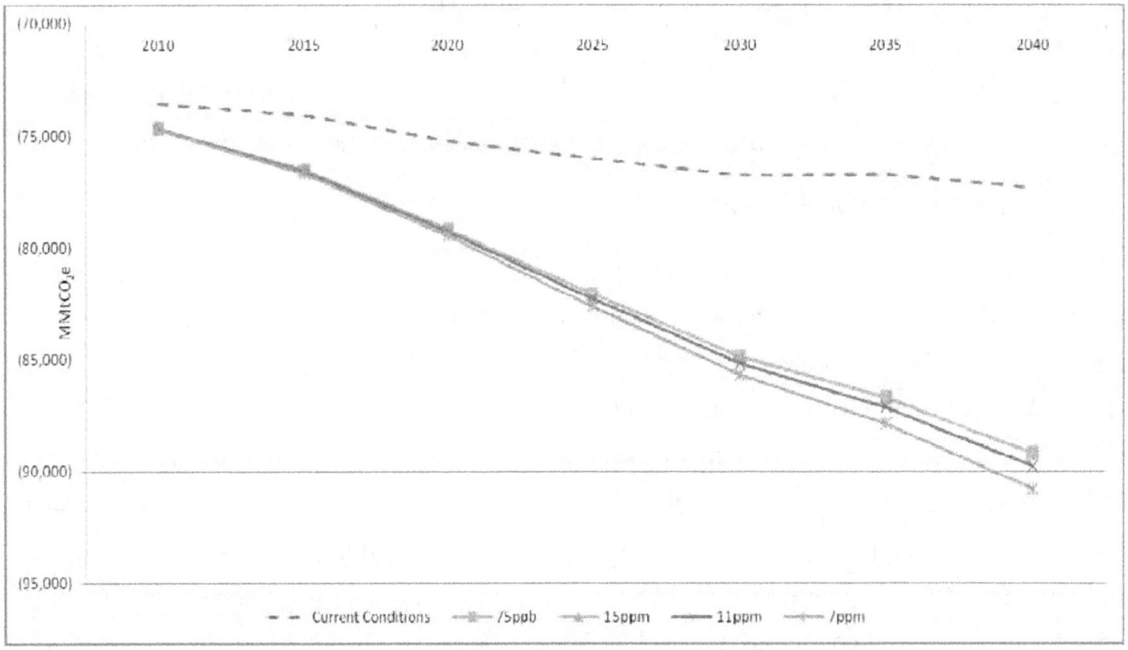

Figure B-4. Carbon Storage in Forestry Sector, MMtCO$_2$e

Changes in forestry sector carbon sequestration are largely driven by changes in forest management, which includes the increases in tree yield in the lower ozone environments. The increased sequestration in this category outweighs losses in sequestration in the other major forestry categories: afforestation and forest soil. Table B-11 presents the detailed changes in forestry carbon sequestration.

Table B-11. Forestry Carbon Sequestration, MMtCO$_2$e

	Policy	2010	2020	2030	2040
Afforestation, trees	Current conditions	−702	−1,579	−1,188	−2,028
	Change with Respect to Current Conditions				
	75 ppb	5	64	388	974
	15 ppm-hr	5	65	389	976
	11 ppm-hr	5	64	394	981
	7 ppm-hr	5	87	394	981
Afforestation, soils	Current conditions	−732	−744	−851	−1,157
	Change with Respect to Current Conditions				
	75 ppb	41	206	478	795
	15 ppm-hr	41	206	480	797
	11 ppm-hr	41	207	482	800
	7 ppm-hr	41	212	485	803
Forest management	Current conditions	−40,174	−38,847	−38,412	−35,366
	Change with Respect to Current Conditions				
	75 ppb	−1,164	−4,175	−8,854	−13,404
	15 ppm-hr	−1,164	−4,174	−8,866	−13,418
	11 ppm-hr	−1,181	−4,278	−9,159	−13,993
	7 ppm-hr	−1,208	−4,509	−9,680	−15,000
Forest soils	Current conditions	−28,270	−27,788	−27,485	−27,505
	Change with Respect to Current Conditions				
	75 ppb	77	222	475	732
	15 ppm-hr	77	222	470	732
	11 ppm-hr	81	229	465	726
	7 ppm-hr	82	239	466	726

B.4   Summary

Impacts to both sectors generally follow similar patterns, although they are generally larger in the forestry sector. Not only are tree species more responsive to changes in ozone, but some of the largest reductions to meet the proposed standards will occur in regions with large forestry sectors: South Central, Southeast, and Rocky Mountains. Reductions in agricultural regions are comparatively moderate. Productivity of both crops and forests is projected to increase at each of the ozone standard levels examined. This increase in supply resulted in decreased prices for forest products and agricultural commodities, which benefits consumer welfare while reducing producer welfare. Unless there are significant changes to wood products markets in particular, producers will be forced to sell at reduced prices in order for the increased supply to be absorbed by the market.

Increased productivity is also projected to affect land use both within and between the agricultural and forest sectors. Within sectors, acreage is generally projected to shift from crops and tree species that are more sensitive to ozone to those that are less sensitive, because productivity in the former will be more substantially affected by reductions in ozone concentrations. For ozone-sensitive crops and species, producers are projected to require less land to produce at the same or higher levels. Forest acreage in particular is projected to decline sharply, driven by declines in both reforestation and afforestation.

Despite reductions in crop and forest area, carbon sequestration is expected to increase over time, led almost entirely by increased forest sequestration. By 2040, the agricultural sector sequestered 8% more carbon under the policy cases and the forestry sector between 15% and 17% more carbon, resulting in gains of nearly 14,000 MMtCO$_2$e.

APPENDIX C
COUNTY-LEVEL AGRICULTURAL WELFARE ANALYSIS: FOCUSING ON COMPARISON WITH CURRENT CONDITIONS

This appendix provides additional information on analyses of changes in county-level agricultural welfare under the ozone standard scenarios relative to current conditions, supplementing the information presented in Section 6. This section uses the comparison of welfare when meeting the current standard versus current conditions as the representative analysis. For incremental effects of the 15 ppm-hr, 11 ppm-hr, and 7 ppm-hr scenarios, see Section 6.

Figure C-1 presents the changes in county-level agricultural welfare for ozone concentrations consistent with the current standard versus current conditions across the modeling periods.

In 2010, the select crop producers' welfare in the Corn Belt region decreased under the 75 ppb scenario—this occurred because the prices of major agricultural commodities decline as yields improve. Soybean production in the Corn Belt region expanded at the expense of corn production. Soybeans have a much larger relative yield gain (RYG) than corn. The revenues lost because of corn production contraction outweighed the revenues gained from expanding soybean production (especially given lower prices for both commodities); therefore, the welfare in the Corn Belt region decreased.

The welfare in the Mississippi River region (located in the South Central region) also decreased because of the corn production contraction and soybean production expansion, accompanied by falling prices for major agricultural products.

Although soybeans are a crop species that would benefit from reduced ozone environments, soybean production in the Great Plains region decreased—this occurred because soybeans have larger RYGs in the Corn Belt region than in the Great Plains region; hence, soybean production shifted to the Corn Belt region. Correspondingly, corn and wheat production in the Great Plains region increased. Nevertheless, the revenues lost because of soybean production contraction are greater than the revenues gained from corn and wheat production expansions; hence, we see welfare decreases in the Great Plains region. In the Rocky Mountains region, the production of various kinds of wheat increased. However, in many counties in the Rocky Mountains region, the effects of price decreases are greater than the effects of production increases on generating revenues. Hence, welfare decreases occurred in many of the counties in the Rocky Mountains region.

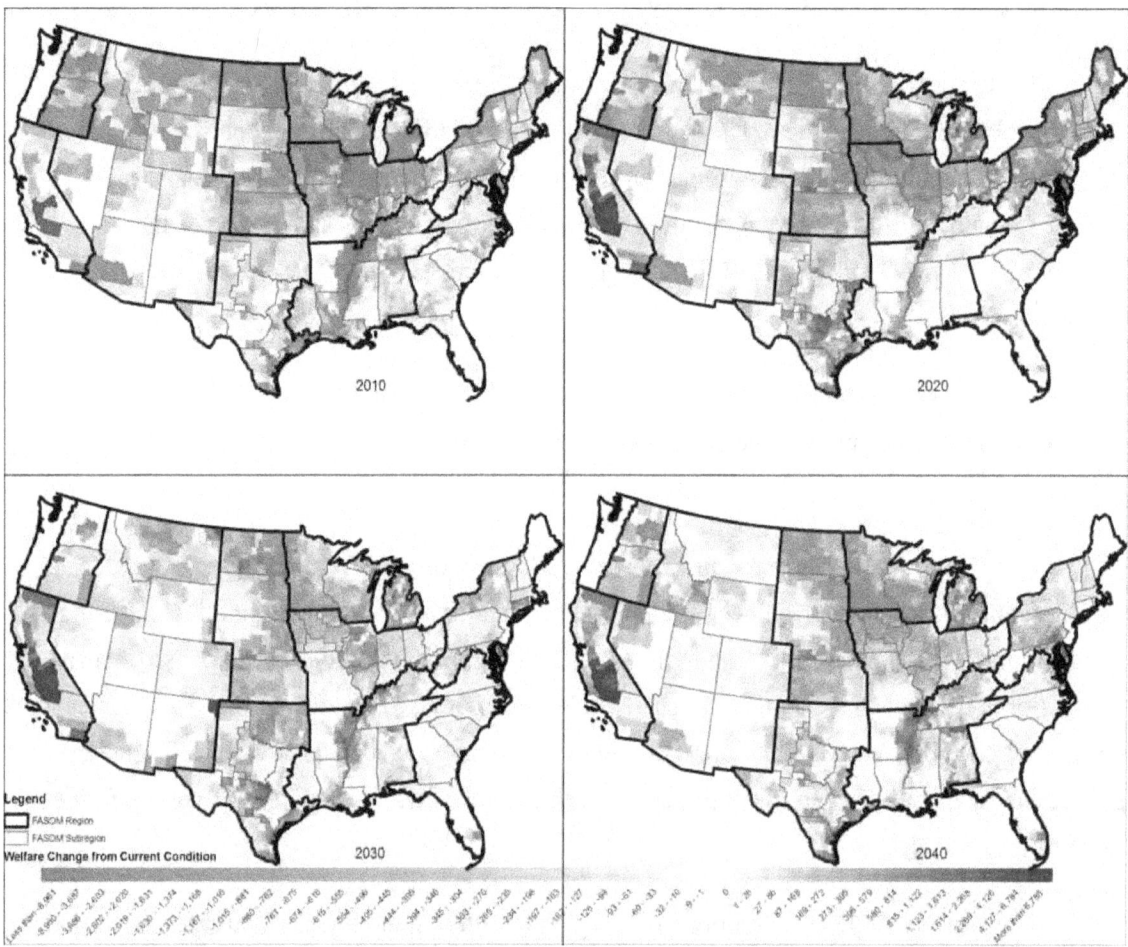

Figure C-1.   Changes in Crop Producer Welfare under 75 ppb with Respect to Current Conditions, 2010-2040

In the Pacific Southwest region, wheat and cotton production increased because these two crops have large RYGs under the 75 ppb scenario. Their production increases were large enough relative to price declines (in part because this region is not a dominant producer of wheat or cotton so has less influence on national prices) that they led to welfare increases in Pacific Southwest region.

Similar patterns of welfare increases and decreases are shown for 2020, as presented in Figure C-1.

By 2030, in the South Central region, welfare increases instead of decreases start to emerge for the majority of this region. This increase occurs because soybean production expansion has gotten larger than it had been in earlier periods. Corn production increases also accompany soybean production expansions as more land is freed from the forestry industry and becomes pasture land for livestock production use, inducing greater demand for feed crop production in the South Central region.

In addition, for the Corn Belt region, welfare decreases still occur in most of this region; however, this time, the underlying reason is because some of the soybean production expansion activities shifted to the South Central and Northeast regions, whereas corn production expanded in part of the Corn Belt region. Overall, lower prices for the primary crops produced in the Corn Belt are still leading to reductions in total producer surplus for the majority of this region.

Similar patterns of welfare increases and decreases are shown for 2040. The changes in magnitudes are attributable to the price changes and production changes under new market equilibriums.

APPENDIX 6C: DISCUSSION OF SUPPLY CURVE SHIFTS

Economists often use consumer and producer surplus as metrics for measuring changes in general welfare. Surplus occurs when the market price differs from what consumers are willing to pay or what producers are willing to accept as payment. As the market price falls below what some consumers are willing to pay this generates a *consumer surplus* for those consumers as they are able to purchase a good or service for less than the maximum amount they are willing to pay. Similarly, if the market price is greater than the minimum level that producers are willing to accept as payment, then a *producer surplus* is generated.

The nature of the impacts on, and the tradeoffs between, consumer and producer surplus is related to the responsiveness of the quantities supplied and demanded to changes in price. The degree of price responsiveness is calculated as the percentage change in that variable divided by the percentage change in price, referred to as the elasticity of the variable. One of the most commonly used elasticities is the own-price elasticity, calculated as the percentage change in quantity demanded for a good or service divided by the percentage change in the price of that good or service. If the own-price elasticity >1, consumers adjust the amount they are willing to purchase by a greater percentage than the change in price and demand for the product is referred to as elastic. When the own-price elasticity <1, on the other hand, demand is inelastic. For many commodities that are relatively low-cost necessities, such as many agricultural goods, consumer demand tends to be inelastic. This is depicted graphically in the steep slope of the demand curve in Figure 6C-1.

At the alternative standard levels considered in this report, crop and forest yields incorporated into FASOMGHG are higher under each of the more stringent alternative standards. The yield increases led to increased quantities of agricultural and forest products, which impact market price and therefore consumer and producer surplus. The precise nature of the tradeoffs depends on a number of factors, including elasticities, but Figure 6C-1 provides an illustrative example that is representative of FASOMGHG commodity projections in many markets.

The initial supply curve, quantity, and price are represented by S_0, Q_0, and P_0, respectively. Consumer surplus is the area of the graph where the price consumers are willing to pay as a function of quantity (D) is greater than the initial market price (P_0), denoted by area A in

Figure 6C-1. Producer surplus is the area where producers are willing to accept (S_0) prices lower than the market price, denoted by area B plus area C. As the supply of agricultural and forest products increases due to increases in yields at more stringent alternative standards, the supply curve shifts to the right (S_1), representing more available product at any given price. The market is now operating at a new equilibrium with a greater quantity produced and sold (Q_1) and lower prices (P_1). Consumer surplus now expands to include areas B, D, and E while producer surplus loses area B and gains F and G. While consumers always benefit from such an outward shift in the supply curve, the net change to producer surplus depends on the magnitude of the lost area B in relation to gained areas F and G. If B is greater than F and G, producer surplus declines and producers are worse off. Conversely, if B is less than what is gained in F and G, producers benefit.

Figure 6C-1. Change in Consumer and Producer Surplus with Outward Supply Shift

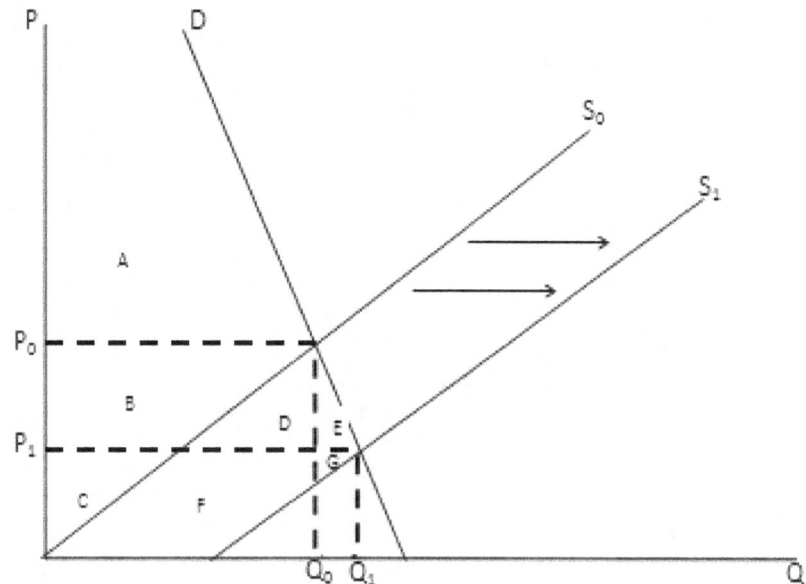

APPENDIX 6D

i-Tree MODEL, METHODOLOGY, and RESULTS

6D.1   i-Tree Model Components

i-Tree version 4.0 offers several urban forest assessment applications including i-Tree Eco, previously known as UFORE. The Urban Forest Effects (UFORE) model was developed to aid in assessing urban forest structure, functions, and values (Nowak and Crane 2000). This model contains protocols to measure and monitor urban forests as well as estimate ecosystem functions and values.

The basic premise behind the UFORE model is that urban forest structure affects forest functions and values. By having an accurate assessment of urban forest structure, better estimates of functions and values can be produced. The model uses a sampling procedure to estimate various measured structural attributes about the forest (e.g., species composition, number of trees, diameter distribution) within a known sampling error. The model uses the measured structural information to estimate other structural attributes (e.g., leaf area, tree and leaf biomass) and incorporates local environmental data to estimate several functional attributes (e.g., air pollution removal, carbon sequestration, building energy effects). Economic data from the literature are used to estimate the value of some of the functions. The model includes the following modules descriptions of which are excerpted from Nowak, 2008.

6D.2   Urban Forest Structure

Urban forest structure is the spatial arrangement and characteristics of vegetation in relation to other objects (e.g., buildings) within urban areas (e.g., Nowak 1994a). This module quantifies urban forest structure (e.g., species composition, tree density, tree health, leaf area, leaf and tree biomass), value, diversity, and potential risk to pests.

6D.2.1   Sampling

Urban Forest Effect model assessments have used two basic types of sampling to quantify urban forest structure: randomized grid and stratified random sampling. With the randomized grid sampling, the study area is divided into equal-area grid cells based on the desired number of plots and then one plot is randomly located within each grid cell. The study

1   area can then be subdivided into smaller units of analysis (i.e., strata) after the plots are

2   distributed (post-stratification). Plot distribution among the strata will be proportional to the

3   strata area. This random sampling approach allows for relatively easy assessment of changes

4   through future measurements (urban forest monitoring), but likely at the cost of increased

5   variance (uncertainty) of the population estimates. With stratified random sampling, the study

6   area is stratified before distributing the plots and plots are randomly distributed within each

7   stratum (e.g., land use). This process allows the user to distribute the plots among the strata to

8   potentially decrease the overall variance of the population estimate. For example, because tree

9   effects are often the primary focus of sampling, the user can distribute more plots into strata that

10 have more trees. The disadvantage of this approach is that it makes long-term change

11 assessments more difficult as a result of the potential for strata to change through time. There is

12 no significant difference in cost or time to establish plots regardless of sampling methods for a

13 fixed number of plots. However, there are likely differences in estimate precision. Pre-

14 stratification, if done properly, can reduce overall variance because it can focus more plots in

15 areas of higher variability. Any plot size can be used in UFORE, but the typical plot size used is

16 0.04 ha (0.1 ac). The number and size of plots will affect total cost of the data collection as well

17 as the variance of the estimates (Nowak et al. 2008).

18 ## 6D.2.2   Data Collection Variables

19   There are four general types of data collected on a UFORE plot: 1) general plot

20 information used to identify the plot and its general characteristics; 2) shrub information used to

21 estimate shrub leaf area/biomass, pollution removal, and volatile organic compound (VOC)

22 emissions by shrubs; 3) tree information used to estimate forest structural attributes, pollution

23 removal, VOC emissions, carbon storage and sequestration, energy conservation effects, and

24 potential pest impacts of trees; and 4) ground cover data used to estimate the amount and

25 distribution of various ground cover types in the study area. Typically, shrubs are defined as

26 woody material with a diameter at breast height (dbh; height at 1.37 m [4.5 ft]) less than 2.54 cm

27 (1 in), whereas trees have a dbh greater than or equal to 2.54 cm (1 in). Trees and shrubs can also

28 be differentiated by species (i.e., certain species are always a tree or always a shrub) or with a

29 different dbh minimum threshold. For example, in densely forested areas, increasing the

30 minimum dbh to 12.7 cm (5 in) can substantially reduce the field work by decreasing the number

1   of trees measured, but less information on trees will be attained. Woody plants that are not 30.5

2   cm (12 in) in height are considered herbaceous cover (e.g., seedlings). Shrub masses within each

3   plot are divided into groups of same species and size, and for each group, appropriate data are

4   collected. Tree variables are collected on every measured tree. Field data are collected during the

5   in-leaf season to help assess crown parameters and health. More detailed information on plot

6   data collection methods and equipment can be found in the i-Tree User's Manual (i-Tree 2008).

7   6D.2.3   Leaf Area and Leaf Biomass

8   Leaf area and leaf biomass of individual open-grown trees (crown light exposure [CLE]

9   of 4 to 5) are calculated using regression equations for deciduous urban species (Nowak 1996). If

10 shading coefficients (percent light intensity intercepted by foliated tree crowns) used in the

11 regression did not exist for an individual species, genus or hardwood averages are used. For

12 deciduous trees that are too large to be used directly in the regression equation, average leaf area

13 index (LAI: m^2 leaf area per m^2 projected ground area of canopy) is calculated by the regression

14 equation for the maximum tree size based on the appropriate height–width ratio and shading

15 coefficient class of the tree. This LAI is applied to the ground area (m^2) projected by the tree's

16 crown to calculate leaf area (m^2). For deciduous trees with height-to-width ratios that are too

17 large or too small to be used directly in the regression equations, tree height or width is scaled

18 downward to allow the crown to the reach maximum (2) or minimum (0.5) height-to-width ratio.

19 Leaf area is calculated using the regression equation with the maximum or minimum ratio; leaf

20 area is then scaled back proportionally to reach the original crown volume. For conifer trees

21 (excluding pines), average LAI per height to-width ratio class for deciduous trees with a shading

22 coefficient of 0.91 is applied to the tree's ground area to calculate leaf area. The 0.91 shading

23 coefficient class is believed to be the best class to represent conifers because conifer forests

24 typically have approximately 1.5 times more LAI than deciduous forests (Barbour et al. 1980)

25 and 1.5 times the average shading coefficient for deciduous trees (0.83; see Nowak 1996) is

26 equivalent to LAI of the 0.91 shading coefficient. Because pines have lower LAI than other

27 conifers and LAI that are comparable to hardwoods (e.g., Jarvis and Leverenz 1983; Leverenz

28 and Hinckley 1990), the average shading coefficient (0.83) is used to estimate pine leaf area.

29 Leaf biomass is calculated by converting leaf area estimates using species-specific

30 measurements of grams of leaf dry weight/m^2 of leaf area. Shrub leaf biomass is calculated as the

1 product of the crown volume occupied by leaves (m^3) and measured leaf biomass factors (g/m^3)

2 for individual species (e.g., Wincr et al. 1983; Nowak 1991). Shrub leaf area is calculated by

3 converting leaf biomass to leaf area based on measured species conversion ratios (m^2/g). As a

4 result of limitations in estimating shrub leaf area by the crown-volume approach, shrub leaf area

5 is not allowed to exceed a LAI of 18. If there are no leaf-biomass to-area or leaf-biomass-to-

6 crown-volume conversion factors for an individual species, genus or hardwood/conifer averages

7 are used. For trees in more forest stand conditions (higher plant competition), LAI for more

8 closed canopy positions (CLE 0–1) is calculated using a forest leaf area formula based on the

9 Beer-Lambert Law:

10
$$LAI = \ln(IIo) - k$$

11 where: I =light intensity beneath canopy; Io = light intensity above canopy; and k=light

12 extinction coefficient (Smith et al. 1991). The light extinction coefficients are 0.52 for conifers

13 and 0.65 for hardwoods (Jarvis and Leverenz 1983). To estimate the tree leaf area (LA):

14
$$LA = [\ln(1 - xs) - k] \times r2$$

15 where xs is average shading coefficient of the species and r is the crown radius. For CLE 2–3:

16 LA is calculated as the average of leaf area from the open-grown (CLE 4–5) and closed canopy

17 equations (CLE 0–1). Estimates of LA and leaf biomass are adjusted downward based on crown

18 leaf dieback (tree condition). Trees are assigned to one of seven condition classes: excellent (less

19 than 1% dieback); good (1% to 10% dieback); fair (11% to 25% dieback); poor (26% to 50%

20 dieback); critical (51% to 75% dieback); dying (76% to 99% dieback); and dead (100% dieback).

21 Condition ratings range between 1 indicating no dieback and 0 indicating 100% dieback (dead

22 tree). Each class between excellent and dead is given a rating between 1 and 0 based on the

23 midvalue of the class (e.g., fair = 11% to 25% dieback is given a rating of 0.82 or 82% healthy

24 crown). Tree leaf area is multiplied by the tree condition factor to produce the final LA estimate.

25 **6D.2.4   Carbon Storage and Annual Sequestration**

26 This module calculates total stored carbon and gross and net carbon sequestered annually

27 by the urban forest. Biomass for each measured tree is calculated using allometric equations

28 from the literature (see Nowak 1994c; Nowak et al. 2002b). Equations that predict aboveground

1 biomass are converted to whole tree biomass based on a root-to-shoot ratio of 0.26 (Cairns et al.

2 1997). Equations that compute fresh weight biomass are multiplied by species- or genus-specific

3 conversion factors to yield dry weight biomass. These conversion factors, derived from average

4 moisture contents of species given in the literature, averaged 0.48 for conifers and 0.56 for

5 hardwoods (see Nowak et al. 2002b). Open-grown, maintained trees tend to have less

6 aboveground biomass than predicted by forest-derived biomass equations for trees of the same

7 dbh (Nowak 1994c). To adjust for this difference, biomass results for urban trees are multiplied

8 by a factor of 0.8 (Nowak 1994c). No adjustment is made for trees found in more natural stand

9 conditions (e.g., on vacant lands or in forest preserves). Because deciduous trees drop their

10 leaves annually, only carbon stored in wood biomass is calculated for these trees. Total tree dry

11 weight biomass is converted to total stored carbon by multiplying by 0.5 (Forest Products

12 Laboratory 1952; Chow and Rolfe 1989). The multiple equations used for individual species

13 were combined to produce one predictive equation for a wide range of diameters for individual

14 species. The process of combining the individual formulas (with limited diameter ranges) into

15 one more general species formula produced results that were typically within 2% of the original

16 estimates for total carbon storage of the urban forest (i.e., the estimates using the multiple

17 equations). Formulas were combined to prevent disjointed sequestration estimates that can occur

18 when calculations switch between individual biomass equations. If no allometric equation could

19 be found for an individual species, the average of results from equations of the same genus is

20 used. If no genus equations are found, the average of results from all broadleaf or conifer

21 equations is used

22 **6D.2.5   Urban Tree Growth and Carbon Sequestration**

23   To determine a base growth rate based on length of growing season, urban street tree

24 (Fleming 1988; Frelich 1992; Nowak 1994c), park tree (deVries 1987), and forest growth

25 estimates (Smith and Shifley 1984) were standardized to growth rates for 153 frost-free days

26 based on: standardized growth = measured growth × (153/number of frost-free days of

27 measurement). Average standardized growth rates for street (open-grown) trees were 0.83

28 cm/year (0.33 in/year). Growth rates of trees of the same species or genera were then compared

29 to determine the average difference between standardized street tree growth and standardized

30 park and forest growth rates. Park growth averaged 1.78 times less than street trees, and forest

1   growth averaged 2.29 times less than street tree growth. Crown light exposure measurements of

2   0 to 1 were used to represent forest growth conditions; 2 to 3 for park conditions; and 4 to 5 for

3   open-grown conditions. Thus, the standardized growth equations are:

4   Standardized growth (SG) _ 0.83 cm/year (0.33 in/year) × number of frost free days/153

5   and for: CLE 0–1: Base growth=SG/2.26; CLE 2–3: base growth=SG /1.78; and CLE 4–5: base

6   growth= SG. Base growth rates are adjusted based on tree condition. For trees in fair to excellent

7   condition, base growth rates are multiplied by 1 (no adjustment), poor trees' growth rates are

8   multiplied by 0.76, critical trees by 0.42, dying trees by 0.15, and dead trees by 0. Adjustment

9   factors are based on percent crown dieback and the assumption that less than 25% crown dieback

10 had a limited effect on dbh growth rates. The difference in estimates of carbon storage between

11 year x and year $x + 1$ is the gross amount of carbon sequestered annually.

12 ### 6D.2.5   Air Pollution Removal

13   This module quantifies the hourly amount of pollution removed by the urban forest, its

14 value, and associated percent improvement in air quality throughout a year. Pollution removal

15 and percent air quality improvement are calculated based on field, pollution concentration, and

16 meteorologic data. This module is used to estimate dry deposition of air pollution (i.e., pollution

17 removal during nonprecipitation periods) to trees and shrubs (Nowak et al. 1998, 2000). This

18 module calculates the hourly dry deposition of ozone (O_3), sulfur dioxide (SO_2), nitrogen dioxide

19 (NO_2), carbon monoxide (CO), and particulate matter less than nominal diameter of 10 μm

20 (PM_{10}) to tree and shrub canopies throughout the year based on tree-cover data, hourly NCDC

21 weather data, and U.S. Environmental Protection Agency pollution concentration monitoring

22 data. The pollutant flux (F; in $g/m^2/s$) is calculated as the product of the deposition velocity (Vd;

23 in m/s) and the pollutant concentration (C; in g/m^3):

24                      $$F = Vd \times C$$

25 Deposition velocity is calculated as the inverse of the sum of the aerodynamic (Ra), quasilaminar

26 boundary layer (Rb), and canopy (Rc) resistances (Baldocchi et al. 1987):

27                      $$Vd = (Ra + Rb + Rc){-}1$$

1   Hourly meteorologic data from the closest weather station (usually airport weather

2   stations) are used in estimating Ra and Rb. In-leaf, hourly tree canopy resistances for O_3, SO_2,

3   and NO_2 are calculated based on a modified hybrid of big leaf and multilayer canopy deposition

4   models (Baldocchi et al. 1987; Baldocchi 1988). Because CO and removal of particulate matter

5   by vegetation are not directly related to transpiration, Rc for CO is set to a constant for in-leaf

6   season (50,000 sec/m [15,240 sec/ft]) and leaf-off season (1,000,000 sec/m [304,800 sec/ft])

7   based on data from Bidwell and Fraser (1972). For particles, the median deposition velocity from

8   the literature (Lovett 1994) is 0.0128 m/s (0.042 ft/s) for the in-leaf season. Base particle Vd is

9   set to 0.064 m/s (0.021 ft/s) based on a LAI of 6 and a 50% resuspension rate of particles back to

10   the atmosphere (Zinke 1967). The base Vd is adjusted according to actual LAI and in-leaf versus

11   leaf-off season parameters. Bounds of total tree removal of O_3, NO_2, SO_2, and PM_{10} are

12   estimated using the typical range of published in-leaf dry deposition velocities (Lovett 1994).

13   Percent air quality improvement is estimated by incorporating local or regional boundary layer

14   height data (height of the pollutant mixing layer). More detailed methods on this module can be

15   found in Nowak et al. (2006a). The monetary value of pollution removal by trees is estimated

16   using the median externality values for the United States for each pollutant. These values, in

17   dollars per metric ton (mt) are: NO_2, $6,752 mt–1; PM_{10}, $4,508 mt–1; SO_2, $1,653 mt–1; and

18   CO, $959 mt–1 (Murray et al. 1994). Values in dollars per short ton, commonly used in the U.S.,

19   are approximated 90% of those per metric ton. Recently, these values were adjusted to 2007

20   values based on the producer's price index (Capital District Planning Commission 2008) and are

21   now: NO_2, $9,906 mt–1; PM_{10}, $6,614 mt–1; SO_2, $2,425 mt–1; and CO, $1,407 mt–1.

22   Externality values for O_3 are set to equal the value for NO_2.

23   **6.D.3 i-Tree Forecast Prototype Model Methods and Results**

24   The i-Tree Forecast Prototype Model was built to simulate future forest structure (e.g.,

25   number of trees and sizes) and various ecosystem services based on annual projections of the

26   current forest structure data. There are 3 main components of the model:

27   1) Tree growth – simulates tree growth to annually project tree diameter, crown size and

28   leaf area for each tree

29   2) Tree mortality – annually removes trees from the projections based on user defined

30   mortality rates

1   3) Tree establishment – annually adds new trees to the projection. These inputs can be used

2   to illustrate the effect of the new trees or determine how many new trees need to be added

3   annually to sustain a certain level of tree cover or benefits.

4   **6D.3.1.   Tree Growth**

5   Annual tree diameter growth is estimated for the region based on: 1) the length of

6   growing season, 2) species average growth rates, 3) tree competition, 4) tree condition, and 5)

7   current tree height relative to maximum tree height.

8   To determine a base growth rate based on length of growing season, urban street tree,

9   park tree, and forest growth estimates were standardized to growth rates for 153 frost free days

10   based on: Standardized growth = measured growth x (153/ number of frost free days of

11   measurement). Growth rates of trees of the same species or genera were also compared to

12   determine the average difference between standardized street tree growth and standardized park

13   and forest growth rates. Park growth averaged 1.78 times less than street trees, and forest growth

14   averaged 2.29 times less than street tree growth.

15   For this study, average standardized growth rates for open-grown (street) trees was input

16   as 0.26 in/yr for slow growing species, 0.39 in/yr for moderate growing species and 0.52 in/yr for

17   fast growing species. Crown light exposure (CLE) measurements of 0-1 were used to represent

18   forest growth conditions; 2-3 for park conditions; and 4-5 for open-grown conditions. Thus, for:

19   CLE 0-1: Base growth = Standardized growth (SG) / 2.26; CLE 2-3: Base growth = SG / 1.78;

20   and CLE 4-5: Base growth = SG. However, as the percent canopy cover increased or decreased,

21   the CLE correction factors were adjusted proportionally to the amount of available greenspace

22   (i.e., as tree cover dropped and available greenspace increased – the CLE adjustment factor

23   dropped; as tree cover increased and available greenspace dropped – the CLE adjustment factor

24   increased).

25   Base growth rates are also adjusted based on tree condition. For trees in fair to excellent

26   condition, base growth rates are multiplied by 1 (no adjustment), Trees in poor condition by

27   0.76, critical trees by 0.42, dying trees by 0.15, and dead trees by 0. Adjustment factors are based

28   on percent crown dieback and the assumption that less than 25 percent crown dieback had a

29   limited effect on dbh growth rates.

1   As trees approach their estimated maximum height, growth rates are reduced. Thus the

2   species growth rates as described above were adjusted based on the ratio between the current

3   height of the tree and the average height at maturity for the species. When a tree's height is over

4   80% of its average height at maturity, the amount of annual dbh growth is proportionally reduced

5   from full growth at 80% of height to ½ growth rate at height at maturity. The growth rate is

6   maintained at ½ growth until the tree is 125% past maximum height, when the growth rate is

7   then reduced to 0 in/yr.

8   Tree height, crown width, crown height and leaf area were then estimated based on tree

9   diameter each year. Height, crown height and crown width are calculated using species, genus,

10   order and family specific equations that were derived from measurements from urban tree data

11   (publication in preparation). If there was no equation for a particular species, then genus

12   equation was used, followed by the family and order equations if necessary. If no order equation

13   could be used, one average equation for all trees was used to estimate these parameters. Leaf area

14   was calculated from the crown height, tree height and crown width estimates based on standard i-

15   Tree methods.

16   Total canopy cover was calculated by summing the crown area of each tree in the

17   population. This estimate of crown area was adjusted to attain the actual tree cover of the study

18   area based on photo-interpretation. As trees often have overlapping crown, the sum of the crown

19   areas will often over estimate total tree cover as determined by aerial estimates. Thus the crown

20   overlap can be determined by comparing the two estimates:

21   % crown overlap = (sum of crown area – actual tree cover area) / sum of crown area

22

23   When future projections predicted an increase in percent canopy cover, the percent crown

24   overlap was held constant However, when 100% canopy cover was attained all new canopy

25   added was considered as overlapping canopy. When there was a projected decrease in percent

26   canopy cover, the percent crown overlap decreased in proportion to the increase in the amount of

27   available greenspace (i.e., as tree cover dropped and available greenspace increased – the crown

28   overlap decreased).

6D.3.2   Tree Mortality Rate

Canopy dieback is the first determinant for tree mortality with trees 50 – 75% dieback having a mortality rate of 13.1% annual mortality rate; trees with 76-99% dieback having a 50% annual mortality rate, and trees with 100% dieback having a 100% annual mortality rate. Trees with less than 50% dieback have a user defined mortality rate that is adjusted based on the tree size class and the current tree dbh.

Trees are placed into species size classes where small trees have an average height at maturity of less than or equal to 40 ft (maximum dbh class = 20+ inches), medium trees have mature tree height of 41- 60 ft (maximum dbh = 30 inches), and large trees have a mature height of greater than 60 ft (maximum dbh = 40 inches). Each size class has a unique set of 7 DBH ranges to which base mortality rates area assigned based on measured tree mortality by dbh class. The same distribution of mortality by dbh class was used for all tree size classes, but the range of the dbh classes differed by size class. The actual mortality rate for each dbh class was adjusted so that the overall average mortality rate for the base population equaled the mortality rates assigned by user. That is, the relative curve of mortality stayed the same among dbh classes, but the actual values would change based on the user defined overall average rate.

6D.3.3.   Tree Establishment

Based on the desired canopy cover level and the number of years desired to reach that canopy level, the program calculates the number of trees needed to be established annually to reach that goal given the model growth and mortality rate. In adding new trees to the model each year, the species composition of new trees was assumed to be proportional to the current species composition. Crown light exposure of newly established trees was also assumed to be proportional to the current growth structure of the canopy. Newly established trees were input with a starting dbh of 1 inch.

6D.4   Ozone Effects Analysis Methods

For this Risk and Exposure Assessment the U.S. Forest Service volunteered their expertise to develop the methodology and run the i-Tree model to project the impact of ozone on carbon sequestration and air pollution removal in selected urban areas. EPA provided CMAQ

1 model generated W126 results for current ambient ozone concentrations and for a model

2 adjusted scenario that just meets the current standard. These methods are described in Chapter 4

3 Air Quality Considerations. For the effects of O_3, we used the concentration-response functions

4 for the 11 tree species analyzed in Chapter 5 to reduce the growth of the trees over a 25 period

5 and compared base model estimates (full-growth) with O_3 effected results (reduced growth).

6 Tree growth was only reduced in analyzed cities for the 11 species that had W126 equations.

7   We used a new forecast model (Nowak, 2012) components of which are described above

8 in the sections on tree growth, mortality, and establishment. This model simulated tree growth,

9 tree influx and mortality annually to estimate annual changes in number of trees, tree cover and

10 stored carbon. For these scenarios, we adjusted the annual mortality (3 or 4%) and influx rate

11 (between 1 and 6 trees / ha / yr) to keep canopy cover as close to current values as possible after

12 25 years. These base assumptions were consistent in both runs (full and reduced growth). Species

13 composition of new trees added annually was proportional to the current species population.

14   Carbon estimates: total carbon storage at the end of the 25 year period was contrasted

15 between the model runs to estimate the impact of reduced growth due to O_3. Differences in

16 number of trees and tree sizes at the end of 25 years will affect the carbon estimate.

17   Pollution removal: pollution removal was based calculating the average tons of air

18 pollutants removed. The forecast model was then used to project differences in estimated tree

19 cover (m^2) between the model runs for each of the 25 years. These annual tons of pollutants

20 removed were summed to estimate the total impact over 25 years.

21   All model runs use the same assumptions, so difference in the estimates are due to

22 reduced growth. However, the magnitude of the impact over 25 years will be affected by the

23 assumptions. As you change the mortality and influx rates, the magnitude of the differences

24 between the model runs will differ. We tried to use reasonable estimates based on limited data on

25 mortality and influx rates. We used Nowak (2012, in press) to help estimate an influx rate and

26 the attached paper on tree mortality to estimate a mortality rate, but we reduced the rate to 3-4%

27 as forest stands are around 1% and the mortality in this paper was around 6%. We have limited

28 mortality data for urban trees, but based on the data and our experience, we believe 3-4% to be

29 reasonable, but it likely varies somewhere between 1% and 5%.

1   **6D.5   i-Tree Results for O₃ Welfare REA**

2   **Table 6D-1   Top Ten Most Common Tree Species (Species with available C-R functions**
3   **highlighted in red)**

Rank	Baltimore	Syracuse	Chicago	Atlanta	Tennessee
1	American beech	European buckthorn	European buckthorn	Sweetgum	Chinese privet
2	Black locust	Sugar maple	Green ash	Loblolly pine	Virginia pine
3	American elm	Black cherry	Boxelder	Flowering dogwood	Eastern red cedar
4	Tree of heaven	Boxelder	Black cherry	Tulip tree	Hackberry
5	White ash	Norway maple	Hardwood	Water oak	Flowering dogwood
6	Black cherry	Northern white cedar	American elm	Boxelder	Amur honeysuckle
7	White mulberry	Norway spruce	Sugar maple	Black cherry	Winged elm
8	Northern red oak	Staghorn sumac	White ash	White oak	Red maple
9	Red maple	Eastern cottonwood	Amur honeysuckle	Red maple	Black tupelo
10	White oak	Eastern hophornbeam	Silver maple	Southern red oak	American beech
Percent of Top 10 species with C-R function	8.5%	18.5%	7.7%	6.6%	9.3%
Percent of Total Tree Species with C-R function	11.2%	20.2%	10.5%	8.9%	17.4%

4

5   Summary: Data from 5 urban areas were simulated to estimate the effect of O₃ (based on
6   the W126 index) on tree ecosystem services of carbon storage and air pollution removal. There
7   were 6 runs: Standard run (i-Tree no adjustments), recent conditions (ozone adjusted), existing
8   standard (75 ppb), adjusted to 15 ppm-hrs, adjusted to 11 ppm-hrs, adjusted to 7 ppm-hs. Runs
9   for 75 ppb (existing standard) and adjusted to 15 ppm-hrs had the same results. The prototype i-
10   Tree Forecast model was used to estimate growth and ecosystem services by trees over a 25-year
11   period. Tree data from the urban areas were loaded in the Forecast model as a base case scenario
12   and simulated for 25 years. The tree growth was then adjusted downward based on the reduced

1 growth factors for 11 species using the W126 protocol and equations (only W126 species had

2 reduced growth). The differences between the scenarios are then contrasted (Standard = base

3 case; O_3 adjusted = W126 reduced growth) for the 25-year period. The model assumed an annual

4 influx of between 1-6 trees/ha/yr and a 3-4% annual mortality rate. These values are updated

5 based on new adjusted RYL values.

6 **Table 6D-2 O_3-Adjusted Canopy Cover for Recent Conditions, Existing and Alternative**
7 **W126 Standard Levels by Region**

Region	Area (ha)	Percent Canopy Cover		O_3-Adjusted Percent Canopy Cover (after 25 years)			
		Recent	After 25yrs	Recent Conditions	Existing Standard & 15 ppm-hrs	11 ppm-hrs	7 ppm-hrs
Atlanta	34,139	52.1	51.9	45.2	49.6	22.212	5.836
Baltimore	20,917	28.5	29.2	26.5	29.1	18.769	0.372
Chicago	993,036	21	20.9	19.1	19.2	7.311	4.582
Syracuse	6,501	26.9	27.8	24.3	27.6	7.812	0.104
Tennessee	630,614	37.7	37.6	34.9	37.1	16.441	4.267

8

9

10 **Table 6D-3   Relative Year Loss Index, Regeneration Rate, and Annual Tree Mortality**
11 **by Region**

Region	Relative Yield Loss Index		Regeneration Rate (trees per ha)	Annual Percent Mortality
	Recent Conditions	Existing Standard & 15 ppm-hrs		
Atlanta	22.212	5.836	2	4%
Baltimore	18.769	0.372	2	3%
Chicago	7.311	4.582	1	3%
Syracuse	7.812	0.104	6	3%
Tennessee	16.441	4.267	1	4%

12

1 **Table 6D-4   Pollution Removal over 25 years for Recent Conditions, Existing and**
2 **Alternative W126 Standards by Region (in metric tons)**

Pollutant and Region	Standard Run	Recent Conditions	Existing Standard & 15 ppm-hrs	11 ppm-hrs	7 ppm-hrs	Relative to Recent Conditions		
						Existing Standard & 15 ppm-hrs	11 ppm-hrs	7 ppm-hrs
CO								
Atlanta	1,482	1,315	1,429	1,432	1,448	113	117	133
Baltimore	186	171	186	186	186	15	15	15
Chicago	8,620	7,916	8,001	8,050	8,143	85	134	227
Syracuse	55	49	55	55	55	6	6	6
Tennessee	12,854	11,731	12,626	12,757	12,916	895	1,026	1,185
NO$_2$								
Atlanta	6,852	6,081	6,605	6,621	6,693	524	540	613
Baltimore	1,968	1,809	1,963	1,963	1,963	155	155	155
Chicago	104,247	95,738	96,766	97,364	98,489	1,028	1,626	2,751
Syracuse	50	45	50	50	50	6	6	6
Tennessee	54,381	49,632	53,419	53,973	54,645	3,788	4,341	5,013
O$_3$								
Atlanta	25,495	22,625	24,574	24,634	24,905	1,949	2,009	2,279
Baltimore	6,262	5,755	6,247	6,247	6,247	492	492	492
Chicago	243,701	223,811	226,214	227,612	230,242	2,403	3,801	6,431
Syracuse	1,544	1,367	1,541	1,541	1,541	173	173	173
Tennessee	393,205	358,861	386,247	390,251	395,107	27,387	31,391	36,246
PM$_{10}$								
Atlanta	20,009	17,756	19,286	19,333	19,545	1,530	1,577	1,789
Baltimore	4,242	3,899	4,232	4,232	4,232	333	333	333
Chicago	171,106	157,140	158,827	159,809	161,655	1,687	2,669	4,515
Syracuse	840	743	838	838	838	94	94	94
Tennessee	175,883	160,521	172,771	174,562	176,734	12,250	14,041	16,213
SO$_2$								
Atlanta	3,380	2,999	3,257	3,265	3,301	258	266	302

Pollutant and Region	Standard Run	Recent Conditions	Existing Standard & 15 ppm-hrs	11 ppm-hrs	7 ppm-hrs	Relative to Recent Conditions		
						Existing Standard & 15 ppm-hrs	11 ppm-hrs	7 ppm-hrs
Baltimore	852	783	850	850	850	67	67	67
Chicago	29,675	27,253	27,546	27,716	28,036	293	463	783
Syracuse	71	63	71	71	71	8	8	8
Tennessee	59,371	54,185	58,320	58,925	59,658	4,135	4,740	5,473
Total								
Atlanta	57,218	50,776	55,151	55,285	55,892	4,374	4,509	5,116
Baltimore	13,510	12,416	13,478	13,478	13,478	1,061	1,061	1,061
Chicago	557,348	511,859	517,354	520,551	526,566	5,495	8,692	14,707
Syracuse	2,561	2,267	2,555	2,555	2,555	287	287	287
Tennessee	695,693	634,929	683,384	690,468	699,059	48,455	55,539	64,130

1

2

3 **Table 6D-5   Carbon Storage after 25 years for Recent Conditions, Existing and**
4 **Alternative W126 Standards by Region (metric tons)**

Region	Current Carbon Storage	Carbon Storage using Standard Growth Rates (normal i-Tree run unadjusted)	Carbon Storage using O_3 Response-Adjusted Growth Rates			
			Recent Conditions	Existing Standard & 15 ppm-hrs	11 ppm-hrs	7 ppm-hrs
Atlanta	1,331,096	1,426,626	1,214,656	1,314,673	1,321,110	1,345,896
Baltimore	598,533	577,824	492,553	570,680	570,680	570,680
Chicago	17,480,805	19,560,361	16,949,766	17,052,562	17,103,025	17,214,633
Syracuse	181,382	169,356	141,145	167,869	167,869	167,869
Tennessee	17,020,383	20,568,155	17,999,081	19,668,486	19,891,847	20,157,865

5

Table 6D-6   Change in Carbon Storage using O_3 Response-Adjusted Growth Rates after 25 years (relative to Unadjusted Rates)

Region	Change in Carbon Storage (metric tons)				Change in Valuation (at $78.5 per metric ton)			
	Recent Conditions	Existing Standard & 15 ppm-hrs	11 ppm-hrs	7 ppm-hrs	Recent Conditions	Existing Standard & 15 ppm-hrs	11 ppm-hrs	7 ppm-hrs
Atlanta	-211,971	-111,954	-105,517	-80,730	-$16,639,715	-$8,788,364	-$8,283,055	-$6,337,302
Baltimore	-85,271	-7,144	-7,144	-7,144	-$6,693,780	-$560,813	-$560,813	-$560,813
Chicago	-2,610,596	-2,507,799	-2,457,336	-2,345,728	-$204,931,749	-$196,862,261	-$192,900,876	-$184,139,650
Syracuse	-28,210	-1,486	-1,486	-1,486	-$2,214,523	-$116,683	-$116,683	-$116,683
Tennessee	-2,569,074	-899,670	-676,308	-410,291	-$201,672,305	-$70,624,078	-$53,090,178	-$32,207,827

Table 6D-7   Change in Carbon Storage using O_3-Response-Adjusted Growth Rates after 25 years (relative to Recent Conditions)

Region	Change in Carbon Storage (metric tons)			Change in Valuation (at $78.5 per metric ton)		
	Existing Standard & 15 ppm-hrs	11 ppm-hrs	7 ppm-hrs	Existing Standard & 15 ppm-hrs	11 ppm-hrs	7 ppm-hrs
Atlanta	100,017	106,454	131,241	$7,851,350	$8,356,660	$10,302,412
Baltimore	78,127	78,127	78,127	$6,132,967	$6,132,967	$6,132,967
Chicago	102,796	153,260	264,868	$8,069,488	$12,030,872	$20,792,099
Syracuse	26,724	26,724	26,724	$2,097,840	$2,097,840	$2,097,840
Tennessee	1,669,404	1,892,766	2,158,783	$131,048,227	$148,582,126	$169,464,478

6D.6 References

Baldocchi, D. (1988). A multi-layer model for estimating sulfur dioxide deposition to a deciduous oak forest canopy. Atmospheric Environment 22:869–884.

Baldocchi, D.D.; Hicks, B.B. and Camara, P. (1987). A canopy stomatal resistance model for gaseous deposition to vegetated surfaces. Atmospheric Environment 21:91–101.

Barbour, M.G.; Burk, J.H. and Pitts, W.D. (1980). Terrestrial Plant Ecology. Benjamin/Cummings, Menlo Park, CA. 604 pp.

Bidwell, R.G.S., and D.E. Fraser. (1972). Carbon monoxide uptake and metabolism by leaves. Canadian Journal of Botany 50:1435–1439.

Cairns, M.A.; Brown, S.; Helmer, E.H and Baumgardner, G.A. (1997). Root biomass allocation in the world's upland forests. Oecologia 111:1–11.

Chow, P., and Rolfe, G.L. (1989). Carbon and hydrogen contents of short-rotation biomass of five hardwood species. Wood and Fiber Science 21:30–36.

deVries, R.E. (1987). A Preliminary Investigation of the Growth and Longevity of Trees in Central Park. M.S. Thesis, Rutgers University, New Brunswick, NJ. 95 pp.

Fleming, L.E. (1988). Growth Estimation of Street Trees in Central New Jersey. M.S. Thesis, Rutgers University, New Brunswick, NJ. 143 pp.

Forest Products Laboratory. (1952). Chemical Analyses of Wood. Tech. Note 235. U.S. Department of Agriculture, Forest Service, Forest Products Laboratory, Madison, WI. 4 pp.

Frelich, L.E. (1992). Predicting Dimensional Relationships for Twin Cities Shade Trees. University of Minnesota, Department of Forest Resources, St. Paul, MN. 33 pp.

Jarvis, P.G., and Leverenz, J.W. (1983). Productivity of temperate, deciduous and evergreen forests, pp. 233–280. In: Lange, O.L., P.S. Nobel, C.B. Osmond, and H. Ziegler (Eds.). Physiological Plant Ecology IV, Encyclopedia of Plant Physiology. Vol. 12D. Springer-Verlag, Berlin, Germany.

1   Leverenz, J.W., and Hinckley, T.M. 1990. Shoot structure, leaf area index and productivity of

2   evergreen conifer stands. Tree Physiology 6:135–149.

3   Lovett, G.M. (1994). Atmospheric deposition of nutrients and pollutants in North America: An

4   ecological perspective. Ecological Applications 4:629–650.

5   Murray, F.J., Marsh, L, and Bradford, P.A. (1994). New York State Energy Plan, Vol. II: Issue

6   Reports. New York State Energy Office, Albany, NY.

7   Nowak, D.J. (2012). Contrasting natural regeneration and tree planting in fourteen North

8   American cities. Urban Forestry and Urban Greening, in press.

9   Nowak, D.J.; Crane, D.E.; Stevens, J.C.; Hoehn, R.E.; Walton, J.T.; Bond, J. (2008). Agriculture

10   and Urban Forestry, 34(6): November.

11   Nowak, D.J., and Crane, D.E. (2000). The urban forest effects (UFORE) model: Quantifying

12   urban forest structure and functions, pp.714–720. In: Hansen M., and T. Burk (Eds.). In:

13   Proceedings Integrated Tools for Natural Resources Inventories in the 21st Century. IUFRO

14   Conference, 16–20 August 1998, Boise, ID. General Technical Report NC-212, U.S. Department

15   of Agriculture, Forest Service, North Central Research Station, St. Paul, MN.

16   Nowak, D.J. (1991). Urban Forest Development and Structure: Analysis of Oakland, California.

17   Ph.D. Dissertation, University of California, Berkeley, CA. 232 pp. 356 Nowak et al.: Assessing

18   Urban Forest Structure and Ecosystem Services ©2008 International Society of Arboriculture

19   ———. (1994a). Understanding the structure of urban forests. Journal of Forestry 92:42–46.

20   ———. (1994b). Urban forest structure: The state of Chicago's urban forest, pp. 3–18, 140–164.

21   In: McPherson, E.G, D.J. Nowak, and R.A. Rowntree (Eds.). Chicago's Urban Forest

22   Ecosystem: Results of the Chicago Urban Forest Climate Project. USDA Forest Service General

23   Technical Report NE-186.

24   ———. (1994c). Atmospheric carbon dioxide reduction by Chicago's urban forest, pp. 83–94.

25   In: McPherson, E.G., D.J. Nowak, and R.A. Rowntree (Eds.). Chicago's Urban Forest

26   Ecosystem: Results of the Chicago Urban Forest Climate Project. Gen. Tech. Rep. NE-186. U.S.

27   Department of Agriculture, Forest Service, Northeastern Forest Experiment Station, Radnor, PA.

1 ———. (1996). Estimating leaf area and leaf biomass of open-grown deciduous urban trees.
2 Forest Science 42:504–507.

3 Nowak, D.J.; Civerolo K.L.; Rao, S.T.; Sistla, G.; Luley, C.J.; Crane. D.E.(2000). A modeling
4 study of the impact of urban trees on ozone. Atmospheric Environment 34:1601–1613.

5 Nowak, D.J.; Crane, D.E.; Stevens, J.C.; Ibarra M. (2002b). Brooklyn's Urban Forest. General
6 Technical Report NE-290, U.S. Department of Agriculture, Forest Service, Northeastern
7 Research Station, Newtown Square, PA. 107 pp.

8 Nowak, D.J., Crane, D.E. and Stevens, J.C. (2006). Air pollution removal by urban trees and
9 shrubs in the United States. Urban Forestry and Urban Greening 4:115–123.

10 Smith, F.W., Sampson, D.A. and Long, J.N. (1991). Comparison of leaf area index estimates
11 from allometrics and measured light interception. Forest Science 37:1682–1688.

12 Smith, W.B., and Shifley, S.R. (19840. Diameter Growth, Survival, and Volume Estimates for
13 Trees in Indiana and Illinois. Res. Pap. NC- 257. U.S. Department of Agriculture, Forest Service,
14 North Central Forest Experiment Station, St. Paul, MN. 10 pp.

15 Winer, A.M., D.R. Fitz, P.R. Miller, R. Atkinson, D.E. Brown, W.P. Carter, M.C. Dodd, C.W.
16 Johnson, M.A. Myers, K.R. Neisess, M.P. Poe, and E.R. Stephens. (1983). Investigation of the
17 Role of Natural Hydrocarbons in Photochemical Smog Formation in California. Statewide Air
18 Pollution Research Center, Riverside, CA.

19 Zinke, P.J. (1967). Forest interception studies in the United States, pp. 137–161. In: Sopper,
20 W.E., and H.W. Lull (Eds.). Forest Hydrology. Pergamon Press, Oxford, UK.

APPENDIX 6E:

Class I Areas and Weighted RBL at Current Standard and Alternative W126 Standard Levels

Class I Area	Number of Grid Cells	W126 (2006 – 2008)	Percent of Basal Area Assessed	Weighted Biomass Loss					
				Recent Conditions	75 ppb	15 ppm-hrs	11 ppm-hrs	7 ppm-hrs
Acadia NP (ME)	4	5.97	9.97	0.14%	<0.01%	<0.01%	<0.01%	<0.01%
Aqua Tibia Wilderness (CA)	4	30.94	0.72	0.37%	0.02%	0.01%	0.01%	0.01%
Alpine Lakes Wilderness (WA)	24	3.67	29.02	0.04%	0.01%	0.01%	0.01%	0.01%
Anaconda Pintler Wilderness (MT)	12	9.38	11.36	<0.01%	<0.01%	<0.01%	<0.01%	<0.01%
Ansel Adams Wilderness (CA)	19	31.22	1.80	0.42%	0.01%	0.01%	0.01%	0.01%
Arches NP (UT)	6	16.69	0.00			No Data		
Badlands/Sage Creek Wilderness, area 1 (ND)	7	9.94	2.98	6.47%	2.21%	2.13%	2.03%	1.43%
Badlands/Sage Creek Wilderness, area 2 (ND)	4	9.80	1.92	4.69%	1.36%	1.27%	1.17%	0.82%
Bandelier Wilderness (NM)	6	12.19	22.53	1.02%	0.21%	0.16%	0.09%	0.08%
Big Bend NP (TX)	32	10.30	0.00			No Data		
Black Canyon of the Gunnison Wilderness (CO)	4	15.27	8.38	0.73%	0.13%	0.10%	0.06%	0.05%
Bob Marshall Wilderness (MT)	46	5.40	24.27	0.02%	<0.01%	<0.01%	<0.01%	<0.01%
Bosque del Apache (NM)	7	12.17	0.00			No Data		
Boundary Waters Canoe Area Wilderness (MN)	53	4.48	27.13	0.45%	0.06%	0.06%	0.06%	0.06%
Bradwell Bay Wilderness (FL)	4	8.92	1.97	0.01%	<0.01%	<0.01%	<0.01%	<0.01%
Bridger Wilderness (WY)	28	12.19	8.67	0.34%	0.06%	0.06%	0.06%	0.05%
Brigantine Wilderness (NJ)	2	18.09	4.16	0.08%	<0.01%	<0.01%	<0.01%	<0.01%
Bryce Canyon NP (UT)	7	18.13	18.02	1.27%	0.26%	0.21%	0.14%	0.12%
Cabinet Mountains Wilderness (MT)	10	4.86	28.86	0.06%	0.01%	0.01%	0.01%	0.01%
Caney Creek Wilderness (AR)	2	7.48	26.41	0.14%	0.03%	0.03%	0.03%	0.03%
Canyonlands NP (UT)	22	17.05	0.50	0.23%	0.05%	0.04%	0.03%	0.02%
Cape Romain Wilderness (SC)	3	11.00	59.09	0.02%	0.02%	0.02%	0.02%	0.02%
Capitol Reef NP (UT)	22	17.83	1.32	0.61%	0.12%	0.10%	0.06%	0.06%
Caribou Wilderness (CA)	4	17.43	5.02	0.33%	0.02%	0.02%	0.02%	0.02%

Class I Area	Number of Grid Cells	W126 (2006–2008)	Percent of Basal Area Assessed	Weighted Biomass Loss				
				Recent Conditions	75 ppb	15 ppm-hrs	11 ppm-hrs	7 ppm-hrs
Carlsbad Caverns NP (NM)	7	13.33	0.10	0.03%	0.01%	0.01%	0.01%	0.01%
Chassahowitzka Wilderness (FL)	3	9.97	5.07	0.04%	0.01%	0.01%	0.01%	0.01%
Chiricahua National Monument (AZ)	7	14.66	3.70	<0.01%	<0.01%	<0.01%	<0.01%	<0.01%
Chiricahua Wilderness (AZ)	9	13.84	9.15	<0.01%	<0.01%	<0.01%	<0.01%	<0.01%
Cohutta Wilderness (TN-GA)	5	13.75	40.00	1.52%	0.17%	0.17%	0.11%	0.05%
Crater Lake NP (OR)	10	6.97	18.94	0.34%	0.07%	0.07%	0.07%	0.07%
Craters of the Moon Wilderness (ID)	8	12.43	14.39	0.03%	0.01%	0.01%	<0.01%	<0.01%
Cucamonga Wilderness (CA)	2	43.55	0.13	0.05%	0.01%	<0.01%	<0.01%	<0.01%
Desolation Wilderness (CA)	5	17.60	0.88	0.05%	<0.01%	<0.01%	<0.01%	<0.01%
Diamond Peak Wilderness (OR)	4	5.68	22.26	0.01%	<0.01%	<0.01%	<0.01%	<0.01%
Dolly Sods Wilderness (WV)	2	9.66	36.08	1.37%	0.24%	0.24%	0.19%	0.13%
Domeland Wilderness (CA)	10	40.82	0.13	0.05%	<0.01%	<0.01%	<0.01%	<0.01%
Eagle Cap Wilderness (OR)	19	6.40	28.76	0.09%	0.03%	0.03%	0.02%	0.02%
Eagles Nest Wilderness (CO)	12	19.88	16.33	1.64%	0.53%	0.38%	0.18%	0.12%
Emigrant Wilderness (CA)	10	26.01	3.37	0.47%	0.02%	0.02%	0.02%	0.01%
Everglades NP (FL)	42	6.27	0.07	0.01%	<0.01%	<0.01%	<0.01%	<0.01%
Fitzpatrick Wilderness (WY)	15	11.87	5.65	0.01%	<0.01%	<0.01%	<0.01%	<0.01%
Flat Top Wilderness (CO)	16	15.85	27.42	2.37%	0.68%	0.53%	0.33%	0.26%
Galiuro Wilderness (AZ)	7	17.01	3.98	0.25%	0.09%	0.07%	0.04%	0.03%
Gates of the Mountains Wilderness (MT)	4	7.23	81.56	0.68%	0.19%	0.17%	0.16%	0.13%
Gearhart Mountain Wilderness (OR)	4	8.77	34.53	1.08%	0.21%	0.19%	0.18%	0.17%
Gila Wilderness (NM)	29	13.26	25.88	1.06%	0.26%	0.20%	0.13%	0.11%
Glacier NP (MT)	40	3.34	26.38	0.05%	0.01%	0.01%	0.01%	0.01%
Glacier Peak Wilderness (WA)	31	2.97	26.11	0.02%	0.01%	0.01%	0.01%	0.01%
Goat Rocks Wilderness (WA)	10	4.03	28.41	0.01%	<0.01%	<0.01%	<0.01%	<0.01%
Grand Canyon NP (AZ)	73	17.90	10.90	2.03%	0.56%	0.44%	0.28%	0.24%
Grand Teton NP (WY)	17	11.67	9.04	0.04%	0.01%	0.01%	0.01%	<0.01%
Great Gulf Wilderness (NH)	2	3.99	8.30	<0.01%	<0.01%	<0.01%	<0.01%	<0.01%
Great Sand Dunes Wilderness (CO)	5	14.24	38.30	1.42%	0.24%	0.18%	0.11%	0.09%
Great Smoky Mountains NP (NC-TN)	26	14.65	30.18	1.24%	0.13%	0.13%	0.08%	0.04%
Guadalupe Mountains NP (TX)	7	12.69	1.08	0.09%	0.02%	0.02%	0.01%	0.01%

6E-2

Class I Area	Number of Grid Cells	W126 (2006–2008)	Percent of Basal Area Assessed	Weighted Biomass Loss				
				Recent Conditions	75 ppb	15 ppm-hrs	11 ppm-hrs	7 ppm-hrs
Hells Canyon Wilderness (ID-OR)	17	6.63	57.18	0.14%	0.04%	0.04%	0.03%	0.03%
Hercules-Glades Wilderness (MO)	4	11.09	0.23	<0.01%	<0.01%	<0.01%	<0.01%	<0.01%
Hoover Wilderness (CA)	9	27.17	3.35	0.56%	0.02%	0.02%	0.02%	0.02%
Isle Royale NP (MI)	4	5.45	15.50	0.32%	0.04%	0.04%	0.04%	0.04%
James River Face Wilderness (VA)	2	8.22	28.30	0.53%	0.02%	0.02%	0.02%	0.01%
Jarbridge Wilderness (NV)	11	16.30	22.47	2.09%	0.32%	0.26%	0.18%	0.16%
John Muir Wilderness (CA)	42	36.36	0.64	0.32%	0.01%	0.01%	0.01%	0.01%
Joshua Tree Wilderness (CA)	38	29.24	0.00		No Data			
Joyce Kilmer-Slickrock Wilderness (NC-TN)	3	14.55	28.26	1.34%	0.09%	0.09%	0.05%	0.02%
Kaiser Wilderness (CA)	4	33.59	5.55	1.02%	0.02%	0.02%	0.02%	0.02%
Kalmiopsis Wilderness (OR)	13	6.02	53.38	<0.01%	<0.01%	<0.01%	<0.01%	<0.01%
Kings Canyon NP (CA)	31	41.68	2.37	0.81%	0.02%	0.02%	0.01%	0.01%
La Garita Wilderness (CO)	11	15.76	18.20	1.17%	0.22%	0.16%	0.09%	0.08%
Lassen Volcanic NP (CA)	11	17.77	11.14	0.67%	0.05%	0.04%	0.04%	0.04%
Lava Beds/Black Lava Flow Wilderness (CA)	4	10.09	32.89	1.65%	0.21%	0.02%	0.18%	0.17%
Lava Beds/Schonchin Wilderness (CA)	2	10.04	29.16	2.22%	0.28%	0.26%	0.24%	0.23%
Linville Gorge Wilderness (NC)	3	10.85	43.58	1.33%	0.05%	0.05%	0.04%	0.03%
Lostwood Wilderness (ND)	2	4.42	2.26	0.09%	0.05%	0.05%	0.05%	0.05%
Lye Brook Wilderness (VT)	4	6.03	35.25	0.17%	<0.01%	<0.01%	<0.01%	<0.01%
Mammoth Cave NP (KY)	6	14.48	19.75	0.81%	0.07%	0.07%	0.04%	0.02%
Marble Mountain Wilderness (CA)	14	6.67	45.64	0.09%	0.01%	0.01%	0.01%	0.01%
Maroon Bells-Snowmass Wilderness (CO)	13	15.39	31.38	2.42%	0.60%	0.45%	0.28%	0.22%
Mazatzal Wilderness (AZ)	18	22.99	6.49	1.54%	0.54%	0.36%	0.16%	0.12%
Medicine Lake Wilderness (MT)	3	6.20	0.00		No Data			
Mesa Verde NP (CO)	4	16.40	0.01	<0.01%	<0.01%	<0.01%	<0.01%	<0.01%
Mingo Wilderness (MO)	4	14.82	4.37	0.06%	0.01%	0.01%	0.01%	0.01%
Mission Mountains Wilderness (MT)	8	5.53	36.13	0.04%	0.01%	0.01%	0.01%	0.01%
Mokelumne Wilderness (CA)	8	20.63	2.45	0.21%	0.01%	0.01%	0.01%	0.01%
Moosehorn Wilderness (ME)	3	2.65	16.07	0.10%	0.01%	0.01%	0.01%	0.01%
Mount Adams Wilderness (WA)	6	3.78	24.95	0.02%	0.01%	0.01%	0.01%	0.01%
Mount Baldy Wilderness (AZ)	3	17.80	54.30	2.72%	0.82%	0.58%	0.31%	0.24%

Class I Area	Number of Grid Cells	W126 (2006 – 2008)	Percent of Basal Area Assessed	Weighted Biomass Loss				
				Recent Conditions	75 ppb	15 ppm-hrs	11 ppm-hrs	7 ppm-hrs
Mount Hood Wilderness (OR)	5	4.00	38.90	0.01%	<0.01%	<0.01%	<0.01%	<0.01%
Mount Jefferson Wilderness (OR)	10	4.32	41.44	0.07%	0.03%	0.03%	0.03%	0.03%
Mount Ranier NP (WA)	14	4.34	26.96	0.01%	<0.01%	<0.01%	<0.01%	<0.01%
Mount Washington Wilderness (OR)	6	4.72	43.29	0.27%	0.11%	0.11%	0.11%	0.10%
Mount Zirkel Wilderness (CO)	13	15.38	14.43	1.19%	0.35%	0.28%	0.22%	0.17%
Mountain Lakes Wilderness (OR)	4	8.15	33.29	0.67%	0.12%	0.11%	0.11%	0.10%
North Absaroka Wilderness (WY)	21	9.25	15.05	<0.01%	<0.01%	<0.01%	<0.01%	<0.01%
North Cascades NP (WA)	27	2.31	18.78	<0.01%	<0.01%	<0.01%	<0.01%	<0.01%
Okefenokee Wilderness (GA)	19	9.83	3.47	0.02%	<0.01%	<0.01%	<0.01%	<0.01%
Olympic NP (WA)	45	1.77	29.16	<0.01%	<0.01%	<0.01%	<0.01%	<0.01%
Otter Creek Wilderness (WV)	3	8.67	42.78	2.36%	0.49%	0.49%	0.37%	0.24%
Pasayten Wilderness (WA)	22	2.42	16.24	0.01%	<0.01%	<0.01%	<0.01%	<0.01%
Pecos Wilderness (NM)	15	12.75	44.38	1.29%	0.26%	0.20%	0.12%	0.10%
Petrified Forest NP (AZ)	16	16.71	0.00			No Data		
Pine Mountain Wilderness (AZ)	2	21.52	9.91	1.83%	0.76%	0.51%	0.23%	0.16%
Pinnacles Wilderness (CA)	4	14.03	0.00			No Data		
Point Reyes NS/Phillip Burton Wilderness 1 (CA)	3	1.42	22.86	<0.01%	<0.01%	<0.01%	<0.01%	<0.01%
Point Reyes NS/Phillip Burton Wilderness 2 (CA)	4	1.44	20.81	<0.01%	<0.01%	<0.01%	<0.01%	<0.01%
Point Reyes NS/Phillip Burton Wilderness 3 (CA)	5	1.57	19.15	<0.01%	<0.01%	<0.01%	<0.01%	<0.01%
Presidential Range-Dry River Wilderness (NH)	5	4.12	12.94	0.02%	<0.01%	<0.01%	<0.01%	<0.01%
Rainbow Lakes Wilderness (WI)	1	4.48	56.42	0.47%	0.05%	0.05%	0.05%	0.05%
Rawah Wilderness (CO)	9	17.59	7.05	0.51%	0.15%	0.11%	0.08%	0.06%
Red Rock Lakes Wilderness (MT)	4	10.50	41.54	0.02%	<0.01%	<0.01%	<0.01%	<0.01%
Redwood NP (CA)	12	6.63	37.56	0.13%	0.02%	0.02%	0.02%	0.02%
Rocky Mountain NP (CO)	17	18.55	14.73	0.60%	0.18%	0.13%	0.06%	0.05%
Saguaro Wilderness (AZ)	13	13.60	3.45	<0.01%	<0.01%	<0.01%	<0.01%	<0.01%
Saint Marks Wilderness (FL)	5	9.71	5.80	0.06%	<0.01%	<0.01%	<0.01%	<0.01%
Salt Creek Wilderness (NM)	4	12.21	0.00			No Data		

Class I Area	Number of Grid Cells	W126 (2006–2008)	Percent of Basal Area Assessed	Weighted Biomass Loss				
				Recent Conditions	75 ppb	15 ppm-hrs	11 ppm-hrs	7 ppm-hrs
San Gabriel Wilderness (CA)	5	31.01	0.02	0.01%	<0.01%	<0.01%	<0.01%	<0.01%
San Gorgonio Wilderness (CA)	7	49.21	0.35	0.09%	0.01%	0.01%	<0.01%	<0.01%
San Jacinto Wilderness (CA)	7	42.32	1.20	0.21%	0.01%	0.01%	0.01%	<0.01%
San Pedro Parks Wilderness (NM)	4	12.38	44.76	1.63%	0.35%	0.26%	0.16%	0.13%
San Rafael Wilderness (CA)	15	12.52	0.00			No Data		
Sawtooth Wilderness (ID)	12	12.54	33.72	0.18%	0.03%	0.03%	0.02%	0.02%
Scapegoat Wilderness (MT)	16	6.07	26.84	0.01%	<0.01%	<0.01%	<0.01%	<0.01%
Selway-Bitterroot Wilderness (MT-ID)	62	7.45	30.59	0.14%	0.03%	0.03%	0.02%	0.02%
Seney Wilderness (MI)	4	6.76	18.25	0.36%	0.04%	0.04%	0.04%	0.04%
Sequioa NP (CA)	31	47.41	2.91	1.25%	0.03%	0.03%	0.02%	0.02%
Shenandoah NP (VA)	22	11.00	25.88	0.98%	0.05%	0.05%	0.05%	0.03%
Shining Rock Wilderness (NC)	4	12.45	23.40	0.97%	0.12%	0.12%	0.11%	0.07%
Sierra Ancha Wilderness (AZ)	4	21.79	27.05	2.86%	0.88%	0.60%	0.28%	0.21%
Sipsey Wilderness (AL)	4	13.99	30.20	1.29%	0.20%	0.20%	0.18%	0.12%
South Warner Wilderness (CA)	6	11.45	11.76	0.54%	0.07%	0.07%	0.06%	0.06%
Strawberry Mountain Wilderness (OR)	6	7.47	53.42	0.67%	0.17%	0.16%	0.15%	0.14%
Superstition Wilderness (AZ)	12	21.73	0.00			No Data		
Swanquarter Wilderness (NC)	1	11.16	24.90	0.18%	0.02%	0.02%	0.02%	0.01%
Sycamore Canyon Wilderness (AZ)	6	20.58	37.58	3.67%	1.44%	1.01%	0.50%	0.37%
Teton Wilderness (WY)	27	10.95	2.11	0.01%	<0.01%	<0.01%	<0.01%	<0.01%
Theodore Roosevelt NP (9)	9	6.08	0.50	0.24%	0.08%	0.08%	0.08%	0.06%
Thousand Lakes Wilderness (CA)	2	15.03	13.31	0.58%	0.05%	0.04%	0.04%	0.04%
Three Sisters Wilderness (OR)	19	5.09	34.51	0.10%	0.04%	0.04%	0.04%	0.04%
UL Bend Wilderness (MT)	5	6.95	42.50	1.01%	0.31%	0.31%	0.31%	0.25%
Upper Buffalo Wilderness (AR)	3	7.95	11.36	0.07%	0.01%	0.01%	0.01%	0.01%
Ventana Wilderness (CA)	15	7.13	0.16	<0.01%	<0.01%	<0.01%	<0.01%	<0.01%
Voyageurs NP (MN)	10	4.54	40.95	0.67%	0.10%	0.10%	0.10%	0.09%
Washakie Wilderness (WY)	41	10.55	9.58	<0.01%	<0.01%	<0.01%	<0.01%	<0.01%
Weminuche Wilderness (CO)	29	16.10	23.51	1.43%	0.29%	0.21%	0.12%	0.10%
West Elk Wilderness (CO)	13	14.82	37.30	2.58%	0.40%	0.31%	0.20%	0.17%
Wheeler Peak Wilderness (NM)	4	13.42	29.72	0.72%	0.14%	0.10%	0.06%	0.05%

Class I Area	Number of Grid Cells	W126 (2006 – 2008)	Percent of Basal Area Assessed	Weighted Biomass Loss					
				Recent Conditions	75 ppb	15 ppm-hrs	11 ppm-hrs	7 ppm-hrs
White Mountain Wilderness (NM)	5	11.46	24.98	0.28%	0.07%	0.05%	0.03%	0.03%
Wichita Mountains (OK)	4	11.89	0.17	0.12%	0.01%	0.01%	0.01%	0.01%
Wind Cave NP (SD)	5	12.47	93.24	4.38%	2.66%	2.65%	2.64%	2.02%
Wolf Island Wilderness (GA)	1	7.46	14.17	0.03%	0.01%	0.01%	0.01%	<0.01%
Yellowstone NP (WY)	84	10.00	4.14	0.01%	<0.01%	<0.01%	<0.01%	<0.01%
Yolla Bolly-Middle Eel Wilderness (CA)	13	10.65	46.51	0.53%	0.03%	0.02%	0.02%	0.02%
Yosemite NP (CA)	35	29.42	5.73	0.90%	0.03%	0.03%	0.02%	0.02%
Zion NP (UT)	13	18.70	1.99	0.12%	0.02%	0.02%	0.01%	0.01%
All Areas Combined	1952	13.59	16.85	0.51%	0.11%	0.09%	0.06%	0.05%

APPENDIX 7A: ADDITIONAL INFORMATION FOR SCREENING-LEVEL ASSESSMENT OF VISIBLE FOLIAR INJURY IN NATIONAL PARKS

This appendix presents the O_3 exposure and soil moisture data used in the assessment of visible foliar injury risk in national parks (Section 7.3) and the park-by-park results of that assessment.

In Figure 7A-1, we provide a plot of the relationship between the percentage of biosites with any visible foliar injury and O_3 exposure; we used this figure to define the base scenario (i.e., 17.7% of biosites at 10.46 ppm-hrs). In Figures 7A-2 through 7A-5, we provide a plot of the relationship between the percentage of biosites with any visible foliar injury and O_3 exposure by soil moisture categorization; we used this figure to define 4 scenarios for any injury (i.e., 5%, 10%, 15%, and 20%). In Figure 7A-6, we provide a plot of the relationship between the percentage of biosites with visible foliar injury measured as biosite index greater than 5 and O_3 exposure by soil moisture category; we used this figure to define 1 scenario for higher injury (i.e., 5% for injury=5).

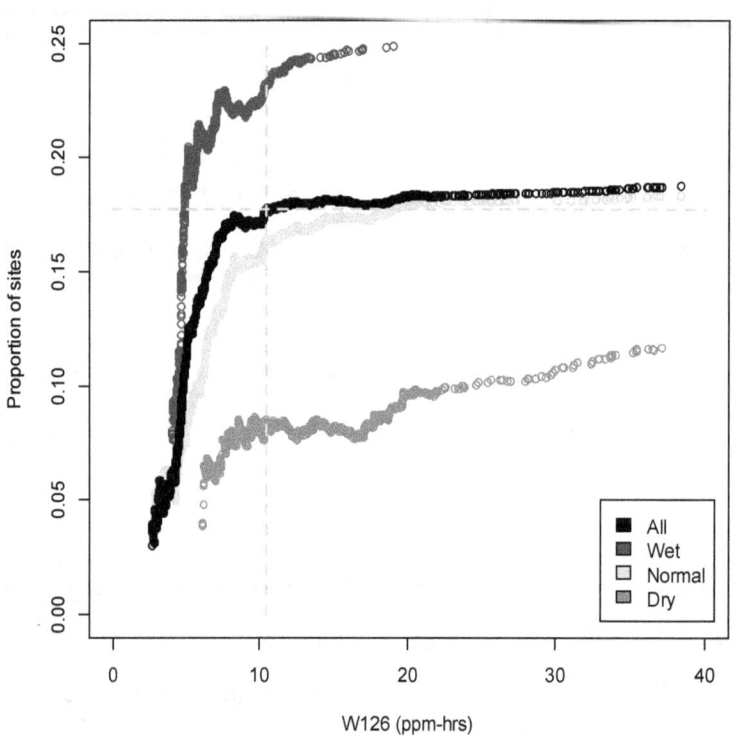

1

2   **Figure 7A-1   Defining Base Scenario (all biosites, any injury)**

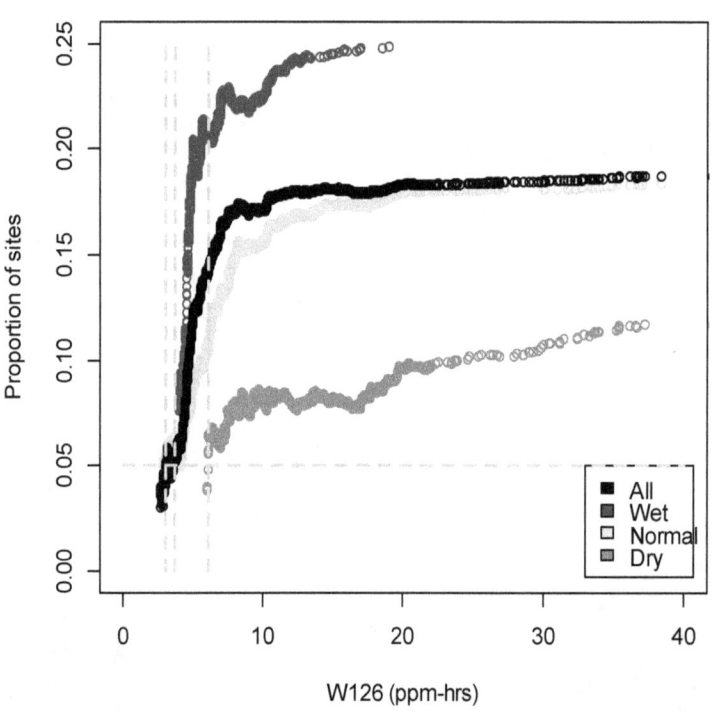

3

1   **Figure 7A-2   Defining Any Injury Scenario (5% of biosites showing injury)**

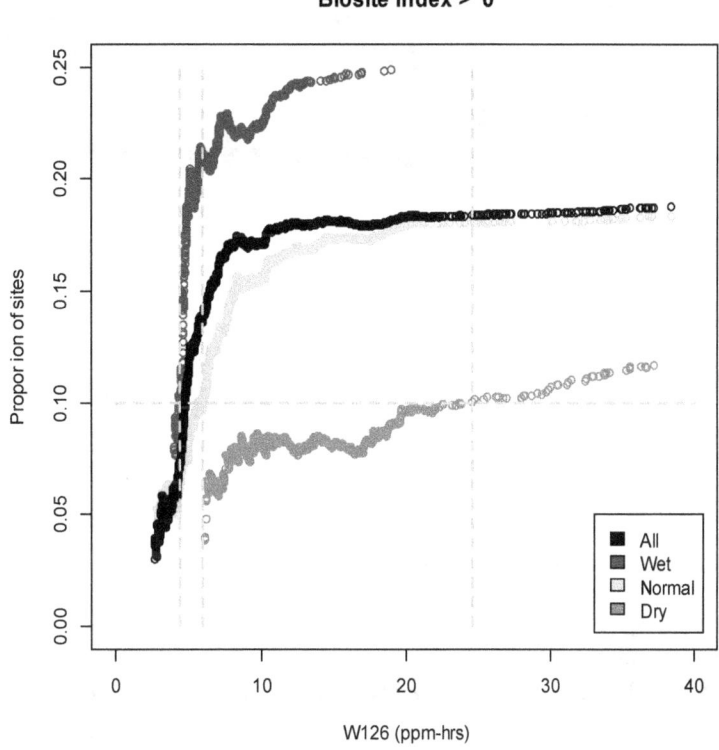

2

3   **Figure 7A-3   Defining Any Injury Scenarios (10% of biosites showing injury)**

4

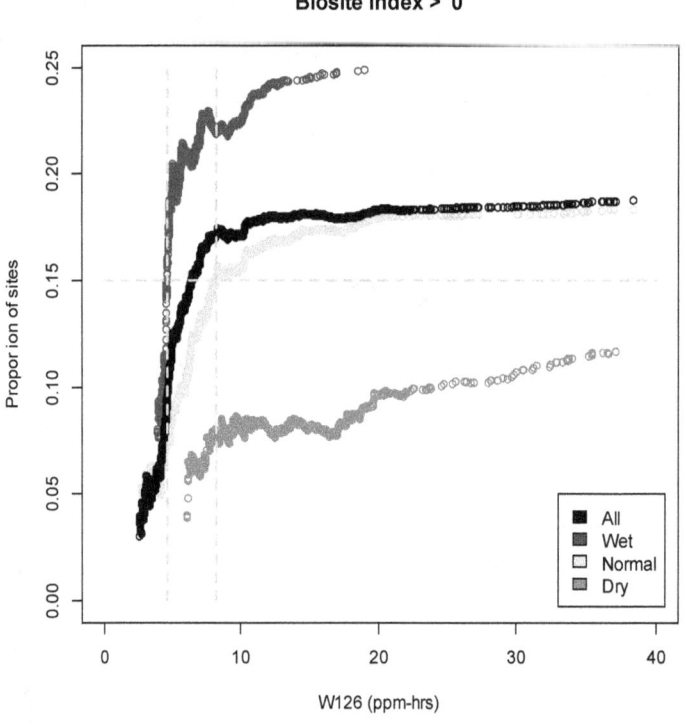

1

2 **Figure 7A-4   Defining Any Injury Scenarios (15% of biosites showing injury)**

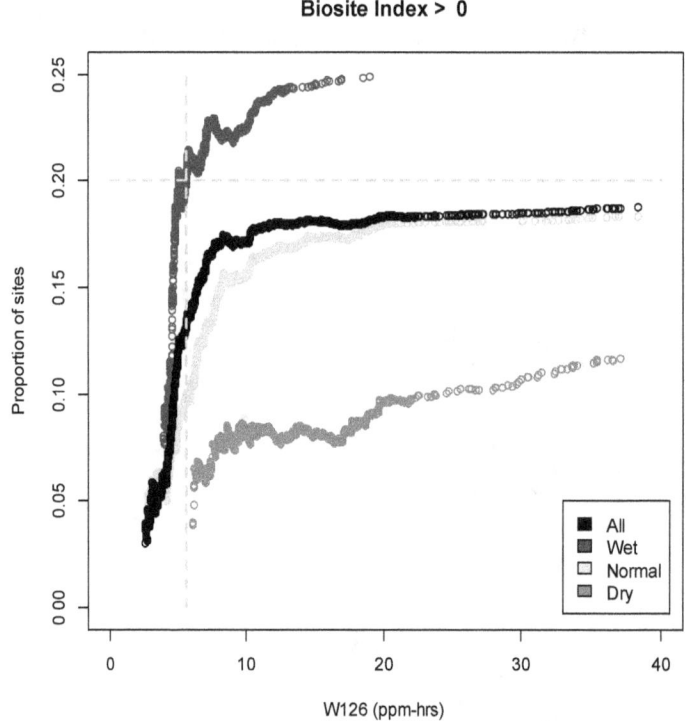

3

4 **Figure 7A-5   Defining Any Injury Scenarios (20% of biosites showing injury)**

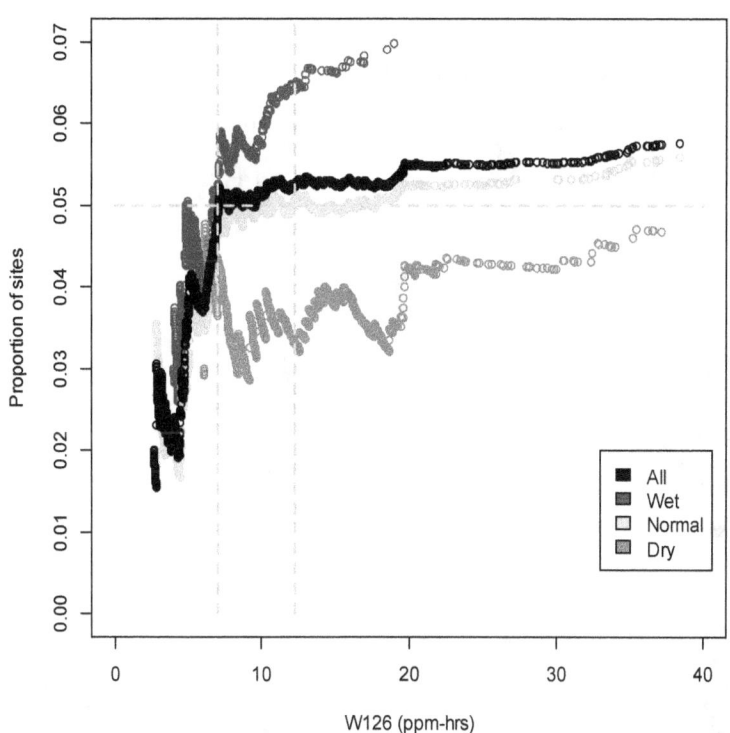

1

2   **Figure 7A-6   Defining Injury = 5 Scenario (5% biosites with biosite Index > 5)**

3

4

1   In Figures 7A-7 through 7A-9, we provide pie charts illustrating the fraction of monitors
2   at parks in locations with different soil moisture categorizations for the 7-month average, 5-
3   month average, and 3-month average, respectively. Soil moisture estimates are based on the
4   Palmer Z index, where estimates above 0 are wetter and estimates below 0 are drier. The soil
5   moisture categories are based on NOAA's classifications for Palmer Z data (NOAA, 2012c).

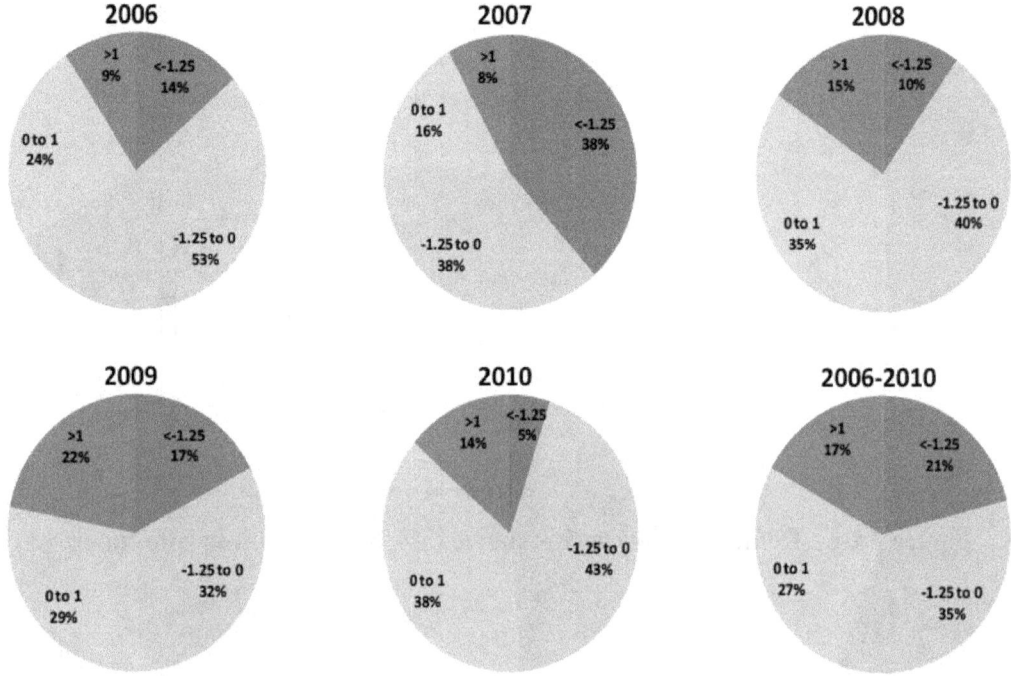

6
7   Figure 7A-7   7-month Palmer Z (March-September) at 57 Monitors at Parks
8
9

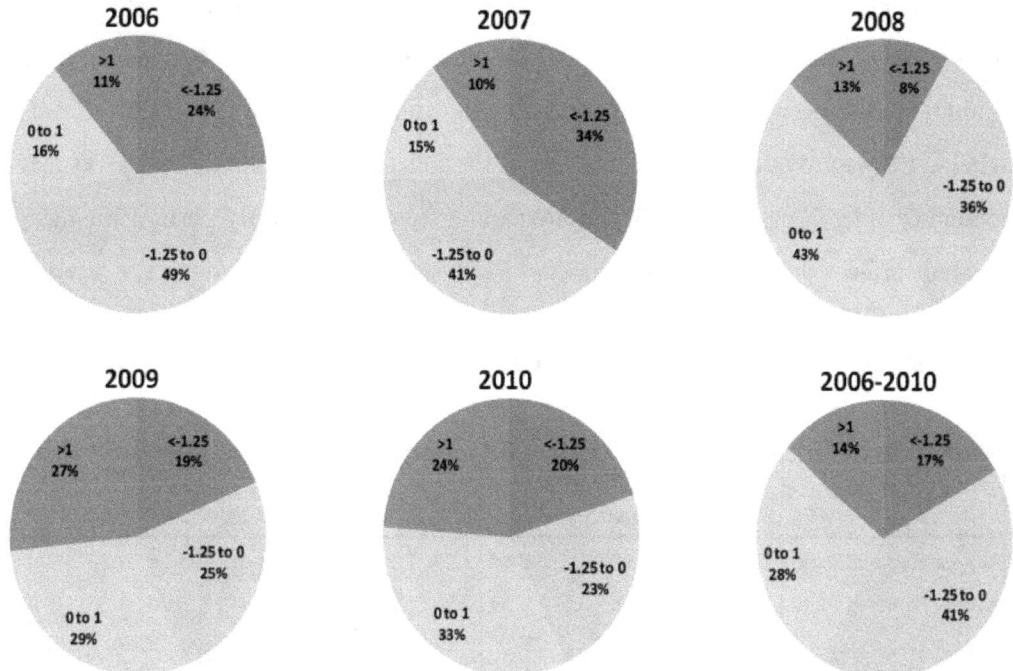

1

2 **Figure 7A-8   5-Month Palmer Z (April-August) at 57 Monitors at Parks**

3

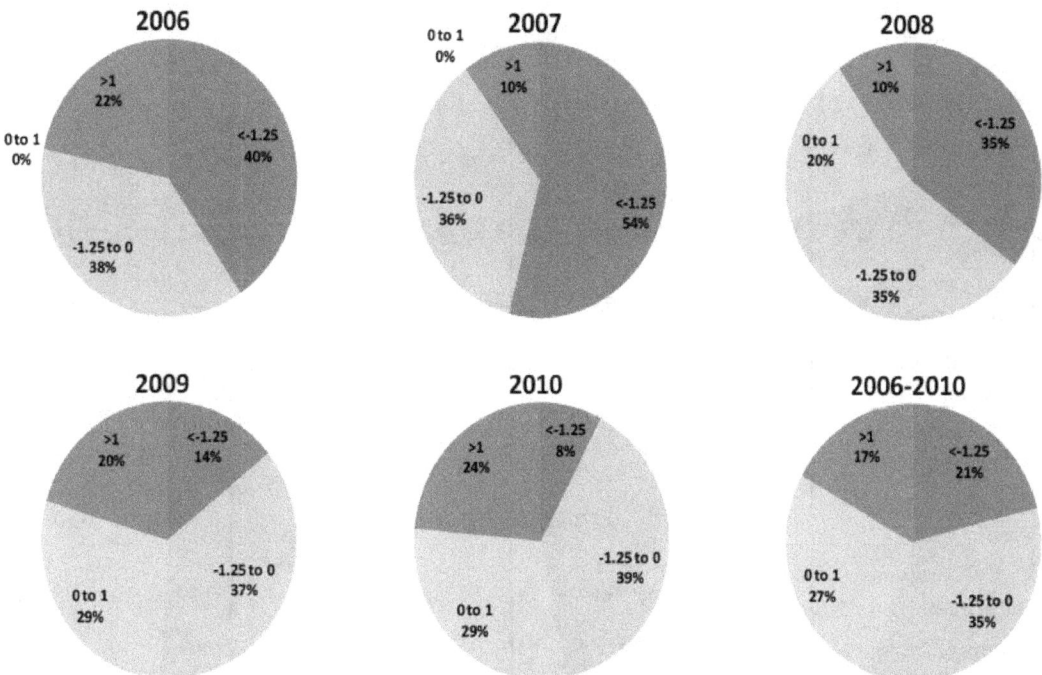

4

5 **Figure 7A-9   3-Month Palmer Z (Monitor-specific) at 57 Monitors at Parks**

6

1   In Table 7A-1, we provide the 7-month, 5-month, and 3-month soil moisture average for

2   each park with an O_3 monitor. In Table 7A-2, we provide the W126 estimates and the timeframe

3   corresponding to those W126 estimates for each park with an O_3 monitor. In Figures 7A-10

4   through 7A-14, we provided larger scale maps of the foliar injury results for 214 parks that are

5   provided in Chapter 7. In Table 7A-3, we provide the O_3 estimates at 214 parks using the

6   interpolated surfaces just meeting the existing 75 ppb standard and alternate W126 standards.

7

8   **Table 7A-1   Average Soil Moisture Data (Palmer Z) by Averaging Time for 57**

9   **Parks with O_3 Monitors**

Monitor Site ID	Park Name	7-Month (Mar-Sept)					5-Month (Apr-Aug)					3-Month (monitor specific)				
		2006	2007	2008	2009	2010	2006	2007	2008	2009	2010	2006	2007	2008	2009	2010
230090102	Acadia National Park	1.69	1.08	1.98	2.58	0.84	2.98	1.28	0.50	3.83	-0.19	5.26	1.83	-0.49	0.81	0.21
230090103	Acadia National Park	1.69	1.08	1.98	2.58	0.84	2.98	1.28	0.50	3.83	-0.19	5.26	1.83	-0.49	0.81	0.21
311651001	Agate Fossil Beds National Monument	-1.40	-1.56	-0.17	1.94	1.95	-2.37	-1.57	-0.31	2.76	2.96	--	-1.83	-0.49	4.06	--
460710001	Badlands National Park	-0.57	-1.41	1.27	2.06	2.15	-1.86	-1.48	1.88	2.17	2.90	--	--	1.31	1.38	2.41
460711001	Badlands National Park	-0.57	-1.41	1.27	2.06	2.15	-1.86	-1.48	1.88	2.17	2.90	-2.46	-1.93	--	--	--
480430101	Big Bend National Park	-0.67	1.78	0.11	-0.25	0.96	-0.65	2.04	-0.07	0.04	1.30	-2.67	2.76	-1.75	-0.63	0.71
370110002	Blue Ridge Parkway	-0.27	-1.95	-1.13	1.14	-0.94	-0.38	-1.92	-1.37	0.64	-1.09	-0.20	-1.80	-1.76	1.28	-0.73
490370101	Canyonlands National Park	-0.91	-0.21	-0.42	-1.39	0.77	-1.87	-0.27	0.05	-1.27	0.97	-2.57	-0.59	0.53	-0.90	0.76
250010002	Cape Cod National Seashore	1.36	0.13	1.06	1.47	0.80	2.83	0.20	0.23	2.06	-0.98	5.37	-0.27	0.49	0.83	-1.06
350153001	Carlsbad Caverns National Park	-0.07	1.12	-0.56	-0.90	1.23	-0.47	0.95	-0.41	-0.69	1.43	--	1.82	-1.69	0.40	0.68
160310001	City of Rocks National Reserve	-0.08	-1.87	-0.68	1.24	-0.12	-0.93	-2.19	-0.66	1.82	0.67	--	--	--	--	0.40
80771001	Colorado National Monument	-0.54	-0.25	-0.26	-0.17	-0.11	-1.24	-0.51	-0.23	0.30	0.24	--	-0.63	-0.01	-1.15	0.14
450790021	Congaree National Park	-0.60	-1.35	-0.04	-0.20	-1.05	-0.38	-1.15	0.09	0.09	-0.95	-1.81	-1.04	-0.09	-0.24	-1.49
450210002	Cowpens National Battlefield	-0.84	-2.18	-1.58	-0.05	-1.24	-0.57	-2.07	-1.55	-0.39	-1.14	-0.54	-1.41	-1.95	-0.25	-1.17
160230101	Craters of the Moon National Monument	0.87	-2.15	-1.50	1.63	-0.25	0.30	-2.51	-1.67	2.33	0.28	--	-2.88	-1.62	0.27	-0.95
210131002	Cumberland Gap National Historical Park	0.73	-1.66	-0.50	1.55	0.09	0.39	-1.43	-0.17	1.82	0.67	--	-1.37	-0.15	2.77	1.37

Monitor Site ID	Park Name	7-Month (Mar-Sept)					5-Month (Apr-Aug)					3-Month (monitor specific)				
		2006	2007	2008	2009	2010	2006	2007	2008	2009	2010	2006	2007	2008	2009	2010
60270101	Death Valley National Park	-1.30	-1.71	-1.15	-1.85	-0.52	-1.59	-1.95	-0.97	-2.01	-0.32	-2.08	-2.11	-0.78	-1.55	-1.31
560111013	Devil's Tower National Monument	-1.77	-0.78	1.41	1.45	0.63	-2.66	-0.71	2.02	1.52	1.60	--	--	0.95	1.63	1.52
490471002	Dinosaur National Monument	-0.57	-1.57	0.50	-0.69	-0.46	-1.52	-1.98	0.38	-0.49	-0.05	--	-1.92	0.92	-0.49	-0.39
300298001	Glacier National Park	-0.16	-1.14	0.15	-0.14	1.01	-0.05	-1.20	0.09	-0.13	1.65	-0.54	-0.48	-0.01	-0.87	1.96
300351001	Glacier National Park	-1.07	-0.88	0.43	-0.50	1.43	-1.68	-0.96	0.57	-0.11	2.12	--	--	--	-0.85	2.15
40058001	Grand Canyon National Park	-0.54	-1.40	0.03	-1.48	0.39	-0.95	-1.41	0.60	-1.42	0.74	-1.74	-2.63	0.27	-2.20	0.41
320330101	Great Basin National Park	0.28	-2.14	-1.62	0.27	-0.35	-0.08	-2.43	-1.64	0.87	-0.23	-0.49	-3.05	-1.58	1.29	0.50
370870036	Great Smoky Mountains National Park	-0.27	-1.95	-1.13	1.14	-0.94	-0.38	-1.92	-1.37	0.64	-1.09	-0.20	-1.80	-1.76	1.28	-0.73
470090102	Great Smoky Mountains National Park	0.65	-1.80	-0.61	1.76	-0.67	0.60	-1.48	-0.47	1.64	-0.68	1.06	-1.69	-0.86	0.75	-0.80
471550101	Great Smoky Mountains National Park	0.65	-1.80	-0.61	1.76	-0.67	0.60	-1.48	-0.47	1.64	-0.68	1.06	-1.03	-0.78	0.75	-0.80
471550102	Great Smoky Mountains National Park	-0.27	-1.95	-1.13	1.14	-0.94	-0.38	-1.92	-1.37	0.64	-1.09	-0.50	-1.71	-2.06	1.14	-0.51
180890022	Indiana Dunes National Lakeshore	0.98	0.22	0.88	0.45	0.37	1.00	0.91	0.06	0.50	0.81	1.59	-0.94	2.47	0.16	1.41
60650008	Joshua Tree National Park	-1.30	-1.71	-1.15	-1.85	-0.52	-1.59	-1.95	-0.97	-2.01	-0.32	-1.84	-2.52	-0.78	-2.34	0.28
60651004	Joshua Tree National Park	-1.30	-1.71	-1.15	-1.85	-0.52	-1.59	-1.95	-0.97	-2.01	-0.32	--	-2.11	-1.27	-1.55	-0.60
60719002	Joshua Tree National Park	-1.30	-1.71	-1.15	-1.85	-0.52	-1.59	-1.95	-0.97	-2.01	-0.32	-2.08	-2.11	-0.76	-1.55	-0.60
60893003	Lassen Volcanic National Park	0.41	-1.07	-1.98	0.14	0.59	0.10	-0.82	-1.81	0.42	1.22	-1.41	-0.80	-1.86	0.64	-0.03
80830101	Mesa Verde National Park	-0.54	-0.25	-0.26	-0.17	-0.11	-1.24	-0.51	-0.23	0.30	0.24	-1.53	-0.48	0.16	0.10	0.14
60711001	Mojave National Preserve	-1.30	-1.71	-1.15	-1.85	-0.52	-1.59	-1.95	-0.97	-2.01	-0.32	--	-2.11	-0.78	-1.50	-0.60
530530012	Mount Rainier Wilderness	-1.25	-0.64	-0.05	-0.78	1.62	-1.00	-1.02	0.25	-0.81	1.75	-1.15	-0.37	-0.07	-1.05	2.84
530090016	Olympic National Park	-0.89	0.07	0.05	0.02	1.55	-0.65	-0.41	0.25	-0.12	1.37	--	--	--	--	1.34

Table 7A-2 Ozone Exposure in 57 Parks with Monitors

Monitor site ID	Park Name	W126					3-Month Timeframe for W126				
		2006	2007	2008	2009	2010	2006	2007	2008	2009	2010
230090102	Acadia National Park	10.59	7.89	7.64	7.02	5.24	MJJ	AMJ	MJJ	MAM	MAM
230090103	Acadia National Park	6.37	6.41	4.72	5.21	4.13	MJJ	AMJ	MJJ	MAM	MAM
311651001	Agate Fossil Beds National Monument	--	8.27	12.76	5.85	--	--	JAS	MJJ	JJA	--
460710001	Badlands National Park	--	--	2.23	2.54	3.85	--	--	JAS	AMJ	JJA
460711001	Badlands National Park	16.74	8.01	--	--	--	JJA	JJA	--	--	--
480430101	Big Bend National Park	11.62	10.60	10.55	8.62	8.47	AMJ	MAM	MAM	MAM	MAM
370110002	Blue Ridge Parkway	9.88	11.46	8.81	4.71	8.19	AMJ	AMJ	AMJ	AMJ	AMJ
490370101	Canyonlands National Park	18.06	16.93	17.06	12.23	13.24	MJJ	MJJ	AMJ	MAM	AMJ
250010002	Cape Cod National Seashore	13.47	13.16	12.89	5.25	7.03	MJJ	MJJ	MJJ	AMJ	MJJ
350153001	Carlsbad Caverns National Park	--	8.65	17.50	11.37	7.09	--	AMJ	AMJ	MJJ	AMJ
160310001	City of Rocks National Reserve	--	--	--	--	6.02	--	--	--	--	JJA
80771001	Colorado National Monument	--	11.61	15.04	4.13	8.75	--	JJA	MJJ	JAS	AMJ
450790021	Congaree National Park	12.31	10.78	9.45	3.97	6.32	MAM	MAM	MAM	FMA	MAM
450210002	Cowpens National Battlefield	14.30	7.87	16.05	3.24	8.81	MJJ	AMJ	JJA	FMA	MAM
160230101	Craters of the Moon National Monument	--	10.17	10.88	5.68	7.82	--	JJA	MJJ	MAM	JAS
210131002	Cumberland Gap National Historical Park	--	18.36	10.12	3.58	7.31	--	MJJ	MJJ	MJJ	MJJ
60270101	Death Valley National Park	29.18	32.55	25.57	15.30	10.61	MJJ	MJJ	MJJ	JJA	JAS
560111013	Devil's Tower National Monument	--	--	7.09	5.42	5.44	--	--	JAS	JAS	JJA
490471002	Dinosaur National Monument	--	10.33	13.34	8.39	13.80	--	MJJ	MJJ	MJJ	MJJ
300298001	Glacier National Park	2.90	2.29	3.98	3.53	2.44	JJA	MAM	MAM	AMJ	AMJ
300351001	Glacier National Park	--	--	--	4.91	3.93	--	--	--	MJJ	MJJ
40058001	Grand Canyon National Park	21.66	18.68	17.02	10.10	14.95	MJJ	AMJ	AMJ	JJA	AMJ
320330101	Great Basin National Park	15.54	15.79	16.94	10.19	11.44	JJA	MJJ	MJJ	AMJ	AMJ
370870036	Great Smoky Mountains National Park	11.46	13.35	11.50	4.59	7.89	AMJ	AMJ	AMJ	AMJ	AMJ
470090102	Great Smoky Mountains National Park	12.97	12.69	10.44	5.31	10.27	AMJ	MAM	AMJ	MAM	MAM
471550101	Great Smoky Mountains National Park	18.87	20.66	14.15	9.03	15.16	AMJ	AMJ	MJJ	MAM	MAM

Monitor site	Park Name	W126					3-Month Timeframe for W126				
471550102	Great Smoky Mountains National Park	19.59	23.51	16.23	7.32	11.94	MJJ	JJA	MJJ	MJJ	ASO
180890022	Indiana Dunes National Lakeshore	8.79	12.21	3.66	2.42	3.91	JJA	AMJ	JAS	MJJ	JJA
60650008	Joshua Tree National Park	24.36	19.97	27.43	19.66	23.39	AMJ	AMJ	MJJ	AMJ	AMJ
60651004	Joshua Tree National Park	--	26.37	30.05	18.81	20.47	--	MJJ	AMJ	JJA	JJA
60719002	Joshua Tree National Park	55.48	52.46	50.99	39.93	43.92	MJJ	MJJ	JJA	JJA	JJA
60893003	Lassen Volcanic National Park	18.97	15.10	18.98	7.64	9.63	JAS	JJA	MJJ	JJA	JAS
80830101	Mesa Verde National Park	23.44	17.57	13.41	15.05	11.94	MJJ	MJJ	AMJ	JJA	AMJ
60711001	Mojave National Preserve	--	28.50	38.92	19.91	19.39	--	MJJ	MJJ	JAS	JJA
530530012	Mount Rainier Wilderness	3.19	3.30	1.18	2.20	1.86	MAM	MAM	JAS	FMA	MAM
530090016	Olympic National Park	--	--	--	--	0.52	--	--	--	--	JAS
530091004	Olympic National Park	--	0.28	0.93	--	--	--	JAS	JAS	--	--
482731001	Padre Island National Seashore	--	8.19	3.66	--	--	--	AMJ	AMJ	--	--
40170119	Petrified Forest National Park	19.16	16.60	19.40	9.04	12.71	AMJ	AMJ	AMJ	AMJ	AMJ
60690003	Pinnacles National Monument	17.14	14.85	19.78	11.41	9.79	JAS	AMJ	MJJ	JAS	JAS
40190021	Saguaro National Park	19.57	17.06	20.13	11.01	15.31	MJJ	MJJ	AMJ	MAM	AMJ
360910004	Saratoga National Historical Park	6.68	10.38	9.26	5.40	5.98	JJA	MJJ	AMJ	MAM	MJJ
311570005	Scotts Bluff National Monument	--	--	--	--	6.20	--	--	--	--	JJA
61070006	Sequoia-Kings Canyon National Park	50.09	53.38	57.24	29.13	26.93	JJA	JJA	JJA	JAS	JAS
61070009	Sequoia-Kings Canyon National Park	66.07	62.88	56.91	55.51	53.79	JAS	JJA	MJJ	JAS	JAS
511130003	Shenandoah National Park	16.43	14.40	12.07	7.63	10.84	AMJ	AMJ	AMJ	MAM	JAS
380070002	Theodore Roosevelt National Park	7.71	5.54	5.55	3.95	4.19	JAS	JJA	AMJ	AMJ	AMJ
380530002	Theodore Roosevelt National Park	9.45	6.29	6.31	4.22	5.17	JJA	JJA	MJJ	AMJ	MAM
40070010	Tonto National Monument	26.39	23.24	25.40	13.67	16.90	MJJ	MJJ	AMJ	AMJ	AMJ
271370034	Voyageurs National Park	5.33	5.19	3.86	4.94	7.66	AMJ	AMJ	MAM	MAM	MAM
460330132	Wind Cave National Park	20.52	12.20	5.92	5.75	5.61	JJA	JJA	JJA	JJA	JAS
560391011	Yellowstone National Park	12.98	9.96	8.84	7.63	11.54	AMJ	AMJ	MAM	MAM	AMJ
60430003	Yosemite National Park	33.78	29.68	42.51	25.70	27.34	JJA	MJJ	JJA	JAS	JAS
60431002	Yosemite National Park	--	12.60	10.03	--	--	--	AMJ	MJJ	--	--
60431003	Yosemite National Park	--	11.61	--	--	--	--	JAS	--	--	--
60431004	Yosemite National Park	--	6.95	15.52	6.58	9.43	--	MJJ	JJA	JAS	JAS
60431005	Yosemite National Park	--	--	27.83	5.18	14.28	--	--	JAS	JAS	JAS

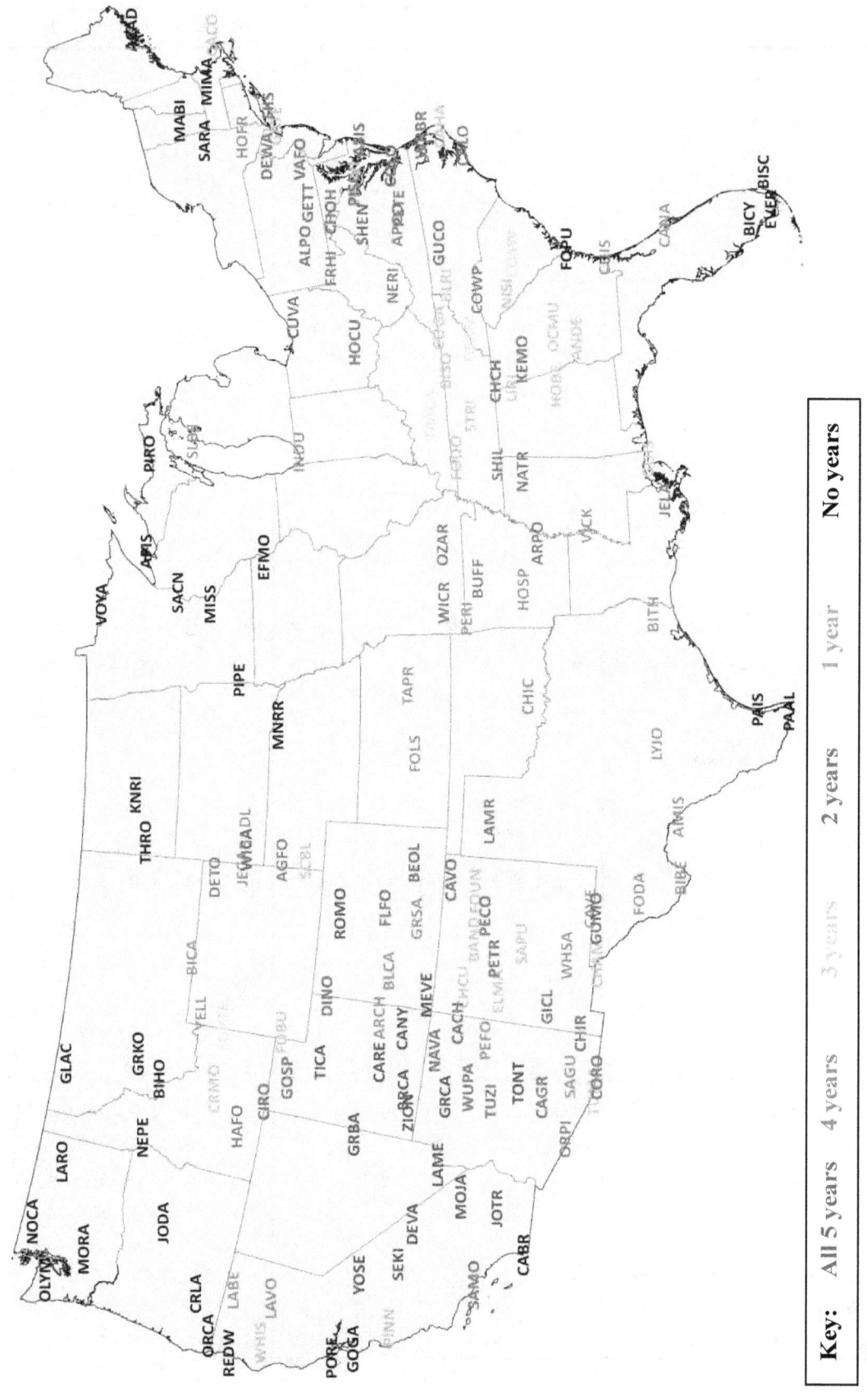

Key: All 5 years   4 years   3 years   2 years   1 year   **No years**

Figure 7A-10 Foliar Injury Results Map for Base Scenario for 214 parks

(Parks identified by park code. Not all park labels shown due to overlap. National Parks are prioritized in mapping.)

7A-12

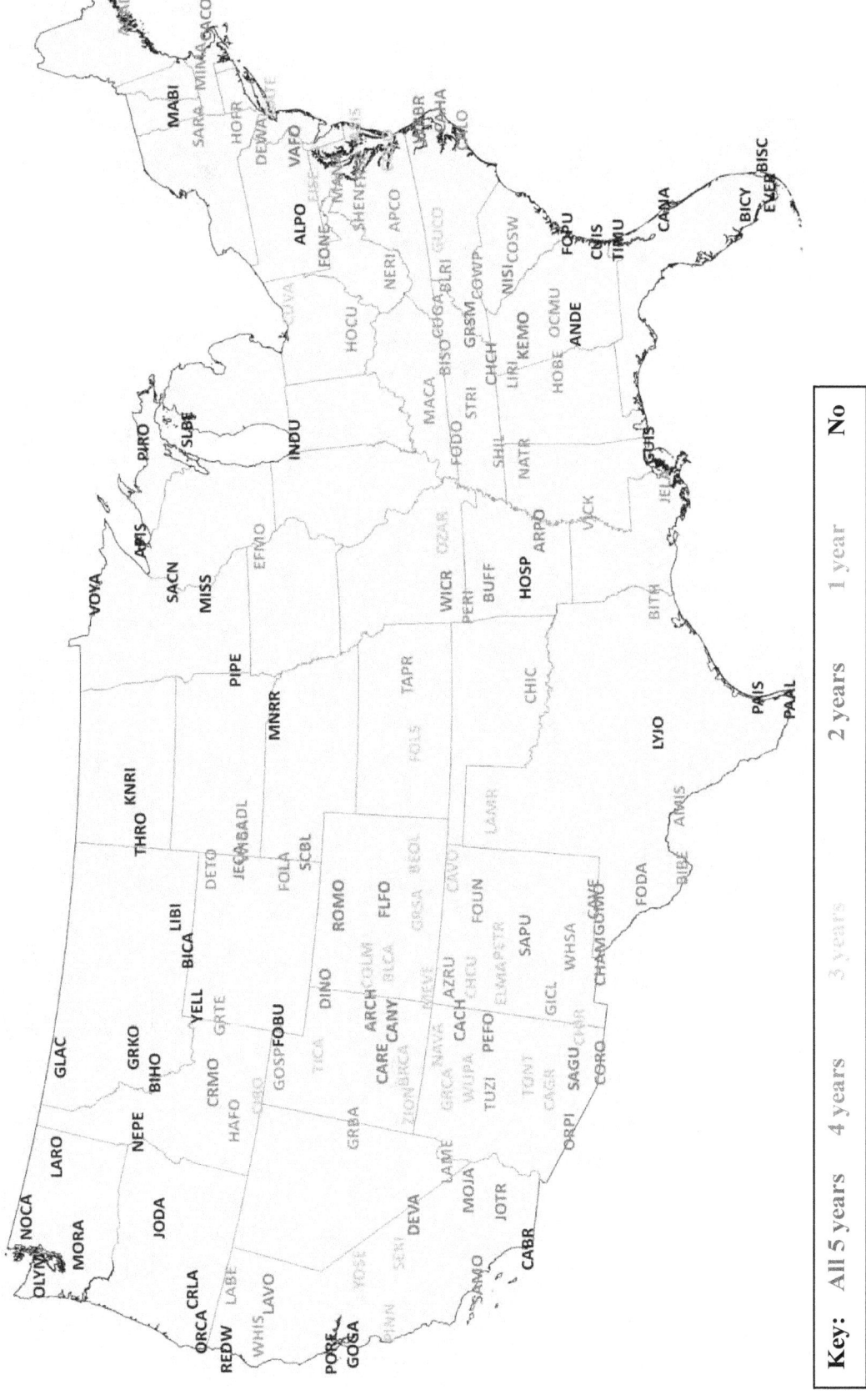

Key: All 5 years   4 years   3 years   2 years   1 year   No

Figure 7A-11 Foliar Injury Results Map for 5% Biosite Scenario (injury=5) for 214 parks

(Parks identified by park code. Not all park labels shown due to overlap. National Parks are prioritized in mapping.)

7A-13

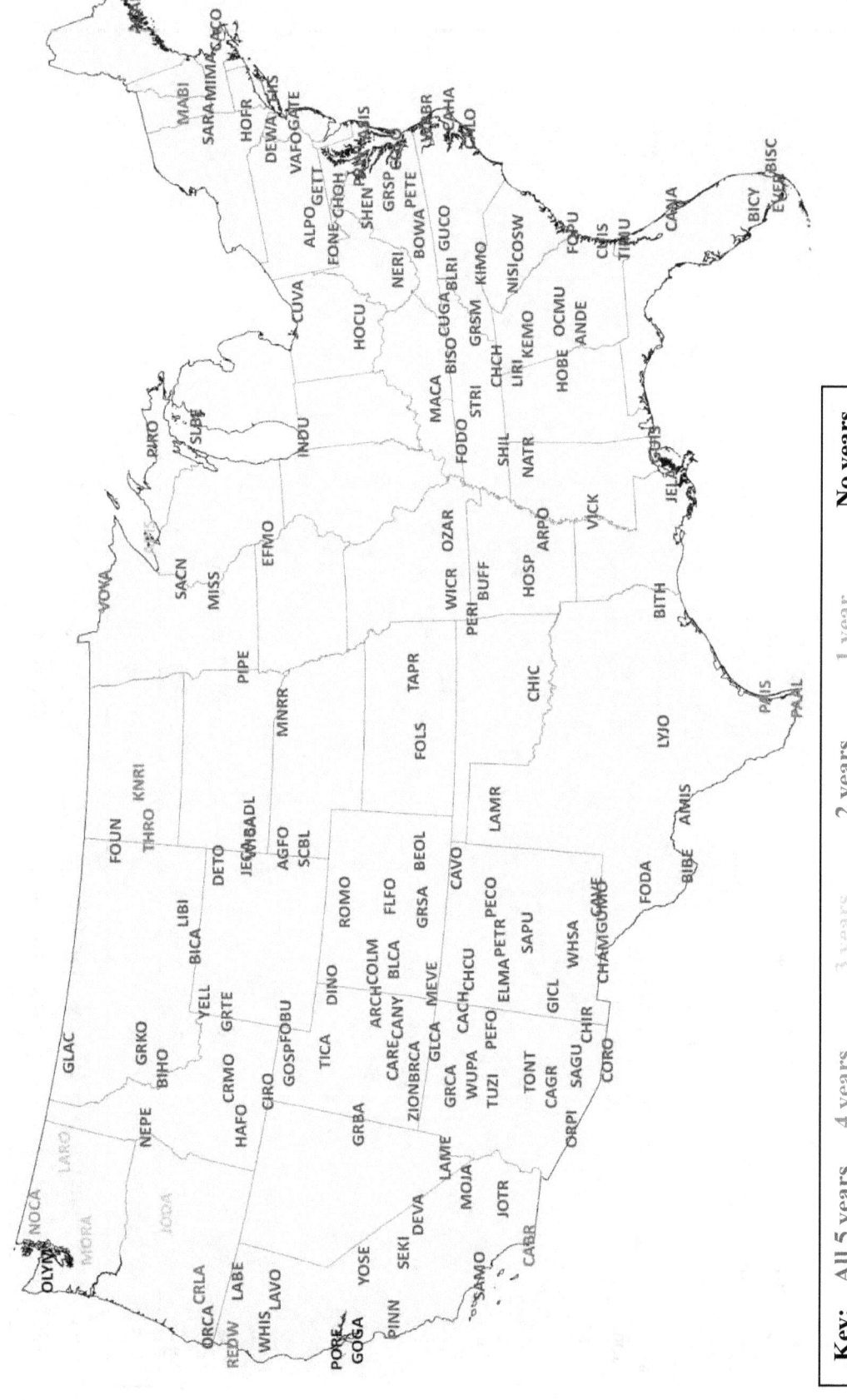

Key: All 5 years   4 years   3 years   2 years   1 year   **No years**

Figure 7A-12 Foliar Injury Results Map for 5% Biosite Scenario (any injury) for 214 parks

(Parks identified by park code. Not all park labels shown due to overlap. National Parks are prioritized in mapping.)

7A-14

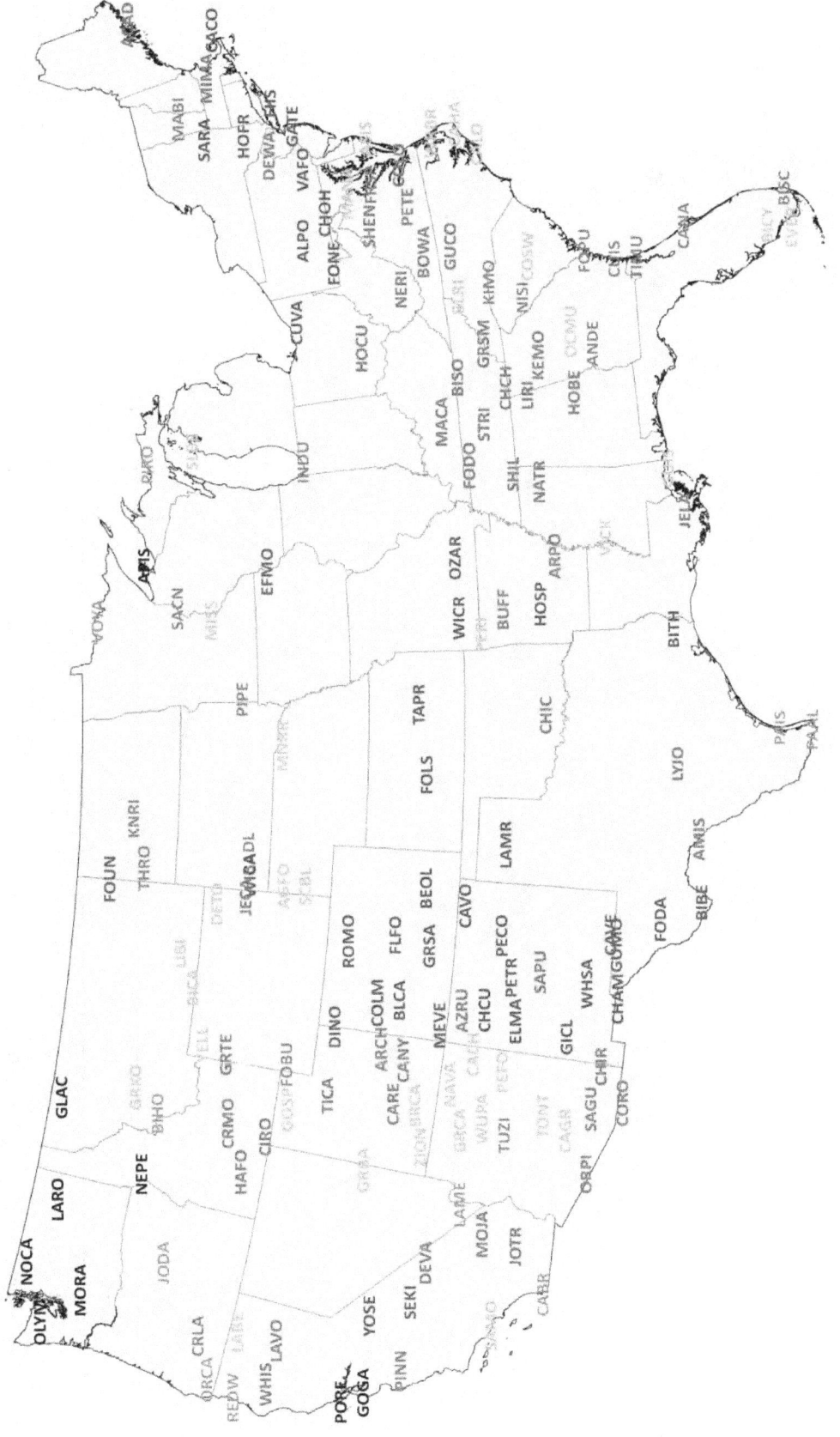

Key: All 5 years   4 years   3 years   2 years   1 year   No years

Figure 7A-13 Foliar Injury Results Map for 10% Biosite Scenario (any injury) for 214 parks
(Parks identified by park code. Not all park labels shown due to overlap. National Parks are prioritized in mapping.)

7A-15

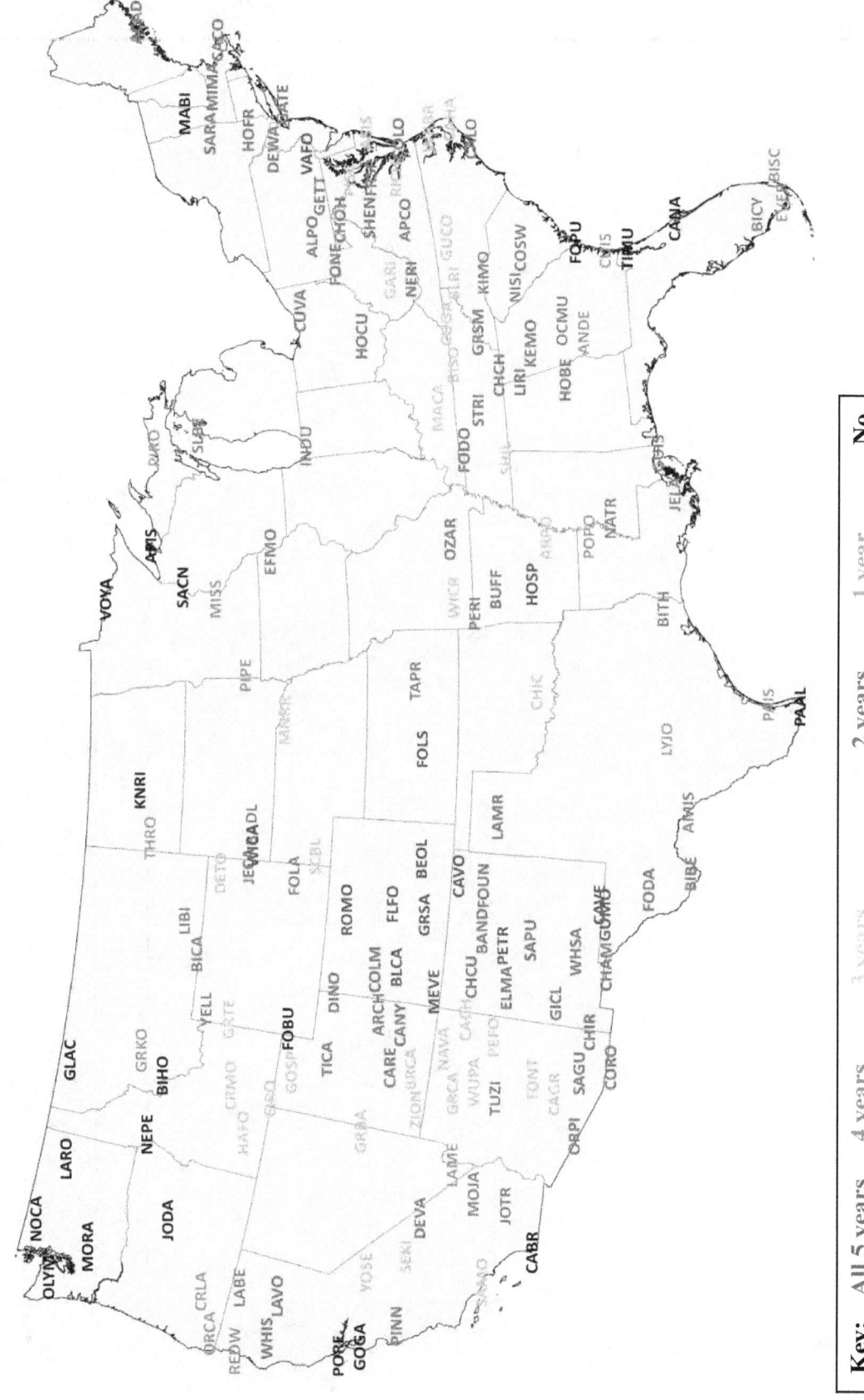

Key: All 5 years   4 years   3 years   2 years   1 year   No

Figure 7A-14 Foliar Injury Results Map for 15% Biosite Scenario (any injury) for 214 parks

(Parks identified by park code. Not all park labels shown due to overlap. National Parks are prioritized in mapping.)

7A-16

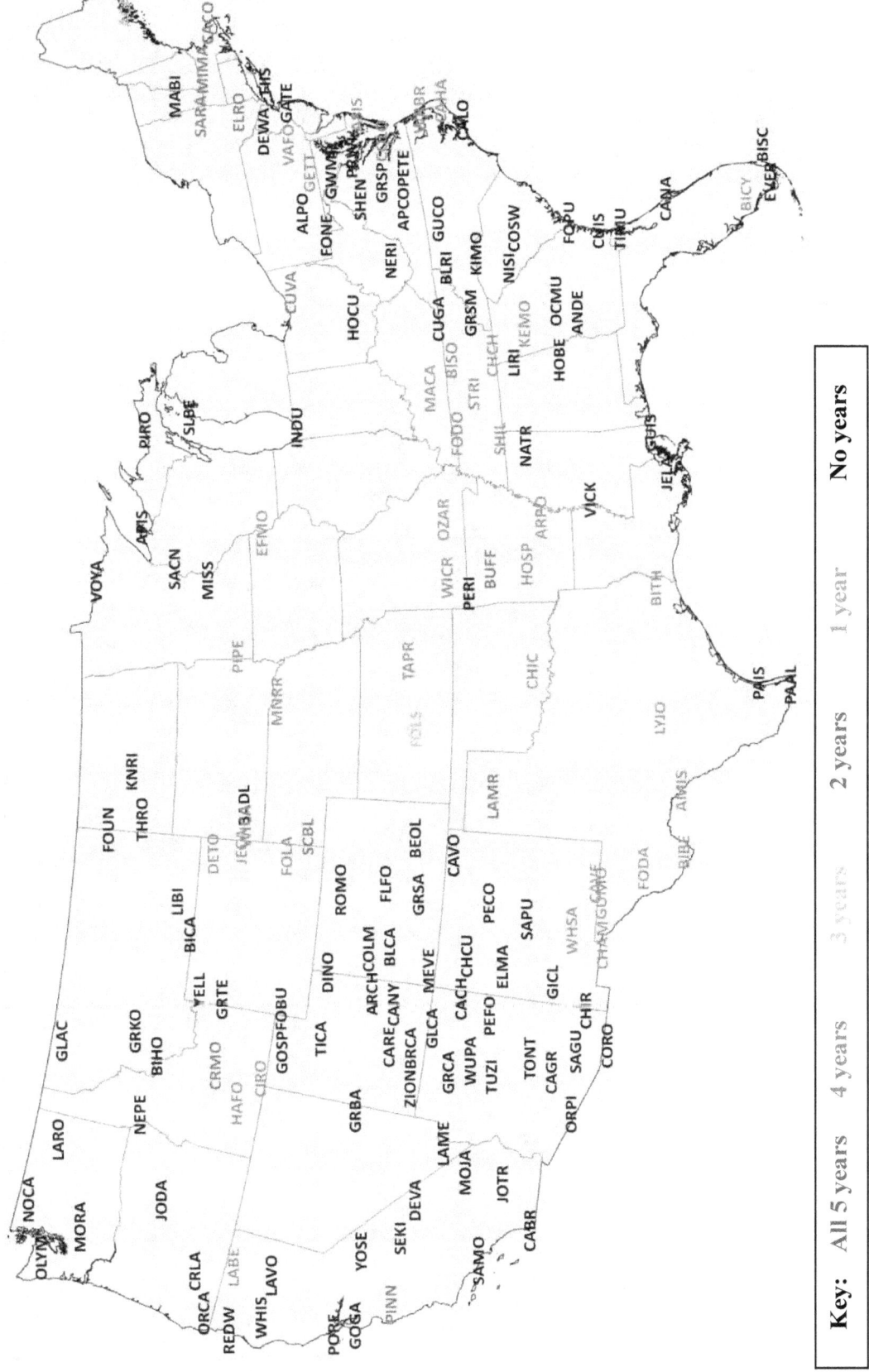

Key: All 5 years   4 years   3 years   2 years   1 year   No years

Figure 7A-12 Foliar Injury Results Map for 20% Biosite Scenario (any injury) for 214 parks

(Parks identified by park code. Not all park labels shown due to overlap. National Parks are prioritized in mapping.)

7A-17

Table 7A-3   Data for 214 Parks Based on Interpolated Ozone Exposure Surface (2006-2010)

Park info		Criteria	2006	2007	2008	2009	2010
		W126 (ppm-hrs)	7.47	6.55	5.68	5.54	4.35
Park Name	Acadia National Park	PZ (7-mo avg)	1.69	1.08	1.98	2.58	0.84
		Exceeds in Base?	No	No	No	No	No
Park Code	ACAD	Exceeds if 5%, injury=5?	Yes	No	No	No	No
Primary State	ME	Exceeds if 5% biosites?	Yes	Yes	Yes	Yes	Yes
Species in park?	Yes	Exceeds if 10% biosites?	Yes	Yes	Yes	Yes	No
# Monitors in park	2	Exceeds if 15% biosites?	Yes	Yes	Yes	Yes	No
Climate Region	Northeast	Exceeds if 20% biosites?	Yes	Yes	Yes	No	No
		W126 (ppm-hrs)	19.16	14.76	10.32	7.44	8.85
Park Name	Agate Fossil Beds National Monument	PZ (7-mo avg)	-1.40	-1.56	-0.17	1.94	1.95
		Exceeds in Base?	Yes	Yes	No	No	No
Park Code	AGFO	Exceeds if 5%, injury=5?	No	No	No	Yes	Yes
Primary State	NE	Exceeds if 5% biosites?	Yes	Yes	Yes	Yes	Yes
Species in park?	Yes	Exceeds if 10% biosites?	No	No	Yes	Yes	Yes
# Monitors in park	1	Exceeds if 15% biosites?	No	No	Yes	Yes	Yes
Climate Region	West North Central	Exceeds if 20% biosites?	No	No	No	Yes	Yes
		W126 (ppm-hrs)	19.52	9.59	11.64	7.93	8.10
Park Name	Alibates Flint Quarries National Monument	PZ (7-mo avg)	-0.72	2.36	-0.27	-0.58	1.40
		Exceeds in Base?	Yes	No	Yes	No	No
Park Code	ALFL	Exceeds if 5%, injury=5?	Yes	Yes	No	No	Yes
Primary State	TX	Exceeds if 5% biosites?	Yes	Yes	Yes	Yes	Yes
Species in park?	Yes	Exceeds if 10% biosites?	Yes	Yes	Yes	Yes	Yes
# Monitors in park	0	Exceeds if 15% biosites?	Yes	Yes	Yes	No	Yes
Climate Region	South	Exceeds if 20% biosites?	No	Yes	No	No	Yes
		W126 (ppm-hrs)	10.96	11.69	10.53	6.01	11.04
Park Name	Allegheny Portage Railroad National Historic Site	PZ (7-mo avg)	-0.27	-0.02	-0.01	-0.23	-0.64
		Exceeds in Base?	Yes	Yes	Yes	No	Yes
Park Code	ALPO	Exceeds if 5%, injury=5?	No	No	No	No	No
Primary State	PA	Exceeds if 5% biosites?	Yes	Yes	Yes	Yes	Yes
Species in park?	Yes	Exceeds if 10% biosites?	Yes	Yes	Yes	Yes	Yes
# Monitors in park	0	Exceeds if 15% biosites?	Yes	Yes	Yes	No	Yes
Climate Region	Northeast	Exceeds if 20% biosites?	No	No	No	No	No
		W126 (ppm-hrs)	12.83	7.15	8.26	7.79	6.46
Park Name	Amistad National Recreation Area	PZ (7-mo avg)	-1.59	4.10	-0.72	-0.68	0.59
		Exceeds in Base?	Yes	No	No	No	No
Park Code	AMIS	Exceeds if 5%, injury=5?	No	Yes	No	No	No
Primary State	TX	Exceeds if 5% biosites?	Yes	Yes	Yes	Yes	Yes
Species in park?	Yes	Exceeds if 10% biosites?	No	Yes	Yes	Yes	Yes
# Monitors in park	0	Exceeds if 15% biosites?	No	Yes	Yes	No	No
Climate Region	South	Exceeds if 20% biosites?	No	Yes	No	No	No
		W126 (ppm-hrs)	15.92	13.34	11.45	4.60	8.01
Park Name	Andersonville National Historic Site	PZ (7-mo avg)	-1.55	-1.92	-0.47	0.76	-1.17
		Exceeds in Base?	Yes	Yes	Yes	No	No
Park Code	ANDE	Exceeds if 5%, injury=5?	No	No	No	No	No

Park info		Criteria	2006	2007	2008	2009	2010
Primary State	GA	Exceeds if 5% biosites?	Yes	Yes	Yes	Yes	Yes
Species in park?	0	Exceeds if 10% biosites?	No	No	Yes	No	Yes
# Monitors in park	0	Exceeds if 15% biosites?	No	No	Yes	No	No
Climate Region	Southeast	Exceeds if 20% biosites?	No	No	No	No	No
Park Name	Antietam National Battlefield	W126 (ppm-hrs)	15.36	15.10	11.85	5.96	15.32
		PZ (7-mo avg)	-0.28	-0.53	1.24	0.05	-1.09
		Exceeds in Base?	Yes	Yes	Yes	No	Yes
Park Code	ANTI	Exceeds if 5%, injury=5?	Yes	Yes	Yes	No	Yes
Primary State	MD	Exceeds if 5% biosites?	Yes	Yes	Yes	Yes	Yes
Species in park?	Yes	Exceeds if 10% biosites?	Yes	Yes	Yes	Yes	Yes
# Monitors in park	0	Exceeds if 15% biosites?	Yes	Yes	Yes	No	Yes
Climate Region	Northeast	Exceeds if 20% biosites?	No	No	Yes	No	No
Park Name	Apostle Islands National Lakeshore	W126 (ppm-hrs)	4.27	5.47	3.79	2.67	3.20
		PZ (7-mo avg)	-1.55	-1.01	-0.09	-1.77	0.80
		Exceeds in Base?	No	No	No	No	No
Park Code	APIS	Exceeds if 5%, injury=5?	No	No	No	No	No
Primary State	WI	Exceeds if 5% biosites?	No	Yes	Yes	No	Yes
Species in park?	Yes	Exceeds if 10% biosites?	No	No	No	No	No
# Monitors in park	0	Exceeds if 15% biosites?	No	No	No	No	No
Climate Region	East North Central	Exceeds if 20% biosites?	No	No	No	No	No
Park Name	Appomattox Court House National Historical Park	W126 (ppm-hrs)	12.87	12.06	9.02	5.15	8.99
		PZ (7-mo avg)	-0.64	-1.19	-0.12	0.36	-0.74
		Exceeds in Base?	Yes	Yes	No	No	No
Park Code	APCO	Exceeds if 5%, injury=5?	Yes	No	No	No	No
Primary State	VA	Exceeds if 5% biosites?	Yes	Yes	Yes	Yes	Yes
Species in park?	Yes	Exceeds if 10% biosites?	Yes	Yes	Yes	No	Yes
# Monitors in park	0	Exceeds if 15% biosites?	Yes	Yes	Yes	No	Yes
Climate Region	Southeast	Exceeds if 20% biosites?	No	No	No	No	No
Park Name	Arches National Park	W126 (ppm-hrs)	19.08	16.53	15.59	9.71	15.20
		PZ (7-mo avg)	-0.91	-0.21	-0.42	-1.39	0.77
		Exceeds in Base?	Yes	Yes	Yes	No	Yes
Park Code	ARCH	Exceeds if 5%, injury=5?	Yes	Yes	Yes	No	Yes
Primary State	UT	Exceeds if 5% biosites?	Yes	Yes	Yes	Yes	Yes
Species in park?	Yes	Exceeds if 10% biosites?	Yes	Yes	Yes	No	Yes
# Monitors in park	0	Exceeds if 15% biosites?	Yes	Yes	Yes	No	Yes
Climate Region	Southwest	Exceeds if 20% biosites?	No	No	No	No	No
Park Name	Arkansas Post National Memorial	W126 (ppm-hrs)	16.65	13.09	7.42	6.43	9.98
		PZ (7-mo avg)	-0.76	-1.16	0.53	2.26	-1.65
		Exceeds in Base?	Yes	Yes	No	No	No
Park Code	ARPO	Exceeds if 5%, injury=5?	Yes	Yes	No	No	No
Primary State	AR	Exceeds if 5% biosites?	Yes	Yes	Yes	Yes	Yes
Species in park?	Yes	Exceeds if 10% biosites?	Yes	Yes	Yes	Yes	No
# Monitors in park	0	Exceeds if 15% biosites?	Yes	Yes	No	Yes	No
Climate Region	South	Exceeds if 20% biosites?	No	No	No	Yes	No
Park Name	Assateague Island National Seashore	W126 (ppm-hrs)	16.15	15.63	16.12	7.34	12.78
		PZ (7-mo avg)	0.20	-1.50	0.27	1.38	-1.70

Park info		Criteria	2006	2007	2008	2009	2010
		Exceeds in Base?	Yes	Yes	Yes	No	Yes
Park Code	ASIS	Exceeds if 5%, injury=5?	Yes	No	Yes	Yes	No
Primary State	MD	Exceeds if 5% biosites?	Yes	Yes	Yes	Yes	Yes
Species in park?	Yes	Exceeds if 10% biosites?	Yes	No	Yes	Yes	No
# Monitors in park	0	Exceeds if 15% biosites?	Yes	No	Yes	Yes	No
Climate Region	Northeast	Exceeds if 20% biosites?	No	No	No	Yes	No
		W126 (ppm-hrs)	11.41	17.91	12.02	5.04	10.86
Park Name	Aztec Ruins National Monument	PZ (7-mo avg)	0.10	0.10	0.02	-0.50	0.75
		Exceeds in Base?	Yes	Yes	Yes	No	Yes
Park Code	AZRU	Exceeds if 5%, injury=5?	No	Yes	No	No	No
Primary State	NM	Exceeds if 5% biosites?	Yes	Yes	Yes	Yes	Yes
Species in park?	Yes	Exceeds if 10% biosites?	Yes	Yes	Yes	No	Yes
# Monitors in park	0	Exceeds if 15% biosites?	Yes	Yes	Yes	No	Yes
Climate Region	Southwest	Exceeds if 20% biosites?	No	No	No	No	No
		W126 (ppm-hrs)	14.98	8.11	4.30	3.85	4.69
Park Name	Badlands National Park	PZ (7-mo avg)	-0.57	-1.41	1.27	2.06	2.15
		Exceeds in Base?	Yes	No	No	No	No
Park Code	BADL	Exceeds if 5%, injury=5?	Yes	No	No	No	No
Primary State	SD	Exceeds if 5% biosites?	Yes	Yes	Yes	Yes	Yes
Species in park?	Yes	Exceeds if 10% biosites?	Yes	No	No	No	Yes
# Monitors in park	2	Exceeds if 15% biosites?	Yes	No	No	No	Yes
Climate Region	West North Central	Exceeds if 20% biosites?	No	No	No	No	No
		W126 (ppm-hrs)	16.50	15.38	12.69	6.88	9.82
Park Name	Bandelier National Monument	PZ (7-mo avg)	-0.91	0.04	-0.84	-0.95	-0.54
		Exceeds in Base?	Yes	Yes	Yes	No	No
Park Code	BAND	Exceeds if 5%, injury=5?	Yes	Yes	Yes	No	No
Primary State	NM	Exceeds if 5% biosites?	Yes	Yes	Yes	Yes	Yes
Species in park?	Yes	Exceeds if 10% biosites?	Yes	Yes	Yes	Yes	Yes
# Monitors in park	0	Exceeds if 15% biosites?	Yes	Yes	Yes	No	Yes
Climate Region	Southwest	Exceeds if 20% biosites?	No	No	No	No	No
		W126 (ppm-hrs)	21.90	14.35	15.44	9.34	11.35
Park Name	Bent's Old Fort National Historic Site	PZ (7-mo avg)	-0.88	0.27	-0.80	-0.62	0.26
		Exceeds in Base?	Yes	Yes	Yes	No	Yes
Park Code	BEOL	Exceeds if 5%, injury=5?	Yes	Yes	Yes	No	No
Primary State	CO	Exceeds if 5% biosites?	Yes	Yes	Yes	Yes	Yes
Species in park?	Yes	Exceeds if 10% biosites?	Yes	Yes	Yes	Yes	Yes
# Monitors in park	0	Exceeds if 15% biosites?	Yes	Yes	Yes	Yes	Yes
Climate Region	Southwest	Exceeds if 20% biosites?	No	No	No	No	No
		W126 (ppm-hrs)	11.95	9.83	9.92	8.29	8.01
Park Name	Big Bend National Park	PZ (7-mo avg)	-0.67	1.78	0.11	-0.25	0.96
		Exceeds in Base?	Yes	No	No	No	No
Park Code	BIBE	Exceeds if 5%, injury=5?	No	Yes	No	No	No
Primary State	TX	Exceeds if 5% biosites?	Yes	Yes	Yes	Yes	Yes
Species in park?	Yes	Exceeds if 10% biosites?	Yes	Yes	Yes	Yes	Yes
# Monitors in park	1	Exceeds if 15% biosites?	Yes	Yes	Yes	Yes	No
Climate Region	South	Exceeds if 20% biosites?	No	Yes	No	No	No

Park info		Criteria	2006	2007	2008	2009	2010
		W126 (ppm-hrs)	9.84	7.73	6.29	3.65	5.64
Park Name	Big Cypress National Preserve	PZ (7-mo avg)	-0.31	-1.10	1.48	-0.28	0.55
		Exceeds in Base?	No	No	No	No	No
Park Code	BICY	Exceeds if 5%, injury=5?	No	No	No	No	No
Primary State	FL	Exceeds if 5% biosites?	Yes	Yes	Yes	Yes	Yes
Species in park?	Yes	Exceeds if 10% biosites?	Yes	Yes	Yes	No	No
# Monitors in park	0	Exceeds if 15% biosites?	Yes	No	Yes	No	No
Climate Region	Southeast	Exceeds if 20% biosites?	No	No	Yes	No	No
		W126 (ppm-hrs)	5.91	7.99	6.91	5.39	5.53
Park Name	Big Hole National Battlefield	PZ (7-mo avg)	-1.10	-1.77	-0.83	-0.06	0.61
		Exceeds in Base?	No	No	No	No	No
Park Code	BIHO	Exceeds if 5%, injury=5?	No	No	No	No	No
Primary State	MT	Exceeds if 5% biosites?	Yes	Yes	Yes	Yes	Yes
Species in park?	Yes	Exceeds if 10% biosites?	No	No	Yes	No	No
# Monitors in park	0	Exceeds if 15% biosites?	No	No	No	No	No
Climate Region	West North Central	Exceeds if 20% biosites?	No	No	No	No	No
	Big South Fork National River and Recreation Area	W126 (ppm-hrs)	12.74	16.75	10.68	5.91	8.12
Park Name		PZ (7-mo avg)	0.16	-1.80	-0.57	1.54	-0.19
		Exceeds in Base?	Yes	Yes	Yes	No	No
Park Code	BISO	Exceeds if 5%, injury=5?	Yes	No	No	No	No
Primary State	TN	Exceeds if 5% biosites?	Yes	Yes	Yes	Yes	Yes
Species in park?	Yes	Exceeds if 10% biosites?	Yes	No	Yes	Yes	Yes
# Monitors in park	0	Exceeds if 15% biosites?	Yes	No	Yes	Yes	No
Climate Region	Central	Exceeds if 20% biosites?	No	No	No	Yes	No
		W126 (ppm-hrs)	11.19	9.17	6.70	6.31	8.13
Park Name	Big Thicket National Preserve	PZ (7-mo avg)	-0.51	1.70	0.18	0.09	-0.60
		Exceeds in Base?	Yes	No	No	No	No
Park Code	BITH	Exceeds if 5%, injury=5?	No	Yes	No	No	No
Primary State	TX	Exceeds if 5% biosites?	Yes	Yes	Yes	Yes	Yes
Species in park?	Yes	Exceeds if 10% biosites?	Yes	Yes	Yes	Yes	Yes
# Monitors in park	0	Exceeds if 15% biosites?	Yes	Yes	No	No	No
Climate Region	South	Exceeds if 20% biosites?	No	Yes	No	No	No
		W126 (ppm-hrs)	11.44	9.08	9.44	6.59	5.07
Park Name	Bighorn Canyon National Recreation Area	PZ (7-mo avg)	-1.83	-0.59	0.55	0.62	0.16
		Exceeds in Base?	Yes	No	No	No	No
Park Code	BICA	Exceeds if 5%, injury=5?	No	No	No	No	No
Primary State	MT	Exceeds if 5% biosites?	Yes	Yes	Yes	Yes	Yes
Species in park?	Yes	Exceeds if 10% biosites?	No	Yes	Yes	Yes	No
# Monitors in park	0	Exceeds if 15% biosites?	No	Yes	Yes	No	No
Climate Region	West North Central	Exceeds if 20% biosites?	No	No	No	No	No
		W126 (ppm-hrs)	9.14	7.41	5.68	3.82	5.40
Park Name	Biscayne National Park	PZ (7-mo avg)	-0.41	0.22	0.59	0.13	0.86
		Exceeds in Base?	No	No	No	No	No
Park Code	BISC	Exceeds if 5%, injury=5?	No	No	No	No	No
Primary State	FL	Exceeds if 5% biosites?	Yes	Yes	Yes	Yes	Yes
Species in park?	Yes	Exceeds if 10% biosites?	Yes	Yes	No	No	No

7A-21

Park info		Criteria	2006	2007	2008	2009	2010
# Monitors in park	0	Exceeds if 15% biosites?	Yes	No	No	No	No
Climate Region	Southeast	Exceeds if 20% biosites?	No	No	No	No	No
Park Name	Black Canyon of the Gunnison National Park	W126 (ppm-hrs)	19.02	15.26	14.70	8.67	11.44
		PZ (7-mo avg)	-0.54	-0.25	-0.26	-0.17	-0.11
		Exceeds in Base?	Yes	Yes	Yes	No	Yes
Park Code	BLCA	Exceeds if 5%, injury=5?	Yes	Yes	Yes	No	No
Primary State	CO	Exceeds if 5% biosites?	Yes	Yes	Yes	Yes	Yes
Species in park?	Yes	Exceeds if 10% biosites?	Yes	Yes	Yes	Yes	Yes
# Monitors in park	0	Exceeds if 15% biosites?	Yes	Yes	Yes	Yes	Yes
Climate Region	Southwest	Exceeds if 20% biosites?	No	No	No	No	No
Park Name	Blue Ridge Parkway	W126 (ppm-hrs)	12.85	14.30	11.53	5.50	9.71
		PZ (7-mo avg)	-0.08	-1.31	-0.33	0.96	-0.81
		Exceeds in Base?	Yes	Yes	Yes	No	No
Park Code	BLRI	Exceeds if 5%, injury=5?	Yes	No	No	No	No
Primary State	NC	Exceeds if 5% biosites?	Yes	Yes	Yes	Yes	Yes
Species in park?	Yes	Exceeds if 10% biosites?	Yes	No	Yes	No	Yes
# Monitors in park	1	Exceeds if 15% biosites?	Yes	No	Yes	No	Yes
Climate Region	Southeast	Exceeds if 20% biosites?	No	No	No	No	No
Park Name	Bluestone National Scenic River	W126 (ppm-hrs)	11.27	14.52	9.78	5.54	8.77
		PZ (7-mo avg)	0.34	-0.54	0.35	1.00	-0.50
		Exceeds in Base?	Yes	Yes	No	No	No
Park Code	BLUE	Exceeds if 5%, injury=5?	No	Yes	No	No	No
Primary State	WV	Exceeds if 5% biosites?	Yes	Yes	Yes	Yes	Yes
Species in park?	Yes	Exceeds if 10% biosites?	Yes	Yes	Yes	Yes	Yes
# Monitors in park	0	Exceeds if 15% biosites?	Yes	Yes	Yes	Yes	Yes
Climate Region	Central	Exceeds if 20% biosites?	No	No	No	No	No
Park Name	Booker T. Washington National Monument	W126 (ppm-hrs)	13.28	14.74	11.57	5.59	9.99
		PZ (7-mo avg)	-0.64	-1.19	-0.12	0.36	-0.74
		Exceeds in Base?	Yes	Yes	Yes	No	No
Park Code	BOWA	Exceeds if 5%, injury=5?	Yes	Yes	No	No	No
Primary State	VA	Exceeds if 5% biosites?	Yes	Yes	Yes	Yes	Yes
Species in park?	Yes	Exceeds if 10% biosites?	Yes	Yes	Yes	No	Yes
# Monitors in park	0	Exceeds if 15% biosites?	Yes	Yes	Yes	No	Yes
Climate Region	Southeast	Exceeds if 20% biosites?	No	No	No	No	No
Park Name	Bryce Canyon National Park	W126 (ppm-hrs)	21.01	18.06	17.01	11.72	14.89
		PZ (7-mo avg)	0.03	-1.53	-0.68	-1.30	0.14
		Exceeds in Base?	Yes	Yes	Yes	Yes	Yes
Park Code	BRCA	Exceeds if 5%, injury=5?	Yes	No	Yes	No	Yes
Primary State	UT	Exceeds if 5% biosites?	Yes	Yes	Yes	Yes	Yes
Species in park?	Yes	Exceeds if 10% biosites?	Yes	No	Yes	No	Yes
# Monitors in park	0	Exceeds if 15% biosites?	Yes	No	Yes	No	Yes
Climate Region	Southwest	Exceeds if 20% biosites?	No	No	No	No	No
Park Name	Buffalo National River	W126 (ppm-hrs)	14.06	12.08	7.03	5.81	8.39
		PZ (7-mo avg)	-0.13	-1.27	3.23	1.23	-0.15
		Exceeds in Base?	Yes	Yes	No	No	No
Park Code	BUFF	Exceeds if 5%, injury=5?	Yes	No	Yes	No	No

Park info		Criteria	2006	2007	2008	2009	2010
Primary State	AR	Exceeds if 5% biosites?	Yes	Yes	Yes	Yes	Yes
Species in park?	Yes	Exceeds if 10% biosites?	Yes	No	Yes	Yes	Yes
# Monitors in park	0	Exceeds if 15% biosites?	Yes	No	Yes	Yes	Yes
Climate Region	South	Exceeds if 20% biosites?	No	No	Yes	Yes	No
Park Name	Cabrillo National Monument	W126 (ppm-hrs)	5.31	6.20	6.65	5.67	5.20
		PZ (7-mo avg)	0.03	-2.53	-1.19	-1.98	0.18
		Exceeds in Base?	No	No	No	No	No
Park Code	CABR	Exceeds if 5%, injury=5?	No	No	No	No	No
Primary State	CA	Exceeds if 5% biosites?	Yes	Yes	Yes	No	Yes
Species in park?	0	Exceeds if 10% biosites?	No	No	Yes	No	No
# Monitors in park	0	Exceeds if 15% biosites?	No	No	No	No	No
Climate Region	West	Exceeds if 20% biosites?	No	No	No	No	No
Park Name	Canaveral National Seashore	W126 (ppm-hrs)	13.47	9.29	7.15	5.28	6.13
		PZ (7-mo avg)	-1.90	-1.26	-0.12	0.06	-0.90
		Exceeds in Base?	Yes	No	No	No	No
Park Code	CANA	Exceeds if 5%, injury=5?	No	No	No	No	No
Primary State	FL	Exceeds if 5% biosites?	Yes	Yes	Yes	Yes	Yes
Species in park?	Yes	Exceeds if 10% biosites?	No	No	Yes	No	Yes
# Monitors in park	0	Exceeds if 15% biosites?	No	No	No	No	No
Climate Region	Southeast	Exceeds if 20% biosites?	No	No	No	No	No
Park Name	Canyon de Chelly National Monument	W126 (ppm-hrs)	17.97	16.13	14.84	7.63	10.91
		PZ (7-mo avg)	-0.54	-1.40	0.03	-1.48	0.39
		Exceeds in Base?	Yes	Yes	Yes	No	Yes
Park Code	CACH	Exceeds if 5%, injury=5?	Yes	No	Yes	No	No
Primary State	AZ	Exceeds if 5% biosites?	Yes	Yes	Yes	Yes	Yes
Species in park?	Yes	Exceeds if 10% biosites?	Yes	No	Yes	No	Yes
# Monitors in park	0	Exceeds if 15% biosites?	Yes	No	Yes	No	Yes
Climate Region	Southwest	Exceeds if 20% biosites?	No	No	No	No	No
Park Name	Canyonlands National Park	W126 (ppm-hrs)	19.25	17.03	16.38	10.82	13.06
		PZ (7-mo avg)	-0.91	-0.21	-0.42	-1.39	0.77
		Exceeds in Base?	Yes	Yes	Yes	Yes	Yes
Park Code	CANY	Exceeds if 5%, injury=5?	Yes	Yes	Yes	No	Yes
Primary State	UT	Exceeds if 5% biosites?	Yes	Yes	Yes	Yes	Yes
Species in park?	Yes	Exceeds if 10% biosites?	Yes	Yes	Yes	No	Yes
# Monitors in park	1	Exceeds if 15% biosites?	Yes	Yes	Yes	No	Yes
Climate Region	Southwest	Exceeds if 20% biosites?	No	No	No	No	No
Park Name	Cape Cod National Seashore	W126 (ppm-hrs)	12.19	10.78	11.83	5.00	7.26
		PZ (7-mo avg)	1.36	0.13	1.06	1.47	0.80
		Exceeds in Base?	Yes	Yes	Yes	No	No
Park Code	CACO	Exceeds if 5%, injury=5?	Yes	No	Yes	No	No
Primary State	MA	Exceeds if 5% biosites?	Yes	Yes	Yes	Yes	Yes
Species in park?	Yes	Exceeds if 10% biosites?	Yes	Yes	Yes	Yes	Yes
# Monitors in park	1	Exceeds if 15% biosites?	Yes	Yes	Yes	Yes	No
Climate Region	Northeast	Exceeds if 20% biosites?	Yes	No	Yes	No	No
Park Name	Cape Hatteras National Seashore	W126 (ppm-hrs)	12.46	12.19	13.06	5.17	10.27
		PZ (7-mo avg)	1.15	-1.60	-0.77	-0.02	-0.43

Park info		Criteria	2006	2007	2008	2009	2010
		Exceeds in Base?	Yes	Yes	Yes	No	No
Park Code	CAHA	Exceeds if 5%, injury=5?	Yes	No	Yes	No	No
Primary State	NC	Exceeds if 5% biosites?	Yes	Yes	Yes	Yes	Yes
Species in park?	Yes	Exceeds if 10% biosites?	Yes	No	Yes	No	Yes
# Monitors in park	0	Exceeds if 15% biosites?	Yes	No	Yes	No	Yes
Climate Region	Southeast	Exceeds if 20% biosites?	Yes	No	No	No	No
		W126 (ppm-hrs)	12.31	9.23	12.08	4.86	6.70
Park Name	Cape Lookout National Seashore	PZ (7-mo avg)	0.20	-1.31	-0.91	-0.06	-0.68
		Exceeds in Base?	Yes	No	Yes	No	No
Park Code	CALO	Exceeds if 5%, injury=5?	Yes	No	No	No	No
Primary State	NC	Exceeds if 5% biosites?	Yes	Yes	Yes	Yes	Yes
Species in park?	Yes	Exceeds if 10% biosites?	Yes	No	Yes	No	Yes
# Monitors in park	0	Exceeds if 15% biosites?	Yes	No	Yes	No	No
Climate Region	Southeast	Exceeds if 20% biosites?	No	No	No	No	No
		W126 (ppm-hrs)	20.45	17.98	16.78	11.33	14.67
Park Name	Capitol Reef National Park	PZ (7-mo avg)	-0.44	-0.87	-0.55	-1.35	0.46
		Exceeds in Base?	Yes	Yes	Yes	Yes	Yes
Park Code	CARE	Exceeds if 5%, injury=5?	Yes	Yes	Yes	No	Yes
Primary State	UT	Exceeds if 5% biosites?	Yes	Yes	Yes	Yes	Yes
Species in park?	Yes	Exceeds if 10% biosites?	Yes	Yes	Yes	No	Yes
# Monitors in park	0	Exceeds if 15% biosites?	Yes	Yes	Yes	No	Yes
Climate Region	Southwest	Exceeds if 20% biosites?	No	No	No	No	No
		W126 (ppm-hrs)	19.33	13.02	13.75	8.35	11.10
Park Name	Capulin Volcano National Monument	PZ (7-mo avg)	-0.91	0.04	-0.84	-0.95	-0.54
		Exceeds in Base?	Yes	Yes	Yes	No	Yes
Park Code	CAVO	Exceeds if 5%, injury=5?	Yes	Yes	Yes	No	No
Primary State	NM	Exceeds if 5% biosites?	Yes	Yes	Yes	Yes	Yes
Species in park?	Yes	Exceeds if 10% biosites?	Yes	Yes	Yes	Yes	Yes
# Monitors in park	0	Exceeds if 15% biosites?	Yes	Yes	Yes	Yes	Yes
Climate Region	Southwest	Exceeds if 20% biosites?	No	No	No	No	No
		W126 (ppm-hrs)	21.24	9.80	10.03	9.32	9.57
Park Name	Carlsbad Caverns National Park	PZ (7-mo avg)	0.63	0.73	0.19	-0.88	1.18
		Exceeds in Base?	Yes	No	No	No	No
Park Code	CAVE	Exceeds if 5%, injury=5?	Yes	No	No	No	Yes
Primary State	NM	Exceeds if 5% biosites?	Yes	Yes	Yes	Yes	Yes
Species in park?	Yes	Exceeds if 10% biosites?	Yes	Yes	Yes	Yes	Yes
# Monitors in park	1	Exceeds if 15% biosites?	Yes	Yes	Yes	Yes	Yes
Climate Region	Southwest	Exceeds if 20% biosites?	No	No	No	No	Yes
		W126 (ppm-hrs)	17.20	11.44	18.53	9.19	12.42
Park Name	Casa Grande Ruins National Monument	PZ (7-mo avg)	-0.01	-1.38	0.30	-1.63	0.02
		Exceeds in Base?	Yes	Yes	Yes	No	Yes
Park Code	CAGR	Exceeds if 5%, injury=5?	Yes	No	Yes	No	Yes
Primary State	NM	Exceeds if 5% biosites?	Yes	Yes	Yes	Yes	Yes
Species in park?	Yes	Exceeds if 10% biosites?	Yes	No	Yes	No	Yes
# Monitors in park	0	Exceeds if 15% biosites?	Yes	No	Yes	No	Yes
Climate Region	Southwest	Exceeds if 20% biosites?	No	No	No	No	No

Park info		Criteria	2006	2007	2008	2009	2010
		W126 (ppm-hrs)	14.08	17.33	13.22	5.97	15.35
Park Name	Catoctin Mountain Park	PZ (7-mo avg)	-0.19	-1.10	0.53	1.08	-0.94
		Exceeds in Base?	Yes	Yes	Yes	No	Yes
Park Code	CATO	Exceeds if 5%, injury=5?	Yes	Yes	Yes	No	Yes
Primary State	MD	Exceeds if 5% biosites?	Yes	Yes	Yes	Yes	Yes
Species in park?	Yes	Exceeds if 10% biosites?	Yes	Yes	Yes	Yes	Yes
# Monitors in park	0	Exceeds if 15% biosites?	Yes	Yes	Yes	Yes	Yes
Climate Region	Northeast	Exceeds if 20% biosites?	No	No	No	Yes	No
		W126 (ppm-hrs)	21.15	17.82	17.53	12.29	15.51
Park Name	Cedar Breaks National Monument	PZ (7-mo avg)	0.03	-1.53	-0.68	-1.30	0.14
		Exceeds in Base?	Yes	Yes	Yes	Yes	Yes
Park Code	CEBR	Exceeds if 5%, injury=5?	Yes	No	Yes	No	Yes
Primary State	UT	Exceeds if 5% biosites?	Yes	Yes	Yes	Yes	Yes
Species in park?	Yes	Exceeds if 10% biosites?	Yes	No	Yes	No	Yes
# Monitors in park	0	Exceeds if 15% biosites?	Yes	No	Yes	No	Yes
Climate Region	Southwest	Exceeds if 20% biosites?	No	No	No	No	No
		W126 (ppm-hrs)	16.04	17.57	13.41	6.16	10.13
Park Name	Chaco Culture National Historical Park	PZ (7-mo avg)	0.10	0.10	0.02	-0.50	0.75
		Exceeds in Base?	Yes	Yes	Yes	No	No
Park Code	CHCU	Exceeds if 5%, injury=5?	Yes	Yes	Yes	No	No
Primary State	NM	Exceeds if 5% biosites?	Yes	Yes	Yes	Yes	Yes
Species in park?	Yes	Exceeds if 10% biosites?	Yes	Yes	Yes	Yes	Yes
# Monitors in park	0	Exceeds if 15% biosites?	Yes	Yes	Yes	No	Yes
Climate Region	Southwest	Exceeds if 20% biosites?	No	No	No	No	No
		W126 (ppm-hrs)	14.30	12.17	10.63	7.14	8.68
Park Name	Chamizal National Memorial	PZ (7-mo avg)	-0.67	1.78	0.11	-0.25	0.96
		Exceeds in Base?	Yes	Yes	Yes	No	No
Park Code	CHAM	Exceeds if 5%, injury=5?	Yes	Yes	No	No	No
Primary State	TX	Exceeds if 5% biosites?	Yes	Yes	Yes	Yes	Yes
Species in park?	0	Exceeds if 10% biosites?	Yes	Yes	Yes	Yes	Yes
# Monitors in park	0	Exceeds if 15% biosites?	Yes	Yes	Yes	No	Yes
Climate Region	South	Exceeds if 20% biosites?	No	Yes	No	No	No
		W126 (ppm-hrs)	26.64	22.60	13.04	9.61	12.26
Park Name	Chattahoochee River National Recreation Area	PZ (7-mo avg)	-0.57	-2.10	-0.45	1.13	0.29
		Exceeds in Base?	Yes	Yes	Yes	No	Yes
Park Code	CHAT	Exceeds if 5%, injury=5?	Yes	No	Yes	Yes	Yes
Primary State	GA	Exceeds if 5% biosites?	Yes	Yes	Yes	Yes	Yes
Species in park?	Yes	Exceeds if 10% biosites?	Yes	No	Yes	Yes	Yes
# Monitors in park	0	Exceeds if 15% biosites?	Yes	No	Yes	Yes	Yes
Climate Region	Southeast	Exceeds if 20% biosites?	No	No	No	Yes	No
	Chesapeake and Ohio Canal National Historical Park	W126 (ppm-hrs)	14.86	14.70	11.88	6.15	14.46
Park Name		PZ (7-mo avg)	-0.23	-0.99	1.06	0.42	-1.14
		Exceeds in Base?	Yes	Yes	Yes	No	Yes
Park Code	CHOH	Exceeds if 5%, injury=5?	Yes	Yes	Yes	No	Yes
Primary State	MD	Exceeds if 5% biosites?	Yes	Yes	Yes	Yes	Yes
Species in park?	Yes	Exceeds if 10% biosites?	Yes	Yes	Yes	Yes	Yes

Park info		Criteria	2006	2007	2008	2009	2010
# Monitors in park	0	Exceeds if 15% biosites?	Yes	Yes	Yes	No	Yes
Climate Region	Northeast	Exceeds if 20% biosites?	No	No	Yes	No	No
	Chickamauga and Chattanooga National Military Park	W126 (ppm-hrs)	18.80	21.34	13.10	6.87	11.09
Park Name		PZ (7-mo avg)	-0.17	-1.96	-0.51	1.24	-0.82
		Exceeds in Base?	Yes	Yes	Yes	No	Yes
Park Code	CHCH	Exceeds if 5%, injury=5?	Yes	No	Yes	No	No
Primary State	GA	Exceeds if 5% biosites?	Yes	Yes	Yes	Yes	Yes
Species in park?	Yes	Exceeds if 10% biosites?	Yes	No	Yes	Yes	Yes
# Monitors in park	0	Exceeds if 15% biosites?	Yes	No	Yes	Yes	Yes
Climate Region	Southeast	Exceeds if 20% biosites?	No	No	No	Yes	No
	Chickasaw National Recreation Area	W126 (ppm-hrs)	20.45	10.37	8.28	9.49	7.85
Park Name		PZ (7-mo avg)	-1.38	2.02	0.58	0.70	-0.10
		Exceeds in Base?	Yes	No	No	No	No
Park Code	CHIC	Exceeds if 5%, injury=5?	No	Yes	No	No	No
Primary State	OK	Exceeds if 5% biosites?	Yes	Yes	Yes	Yes	Yes
Species in park?	Yes	Exceeds if 10% biosites?	No	Yes	Yes	Yes	Yes
# Monitors in park	0	Exceeds if 15% biosites?	No	Yes	Yes	Yes	No
Climate Region	South	Exceeds if 20% biosites?	No	Yes	No	No	No
	Chiricahua National Monument Wilderness	W126 (ppm-hrs)	16.29	12.64	14.77	10.24	11.96
Park Name		PZ (7-mo avg)	-0.31	-0.84	0.17	-1.62	0.29
		Exceeds in Base?	Yes	Yes	Yes	No	Yes
Park Code	CHIR	Exceeds if 5%, injury=5?	Yes	Yes	Yes	No	No
Primary State	AZ	Exceeds if 5% biosites?	Yes	Yes	Yes	Yes	Yes
Species in park?	Yes	Exceeds if 10% biosites?	Yes	Yes	Yes	No	Yes
# Monitors in park	0	Exceeds if 15% biosites?	Yes	Yes	Yes	No	Yes
Climate Region	Southwest	Exceeds if 20% biosites?	No	No	No	No	No
	City of Rocks National Reserve	W126 (ppm-hrs)	16.49	18.96	13.85	10.48	6.46
Park Name		PZ (7-mo avg)	-0.08	-1.87	-0.68	1.24	-0.12
		Exceeds in Base?	Yes	Yes	Yes	Yes	No
Park Code	CIRO	Exceeds if 5%, injury=5?	Yes	No	Yes	Yes	No
Primary State	ID	Exceeds if 5% biosites?	Yes	Yes	Yes	Yes	Yes
Species in park?	Yes	Exceeds if 10% biosites?	Yes	No	Yes	Yes	Yes
# Monitors in park	1	Exceeds if 15% biosites?	Yes	No	Yes	Yes	No
Climate Region	Northwest	Exceeds if 20% biosites?	No	No	No	Yes	No
	Colonial National Historical Park	W126 (ppm-hrs)	15.27	15.17	16.46	6.01	13.43
Park Name		PZ (7-mo avg)	0.99	-1.35	0.22	1.02	-0.90
		Exceeds in Base?	Yes	Yes	Yes	No	Yes
Park Code	COLO	Exceeds if 5%, injury=5?	Yes	No	Yes	No	Yes
Primary State	VA	Exceeds if 5% biosites?	Yes	Yes	Yes	Yes	Yes
Species in park?	Yes	Exceeds if 10% biosites?	Yes	No	Yes	Yes	Yes
# Monitors in park	0	Exceeds if 15% biosites?	Yes	No	Yes	Yes	Yes
Climate Region	Southeast	Exceeds if 20% biosites?	No	No	No	Yes	No
	Colorado National Monument	W126 (ppm-hrs)	19.38	15.48	15.14	6.40	10.38
Park Name		PZ (7-mo avg)	-0.54	-0.25	-0.26	-0.17	-0.11
		Exceeds in Base?	Yes	Yes	Yes	No	No
Park Code	COLM	Exceeds if 5%, injury=5?	Yes	Yes	Yes	No	No

Park info		Criteria	2006	2007	2008	2009	2010
Primary State	CO	Exceeds if 5% biosites?	Yes	Yes	Yes	Yes	Yes
Species in park?	Yes	Exceeds if 10% biosites?	Yes	Yes	Yes	Yes	Yes
# Monitors in park	1	Exceeds if 15% biosites?	Yes	Yes	Yes	No	Yes
Climate Region	Southwest	Exceeds if 20% biosites?	No	No	No	No	No
Park Name	Congaree National Park	W126 (ppm-hrs)	13.67	12.11	11.41	4.62	7.18
		PZ (7-mo avg)	-0.60	-1.35	-0.04	-0.20	-1.05
		Exceeds in Base?	Yes	Yes	Yes	No	No
Park Code	COSW	Exceeds if 5%, injury=5?	Yes	No	No	No	No
Primary State	SC	Exceeds if 5% biosites?	Yes	Yes	Yes	Yes	Yes
Species in park?	Yes	Exceeds if 10% biosites?	Yes	No	Yes	No	Yes
# Monitors in park	1	Exceeds if 15% biosites?	Yes	No	Yes	No	No
Climate Region	Southeast	Exceeds if 20% biosites?	No	No	No	No	No
Park Name	Coronado National	W126 (ppm-hrs)	13.40	11.23	14.44	8.63	10.77
		PZ (7-mo avg)	-0.31	-0.84	0.17	-1.62	0.29
		Exceeds in Base?	Yes	Yes	Yes	No	Yes
Park Code	CORO	Exceeds if 5%, injury=5?	Yes	No	Yes	No	No
Primary State	AZ	Exceeds if 5% biosites?	Yes	Yes	Yes	Yes	Yes
Species in park?	Yes	Exceeds if 10% biosites?	Yes	Yes	Yes	No	Yes
# Monitors in park	0	Exceeds if 15% biosites?	Yes	Yes	Yes	No	Yes
Climate Region	Southwest	Exceeds if 20% biosites?	No	No	No	No	No
Park Name	Cowpens National Battlefield	W126 (ppm-hrs)	15.34	9.62	16.82	3.71	9.29
		PZ (7-mo avg)	-0.84	-2.18	-1.58	-0.05	-1.24
		Exceeds in Base?	Yes	No	Yes	No	No
Park Code	COWP	Exceeds if 5%, injury=5?	Yes	No	No	No	No
Primary State	SC	Exceeds if 5% biosites?	Yes	Yes	Yes	Yes	Yes
Species in park?	Yes	Exceeds if 10% biosites?	Yes	No	No	No	Yes
# Monitors in park	1	Exceeds if 15% biosites?	Yes	No	No	No	Yes
Climate Region	Southeast	Exceeds if 20% biosites?	No	No	No	No	No
Park Name	Crater Lake National Park	W126 (ppm-hrs)	8.31	5.10	6.14	4.28	4.51
		PZ (7-mo avg)	-0.22	-1.42	-1.00	-0.42	0.70
		Exceeds in Base?	No	No	No	No	No
Park Code	CRLA	Exceeds if 5%, injury=5?	No	No	No	No	No
Primary State	OR	Exceeds if 5% biosites?	Yes	No	Yes	Yes	Yes
Species in park?	Yes	Exceeds if 10% biosites?	Yes	No	Yes	No	No
# Monitors in park	1	Exceeds if 15% biosites?	Yes	No	No	No	No
Climate Region	Northwest	Exceeds if 20% biosites?	No	No	No	No	No
Park Name	Craters of the Moon National Monument	W126 (ppm-hrs)	11.29	13.23	11.64	7.57	7.72
		PZ (7-mo avg)	1.06	-1.91	-1.22	1.23	0.06
		Exceeds in Base?	Yes	Yes	Yes	No	No
Park Code	CRMO	Exceeds if 5%, injury=5?	Yes	No	No	Yes	No
Primary State	ID	Exceeds if 5% biosites?	Yes	Yes	Yes	Yes	Yes
Species in park?	Yes	Exceeds if 10% biosites?	Yes	No	Yes	Yes	Yes
# Monitors in park	0	Exceeds if 15% biosites?	Yes	No	Yes	Yes	No
Climate Region	Northwest	Exceeds if 20% biosites?	Yes	No	No	Yes	No
Park Name	Cumberland Gap National Historical Park	W126 (ppm-hrs)	13.39	15.24	10.96	5.13	8.06
		PZ (7-mo avg)	0.56	-1.66	-0.46	1.51	-0.35

Park info		Criteria	2006	2007	2008	2009	2010
		Exceeds in Base?	Yes	Yes	Yes	No	No
Park Code	CUGA	Exceeds if 5%, injury=5?	Yes	No	No	No	No
Primary State	KY	Exceeds if 5% biosites?	Yes	Yes	Yes	Yes	Yes
Species in park?	Yes	Exceeds if 10% biosites?	Yes	No	Yes	Yes	Yes
# Monitors in park	1	Exceeds if 15% biosites?	Yes	No	Yes	Yes	No
Climate Region	Central	Exceeds if 20% biosites?	No	No	No	No	No
Park Name	Cumberland Island National Seashore	W126 (ppm-hrs)	11.26	9.74	7.12	4.50	6.33
		PZ (7-mo avg)	-1.76	-0.98	-0.84	0.98	-1.43
		Exceeds in Base?	Yes	No	No	No	No
Park Code	CUIS	Exceeds if 5%, injury=5?	No	No	No	No	No
Primary State	GA	Exceeds if 5% biosites?	Yes	Yes	Yes	Yes	Yes
Species in park?	Yes	Exceeds if 10% biosites?	No	Yes	Yes	No	No
# Monitors in park	0	Exceeds if 15% biosites?	No	Yes	No	No	No
Climate Region	Southeast	Exceeds if 20% biosites?	No	No	No	No	No
Park Name	Curecanti National Recreation Area	W126 (ppm-hrs)	19.63	16.32	14.33	9.78	10.76
		PZ (7-mo avg)	-0.54	-0.25	-0.26	-0.17	-0.11
		Exceeds in Base?	Yes	Yes	Yes	No	Yes
Park Code	CURE	Exceeds if 5%, injury=5?	Yes	Yes	Yes	No	No
Primary State	CO	Exceeds if 5% biosites?	Yes	Yes	Yes	Yes	Yes
Species in park?	Yes	Exceeds if 10% biosites?	Yes	Yes	Yes	Yes	Yes
# Monitors in park	0	Exceeds if 15% biosites?	Yes	Yes	Yes	Yes	Yes
Climate Region	Southwest	Exceeds if 20% biosites?	No	No	No	No	No
Park Name	Cuyahoga Valley National Park	W126 (ppm-hrs)	11.02	13.64	9.86	6.14	9.42
		PZ (7-mo avg)	1.76	-0.22	1.23	0.51	-0.64
		Exceeds in Base?	Yes	Yes	No	No	No
Park Code	CUVA	Exceeds if 5%, injury=5?	Yes	Yes	Yes	No	No
Primary State	OH	Exceeds if 5% biosites?	Yes	Yes	Yes	Yes	Yes
Species in park?	Yes	Exceeds if 10% biosites?	Yes	Yes	Yes	Yes	Yes
# Monitors in park	0	Exceeds if 15% biosites?	Yes	Yes	Yes	No	Yes
Climate Region	Central	Exceeds if 20% biosites?	Yes	No	Yes	No	No
Park Name	Death Valley National Park	W126 (ppm-hrs)	30.56	30.77	28.53	17.99	16.37
		PZ (7-mo avg)	-0.67	-1.22	-0.99	-1.37	-0.06
		Exceeds in Base?	Yes	Yes	Yes	Yes	Yes
Park Code	DEVA	Exceeds if 5%, injury=5?	Yes	Yes	Yes	No	Yes
Primary State	CA	Exceeds if 5% biosites?	Yes	Yes	Yes	Yes	Yes
Species in park?	Yes	Exceeds if 10% biosites?	Yes	Yes	Yes	No	Yes
# Monitors in park	1	Exceeds if 15% biosites?	Yes	Yes	Yes	No	Yes
Climate Region	West	Exceeds if 20% biosites?	No	No	No	No	No
Park Name	Delaware Water Gap National Recreation Area	W126 (ppm-hrs)	12.57	11.69	10.59	4.57	10.91
		PZ (7-mo avg)	0.54	0.30	-0.03	0.62	-0.62
		Exceeds in Base?	Yes	Yes	Yes	No	Yes
Park Code	DEWA	Exceeds if 5%, injury=5?	Yes	No	No	No	No
Primary State	PA	Exceeds if 5% biosites?	Yes	Yes	Yes	Yes	Yes
Species in park?	Yes	Exceeds if 10% biosites?	Yes	Yes	Yes	No	Yes
# Monitors in park	0	Exceeds if 15% biosites?	Yes	Yes	Yes	No	Yes
Climate Region	Northeast	Exceeds if 20% biosites?	No	No	No	No	No

Park info		Criteria	2006	2007	2008	2009	2010
Park Name	Devil's Tower National Monument	W126 (ppm-hrs)	14.08	10.69	8.72	5.38	5.93
		PZ (7-mo avg)	-1.77	-0.78	1.41	1.45	0.63
		Exceeds in Base?	Yes	Yes	No	No	No
Park Code	DETO	Exceeds if 5%, injury=5?	No	No	Yes	No	No
Primary State	WY	Exceeds if 5% biosites?	Yes	Yes	Yes	Yes	Yes
Species in park?	Yes	Exceeds if 10% biosites?	No	Yes	Yes	Yes	No
# Monitors in park	1	Exceeds if 15% biosites?	No	Yes	Yes	Yes	No
Climate Region	West North Central	Exceeds if 20% biosites?	No	No	Yes	No	No
Park Name	Dinosaur National Monument	W126 (ppm-hrs)	18.30	11.79	11.89	7.64	15.39
		PZ (7-mo avg)	-0.55	-0.91	0.12	-0.43	-0.29
		Exceeds in Base?	Yes	Yes	Yes	No	Yes
Park Code	DINO	Exceeds if 5%, injury=5?	Yes	No	No	No	Yes
Primary State	CO	Exceeds if 5% biosites?	Yes	Yes	Yes	Yes	Yes
Species in park?	Yes	Exceeds if 10% biosites?	Yes	Yes	Yes	Yes	Yes
# Monitors in park	1	Exceeds if 15% biosites?	Yes	Yes	Yes	No	Yes
Climate Region	Southwest	Exceeds if 20% biosites?	No	No	No	No	No
Park Name	Effigy Mounds National Monument	W126 (ppm-hrs)	6.85	9.26	5.25	4.96	5.10
		PZ (7-mo avg)	0.44	1.50	2.41	1.01	1.76
		Exceeds in Base?	No	No	No	No	No
Park Code	EFMO	Exceeds if 5%, injury=5?	No	Yes	No	No	No
Primary State	IA	Exceeds if 5% biosites?	Yes	Yes	Yes	Yes	Yes
Species in park?	Yes	Exceeds if 10% biosites?	Yes	Yes	Yes	Yes	Yes
# Monitors in park	0	Exceeds if 15% biosites?	No	Yes	Yes	Yes	Yes
Climate Region	East North Central	Exceeds if 20% biosites?	No	Yes	No	No	No
Park Name	Eisenhower National Historic Site	W126 (ppm-hrs)	11.82	16.79	13.00	6.01	14.02
		PZ (7-mo avg)	-0.06	-0.80	1.07	0.89	-0.60
		Exceeds in Base?	Yes	Yes	Yes	No	Yes
Park Code	EISE	Exceeds if 5%, injury=5?	No	Yes	Yes	No	Yes
Primary State	PA	Exceeds if 5% biosites?	Yes	Yes	Yes	Yes	Yes
Species in park?	Yes	Exceeds if 10% biosites?	Yes	Yes	Yes	Yes	Yes
# Monitors in park	0	Exceeds if 15% biosites?	Yes	Yes	Yes	No	Yes
Climate Region	Northeast	Exceeds if 20% biosites?	No	No	Yes	No	No
Park Name	El Malpais National Monument	W126 (ppm-hrs)	16.53	14.25	12.29	7.02	9.89
		PZ (7-mo avg)	0.44	-0.35	0.13	-0.75	0.44
		Exceeds in Base?	Yes	Yes	Yes	No	No
Park Code	ELMA	Exceeds if 5%, injury=5?	Yes	Yes	Yes	No	No
Primary State	NM	Exceeds if 5% biosites?	Yes	Yes	Yes	Yes	Yes
Species in park?	Yes	Exceeds if 10% biosites?	Yes	Yes	Yes	Yes	Yes
# Monitors in park	0	Exceeds if 15% biosites?	Yes	Yes	Yes	No	Yes
Climate Region	Southwest	Exceeds if 20% biosites?	No	No	No	No	No
Park Name	El Morro National Monument	W126 (ppm-hrs)	16.49	14.40	12.61	6.89	10.00
		PZ (7-mo avg)	0.10	0.10	0.02	-0.50	0.75
		Exceeds in Base?	Yes	Yes	Yes	No	No
Park Code	ELMO	Exceeds if 5%, injury=5?	Yes	Yes	Yes	No	No
Primary State	NM	Exceeds if 5% biosites?	Yes	Yes	Yes	Yes	Yes
Species in park?	Yes	Exceeds if 10% biosites?	Yes	Yes	Yes	Yes	Yes

Park info		Criteria	2006	2007	2008	2009	2010
# Monitors in park	0	Exceeds if 15% biosites?	Yes	Yes	Yes	No	Yes
Climate Region	Southwest	Exceeds if 20% biosites?	No	No	No	No	No
Park Name	Eleanor Roosevelt National Historic Site	W126 (ppm-hrs)	6.69	10.69	8.84	5.65	8.10
		PZ (7-mo avg)	1.09	0.44	1.13	1.42	-0.90
		Exceeds in Base?	No	Yes	No	No	No
Park Code	ELRO	Exceeds if 5%, injury=5?	No	No	Yes	No	No
Primary State	NY	Exceeds if 5% biosites?	Yes	Yes	Yes	Yes	Yes
Species in park?	Yes	Exceeds if 10% biosites?	Yes	Yes	Yes	Yes	Yes
# Monitors in park	0	Exceeds if 15% biosites?	Yes	Yes	Yes	Yes	No
Climate Region	Northeast	Exceeds if 20% biosites?	Yes	No	Yes	No	No
Park Name	Everglades National Park	W126 (ppm-hrs)	9.74	8.01	6.36	3.72	5.52
		PZ (7-mo avg)	-0.38	-0.23	0.49	-0.36	0.73
		Exceeds in Base?	No	No	No	No	No
Park Code	EVER	Exceeds if 5%, injury=5?	No	No	No	No	No
Primary State	FL	Exceeds if 5% biosites?	Yes	Yes	Yes	Yes	Yes
Species in park?	Yes	Exceeds if 10% biosites?	Yes	Yes	Yes	No	No
# Monitors in park	0	Exceeds if 15% biosites?	Yes	No	No	No	No
Climate Region	Southeast	Exceeds if 20% biosites?	No	No	No	No	No
Park Name	Fire Island National Seashore	W126 (ppm-hrs)	14.41	13.72	16.62	9.02	15.06
		PZ (7-mo avg)	0.50	0.76	0.13	0.56	-0.90
		Exceeds in Base?	Yes	Yes	Yes	No	Yes
Park Code	FIIS	Exceeds if 5%, injury=5?	Yes	Yes	Yes	No	Yes
Primary State	NY	Exceeds if 5% biosites?	Yes	Yes	Yes	Yes	Yes
Species in park?	Yes	Exceeds if 10% biosites?	Yes	Yes	Yes	Yes	Yes
# Monitors in park	0	Exceeds if 15% biosites?	Yes	Yes	Yes	Yes	Yes
Climate Region	Northeast	Exceeds if 20% biosites?	No	No	No	No	No
Park Name	Florissant Fossil Beds National Monument	W126 (ppm-hrs)	23.55	19.10	19.14	9.94	13.30
		PZ (7-mo avg)	-1.14	0.53	-0.32	0.54	0.78
		Exceeds in Base?	Yes	Yes	Yes	No	Yes
Park Code	FLFO	Exceeds if 5%, injury=5?	Yes	Yes	Yes	No	Yes
Primary State	CO	Exceeds if 5% biosites?	Yes	Yes	Yes	Yes	Yes
Species in park?	Yes	Exceeds if 10% biosites?	Yes	Yes	Yes	Yes	Yes
# Monitors in park	0	Exceeds if 15% biosites?	Yes	Yes	Yes	Yes	Yes
Climate Region	Southwest	Exceeds if 20% biosites?	No	No	No	No	No
Park Name	Fort Bowie National Historic Site	W126 (ppm-hrs)	18.43	14.42	17.48	11.58	13.33
		PZ (7-mo avg)	-0.31	-0.84	0.17	-1.62	0.29
		Exceeds in Base?	Yes	Yes	Yes	Yes	Yes
Park Code	FOBO	Exceeds if 5%, injury=5?	Yes	Yes	Yes	No	Yes
Primary State	AZ	Exceeds if 5% biosites?	Yes	Yes	Yes	Yes	Yes
Species in park?	Yes	Exceeds if 10% biosites?	Yes	Yes	Yes	No	Yes
# Monitors in park	0	Exceeds if 15% biosites?	Yes	Yes	Yes	No	Yes
Climate Region	Southwest	Exceeds if 20% biosites?	No	No	No	No	No
Park Name	Fort Davis National Historic Site	W126 (ppm-hrs)	17.77	9.85	9.88	8.34	7.55
		PZ (7-mo avg)	-0.67	1.78	0.11	-0.25	0.96
		Exceeds in Base?	Yes	No	No	No	No
Park Code	FODA	Exceeds if 5%, injury=5?	Yes	Yes	No	No	No

Park info		Criteria	2006	2007	2008	2009	2010
Primary State	TX	Exceeds if 5% biosites?	Yes	Yes	Yes	Yes	Yes
Species in park?	Yes	Exceeds if 10% biosites?	Yes	Yes	Yes	Yes	Yes
# Monitors in park	0	Exceeds if 15% biosites?	Yes	Yes	Yes	Yes	No
Climate Region	South	Exceeds if 20% biosites?	No	Yes	No	No	No
Park Name	Fort Donelson National Battlefield	W126 (ppm-hrs)	13.87	21.85	9.42	6.51	10.98
		PZ (7-mo avg)	-0.42	-2.13	-0.15	1.50	0.14
		Exceeds in Base?	Yes	Yes	No	No	Yes
Park Code	FODO	Exceeds if 5%, injury=5?	Yes	No	No	No	No
Primary State	TN	Exceeds if 5% biosites?	Yes	Yes	Yes	Yes	Yes
Species in park?	Yes	Exceeds if 10% biosites?	Yes	No	Yes	Yes	Yes
# Monitors in park	0	Exceeds if 15% biosites?	Yes	No	Yes	Yes	Yes
Climate Region	Central	Exceeds if 20% biosites?	No	No	No	Yes	No
Park Name	Fort Laramie National Historic Site	W126 (ppm-hrs)	19.12	15.54	11.83	7.98	9.60
		PZ (7-mo avg)	-1.78	-1.26	-0.08	0.95	1.79
		Exceeds in Base?	Yes	Yes	Yes	No	No
Park Code	FOLA	Exceeds if 5%, injury=5?	No	No	No	No	Yes
Primary State	WY	Exceeds if 5% biosites?	Yes	Yes	Yes	Yes	Yes
Species in park?	Yes	Exceeds if 10% biosites?	No	No	Yes	Yes	Yes
# Monitors in park	0	Exceeds if 15% biosites?	No	No	Yes	No	Yes
Climate Region	West North Central	Exceeds if 20% biosites?	No	No	No	No	Yes
Park Name	Fort Larned National Historic Site	W126 (ppm-hrs)	17.88	6.83	7.59	8.67	10.41
		PZ (7-mo avg)	-0.55	1.75	1.66	1.23	0.71
		Exceeds in Base?	Yes	No	No	No	No
Park Code	FOLS	Exceeds if 5%, injury=5?	Yes	No	Yes	Yes	No
Primary State	KS	Exceeds if 5% biosites?	Yes	Yes	Yes	Yes	Yes
Species in park?	Yes	Exceeds if 10% biosites?	Yes	Yes	Yes	Yes	Yes
# Monitors in park	0	Exceeds if 15% biosites?	Yes	Yes	Yes	Yes	Yes
Climate Region	South	Exceeds if 20% biosites?	No	Yes	Yes	Yes	No
Park Name	Fort Necessity National Battlefield	W126 (ppm-hrs)	13.24	12.73	8.45	6.06	10.61
		PZ (7-mo avg)	0.27	0.24	0.35	-0.14	-0.88
		Exceeds in Base?	Yes	Yes	No	No	Yes
Park Code	FONE	Exceeds if 5%, injury=5?	Yes	Yes	No	No	No
Primary State	PA	Exceeds if 5% biosites?	Yes	Yes	Yes	Yes	Yes
Species in park?	Yes	Exceeds if 10% biosites?	Yes	Yes	Yes	Yes	Yes
# Monitors in park	0	Exceeds if 15% biosites?	Yes	Yes	Yes	No	Yes
Climate Region	Northeast	Exceeds if 20% biosites?	No	No	No	No	No
Park Name	Fort Pulaski National Monument	W126 (ppm-hrs)	8.24	6.70	6.16	4.29	5.58
		PZ (7-mo avg)	-1.51	-1.13	-0.74	0.43	-1.06
		Exceeds in Base?	No	No	No	No	No
Park Code	FOPU	Exceeds if 5%, injury=5?	No	No	No	No	No
Primary State	GA	Exceeds if 5% biosites?	Yes	Yes	Yes	Yes	Yes
Species in park?	Yes	Exceeds if 10% biosites?	No	Yes	Yes	No	No
# Monitors in park	0	Exceeds if 15% biosites?	No	No	No	No	No
Climate Region	Southeast	Exceeds if 20% biosites?	No	No	No	No	No
Park Name	Fort Raleigh National Historic Site	W126 (ppm-hrs)	12.57	12.38	13.19	5.11	10.53
		PZ (7-mo avg)	1.15	-1.60	-0.77	-0.02	-0.43

Park info		Criteria	2006	2007	2008	2009	2010
		Exceeds in Base?	Yes	Yes	Yes	No	Yes
Park Code	FORA	Exceeds if 5%, injury=5?	Yes	No	Yes	No	No
Primary State	NC	Exceeds if 5% biosites?	Yes	Yes	Yes	Yes	Yes
Species in park?	Yes	Exceeds if 10% biosites?	Yes	No	Yes	No	Yes
# Monitors in park	0	Exceeds if 15% biosites?	Yes	No	Yes	No	Yes
Climate Region	Southeast	Exceeds if 20% biosites?	Yes	No	No	No	No
Park Name	Fort Union National Monument	W126 (ppm-hrs)	15.09	10.60	10.97	7.54	8.57
		PZ (7-mo avg)	-0.93	-0.08	-0.91	-0.17	0.69
		Exceeds in Base?	Yes	Yes	Yes	No	No
Park Code	FOUN	Exceeds if 5%, injury=5?	Yes	No	No	No	No
Primary State	NM	Exceeds if 5% biosites?	Yes	Yes	Yes	Yes	Yes
Species in park?	Yes	Exceeds if 10% biosites?	Yes	Yes	Yes	Yes	Yes
# Monitors in park	0	Exceeds if 15% biosites?	Yes	Yes	Yes	No	Yes
Climate Region	Southwest	Exceeds if 20% biosites?	No	No	No	No	No
Park Name	Fort Washington Park	W126 (ppm-hrs)	21.90	20.65	15.16	8.28	18.03
		PZ (7-mo avg)	-0.25	-1.67	1.02	1.04	-1.06
		Exceeds in Base?	Yes	Yes	Yes	No	Yes
Park Code	FOWA	Exceeds if 5%, injury=5?	Yes	No	Yes	Yes	Yes
Primary State	MD	Exceeds if 5% biosites?	Yes	Yes	Yes	Yes	Yes
Species in park?	Yes	Exceeds if 10% biosites?	Yes	No	Yes	Yes	Yes
# Monitors in park	0	Exceeds if 15% biosites?	Yes	No	Yes	Yes	Yes
Climate Region	Northeast	Exceeds if 20% biosites?	No	No	Yes	Yes	No
Park Name	Fossil Butte National Monument	W126 (ppm-hrs)	19.10	15.76	11.89	6.84	8.10
		PZ (7-mo avg)	-2.29	-2.32	-1.39	-0.77	-1.22
		Exceeds in Base?	Yes	Yes	Yes	No	No
Park Code	FOBU	Exceeds if 5%, injury=5?	No	No	No	No	No
Primary State	WY	Exceeds if 5% biosites?	Yes	Yes	Yes	Yes	Yes
Species in park?	Yes	Exceeds if 10% biosites?	No	No	No	Yes	Yes
# Monitors in park	0	Exceeds if 15% biosites?	No	No	No	No	No
Climate Region	West North Central	Exceeds if 20% biosites?	No	No	No	No	No
Park Name	Fredericksburg and Spotsylvania Co. Battlefields Memorial National Military Park	W126 (ppm-hrs)	13.71	13.91	9.63	4.91	10.49
		PZ (7-mo avg)	-0.14	-1.32	0.58	0.23	-1.36
		Exceeds in Base?	Yes	Yes	No	No	Yes
Park Code	FRSP	Exceeds if 5%, injury=5?	Yes	No	No	No	No
Primary State	VA	Exceeds if 5% biosites?	Yes	Yes	Yes	Yes	Yes
Species in park?	Yes	Exceeds if 10% biosites?	Yes	No	Yes	No	No
# Monitors in park	0	Exceeds if 15% biosites?	Yes	No	Yes	No	No
Climate Region	Southeast	Exceeds if 20% biosites?	No	No	No	No	No
Park Name	Friendship Hill National Historic Site	W126 (ppm-hrs)	13.34	14.84	8.34	6.07	10.24
		PZ (7-mo avg)	0.27	0.24	0.35	-0.14	-0.88
		Exceeds in Base?	Yes	Yes	No	No	No
Park Code	FRHI	Exceeds if 5%, injury=5?	Yes	Yes	No	No	No
Primary State	PA	Exceeds if 5% biosites?	Yes	Yes	Yes	Yes	Yes
Species in park?	Yes	Exceeds if 10% biosites?	Yes	Yes	Yes	Yes	Yes
# Monitors in park	0	Exceeds if 15% biosites?	Yes	Yes	Yes	No	Yes
Climate Region	Northeast	Exceeds if 20% biosites?	No	No	No	No	No

Park info		Criteria	2006	2007	2008	2009	2010
Park Name	Gateway National Recreation Area	W126 (ppm-hrs)	14.68	12.72	9.77	6.06	15.33
		PZ (7-mo avg)	0.50	0.76	0.13	0.56	-0.90
		Exceeds in Base?	Yes	Yes	No	No	Yes
Park Code	GATE	Exceeds if 5%, injury=5?	Yes	Yes	No	No	Yes
Primary State	NY	Exceeds if 5% biosites?	Yes	Yes	Yes	Yes	Yes
Species in park?	Yes	Exceeds if 10% biosites?	Yes	Yes	Yes	Yes	Yes
# Monitors in park	0	Exceeds if 15% biosites?	Yes	Yes	Yes	No	Yes
Climate Region	Northeast	Exceeds if 20% biosites?	No	No	No	No	No
Park Name	Gauley River National Recreation Area	W126 (ppm-hrs)	10.65	13.40	9.14	4.94	7.47
		PZ (7-mo avg)	0.05	-0.15	0.33	0.26	-1.00
		Exceeds in Base?	Yes	Yes	No	No	No
Park Code	GARI	Exceeds if 5%, injury=5?	No	Yes	No	No	No
Primary State	WV	Exceeds if 5% biosites?	Yes	Yes	Yes	Yes	Yes
Species in park?	Yes	Exceeds if 10% biosites?	Yes	Yes	Yes	No	Yes
# Monitors in park	0	Exceeds if 15% biosites?	Yes	Yes	Yes	No	No
Climate Region	Central	Exceeds if 20% biosites?	No	No	No	No	No
Park Name	George Washington Birthplace National Monument	W126 (ppm-hrs)	17.89	18.11	13.42	6.97	14.87
		PZ (7-mo avg)	0.99	-1.35	0.22	1.02	-0.90
		Exceeds in Base?	Yes	Yes	Yes	No	Yes
Park Code	GEWA	Exceeds if 5%, injury=5?	Yes	No	Yes	No	Yes
Primary State	VA	Exceeds if 5% biosites?	Yes	Yes	Yes	Yes	Yes
Species in park?	Yes	Exceeds if 10% biosites?	Yes	No	Yes	Yes	Yes
# Monitors in park	0	Exceeds if 15% biosites?	Yes	No	Yes	Yes	Yes
Climate Region	Southeast	Exceeds if 20% biosites?	No	No	No	Yes	No
Park Name	George Washington Memorial Parkway	W126 (ppm-hrs)	18.03	18.23	15.37	6.82	18.65
		PZ (7-mo avg)	-0.24	-1.56	0.94	0.31	-1.19
		Exceeds in Base?	Yes	Yes	Yes	No	Yes
Park Code	GWMP	Exceeds if 5%, injury=5?	Yes	No	Yes	No	Yes
Primary State	VA	Exceeds if 5% biosites?	Yes	Yes	Yes	Yes	Yes
Species in park?	Yes	Exceeds if 10% biosites?	Yes	No	Yes	Yes	Yes
# Monitors in park	0	Exceeds if 15% biosites?	Yes	No	Yes	No	Yes
Climate Region	Southeast	Exceeds if 20% biosites?	No	No	No	No	No
Park Name	Gettysburg National Military Park	W126 (ppm-hrs)	12.33	16.75	13.25	6.10	14.38
		PZ (7-mo avg)	-0.06	-0.80	1.07	0.89	-0.60
		Exceeds in Base?	Yes	Yes	Yes	No	Yes
Park Code	GETT	Exceeds if 5%, injury=5?	Yes	Yes	Yes	No	Yes
Primary State	PA	Exceeds if 5% biosites?	Yes	Yes	Yes	Yes	Yes
Species in park?	Yes	Exceeds if 10% biosites?	Yes	Yes	Yes	Yes	Yes
# Monitors in park	0	Exceeds if 15% biosites?	Yes	Yes	Yes	No	Yes
Climate Region	Northeast	Exceeds if 20% biosites?	No	No	Yes	No	No
Park Name	Gila Cliff Dwellings National Monument	W126 (ppm-hrs)	15.37	11.66	12.00	7.41	11.50
		PZ (7-mo avg)	0.44	-0.35	0.13	-0.75	0.44
		Exceeds in Base?	Yes	Yes	Yes	No	Yes
Park Code	GICL	Exceeds if 5%, injury=5?	Yes	No	No	No	No
Primary State	NM	Exceeds if 5% biosites?	Yes	Yes	Yes	Yes	Yes
Species in park?	Yes	Exceeds if 10% biosites?	Yes	Yes	Yes	Yes	Yes

Park info		Criteria	2006	2007	2008	2009	2010
# Monitors in park	0	Exceeds if 15% biosites?	Yes	Yes	Yes	No	Yes
Climate Region	Southwest	Exceeds if 20% biosites?	No	No	No	No	No
Park Name	Glacier National Park	W126 (ppm-hrs)	3.66	3.23	4.06	4.22	3.22
		PZ (7-mo avg)	-0.62	-1.01	0.29	-0.32	1.22
		Exceeds in Base?	No	No	No	No	No
Park Code	GLAC	Exceeds if 5%, injury=5?	No	No	No	No	No
Primary State	MT	Exceeds if 5% biosites?	Yes	Yes	Yes	Yes	Yes
Species in park?	Yes	Exceeds if 10% biosites?	No	No	No	No	No
# Monitors in park	1	Exceeds if 15% biosites?	No	No	No	No	No
Climate Region	West North Central	Exceeds if 20% biosites?	No	No	No	No	No
Park Name	Glen Canyon National Recreation Area	W126 (ppm-hrs)	20.49	17.62	17.04	10.73	13.34
		PZ (7-mo avg)	-0.47	-1.05	-0.36	-1.39	0.43
		Exceeds in Base?	Yes	Yes	Yes	Yes	Yes
Park Code	GLCA	Exceeds if 5%, injury=5?	Yes	Yes	Yes	No	Yes
Primary State	UT	Exceeds if 5% biosites?	Yes	Yes	Yes	Yes	Yes
Species in park?	Yes	Exceeds if 10% biosites?	Yes	Yes	Yes	No	Yes
# Monitors in park	0	Exceeds if 15% biosites?	Yes	Yes	Yes	No	Yes
Climate Region	Southwest	Exceeds if 20% biosites?	No	No	No	No	No
Park Name	Golden Gate National Recreation Area	W126 (ppm-hrs)	1.67	1.36	2.25	1.69	1.68
		PZ (7-mo avg)	1.15	-1.17	-1.68	-0.20	1.17
		Exceeds in Base?	No	No	No	No	No
Park Code	GOGA	Exceeds if 5%, injury=5?	No	No	No	No	No
Primary State	CA	Exceeds if 5% biosites?	No	No	No	No	No
Species in park?	Yes	Exceeds if 10% biosites?	No	No	No	No	No
# Monitors in park	0	Exceeds if 15% biosites?	No	No	No	No	No
Climate Region	West	Exceeds if 20% biosites?	No	No	No	No	No
Park Name	Golden Spike National Historic Site	W126 (ppm-hrs)	20.72	21.66	15.13	11.02	8.43
		PZ (7-mo avg)	-0.33	-2.36	-1.34	-0.07	-0.43
		Exceeds in Base?	Yes	Yes	Yes	Yes	No
Park Code	GOSP	Exceeds if 5%, injury=5?	Yes	No	No	No	No
Primary State	UT	Exceeds if 5% biosites?	Yes	Yes	Yes	Yes	Yes
Species in park?	Yes	Exceeds if 10% biosites?	Yes	No	No	Yes	Yes
# Monitors in park	0	Exceeds if 15% biosites?	Yes	No	No	Yes	Yes
Climate Region	Southwest	Exceeds if 20% biosites?	No	No	No	No	No
Park Name	Grand Canyon National Park	W126 (ppm-hrs)	21.51	17.08	17.30	10.68	14.74
		PZ (7-mo avg)	-0.99	-1.57	-0.54	-1.59	0.03
		Exceeds in Base?	Yes	Yes	Yes	Yes	Yes
Park Code	GRCA	Exceeds if 5%, injury=5?	Yes	No	Yes	No	Yes
Primary State	AZ	Exceeds if 5% biosites?	Yes	Yes	Yes	Yes	Yes
Species in park?	Yes	Exceeds if 10% biosites?	Yes	No	Yes	No	Yes
# Monitors in park	1	Exceeds if 15% biosites?	Yes	No	Yes	No	Yes
Climate Region	Southwest	Exceeds if 20% biosites?	No	No	No	No	No
Park Name	Grand Teton National Park	W126 (ppm-hrs)	12.78	12.69	11.48	7.56	9.21
		PZ (7-mo avg)	-0.80	-1.51	-0.26	0.92	0.93
		Exceeds in Base?	Yes	Yes	Yes	No	No
Park Code	GRTE	Exceeds if 5%, injury=5?	Yes	No	No	No	No

Park info		Criteria	2006	2007	2008	2009	2010
Primary State	WY	Exceeds if 5% biosites?	Yes	Yes	Yes	Yes	Yes
Species in park?	Yes	Exceeds if 10% biosites?	Yes	No	Yes	Yes	Yes
# Monitors in park	0	Exceeds if 15% biosites?	Yes	No	Yes	No	Yes
Climate Region	West North Central	Exceeds if 20% biosites?	No	No	No	No	No
Park Name	Grant-Kohrs Ranch National Historic Site	W126 (ppm-hrs)	5.29	7.55	6.50	5.23	4.95
		PZ (7-mo avg)	-0.16	-1.14	0.15	-0.14	1.01
		Exceeds in Base?	No	No	No	No	No
Park Code	GRKO	Exceeds if 5%, injury=5?	No	No	No	No	No
Primary State	MT	Exceeds if 5% biosites?	Yes	Yes	Yes	Yes	Yes
Species in park?	Yes	Exceeds if 10% biosites?	No	Yes	Yes	No	Yes
# Monitors in park	0	Exceeds if 15% biosites?	No	No	No	No	Yes
Climate Region	West North Central	Exceeds if 20% biosites?	No	No	No	No	No
Park Name	Great Basin National Park	W126 (ppm-hrs)	16.78	16.47	16.81	10.51	11.84
		PZ (7-mo avg)	0.28	-2.14	-1.62	0.27	-0.35
		Exceeds in Base?	Yes	Yes	Yes	Yes	Yes
Park Code	GRBA	Exceeds if 5%, injury=5?	Yes	No	No	No	No
Primary State	NV	Exceeds if 5% biosites?	Yes	Yes	Yes	Yes	Yes
Species in park?	Yes	Exceeds if 10% biosites?	Yes	No	No	Yes	Yes
# Monitors in park	1	Exceeds if 15% biosites?	Yes	No	No	Yes	Yes
Climate Region	West	Exceeds if 20% biosites?	No	No	No	No	No
Park Name	Great Sand Dunes National Park	W126 (ppm-hrs)	19.35	17.15	14.87	9.61	12.16
		PZ (7-mo avg)	-0.98	0.17	-1.06	-0.67	-0.21
		Exceeds in Base?	Yes	Yes	Yes	No	Yes
Park Code	GRSA	Exceeds if 5%, injury=5?	Yes	Yes	Yes	No	No
Primary State	CO	Exceeds if 5% biosites?	Yes	Yes	Yes	Yes	Yes
Species in park?	Yes	Exceeds if 10% biosites?	Yes	Yes	Yes	Yes	Yes
# Monitors in park	0	Exceeds if 15% biosites?	Yes	Yes	Yes	Yes	Yes
Climate Region	Southwest	Exceeds if 20% biosites?	No	No	No	No	No
Park Name	Great Smoky Mountains National Park	W126 (ppm-hrs)	16.53	15.41	13.15	5.61	10.16
		PZ (7-mo avg)	0.19	-1.87	-0.87	1.45	-0.80
		Exceeds in Base?	Yes	Yes	Yes	No	No
Park Code	GRSM	Exceeds if 5%, injury=5?	Yes	No	Yes	No	No
Primary State	TN	Exceeds if 5% biosites?	Yes	Yes	Yes	Yes	Yes
Species in park?	Yes	Exceeds if 10% biosites?	Yes	No	Yes	Yes	Yes
# Monitors in park	4	Exceeds if 15% biosites?	Yes	No	Yes	Yes	Yes
Climate Region	Central	Exceeds if 20% biosites?	No	No	No	No	No
Park Name	Green Springs National Historic Landmark District	W126 (ppm-hrs)	13.49	12.60	11.04	5.58	10.34
		PZ (7-mo avg)	-0.05	-1.08	0.22	0.14	-1.52
		Exceeds in Base?	Yes	Yes	Yes	No	No
Park Code	GRSP	Exceeds if 5%, injury=5?	Yes	Yes	No	No	No
Primary State	VA	Exceeds if 5% biosites?	Yes	Yes	Yes	Yes	Yes
Species in park?	0	Exceeds if 10% biosites?	Yes	Yes	Yes	No	No
# Monitors in park	0	Exceeds if 15% biosites?	Yes	Yes	Yes	No	No
Climate Region	Southeast	Exceeds if 20% biosites?	No	No	No	No	No
Park Name	Greenbelt Park	W126 (ppm-hrs)	23.81	19.77	16.65	7.83	21.15
		PZ (7-mo avg)	-0.25	-1.67	1.02	1.04	-1.06

Park info		Criteria	2006	2007	2008	2009	2010
		Exceeds in Base?	Yes	Yes	Yes	No	Yes
Park Code	GREE	Exceeds if 5%, injury=5?	Yes	No	Yes	Yes	Yes
Primary State	MD	Exceeds if 5% biosites?	Yes	Yes	Yes	Yes	Yes
Species in park?	0	Exceeds if 10% biosites?	Yes	No	Yes	Yes	Yes
# Monitors in park	0	Exceeds if 15% biosites?	Yes	No	Yes	Yes	Yes
Climate Region	Northeast	Exceeds if 20% biosites?	No	No	Yes	Yes	No
Park Name	Guadalupe Mountains National Park	W126 (ppm-hrs)	18.72	10.66	9.99	9.09	8.79
		PZ (7-mo avg)	-0.67	1.78	0.11	-0.25	0.96
		Exceeds in Base?	Yes	Yes	No	No	No
Park Code	GUMO	Exceeds if 5%, injury=5?	Yes	Yes	No	No	No
Primary State	TX	Exceeds if 5% biosites?	Yes	Yes	Yes	Yes	Yes
Species in park?	Yes	Exceeds if 10% biosites?	Yes	Yes	Yes	Yes	Yes
# Monitors in park	0	Exceeds if 15% biosites?	Yes	Yes	Yes	Yes	Yes
Climate Region	South	Exceeds if 20% biosites?	No	Yes	No	No	No
Park Name	Guilford Courthouse National Military Park	W126 (ppm-hrs)	17.70	23.09	16.91	8.05	14.46
		PZ (7-mo avg)	0.76	-1.85	0.71	-0.01	-0.79
		Exceeds in Base?	Yes	Yes	Yes	No	Yes
Park Code	GUCO	Exceeds if 5%, injury=5?	Yes	No	Yes	No	Yes
Primary State	NC	Exceeds if 5% biosites?	Yes	Yes	Yes	Yes	Yes
Species in park?	Yes	Exceeds if 10% biosites?	Yes	No	Yes	Yes	Yes
# Monitors in park	0	Exceeds if 15% biosites?	Yes	No	Yes	No	Yes
Climate Region	Southeast	Exceeds if 20% biosites?	No	No	No	No	No
Park Name	Gulf Islands National Seashore	W126 (ppm-hrs)	18.13	16.89	11.16	7.32	8.54
		PZ (7-mo avg)	-1.99	-2.41	-0.49	0.71	-0.71
		Exceeds in Base?	Yes	Yes	Yes	No	No
Park Code	GUIS	Exceeds if 5%, injury=5?	No	No	No	No	No
Primary State	FL	Exceeds if 5% biosites?	Yes	Yes	Yes	Yes	Yes
Species in park?	Yes	Exceeds if 10% biosites?	No	No	Yes	Yes	Yes
# Monitors in park	0	Exceeds if 15% biosites?	No	No	Yes	No	Yes
Climate Region	Southeast	Exceeds if 20% biosites?	No	No	No	No	No
Park Name	Hagerman Fossil Beds National Monument	W126 (ppm-hrs)	9.95	15.73	11.00	9.78	7.84
		PZ (7-mo avg)	1.25	-1.67	-0.94	0.82	0.38
		Exceeds in Base?	No	Yes	Yes	No	No
Park Code	HAFO	Exceeds if 5%, injury=5?	Yes	No	No	No	No
Primary State	ID	Exceeds if 5% biosites?	Yes	Yes	Yes	Yes	Yes
Species in park?	Yes	Exceeds if 10% biosites?	Yes	No	Yes	Yes	Yes
# Monitors in park	0	Exceeds if 15% biosites?	Yes	No	Yes	Yes	No
Climate Region	Northwest	Exceeds if 20% biosites?	Yes	No	No	No	No
Park Name	Harpers Ferry National Historical Park	W126 (ppm-hrs)	15.43	15.28	12.08	5.86	14.76
		PZ (7-mo avg)	-0.21	-0.94	1.02	0.36	-1.16
		Exceeds in Base?	Yes	Yes	Yes	No	Yes
Park Code	HAFE	Exceeds if 5%, injury=5?	Yes	Yes	Yes	No	Yes
Primary State	WV	Exceeds if 5% biosites?	Yes	Yes	Yes	Yes	Yes
Species in park?	Yes	Exceeds if 10% biosites?	Yes	Yes	Yes	No	Yes
# Monitors in park	0	Exceeds if 15% biosites?	Yes	Yes	Yes	No	Yes
Climate Region	Central	Exceeds if 20% biosites?	No	No	Yes	No	No

Park info		Criteria	2006	2007	2008	2009	2010
Park Name	Home of F. D. Roosevelt National Historic Site	W126 (ppm-hrs)	7.01	10.69	8.77	5.70	8.14
		PZ (7-mo avg)	1.09	0.44	1.13	1.42	-0.90
		Exceeds in Base?	No	Yes	No	No	No
Park Code	HOFR	Exceeds if 5%, injury=5?	No	No	Yes	No	No
Primary State	NY	Exceeds if 5% biosites?	Yes	Yes	Yes	Yes	Yes
Species in park?	Yes	Exceeds if 10% biosites?	Yes	Yes	Yes	Yes	Yes
# Monitors in park	0	Exceeds if 15% biosites?	Yes	Yes	Yes	Yes	No
Climate Region	Northeast	Exceeds if 20% biosites?	Yes	No	Yes	Yes	No
Park Name	Hopewell Culture National Historical Park	W126 (ppm-hrs)	14.30	17.10	11.01	8.23	10.48
		PZ (7-mo avg)	0.32	-1.26	0.83	0.81	-0.11
		Exceeds in Base?	Yes	Yes	Yes	No	Yes
Park Code	HOCU	Exceeds if 5%, injury=5?	Yes	No	No	No	No
Primary State	OH	Exceeds if 5% biosites?	Yes	Yes	Yes	Yes	Yes
Species in park?	Yes	Exceeds if 10% biosites?	Yes	No	Yes	Yes	Yes
# Monitors in park	0	Exceeds if 15% biosites?	Yes	No	Yes	Yes	Yes
Climate Region	Central	Exceeds if 20% biosites?	No	No	No	No	No
Park Name	Hopewell Furnace National Historic Park	W126 (ppm-hrs)	16.24	16.31	16.49	5.73	15.07
		PZ (7-mo avg)	0.84	0.12	-0.10	1.02	-0.60
		Exceeds in Base?	Yes	Yes	Yes	No	Yes
Park Code	HOFU	Exceeds if 5%, injury=5?	Yes	Yes	Yes	No	Yes
Primary State	PA	Exceeds if 5% biosites?	Yes	Yes	Yes	Yes	Yes
Species in park?	Yes	Exceeds if 10% biosites?	Yes	Yes	Yes	Yes	Yes
# Monitors in park	0	Exceeds if 15% biosites?	Yes	Yes	Yes	Yes	Yes
Climate Region	Northeast	Exceeds if 20% biosites?	No	No	No	Yes	No
Park Name	Horseshoe Bend National Military Park	W126 (ppm-hrs)	16.37	15.38	10.80	5.05	8.31
		PZ (7-mo avg)	-1.07	-2.36	-0.11	1.37	-0.82
		Exceeds in Base?	Yes	Yes	Yes	No	No
Park Code	HOBE	Exceeds if 5%, injury=5?	Yes	No	No	No	No
Primary State	AL	Exceeds if 5% biosites?	Yes	Yes	Yes	Yes	Yes
Species in park?	Yes	Exceeds if 10% biosites?	Yes	No	Yes	Yes	Yes
# Monitors in park	0	Exceeds if 15% biosites?	Yes	No	Yes	Yes	Yes
Climate Region	Southeast	Exceeds if 20% biosites?	No	No	No	No	No
Park Name	Hot Springs National Park	W126 (ppm-hrs)	11.89	9.03	5.67	4.81	9.53
		PZ (7-mo avg)	-0.51	-0.61	1.96	2.38	-1.15
		Exceeds in Base?	Yes	No	No	No	No
Park Code	HOSP	Exceeds if 5%, injury=5?	No	No	No	No	No
Primary State	AR	Exceeds if 5% biosites?	Yes	Yes	Yes	Yes	Yes
Species in park?	Yes	Exceeds if 10% biosites?	Yes	Yes	Yes	Yes	Yes
# Monitors in park	0	Exceeds if 15% biosites?	Yes	Yes	Yes	Yes	Yes
Climate Region	South	Exceeds if 20% biosites?	No	No	Yes	No	No
Park Name	Hovenweep National Monument	W126 (ppm-hrs)	20.25	16.77	15.04	9.91	10.81
		PZ (7-mo avg)	-0.72	-0.23	-0.34	-0.78	0.33
		Exceeds in Base?	Yes	Yes	Yes	No	Yes
Park Code	HOVE	Exceeds if 5%, injury=5?	Yes	Yes	Yes	No	No
Primary State	CO	Exceeds if 5% biosites?	Yes	Yes	Yes	Yes	Yes
Species in park?	Yes	Exceeds if 10% biosites?	Yes	Yes	Yes	Yes	Yes

Park info		Criteria	2006	2007	2008	2009	2010
# Monitors in park	0	Exceeds if 15% biosites?	Yes	Yes	Yes	Yes	Yes
Climate Region	Southwest	Exceeds if 20% biosites?	No	No	No	No	No
Park Name	Indiana Dunes National Lakeshore	W126 (ppm-hrs)	9.05	11.48	4.92	4.83	5.73
		PZ (7-mo avg)	0.98	0.22	0.88	0.45	0.37
		Exceeds in Base?	No	Yes	No	No	No
Park Code	INDU	Exceeds if 5%, injury=5?	No	No	No	No	No
Primary State	IN	Exceeds if 5% biosites?	Yes	Yes	Yes	Yes	Yes
Species in park?	Yes	Exceeds if 10% biosites?	Yes	Yes	No	No	No
# Monitors in park	1	Exceeds if 15% biosites?	Yes	Yes	No	No	No
Climate Region	Central	Exceeds if 20% biosites?	No	No	No	No	No
Park Name	Jean Lafitte National Historical Park and Preserve	W126 (ppm-hrs)	17.22	13.43	6.69	7.20	8.93
		PZ (7-mo avg)	-1.65	-0.36	0.41	-1.16	-0.05
		Exceeds in Base?	Yes	Yes	No	No	No
Park Code	JELA	Exceeds if 5%, injury=5?	No	Yes	No	No	No
Primary State	LA	Exceeds if 5% biosites?	Yes	Yes	Yes	Yes	Yes
Species in park?	Yes	Exceeds if 10% biosites?	No	Yes	Yes	Yes	Yes
# Monitors in park	0	Exceeds if 15% biosites?	No	Yes	No	No	Yes
Climate Region	South	Exceeds if 20% biosites?	No	No	No	No	No
Park Name	Jewel Cave National Monument	W126 (ppm-hrs)	14.59	9.74	7.22	5.26	5.73
		PZ (7-mo avg)	0.01	-0.46	1.73	1.13	1.00
		Exceeds in Base?	Yes	No	No	No	No
Park Code	JECA	Exceeds if 5%, injury=5?	Yes	No	Yes	No	No
Primary State	SD	Exceeds if 5% biosites?	Yes	Yes	Yes	Yes	Yes
Species in park?	Yes	Exceeds if 10% biosites?	Yes	Yes	Yes	Yes	No
# Monitors in park	0	Exceeds if 15% biosites?	Yes	Yes	Yes	Yes	No
Climate Region	West North Central	Exceeds if 20% biosites?	No	No	Yes	No	No
Park Name	John D. Rockefeller Jr. Memorial Parkway	W126 (ppm-hrs)	12.49	12.13	10.89	7.56	9.58
		PZ (7-mo avg)	-0.80	-1.51	-0.26	0.92	0.93
		Exceeds in Base?	Yes	Yes	Yes	No	No
Park Code	JODR	Exceeds if 5%, injury=5?	Yes	No	No	No	No
Primary State	WY	Exceeds if 5% biosites?	Yes	Yes	Yes	Yes	Yes
Species in park?	0	Exceeds if 10% biosites?	Yes	No	Yes	Yes	Yes
# Monitors in park	0	Exceeds if 15% biosites?	Yes	No	Yes	No	Yes
Climate Region	West North Central	Exceeds if 20% biosites?	No	No	No	No	No
Park Name	John Day Fossil Beds National Monument	W126 (ppm-hrs)	6.16	4.57	4.60	3.86	4.04
		PZ (7-mo avg)	0.48	-2.24	-1.75	-0.96	0.31
		Exceeds in Base?	No	No	No	No	No
Park Code	JODA	Exceeds if 5%, injury=5?	No	No	No	No	No
Primary State	OR	Exceeds if 5% biosites?	Yes	No	No	Yes	Yes
Species in park?	Yes	Exceeds if 10% biosites?	Yes	No	No	No	No
# Monitors in park	0	Exceeds if 15% biosites?	No	No	No	No	No
Climate Region	Northwest	Exceeds if 20% biosites?	No	No	No	No	No
Park Name	John Muir National Historic Site	W126 (ppm-hrs)	6.72	2.99	5.31	3.95	3.63
		PZ (7-mo avg)	1.87	-1.25	-1.57	-0.20	1.06
		Exceeds in Base?	No	No	No	No	No
Park Code	JOMU	Exceeds if 5%, injury=5?	No	No	No	No	No

Park info		Criteria	2006	2007	2008	2009	2010
Primary State	CA	Exceeds if 5% biosites?	Yes	No	No	Yes	Yes
Species in park?	Yes	Exceeds if 10% biosites?	Yes	No	No	No	No
# Monitors in park	0	Exceeds if 15% biosites?	Yes	No	No	No	No
Climate Region	West	Exceeds if 20% biosites?	Yes	No	No	No	No
Park Name	Joshua Tree National Park	W126 (ppm-hrs)	32.12	28.20	30.70	23.82	25.07
		PZ (7-mo avg)	-1.30	-1.71	-1.15	-1.85	-0.52
		Exceeds in Base?	Yes	Yes	Yes	Yes	Yes
Park Code	JOTR	Exceeds if 5%, injury=5?	No	No	Yes	No	Yes
Primary State	CA	Exceeds if 5% biosites?	Yes	Yes	Yes	Yes	Yes
Species in park?	Yes	Exceeds if 10% biosites?	Yes	Yes	Yes	No	Yes
# Monitors in park	3	Exceeds if 15% biosites?	No	No	Yes	No	Yes
Climate Region	West	Exceeds if 20% biosites?	No	No	No	No	No
Park Name	Kennesaw Mountain National Battlefield Park	W126 (ppm-hrs)	26.52	22.62	14.10	9.05	13.61
		PZ (7-mo avg)	-0.57	-2.10	-0.45	1.13	0.29
		Exceeds in Base?	Yes	Yes	Yes	No	Yes
Park Code	KEMO	Exceeds if 5%, injury=5?	Yes	No	Yes	Yes	Yes
Primary State	GA	Exceeds if 5% biosites?	Yes	Yes	Yes	Yes	Yes
Species in park?	Yes	Exceeds if 10% biosites?	Yes	No	Yes	Yes	Yes
# Monitors in park	0	Exceeds if 15% biosites?	Yes	No	Yes	Yes	Yes
Climate Region	Southeast	Exceeds if 20% biosites?	No	No	No	Yes	No
Park Name	Kings Mountain National Military Park	W126 (ppm-hrs)	15.10	15.78	15.44	4.66	8.54
		PZ (7-mo avg)	-0.56	-1.98	-1.06	-0.08	-1.26
		Exceeds in Base?	Yes	Yes	Yes	No	No
Park Code	KIMO	Exceeds if 5%, injury=5?	Yes	No	Yes	No	No
Primary State	SC	Exceeds if 5% biosites?	Yes	Yes	Yes	Yes	Yes
Species in park?	Yes	Exceeds if 10% biosites?	Yes	No	Yes	No	No
# Monitors in park	0	Exceeds if 15% biosites?	Yes	No	Yes	No	No
Climate Region	Southeast	Exceeds if 20% biosites?	No	No	No	No	No
Park Name	Knife River Indian Villages National Historic Site	W126 (ppm-hrs)	7.25	3.08	4.87	3.94	4.68
		PZ (7-mo avg)	-0.68	-0.01	-1.55	0.68	2.76
		Exceeds in Base?	No	No	No	No	No
Park Code	KNRI	Exceeds if 5%, injury=5?	No	No	No	No	No
Primary State	ND	Exceeds if 5% biosites?	Yes	Yes	No	Yes	Yes
Species in park?	Yes	Exceeds if 10% biosites?	Yes	No	No	No	Yes
# Monitors in park	0	Exceeds if 15% biosites?	No	No	No	No	No
Climate Region	West North Central	Exceeds if 20% biosites?	No	No	No	No	No
Park Name	Lake Mead National Recreation Area	W126 (ppm-hrs)	23.51	17.81	17.64	12.43	15.30
		PZ (7-mo avg)	-1.32	-1.49	-1.31	-1.79	-0.33
		Exceeds in Base?	Yes	Yes	Yes	Yes	Yes
Park Code	LAME	Exceeds if 5%, injury=5?	No	No	No	No	Yes
Primary State	NV	Exceeds if 5% biosites?	Yes	Yes	Yes	Yes	Yes
Species in park?	Yes	Exceeds if 10% biosites?	No	No	No	No	Yes
# Monitors in park	0	Exceeds if 15% biosites?	No	No	No	No	Yes
Climate Region	West	Exceeds if 20% biosites?	No	No	No	No	No
Park Name	Lake Meredith National Recreation Area	W126 (ppm-hrs)	19.51	9.60	11.66	7.93	8.10
		PZ (7-mo avg)	-0.72	2.36	-0.27	-0.58	1.40

Park info		Criteria	2006	2007	2008	2009	2010
		Exceeds in Base?	Yes	No	Yes	No	No
Park Code	LAMR	Exceeds if 5%, injury=5?	Yes	Yes	No	No	Yes
Primary State	TX	Exceeds if 5% biosites?	Yes	Yes	Yes	Yes	Yes
Species in park?	Yes	Exceeds if 10% biosites?	Yes	Yes	Yes	Yes	Yes
# Monitors in park	0	Exceeds if 15% biosites?	Yes	Yes	Yes	No	Yes
Climate Region	South	Exceeds if 20% biosites?	No	Yes	No	No	Yes
Park Name	Lake Roosevelt National Recreation Area	W126 (ppm-hrs)	5.47	3.85	2.89	3.86	2.72
		PZ (7-mo avg)	0.08	-1.20	-0.78	-0.67	1.00
		Exceeds in Base?	No	No	No	No	No
Park Code	LARO	Exceeds if 5%, injury=5?	No	No	No	No	No
Primary State	WA	Exceeds if 5% biosites?	Yes	Yes	No	Yes	No
Species in park?	Yes	Exceeds if 10% biosites?	No	No	No	No	No
# Monitors in park	0	Exceeds if 15% biosites?	No	No	No	No	No
Climate Region	Northwest	Exceeds if 20% biosites?	No	No	No	No	No
Park Name	Lassen Volcanic National Park	W126 (ppm-hrs)	21.31	15.29	18.03	10.42	11.59
		PZ (7-mo avg)	0.41	-1.07	-1.98	0.14	0.59
		Exceeds in Base?	Yes	Yes	Yes	No	Yes
Park Code	LAVO	Exceeds if 5%, injury=5?	Yes	Yes	No	No	No
Primary State	CA	Exceeds if 5% biosites?	Yes	Yes	Yes	Yes	Yes
Species in park?	Yes	Exceeds if 10% biosites?	Yes	Yes	No	Yes	Yes
# Monitors in park	1	Exceeds if 15% biosites?	Yes	Yes	No	Yes	Yes
Climate Region	West	Exceeds if 20% biosites?	No	No	No	No	No
Park Name	Lava Beds National Monument	W126 (ppm-hrs)	11.08	8.09	9.78	5.59	7.59
		PZ (7-mo avg)	0.43	-1.08	-1.78	-0.20	1.27
		Exceeds in Base?	Yes	No	No	No	No
Park Code	LABE	Exceeds if 5%, injury=5?	No	No	No	No	Yes
Primary State	CA	Exceeds if 5% biosites?	Yes	Yes	Yes	Yes	Yes
Species in park?	Yes	Exceeds if 10% biosites?	Yes	Yes	No	No	Yes
# Monitors in park	0	Exceeds if 15% biosites?	Yes	No	No	No	Yes
Climate Region	West	Exceeds if 20% biosites?	No	No	No	No	Yes
Park Name	Little Bighorn Battlefield National Monument	W126 (ppm-hrs)	11.16	8.71	9.09	6.15	4.67
		PZ (7-mo avg)	-1.41	-0.03	0.22	0.60	0.27
		Exceeds in Base?	Yes	No	No	No	No
Park Code	LIBI	Exceeds if 5%, injury=5?	No	No	No	No	No
Primary State	IN	Exceeds if 5% biosites?	Yes	Yes	Yes	Yes	Yes
Species in park?	Yes	Exceeds if 10% biosites?	No	Yes	Yes	Yes	No
# Monitors in park	0	Exceeds if 15% biosites?	No	Yes	Yes	No	No
Climate Region	Central	Exceeds if 20% biosites?	No	No	No	No	No
Park Name	Little River Canyon National Preserve	W126 (ppm-hrs)	17.64	18.72	10.94	5.39	9.72
		PZ (7-mo avg)	-1.06	-2.50	0.12	1.02	-0.80
		Exceeds in Base?	Yes	Yes	Yes	No	No
Park Code	LIRI	Exceeds if 5%, injury=5?	Yes	No	No	No	No
Primary State	MT	Exceeds if 5% biosites?	Yes	Yes	Yes	Yes	Yes
Species in park?	Yes	Exceeds if 10% biosites?	Yes	No	Yes	Yes	Yes
# Monitors in park	0	Exceeds if 15% biosites?	Yes	No	Yes	Yes	Yes
Climate Region	West North Central	Exceeds if 20% biosites?	No	No	No	No	No

Park info		Criteria	2006	2007	2008	2009	2010
Park Name	Lyndon B. Johnson National Historical Park	W126 (ppm-hrs)	13.69	6.58	7.90	8.17	7.27
		PZ (7-mo avg)	-1.59	4.10	-0.72	-0.68	0.59
		Exceeds in Base?	Yes	No	No	No	No
Park Code	LYJO	Exceeds if 5%, injury=5?	No	No	No	No	No
Primary State	TX	Exceeds if 5% biosites?	Yes	Yes	Yes	Yes	Yes
Species in park?	Yes	Exceeds if 10% biosites?	No	Yes	Yes	Yes	Yes
# Monitors in park	0	Exceeds if 15% biosites?	No	Yes	No	No	No
Climate Region	South	Exceeds if 20% biosites?	No	Yes	No	No	No
Park Name	Mammoth Cave National Park	W126 (ppm-hrs)	12.09	20.19	11.44	6.23	10.03
		PZ (7-mo avg)	1.00	-1.39	-0.28	0.98	-0.33
		Exceeds in Base?	Yes	Yes	Yes	No	No
Park Code	MACA	Exceeds if 5%, injury=5?	Yes	No	No	No	No
Primary State	KY	Exceeds if 5% biosites?	Yes	Yes	Yes	Yes	Yes
Species in park?	Yes	Exceeds if 10% biosites?	Yes	No	Yes	Yes	Yes
# Monitors in park	0	Exceeds if 15% biosites?	Yes	No	Yes	No	Yes
Climate Region	Central	Exceeds if 20% biosites?	Yes	No	No	No	No
Park Name	Manassas National Battlefield Park	W126 (ppm-hrs)	13.38	14.37	11.44	5.09	12.08
		PZ (7-mo avg)	-0.24	-1.56	0.94	0.31	-1.19
		Exceeds in Base?	Yes	Yes	Yes	No	Yes
Park Code	MANA	Exceeds if 5%, injury=5?	Yes	No	No	No	No
Primary State	VA	Exceeds if 5% biosites?	Yes	Yes	Yes	Yes	Yes
Species in park?	Yes	Exceeds if 10% biosites?	Yes	No	Yes	No	Yes
# Monitors in park	0	Exceeds if 15% biosites?	Yes	No	Yes	No	Yes
Climate Region	Southeast	Exceeds if 20% biosites?	No	No	No	No	No
Park Name	Manzanar National Historic Site	W126 (ppm-hrs)	43.41	37.02	39.34	30.02	29.02
		PZ (7-mo avg)	-1.30	-1.71	-1.15	-1.85	-0.52
		Exceeds in Base?	Yes	Yes	Yes	Yes	Yes
Park Code	MANZ	Exceeds if 5%, injury=5?	No	No	Yes	No	Yes
Primary State	CA	Exceeds if 5% biosites?	Yes	Yes	Yes	Yes	Yes
Species in park?	0	Exceeds if 10% biosites?	Yes	Yes	Yes	Yes	Yes
# Monitors in park	0	Exceeds if 15% biosites?	No	No	Yes	No	Yes
Climate Region	West	Exceeds if 20% biosites?	No	No	No	No	No
Park Name	Marsh-Billings-Rockefeller National Historical Park	W126 (ppm-hrs)	4.23	6.27	6.21	4.06	4.45
		PZ (7-mo avg)	1.18	-0.26	0.42	0.34	-1.25
		Exceeds in Base?	No	No	No	No	No
Park Code	MABI	Exceeds if 5%, injury=5?	No	No	No	No	No
Primary State	VT	Exceeds if 5% biosites?	Yes	Yes	Yes	Yes	No
Species in park?	Yes	Exceeds if 10% biosites?	No	Yes	Yes	No	No
# Monitors in park	0	Exceeds if 15% biosites?	No	No	No	No	No
Climate Region	Northeast	Exceeds if 20% biosites?	No	No	No	No	No
Park Name	Mesa Verde National Park	W126 (ppm-hrs)	20.79	17.13	13.92	12.06	11.62
		PZ (7-mo avg)	-0.54	-0.25	-0.26	-0.17	-0.11
		Exceeds in Base?	Yes	Yes	Yes	Yes	Yes
Park Code	MEVE	Exceeds if 5%, injury=5?	Yes	Yes	Yes	No	No
Primary State	CO	Exceeds if 5% biosites?	Yes	Yes	Yes	Yes	Yes
Species in park?	Yes	Exceeds if 10% biosites?	Yes	Yes	Yes	Yes	Yes

Park info		Criteria	2006	2007	2008	2009	2010
# Monitors in park	1	Exceeds if 15% biosites?	Yes	Yes	Yes	Yes	Yes
Climate Region	Southwest	Exceeds if 20% biosites?	No	No	No	No	No
Park Name	Minute Man National Historical Park	W126 (ppm-hrs)	8.26	9.66	6.40	4.71	5.88
		PZ (7-mo avg)	1.06	0.34	1.69	1.22	-0.17
		Exceeds in Base?	No	No	No	No	No
Park Code	MIMA	Exceeds if 5%, injury=5?	Yes	No	No	No	No
Primary State	MA	Exceeds if 5% biosites?	Yes	Yes	Yes	Yes	Yes
Species in park?	Yes	Exceeds if 10% biosites?	Yes	Yes	Yes	Yes	No
# Monitors in park	0	Exceeds if 15% biosites?	Yes	Yes	Yes	Yes	No
Climate Region	Northeast	Exceeds if 20% biosites?	Yes	No	Yes	No	No
Park Name	Mississippi National River And Recreation Area	W126 (ppm-hrs)	7.48	8.20	4.61	4.72	4.55
		PZ (7-mo avg)	-0.29	0.27	0.36	-0.34	1.15
		Exceeds in Base?	No	No	No	No	No
Park Code	MISS	Exceeds if 5%, injury=5?	No	No	No	No	No
Primary State	MN	Exceeds if 5% biosites?	Yes	Yes	Yes	Yes	Yes
Species in park?	Yes	Exceeds if 10% biosites?	Yes	Yes	No	No	Yes
# Monitors in park	0	Exceeds if 15% biosites?	No	Yes	No	No	No
Climate Region	East North Central	Exceeds if 20% biosites?	No	No	No	No	No
Park Name	Missouri National Recreational River	W126 (ppm-hrs)	9.89	6.06	4.12	4.25	5.26
		PZ (7-mo avg)	-0.06	1.31	1.16	1.13	2.63
		Exceeds in Base?	No	No	No	No	No
Park Code	MNRR	Exceeds if 5%, injury=5?	No	No	No	No	No
Primary State	NE	Exceeds if 5% biosites?	Yes	Yes	Yes	Yes	Yes
Species in park?	Yes	Exceeds if 10% biosites?	Yes	Yes	No	No	Yes
# Monitors in park	0	Exceeds if 15% biosites?	Yes	Yes	No	No	Yes
Climate Region	West North Central	Exceeds if 20% biosites?	No	Yes	No	No	No
Park Name	Mojave National Preserve	W126 (ppm-hrs)	28.70	26.92	25.28	19.31	19.85
		PZ (7-mo avg)	-1.24	-1.47	-1.33	-1.86	-0.43
		Exceeds in Base?	Yes	Yes	Yes	Yes	Yes
Park Code	MOJA	Exceeds if 5%, injury=5?	Yes	No	No	No	Yes
Primary State	CA	Exceeds if 5% biosites?	Yes	Yes	Yes	Yes	Yes
Species in park?	Yes	Exceeds if 10% biosites?	Yes	Yes	Yes	No	Yes
# Monitors in park	1	Exceeds if 15% biosites?	Yes	No	No	No	Yes
Climate Region	West	Exceeds if 20% biosites?	No	No	No	No	No
Park Name	Monocacy National Battlefield	W126 (ppm-hrs)	17.17	17.44	14.47	6.47	16.84
		PZ (7-mo avg)	-0.19	-1.10	0.53	1.08	-0.94
		Exceeds in Base?	Yes	Yes	Yes	No	Yes
Park Code	MONO	Exceeds if 5%, injury=5?	Yes	Yes	Yes	No	Yes
Primary State	MD	Exceeds if 5% biosites?	Yes	Yes	Yes	Yes	Yes
Species in park?	Yes	Exceeds if 10% biosites?	Yes	Yes	Yes	Yes	Yes
# Monitors in park	0	Exceeds if 15% biosites?	Yes	Yes	Yes	Yes	Yes
Climate Region	Northeast	Exceeds if 20% biosites?	No	No	No	Yes	No
Park Name	Montezuma Castle National Monument	W126 (ppm-hrs)	25.01	19.96	23.10	10.27	12.48
		PZ (7-mo avg)	-1.45	-1.95	-0.31	-1.56	-0.31
		Exceeds in Base?	Yes	Yes	Yes	No	Yes
Park Code	MOCA	Exceeds if 5%, injury=5?	No	No	Yes	No	Yes

Park info		Criteria	2006	2007	2008	2009	2010
Primary State	AZ	Exceeds if 5% biosites?	Yes	Yes	Yes	Yes	Yes
Species in park?	Yes	Exceeds if 10% biosites?	Yes	No	Yes	No	Yes
# Monitors in park	0	Exceeds if 15% biosites?	No	No	Yes	No	Yes
Climate Region	Southwest	Exceeds if 20% biosites?	No	No	No	No	No
Park Name	Morristown National Historical Park	W126 (ppm-hrs)	19.03	18.20	14.78	5.66	14.80
		PZ (7-mo avg)	0.15	0.91	-0.31	0.53	-0.62
		Exceeds in Base?	Yes	Yes	Yes	No	Yes
Park Code	MORR	Exceeds if 5%, injury=5?	Yes	Yes	Yes	No	Yes
Primary State	NJ	Exceeds if 5% biosites?	Yes	Yes	Yes	Yes	Yes
Species in park?	Yes	Exceeds if 10% biosites?	Yes	Yes	Yes	No	Yes
# Monitors in park	0	Exceeds if 15% biosites?	Yes	Yes	Yes	No	Yes
Climate Region	Northeast	Exceeds if 20% biosites?	No	No	No	No	No
Park Name	Mount Rainier Wilderness	W126 (ppm-hrs)	5.43	2.28	3.28	3.96	2.34
		PZ (7-mo avg)	-1.01	-0.26	0.08	-0.52	1.47
		Exceeds in Base?	No	No	No	No	No
Park Code	MORA	Exceeds if 5%, injury=5?	No	No	No	No	No
Primary State	WA	Exceeds if 5% biosites?	Yes	No	Yes	Yes	No
Species in park?	Yes	Exceeds if 10% biosites?	No	No	No	No	No
# Monitors in park	1	Exceeds if 15% biosites?	No	No	No	No	No
Climate Region	Northwest	Exceeds if 20% biosites?	No	No	No	No	No
Park Name	Mount Rushmore National Memorial	W126 (ppm-hrs)	11.68	7.59	5.99	4.54	4.80
		PZ (7-mo avg)	0.01	-0.46	1.73	1.13	1.00
		Exceeds in Base?	Yes	No	No	No	No
Park Code	MORU	Exceeds if 5%, injury=5?	No	No	No	No	No
Primary State	SD	Exceeds if 5% biosites?	Yes	Yes	Yes	Yes	Yes
Species in park?	Yes	Exceeds if 10% biosites?	Yes	Yes	Yes	Yes	No
# Monitors in park	0	Exceeds if 15% biosites?	Yes	No	Yes	No	No
Climate Region	West North Central	Exceeds if 20% biosites?	No	No	Yes	No	No
Park Name	Muir Woods National Monument	W126 (ppm-hrs)	0.99	1.14	1.80	1.27	1.22
		PZ (7-mo avg)	0.43	-1.08	-1.78	-0.20	1.27
		Exceeds in Base?	No	No	No	No	No
Park Code	MUWO	Exceeds if 5%, injury=5?	No	No	No	No	No
Primary State	CA	Exceeds if 5% biosites?	No	No	No	No	No
Species in park?	Yes	Exceeds if 10% biosites?	No	No	No	No	No
# Monitors in park	0	Exceeds if 15% biosites?	No	No	No	No	No
Climate Region	West	Exceeds if 20% biosites?	No	No	No	No	No
Park Name	Natchez Trace Parkway	W126 (ppm-hrs)	15.19	12.22	6.95	4.81	7.31
		PZ (7-mo avg)	-1.25	-1.21	0.78	0.97	-0.87
		Exceeds in Base?	Yes	Yes	No	No	No
Park Code	NATR	Exceeds if 5%, injury=5?	Yes	No	No	No	No
Primary State	AL	Exceeds if 5% biosites?	Yes	Yes	Yes	Yes	Yes
Species in park?	Yes	Exceeds if 10% biosites?	Yes	Yes	Yes	No	Yes
# Monitors in park	0	Exceeds if 15% biosites?	Yes	Yes	No	No	No
Climate Region	Southeast	Exceeds if 20% biosites?	No	No	No	No	No
Park Name	National Mall & Memorial Parks	W126 (ppm-hrs)	21.89	20.39	16.18	7.71	20.38
		PZ (7-mo avg)	-0.25	-1.67	1.02	1.04	-1.06

Park info		Criteria	2006	2007	2008	2009	2010
		Exceeds in Base?	Yes	Yes	Yes	No	Yes
Park Code	NACC	Exceeds if 5%, injury=5?	Yes	No	Yes	Yes	Yes
Primary State	DC	Exceeds if 5% biosites?	Yes	Yes	Yes	Yes	Yes
Species in park?	0	Exceeds if 10% biosites?	Yes	No	Yes	Yes	Yes
# Monitors in park	0	Exceeds if 15% biosites?	Yes	No	Yes	Yes	Yes
Climate Region	Southeast	Exceeds if 20% biosites?	No	No	Yes	Yes	No
	Natural Bridges National Monument	W126 (ppm-hrs)	20.14	17.02	16.82	9.95	12.50
Park Name		PZ (7-mo avg)	-0.91	-0.21	-0.42	-1.39	0.77
		Exceeds in Base?	Yes	Yes	Yes	No	Yes
Park Code	NABR	Exceeds if 5%, injury=5?	Yes	Yes	Yes	No	Yes
Primary State	UT	Exceeds if 5% biosites?	Yes	Yes	Yes	Yes	Yes
Species in park?	Yes	Exceeds if 10% biosites?	Yes	Yes	Yes	No	Yes
# Monitors in park	0	Exceeds if 15% biosites?	Yes	Yes	Yes	No	Yes
Climate Region	Southwest	Exceeds if 20% biosites?	No	No	No	No	No
	Navajo National Monument	W126 (ppm-hrs)	20.51	17.19	17.49	10.00	13.08
Park Name		PZ (7-mo avg)	-0.54	-1.40	0.03	-1.48	0.39
		Exceeds in Base?	Yes	Yes	Yes	No	Yes
Park Code	NAVA	Exceeds if 5%, injury=5?	Yes	No	Yes	No	Yes
Primary State	AZ	Exceeds if 5% biosites?	Yes	Yes	Yes	Yes	Yes
Species in park?	Yes	Exceeds if 10% biosites?	Yes	No	Yes	No	Yes
# Monitors in park	0	Exceeds if 15% biosites?	Yes	No	Yes	No	Yes
Climate Region	Southwest	Exceeds if 20% biosites?	No	No	No	No	No
	New River Gorge National River	W126 (ppm-hrs)	10.90	14.07	9.44	5.33	8.20
Park Name		PZ (7-mo avg)	0.05	-0.15	0.33	0.26	-1.00
		Exceeds in Base?	Yes	Yes	No	No	No
Park Code	NERI	Exceeds if 5%, injury=5?	No	Yes	No	No	No
Primary State	WV	Exceeds if 5% biosites?	Yes	Yes	Yes	Yes	Yes
Species in park?	Yes	Exceeds if 10% biosites?	Yes	Yes	Yes	No	Yes
# Monitors in park	0	Exceeds if 15% biosites?	Yes	Yes	Yes	No	Yes
Climate Region	Central	Exceeds if 20% biosites?	No	No	No	No	No
	Nez Perce National Historical Park	W126 (ppm-hrs)	4.00	7.25	5.29	5.35	4.23
Park Name		PZ (7-mo avg)	-0.41	-1.51	-0.14	0.21	0.71
		Exceeds in Base?	No	No	No	No	No
Park Code	NEPE	Exceeds if 5%, injury=5?	No	No	No	No	No
Primary State	ID	Exceeds if 5% biosites?	Yes	Yes	Yes	Yes	Yes
Species in park?	Yes	Exceeds if 10% biosites?	No	No	No	No	No
# Monitors in park	0	Exceeds if 15% biosites?	No	No	No	No	No
Climate Region	Northwest	Exceeds if 20% biosites?	No	No	No	No	No
	Ninety Six National Historic Site	W126 (ppm-hrs)	15.07	14.88	14.71	4.66	7.20
Park Name		PZ (7-mo avg)	-1.05	-1.67	-1.08	0.12	-1.33
		Exceeds in Base?	Yes	Yes	Yes	No	No
Park Code	NISI	Exceeds if 5%, injury=5?	Yes	No	Yes	No	No
Primary State	SC	Exceeds if 5% biosites?	Yes	Yes	Yes	Yes	Yes
Species in park?	Yes	Exceeds if 10% biosites?	Yes	No	Yes	No	No
# Monitors in park	0	Exceeds if 15% biosites?	Yes	No	Yes	No	No
Climate Region	Southeast	Exceeds if 20% biosites?	No	No	No	No	No

Park info		Criteria	2006	2007	2008	2009	2010
	North Cascades National Park	W126 (ppm-hrs)	2.99	1.86	2.14	3.31	2.31
Park Name		PZ (7-mo avg)	-0.78	-0.56	-0.23	-0.40	1.41
		Exceeds in Base?	No	No	No	No	No
Park Code	NOCA	Exceeds if 5%, injury=5?	No	No	No	No	No
Primary State	WA	Exceeds if 5% biosites?	No	No	No	Yes	No
Species in park?	Yes	Exceeds if 10% biosites?	No	No	No	No	No
# Monitors in park	0	Exceeds if 15% biosites?	No	No	No	No	No
Climate Region	Northwest	Exceeds if 20% biosites?	No	No	No	No	No
	Ocmulgee National Monument	W126 (ppm-hrs)	18.94	16.03	14.33	8.10	8.97
Park Name		PZ (7-mo avg)	-1.47	-0.59	-1.32	1.00	-0.07
		Exceeds in Base?	Yes	Yes	Yes	No	No
Park Code	OCMU	Exceeds if 5%, injury=5?	No	Yes	No	No	No
Primary State	GA	Exceeds if 5% biosites?	Yes	Yes	Yes	Yes	Yes
Species in park?	Yes	Exceeds if 10% biosites?	No	Yes	No	Yes	Yes
# Monitors in park	0	Exceeds if 15% biosites?	No	Yes	No	No	Yes
Climate Region	Southeast	Exceeds if 20% biosites?	No	No	No	No	No
	Olympic National Park	W126 (ppm-hrs)	2.74	1.49	1.73	2.06	1.79
Park Name		PZ (7-mo avg)	-0.90	0.23	0.03	-0.40	1.17
		Exceeds in Base?	No	No	No	No	No
Park Code	OLYM	Exceeds if 5%, injury=5?	No	No	No	No	No
Primary State	WA	Exceeds if 5% biosites?	No	No	No	No	No
Species in park?	Yes	Exceeds if 10% biosites?	No	No	No	No	No
# Monitors in park	2	Exceeds if 15% biosites?	No	No	No	No	No
Climate Region	Northwest	Exceeds if 20% biosites?	No	No	No	No	No
	Oregon Caves National Monument	W126 (ppm-hrs)	8.91	3.43	5.44	4.09	3.38
Park Name		PZ (7-mo avg)	0.13	-0.39	-0.43	-0.05	1.95
		Exceeds in Base?	No	No	No	No	No
Park Code	ORCA	Exceeds if 5%, injury=5?	No	No	No	No	No
Primary State	OR	Exceeds if 5% biosites?	Yes	Yes	Yes	Yes	Yes
Species in park?	Yes	Exceeds if 10% biosites?	Yes	No	No	No	No
# Monitors in park	0	Exceeds if 15% biosites?	Yes	No	No	No	No
Climate Region	Northwest	Exceeds if 20% biosites?	No	No	No	No	No
	Organ Pipe Cactus National Monument	W126 (ppm-hrs)	12.40	8.76	15.96	7.16	9.83
Park Name		PZ (7-mo avg)	-0.31	-0.84	0.17	-1.62	0.29
		Exceeds in Base?	Yes	No	Yes	No	No
Park Code	ORPI	Exceeds if 5%, injury=5?	Yes	No	Yes	No	No
Primary State	AZ	Exceeds if 5% biosites?	Yes	Yes	Yes	Yes	Yes
Species in park?	Yes	Exceeds if 10% biosites?	Yes	Yes	Yes	No	Yes
# Monitors in park	0	Exceeds if 15% biosites?	Yes	Yes	Yes	No	Yes
Climate Region	Southwest	Exceeds if 20% biosites?	No	No	No	No	No
	Ozark National Scenic Riverways	W126 (ppm-hrs)	16.63	17.40	8.88	7.01	9.23
Park Name		PZ (7-mo avg)	0.12	-1.01	2.40	0.72	-0.17
		Exceeds in Base?	Yes	Yes	No	No	No
Park Code	OZAR	Exceeds if 5%, injury=5?	Yes	Yes	Yes	No	No
Primary State	MO	Exceeds if 5% biosites?	Yes	Yes	Yes	Yes	Yes
Species in park?	Yes	Exceeds if 10% biosites?	Yes	Yes	Yes	Yes	Yes

Park info		Criteria	2006	2007	2008	2009	2010
# Monitors in park	0	Exceeds if 15% biosites?	Yes	Yes	Yes	No	Yes
Climate Region	Central	Exceeds if 20% biosites?	No	No	Yes	No	No
Park Name	Padre Island National Seashore	W126 (ppm-hrs)	5.01	4.37	5.79	5.04	5.19
		PZ (7-mo avg)	-0.98	4.66	0.31	-1.53	3.02
		Exceeds in Base?	No	No	No	No	No
Park Code	PAIS	Exceeds if 5%, injury=5?	No	No	No	No	No
Primary State	TX	Exceeds if 5% biosites?	Yes	Yes	Yes	No	Yes
Species in park?	Yes	Exceeds if 10% biosites?	No	No	No	No	Yes
# Monitors in park	1	Exceeds if 15% biosites?	No	No	No	No	Yes
Climate Region	South	Exceeds if 20% biosites?	No	No	No	No	No
Park Name	Palo Alto Battlefield National Historic Site	W126 (ppm-hrs)	4.27	3.84	4.51	3.82	3.47
		PZ (7-mo avg)	-1.08	1.18	1.63	-1.42	2.29
		Exceeds in Base?	No	No	No	No	No
Park Code	PAAL	Exceeds if 5%, injury=5?	No	No	No	No	No
Primary State	TX	Exceeds if 5% biosites?	Yes	Yes	Yes	No	Yes
Species in park?	0	Exceeds if 10% biosites?	No	No	Yes	No	No
# Monitors in park	0	Exceeds if 15% biosites?	No	No	No	No	No
Climate Region	South	Exceeds if 20% biosites?	No	No	No	No	No
Park Name	Pea Ridge National Military Park	W126 (ppm-hrs)	17.70	8.18	5.47	5.71	7.83
		PZ (7-mo avg)	-0.42	-1.41	3.29	0.88	-0.15
		Exceeds in Base?	Yes	No	No	No	No
Park Code	PERI	Exceeds if 5%, injury=5?	Yes	No	No	No	No
Primary State	AR	Exceeds if 5% biosites?	Yes	Yes	Yes	Yes	Yes
Species in park?	Yes	Exceeds if 10% biosites?	Yes	No	Yes	No	Yes
# Monitors in park	0	Exceeds if 15% biosites?	Yes	No	Yes	No	No
Climate Region	South	Exceeds if 20% biosites?	No	No	No	No	No
Park Name	Pecos National Historical Park	W126 (ppm-hrs)	17.84	14.01	13.48	9.27	10.48
		PZ (7-mo avg)	-0.91	0.04	-0.84	-0.95	-0.54
		Exceeds in Base?	Yes	Yes	Yes	No	Yes
Park Code	PECO	Exceeds if 5%, injury=5?	Yes	Yes	Yes	No	No
Primary State	NM	Exceeds if 5% biosites?	Yes	Yes	Yes	Yes	Yes
Species in park?	Yes	Exceeds if 10% biosites?	Yes	Yes	Yes	Yes	Yes
# Monitors in park	0	Exceeds if 15% biosites?	Yes	Yes	Yes	Yes	Yes
Climate Region	Southwest	Exceeds if 20% biosites?	No	No	No	No	No
Park Name	Petersburg National Battlefield	W126 (ppm-hrs)	14.41	14.70	15.51	5.62	11.19
		PZ (7-mo avg)	0.47	-1.22	0.22	0.58	-1.21
		Exceeds in Base?	Yes	Yes	Yes	No	Yes
Park Code	PETE	Exceeds if 5%, injury=5?	Yes	Yes	Yes	No	No
Primary State	VA	Exceeds if 5% biosites?	Yes	Yes	Yes	Yes	Yes
Species in park?	Yes	Exceeds if 10% biosites?	Yes	Yes	Yes	No	Yes
# Monitors in park	0	Exceeds if 15% biosites?	Yes	Yes	Yes	No	Yes
Climate Region	Southeast	Exceeds if 20% biosites?	No	No	No	No	No
Park Name	Petrified Forest National Park	W126 (ppm-hrs)	18.54	16.00	17.49	8.56	12.05
		PZ (7-mo avg)	-0.54	-1.40	0.03	-1.48	0.39
		Exceeds in Base?	Yes	Yes	Yes	No	Yes
Park Code	PEFO	Exceeds if 5%, injury=5?	Yes	No	Yes	No	No

Park info		Criteria	2006	2007	2008	2009	2010
Primary State	AZ	Exceeds if 5% biosites?	Yes	Yes	Yes	Yes	Yes
Species in park?	Yes	Exceeds if 10% biosites?	Yes	No	Yes	No	Yes
# Monitors in park	1	Exceeds if 15% biosites?	Yes	No	Yes	No	Yes
Climate Region	Southwest	Exceeds if 20% biosites?	No	No	No	No	No
Park Name	Petroglyph National Monument	W126 (ppm-hrs)	16.66	15.54	12.68	10.35	10.84
		PZ (7-mo avg)	1.31	1.10	0.16	-0.33	0.47
		Exceeds in Base?	Yes	Yes	Yes	No	Yes
Park Code	PETR	Exceeds if 5%, injury=5?	Yes	Yes	Yes	No	No
Primary State	NM	Exceeds if 5% biosites?	Yes	Yes	Yes	Yes	Yes
Species in park?	Yes	Exceeds if 10% biosites?	Yes	Yes	Yes	Yes	Yes
# Monitors in park	0	Exceeds if 15% biosites?	Yes	Yes	Yes	Yes	Yes
Climate Region	Southwest	Exceeds if 20% biosites?	Yes	Yes	No	No	No
Park Name	Pictured Rocks National Lakeshore	W126 (ppm-hrs)	6.41	9.18	4.55	3.23	4.24
		PZ (7-mo avg)	-1.53	-1.15	0.10	-0.48	0.11
		Exceeds in Base?	No	No	No	No	No
Park Code	PIRO	Exceeds if 5%, injury=5?	No	No	No	No	No
Primary State	MI	Exceeds if 5% biosites?	Yes	Yes	Yes	Yes	Yes
Species in park?	Yes	Exceeds if 10% biosites?	No	Yes	No	No	No
# Monitors in park	0	Exceeds if 15% biosites?	No	Yes	No	No	No
Climate Region	East North Central	Exceeds if 20% biosites?	No	No	No	No	No
Park Name	Pinnacles National Monument	W126 (ppm-hrs)	14.06	12.79	16.44	9.83	8.87
		PZ (7-mo avg)	1.87	-1.25	-1.57	-0.20	1.06
		Exceeds in Base?	Yes	Yes	Yes	No	No
Park Code	PINN	Exceeds if 5%, injury=5?	Yes	Yes	No	No	Yes
Primary State	CA	Exceeds if 5% biosites?	Yes	Yes	Yes	Yes	Yes
Species in park?	Yes	Exceeds if 10% biosites?	Yes	Yes	No	Yes	Yes
# Monitors in park	1	Exceeds if 15% biosites?	Yes	Yes	No	Yes	Yes
Climate Region	West	Exceeds if 20% biosites?	Yes	No	No	No	Yes
Park Name	Pipe Spring National Monument	W126 (ppm-hrs)	21.00	16.32	16.86	11.30	16.17
		PZ (7-mo avg)	-1.45	-1.74	-1.11	-1.71	-0.34
		Exceeds in Base?	Yes	Yes	Yes	Yes	Yes
Park Code	PISP	Exceeds if 5%, injury=5?	No	No	Yes	No	Yes
Primary State	AZ	Exceeds if 5% biosites?	Yes	Yes	Yes	Yes	Yes
Species in park?	Yes	Exceeds if 10% biosites?	No	No	Yes	No	Yes
# Monitors in park	0	Exceeds if 15% biosites?	No	No	Yes	No	Yes
Climate Region	Southwest	Exceeds if 20% biosites?	No	No	No	No	No
Park Name	Pipestone National Monument	W126 (ppm-hrs)	6.94	5.37	4.36	4.43	5.66
		PZ (7-mo avg)	0.43	0.30	0.22	-0.33	2.42
		Exceeds in Base?	No	No	No	No	No
Park Code	PIPE	Exceeds if 5%, injury=5?	No	No	No	No	No
Primary State	MN	Exceeds if 5% biosites?	Yes	Yes	Yes	Yes	Yes
Species in park?	Yes	Exceeds if 10% biosites?	Yes	No	No	No	Yes
# Monitors in park	0	Exceeds if 15% biosites?	No	No	No	No	Yes
Climate Region	East North Central	Exceeds if 20% biosites?	No	No	No	No	Yes
Park Name	Piscataway Park	W126 (ppm-hrs)	20.83	19.99	14.57	7.95	17.35
		PZ (7-mo avg)	0.00	-1.82	0.82	0.69	-1.11

Park info		Criteria	2006	2007	2008	2009	2010
		Exceeds in Base?	Yes	Yes	Yes	No	Yes
Park Code	PISC	Exceeds if 5%, injury=5?	Yes	No	Yes	No	Yes
Primary State	MD	Exceeds if 5% biosites?	Yes	Yes	Yes	Yes	Yes
Species in park?	Yes	Exceeds if 10% biosites?	Yes	No	Yes	Yes	Yes
# Monitors in park	0	Exceeds if 15% biosites?	Yes	No	Yes	No	Yes
Climate Region	Northeast	Exceeds if 20% biosites?	No	No	No	No	No
Park Name	Point Reyes National Seashore	W126 (ppm-hrs)	1.61	1.36	2.29	1.58	1.71
		PZ (7-mo avg)	0.43	-1.08	-1.78	-0.20	1.27
		Exceeds in Base?	No	No	No	No	No
Park Code	PORE	Exceeds if 5%, injury=5?	No	No	No	No	No
Primary State	CA	Exceeds if 5% biosites?	No	No	No	No	No
Species in park?	Yes	Exceeds if 10% biosites?	No	No	No	No	No
# Monitors in park	0	Exceeds if 15% biosites?	No	No	No	No	No
Climate Region	West	Exceeds if 20% biosites?	No	No	No	No	No
Park Name	Poverty Point National Monument	W126 (ppm-hrs)	13.76	7.39	4.48	4.68	8.22
		PZ (7-mo avg)	-0.99	-0.23	1.47	0.38	-1.88
		Exceeds in Base?	Yes	No	No	No	No
Park Code	POPO	Exceeds if 5%, injury=5?	Yes	No	No	No	No
Primary State	LA	Exceeds if 5% biosites?	Yes	Yes	Yes	Yes	Yes
Species in park?	0	Exceeds if 10% biosites?	Yes	Yes	Yes	No	No
# Monitors in park	0	Exceeds if 15% biosites?	Yes	No	No	No	No
Climate Region	South	Exceeds if 20% biosites?	No	No	No	No	No
Park Name	Prince William Forest Park	W126 (ppm-hrs)	15.40	15.91	10.16	5.71	12.94
		PZ (7-mo avg)	0.37	-1.46	0.58	0.67	-1.05
		Exceeds in Base?	Yes	Yes	No	No	Yes
Park Code	PRWI	Exceeds if 5%, injury=5?	Yes	No	No	No	Yes
Primary State	VA	Exceeds if 5% biosites?	Yes	Yes	Yes	Yes	Yes
Species in park?	Yes	Exceeds if 10% biosites?	Yes	No	Yes	No	Yes
# Monitors in park	0	Exceeds if 15% biosites?	Yes	No	Yes	No	Yes
Climate Region	Southeast	Exceeds if 20% biosites?	No	No	No	No	No
Park Name	Rainbow Bridge National Monument	W126 (ppm-hrs)	20.61	17.32	17.53	10.27	13.43
		PZ (7-mo avg)	-0.91	-0.21	-0.42	-1.39	0.77
		Exceeds in Base?	Yes	Yes	Yes	No	Yes
Park Code	RABR	Exceeds if 5%, injury=5?	Yes	Yes	Yes	No	Yes
Primary State	UT	Exceeds if 5% biosites?	Yes	Yes	Yes	Yes	Yes
Species in park?	Yes	Exceeds if 10% biosites?	Yes	Yes	Yes	No	Yes
# Monitors in park	0	Exceeds if 15% biosites?	Yes	Yes	Yes	No	Yes
Climate Region	Southwest	Exceeds if 20% biosites?	No	No	No	No	No
Park Name	Redwood National Park	W126 (ppm-hrs)	9.21	3.32	4.86	3.60	4.23
		PZ (7-mo avg)	0.43	-1.08	-1.78	-0.20	1.27
		Exceeds in Base?	No	No	No	No	No
Park Code	REDW	Exceeds if 5%, injury=5?	No	No	No	No	No
Primary State	CA	Exceeds if 5% biosites?	Yes	Yes	No	Yes	Yes
Species in park?	Yes	Exceeds if 10% biosites?	Yes	No	No	No	No
# Monitors in park	0	Exceeds if 15% biosites?	Yes	No	No	No	No
Climate Region	West	Exceeds if 20% biosites?	No	No	No	No	No

Park info		Criteria	2006	2007	2008	2009	2010
	Richmond National Battlefield Park	W126 (ppm-hrs)	16.76	17.16	17.47	6.33	13.63
Park Name		PZ (7-mo avg)	-0.05	-1.08	0.22	0.14	-1.52
		Exceeds in Base?	Yes	Yes	Yes	No	Yes
Park Code	RICH	Exceeds if 5%, injury=5?	Yes	Yes	Yes	No	No
Primary State	VA	Exceeds if 5% biosites?	Yes	Yes	Yes	Yes	Yes
Species in park?	Yes	Exceeds if 10% biosites?	Yes	Yes	Yes	Yes	No
# Monitors in park	0	Exceeds if 15% biosites?	Yes	Yes	Yes	No	No
Climate Region	Southeast	Exceeds if 20% biosites?	No	No	No	No	No
	Rock Creek Park	W126 (ppm-hrs)	21.83	19.69	16.29	7.74	19.67
Park Name		PZ (7-mo avg)	-0.22	-1.38	0.78	1.06	-1.00
		Exceeds in Base?	Yes	Yes	Yes	No	Yes
Park Code	ROCR	Exceeds if 5%, injury=5?	Yes	No	Yes	Yes	Yes
Primary State	DC	Exceeds if 5% biosites?	Yes	Yes	Yes	Yes	Yes
Species in park?	Yes	Exceeds if 10% biosites?	Yes	No	Yes	Yes	Yes
# Monitors in park	0	Exceeds if 15% biosites?	Yes	No	Yes	Yes	Yes
Climate Region	Southeast	Exceeds if 20% biosites?	No	No	No	Yes	No
	Rocky Mountain National Park	W126 (ppm-hrs)	19.31	17.74	18.51	10.54	15.48
Park Name		PZ (7-mo avg)	-0.97	0.27	-0.05	0.77	0.60
		Exceeds in Base?	Yes	Yes	Yes	Yes	Yes
Park Code	ROMO	Exceeds if 5%, injury=5?	Yes	Yes	Yes	No	Yes
Primary State	CO	Exceeds if 5% biosites?	Yes	Yes	Yes	Yes	Yes
Species in park?	Yes	Exceeds if 10% biosites?	Yes	Yes	Yes	Yes	Yes
# Monitors in park	0	Exceeds if 15% biosites?	Yes	Yes	Yes	Yes	Yes
Climate Region	Southwest	Exceeds if 20% biosites?	No	No	No	No	No
	Saguaro National Park	W126 (ppm-hrs)	16.16	12.93	15.59	9.60	12.26
Park Name		PZ (7-mo avg)	-0.31	-0.84	0.17	-1.62	0.29
		Exceeds in Base?	Yes	Yes	Yes	No	Yes
Park Code	SAGU	Exceeds if 5%, injury=5?	Yes	Yes	Yes	No	Yes
Primary State	AZ	Exceeds if 5% biosites?	Yes	Yes	Yes	Yes	Yes
Species in park?	Yes	Exceeds if 10% biosites?	Yes	Yes	Yes	No	Yes
# Monitors in park	1	Exceeds if 15% biosites?	Yes	Yes	Yes	No	Yes
Climate Region	Southwest	Exceeds if 20% biosites?	No	No	No	No	No
	Saint Croix National Scenic Riverway	W126 (ppm-hrs)	6.40	5.94	4.21	3.89	4.02
Park Name		PZ (7-mo avg)	-0.66	-0.03	0.18	-1.04	0.87
		Exceeds in Base?	No	No	No	No	No
Park Code	SACN	Exceeds if 5%, injury=5?	No	No	No	No	No
Primary State	WI	Exceeds if 5% biosites?	Yes	Yes	Yes	Yes	Yes
Species in park?	Yes	Exceeds if 10% biosites?	Yes	Yes	No	No	No
# Monitors in park	0	Exceeds if 15% biosites?	No	No	No	No	No
Climate Region	East North Central	Exceeds if 20% biosites?	No	No	No	No	No
	Salinas Pueblo Missions National Monument	W126 (ppm-hrs)	16.46	12.77	11.93	8.88	9.93
Park Name		PZ (7-mo avg)	0.03	0.62	-0.20	-1.50	-0.03
		Exceeds in Base?	Yes	Yes	Yes	No	No
Park Code	SAPU	Exceeds if 5%, injury=5?	Yes	Yes	No	No	No
Primary State	NM	Exceeds if 5% biosites?	Yes	Yes	Yes	Yes	Yes
Species in park?	Yes	Exceeds if 10% biosites?	Yes	Yes	Yes	No	Yes

Park info		Criteria	2006	2007	2008	2009	2010
# Monitors in park	0	Exceeds if 15% biosites?	Yes	Yes	Yes	No	Yes
Climate Region	Southwest	Exceeds if 20% biosites?	No	No	No	No	No
Park Name	Santa Monica Mountains National Recreation Area	W126 (ppm-hrs)	15.95	13.68	16.41	13.86	10.36
		PZ (7-mo avg)	0.03	-2.53	-1.19	-1.98	0.18
		Exceeds in Base?	Yes	Yes	Yes	Yes	No
Park Code	SAMO	Exceeds if 5%, injury=5?	Yes	No	Yes	No	No
Primary State	CA	Exceeds if 5% biosites?	Yes	Yes	Yes	Yes	Yes
Species in park?	Yes	Exceeds if 10% biosites?	Yes	No	Yes	No	Yes
# Monitors in park	0	Exceeds if 15% biosites?	Yes	No	Yes	No	Yes
Climate Region	West	Exceeds if 20% biosites?	No	No	No	No	No
Park Name	Saratoga National Historical Park	W126 (ppm-hrs)	5.96	9.14	7.78	5.24	6.07
		PZ (7-mo avg)	1.09	0.44	1.13	1.42	-0.90
		Exceeds in Base?	No	No	No	No	No
Park Code	SARA	Exceeds if 5%, injury=5?	No	No	Yes	No	No
Primary State	NY	Exceeds if 5% biosites?	Yes	Yes	Yes	Yes	Yes
Species in park?	Yes	Exceeds if 10% biosites?	Yes	Yes	Yes	Yes	Yes
# Monitors in park	1	Exceeds if 15% biosites?	Yes	Yes	Yes	Yes	No
Climate Region	Northeast	Exceeds if 20% biosites?	Yes	No	Yes	No	No
Park Name	Scotts Bluff National Monument	W126 (ppm-hrs)	19.08	15.43	11.45	8.08	10.16
		PZ (7-mo avg)	-1.40	-1.56	-0.17	1.94	1.95
		Exceeds in Base?	Yes	Yes	Yes	No	No
Park Code	SCBL	Exceeds if 5%, injury=5?	No	No	No	Yes	Yes
Primary State	NE	Exceeds if 5% biosites?	Yes	Yes	Yes	Yes	Yes
Species in park?	Yes	Exceeds if 10% biosites?	No	No	Yes	Yes	Yes
# Monitors in park	1	Exceeds if 15% biosites?	No	No	Yes	Yes	Yes
Climate Region	West North Central	Exceeds if 20% biosites?	No	No	No	Yes	Yes
Park Name	Sequoia-Kings Canyon National Park	W126 (ppm-hrs)	44.59	38.05	43.95	31.74	30.55
		PZ (7-mo avg)	-0.12	-1.72	-1.43	-1.18	-0.02
		Exceeds in Base?	Yes	Yes	Yes	Yes	Yes
Park Code	SEKI	Exceeds if 5%, injury=5?	Yes	No	No	Yes	Yes
Primary State	CA	Exceeds if 5% biosites?	Yes	Yes	Yes	Yes	Yes
Species in park?	Yes	Exceeds if 10% biosites?	Yes	Yes	Yes	Yes	Yes
# Monitors in park	2	Exceeds if 15% biosites?	Yes	No	No	Yes	Yes
Climate Region	West	Exceeds if 20% biosites?	No	No	No	No	No
Park Name	Shenandoah National Park	W126 (ppm-hrs)	13.11	11.74	10.07	5.62	9.46
		PZ (7-mo avg)	-0.23	-1.02	0.26	0.51	-0.98
		Exceeds in Base?	Yes	Yes	No	No	No
Park Code	SHEN	Exceeds if 5%, injury=5?	Yes	No	No	No	No
Primary State	VA	Exceeds if 5% biosites?	Yes	Yes	Yes	Yes	Yes
Species in park?	Yes	Exceeds if 10% biosites?	Yes	Yes	Yes	No	Yes
# Monitors in park	1	Exceeds if 15% biosites?	Yes	Yes	Yes	No	Yes
Climate Region	Southeast	Exceeds if 20% biosites?	No	No	No	No	No
Park Name	Shiloh National Military Park	W126 (ppm-hrs)	13.82	18.18	7.78	6.16	9.33
		PZ (7-mo avg)	-0.42	-2.13	-0.15	1.50	0.14
		Exceeds in Base?	Yes	Yes	No	No	No
Park Code	SHIL	Exceeds if 5%, injury=5?	Yes	No	No	No	No

Park info		Criteria	2006	2007	2008	2009	2010
Primary State	TN	Exceeds if 5% biosites?	Yes	Yes	Yes	Yes	Yes
Species in park?	Yes	Exceeds if 10% biosites?	Yes	No	Yes	Yes	Yes
# Monitors in park	0	Exceeds if 15% biosites?	Yes	No	No	Yes	Yes
Climate Region	Central	Exceeds if 20% biosites?	No	No	No	Yes	No
Park Name	Sleeping Bear Dunes National Lakeshore	W126 (ppm-hrs)	8.84	11.09	5.58	5.37	6.01
		PZ (7-mo avg)	-0.12	-1.20	0.55	0.22	0.04
		Exceeds in Base?	No	Yes	No	No	No
Park Code	SLBE	Exceeds if 5%, injury=5?	No	No	No	No	No
Primary State	MI	Exceeds if 5% biosites?	Yes	Yes	Yes	Yes	Yes
Species in park?	Yes	Exceeds if 10% biosites?	Yes	Yes	No	No	Yes
# Monitors in park	0	Exceeds if 15% biosites?	Yes	Yes	No	No	No
Climate Region	East North Central	Exceeds if 20% biosites?	No	No	No	No	No
Park Name	Stones River National Battlefield	W126 (ppm-hrs)	14.70	20.43	11.59	6.33	10.02
		PZ (7-mo avg)	-0.42	-2.13	-0.15	1.50	0.14
		Exceeds in Base?	Yes	Yes	No	No	No
Park Code	STRI	Exceeds if 5%, injury=5?	Yes	No	No	No	No
Primary State	TN	Exceeds if 5% biosites?	Yes	Yes	Yes	Yes	Yes
Species in park?	Yes	Exceeds if 10% biosites?	Yes	No	Yes	Yes	Yes
# Monitors in park	0	Exceeds if 15% biosites?	Yes	No	Yes	Yes	Yes
Climate Region	Central	Exceeds if 20% biosites?	No	No	No	Yes	No
Park Name	Sunset Crater Volcano National Monument	W126 (ppm-hrs)	22.26	18.36	21.05	10.72	13.38
		PZ (7-mo avg)	-0.54	-1.40	0.03	-1.48	0.39
		Exceeds in Base?	Yes	Yes	Yes	Yes	Yes
Park Code	SUCR	Exceeds if 5%, injury=5?	Yes	No	Yes	No	Yes
Primary State	AZ	Exceeds if 5% biosites?	Yes	Yes	Yes	Yes	Yes
Species in park?	Yes	Exceeds if 10% biosites?	Yes	No	Yes	No	Yes
# Monitors in park	0	Exceeds if 15% biosites?	Yes	No	Yes	No	Yes
Climate Region	Southwest	Exceeds if 20% biosites?	No	No	No	No	No
Park Name	Tallgrass Prairie National Preserve	W126 (ppm-hrs)	16.18	6.58	6.22	6.92	7.81
		PZ (7-mo avg)	-0.71	0.52	0.93	1.49	0.83
		Exceeds in Base?	Yes	No	No	No	No
Park Code	TAPR	Exceeds if 5%, injury=5?	Yes	No	No	No	No
Primary State	KS	Exceeds if 5% biosites?	Yes	Yes	Yes	Yes	Yes
Species in park?	Yes	Exceeds if 10% biosites?	Yes	Yes	Yes	Yes	Yes
# Monitors in park	0	Exceeds if 15% biosites?	Yes	No	No	Yes	No
Climate Region	South	Exceeds if 20% biosites?	No	No	No	Yes	No
Park Name	Theodore Roosevelt National Park	W126 (ppm-hrs)	8.34	5.68	5.99	4.13	4.58
		PZ (7-mo avg)	-0.55	-0.16	-1.71	0.93	2.08
		Exceeds in Base?	No	No	No	No	No
Park Code	THRO	Exceeds if 5%, injury=5?	No	No	No	No	No
Primary State	ND	Exceeds if 5% biosites?	Yes	Yes	No	Yes	Yes
Species in park?	Yes	Exceeds if 10% biosites?	Yes	No	No	No	Yes
# Monitors in park	2	Exceeds if 15% biosites?	Yes	No	No	No	No
Climate Region	West North Central	Exceeds if 20% biosites?	No	No	No	No	No
Park Name	Timpanogos Cave National Monument	W126 (ppm-hrs)	21.45	19.71	15.96	11.75	13.49
		PZ (7-mo avg)	0.01	-2.12	-0.69	0.62	-0.35

Park info		Criteria	2006	2007	2008	2009	2010
		Exceeds in Base?	Yes	Yes	Yes	Yes	Yes
Park Code	TICA	Exceeds if 5%, injury=5?	Yes	No	Yes	No	Yes
Primary State	UT	Exceeds if 5% biosites?	Yes	Yes	Yes	Yes	Yes
Species in park?	Yes	Exceeds if 10% biosites?	Yes	No	Yes	Yes	Yes
# Monitors in park	0	Exceeds if 15% biosites?	Yes	No	Yes	Yes	Yes
Climate Region	Southwest	Exceeds if 20% biosites?	No	No	No	No	No
Park Name	Timucuan Ecological And Historic Preserve	W126 (ppm-hrs)	13.28	11.63	7.84	5.48	6.67
		PZ (7-mo avg)	-2.28	-1.29	-0.16	0.79	-1.01
		Exceeds in Base?	Yes	Yes	No	No	No
Park Code	TIMU	Exceeds if 5%, injury=5?	No	No	No	No	No
Primary State	FL	Exceeds if 5% biosites?	Yes	Yes	Yes	Yes	Yes
Species in park?	Yes	Exceeds if 10% biosites?	No	No	Yes	No	Yes
# Monitors in park	0	Exceeds if 15% biosites?	No	No	No	No	No
Climate Region	Southeast	Exceeds if 20% biosites?	No	No	No	No	No
Park Name	Tonto National Monument	W126 (ppm-hrs)	24.01	21.11	24.48	13.32	16.29
		PZ (7-mo avg)	-1.33	-1.49	0.88	-0.67	0.21
		Exceeds in Base?	Yes	Yes	Yes	Yes	Yes
Park Code	TONT	Exceeds if 5%, injury=5?	No	No	Yes	Yes	Yes
Primary State	AZ	Exceeds if 5% biosites?	Yes	Yes	Yes	Yes	Yes
Species in park?	Yes	Exceeds if 10% biosites?	No	No	Yes	Yes	Yes
# Monitors in park	1	Exceeds if 15% biosites?	No	No	Yes	Yes	Yes
Climate Region	Southwest	Exceeds if 20% biosites?	No	No	No	No	No
Park Name	Tumacacori National Historical Park	W126 (ppm-hrs)	12.86	10.25	14.17	8.67	11.39
		PZ (7-mo avg)	-0.31	-0.84	0.17	-1.62	0.29
		Exceeds in Base?	Yes	No	Yes	No	Yes
Park Code	TUMA	Exceeds if 5%, injury=5?	Yes	No	Yes	No	No
Primary State	AZ	Exceeds if 5% biosites?	Yes	Yes	Yes	Yes	Yes
Species in park?	Yes	Exceeds if 10% biosites?	Yes	Yes	Yes	No	Yes
# Monitors in park	0	Exceeds if 15% biosites?	Yes	Yes	Yes	No	Yes
Climate Region	Southwest	Exceeds if 20% biosites?	No	No	No	No	No
Park Name	Tuzigoot National Monument	W126 (ppm-hrs)	22.90	18.12	20.55	9.75	12.22
		PZ (7-mo avg)	-1.45	-1.95	-0.31	-1.56	-0.31
		Exceeds in Base?	Yes	Yes	Yes	No	Yes
Park Code	TUZI	Exceeds if 5%, injury=5?	No	No	Yes	No	No
Primary State	AZ	Exceeds if 5% biosites?	Yes	Yes	Yes	Yes	Yes
Species in park?	Yes	Exceeds if 10% biosites?	No	No	Yes	No	Yes
# Monitors in park	0	Exceeds if 15% biosites?	No	No	Yes	No	Yes
Climate Region	Southwest	Exceeds if 20% biosites?	No	No	No	No	No
Park Name	Valley Forge National Historical Park	W126 (ppm-hrs)	17.35	16.86	17.33	6.10	15.54
		PZ (7-mo avg)	0.84	0.12	-0.10	1.02	-0.60
		Exceeds in Base?	Yes	Yes	Yes	No	Yes
Park Code	VAFO	Exceeds if 5%, injury=5?	Yes	Yes	Yes	No	Yes
Primary State	PA	Exceeds if 5% biosites?	Yes	Yes	Yes	Yes	Yes
Species in park?	Yes	Exceeds if 10% biosites?	Yes	Yes	Yes	Yes	Yes
# Monitors in park	0	Exceeds if 15% biosites?	Yes	Yes	Yes	Yes	Yes
Climate Region	Northeast	Exceeds if 20% biosites?	No	No	No	Yes	No

Park info		Criteria	2006	2007	2008	2009	2010
Park Name	Vanderbilt Mansion National Historic Site	W126 (ppm-hrs)	7.32	10.68	8.69	5.75	8.19
		PZ (7-mo avg)	1.09	0.44	1.13	1.42	-0.90
		Exceeds in Base?	No	Yes	No	No	No
Park Code	VAMA	Exceeds if 5%, injury=5?	Yes	No	Yes	No	No
Primary State	NY	Exceeds if 5% biosites?	Yes	Yes	Yes	Yes	Yes
Species in park?	Yes	Exceeds if 10% biosites?	Yes	Yes	Yes	Yes	Yes
# Monitors in park	0	Exceeds if 15% biosites?	Yes	Yes	Yes	Yes	Yes
Climate Region	Northeast	Exceeds if 20% biosites?	Yes	No	Yes	Yes	No
Park Name	Vicksburg National Military Park	W126 (ppm-hrs)	14.37	8.55	5.50	4.86	7.61
		PZ (7-mo avg)	-1.23	-0.81	1.12	0.02	-1.48
		Exceeds in Base?	Yes	No	No	No	No
Park Code	VICK	Exceeds if 5%, injury=5?	Yes	No	No	No	No
Primary State	MS	Exceeds if 5% biosites?	Yes	Yes	Yes	Yes	Yes
Species in park?	Yes	Exceeds if 10% biosites?	Yes	Yes	Yes	No	No
# Monitors in park	0	Exceeds if 15% biosites?	Yes	Yes	Yes	No	No
Climate Region	South	Exceeds if 20% biosites?	No	No	No	No	No
Park Name	Voyageurs National Park	W126 (ppm-hrs)	5.15	5.02	3.52	4.34	6.44
		PZ (7-mo avg)	-1.39	-0.23	0.41	-0.26	0.13
		Exceeds in Base?	No	No	No	No	No
Park Code	VOYA	Exceeds if 5%, injury=5?	No	No	No	No	No
Primary State	MN	Exceeds if 5% biosites?	No	Yes	Yes	Yes	Yes
Species in park?	Yes	Exceeds if 10% biosites?	No	No	No	No	Yes
# Monitors in park	1	Exceeds if 15% biosites?	No	No	No	No	No
Climate Region	East North Central	Exceeds if 20% biosites?	No	No	No	No	No
Park Name	Walnut Canyon National Monument	W126 (ppm-hrs)	22.79	18.83	21.78	11.04	13.39
		PZ (7-mo avg)	-0.54	-1.40	0.03	-1.48	0.39
		Exceeds in Base?	Yes	No	Yes	Yes	Yes
Park Code	WACA	Exceeds if 5%, injury=5?	Yes	No	Yes	No	Yes
Primary State	AZ	Exceeds if 5% biosites?	Yes	Yes	Yes	Yes	Yes
Species in park?	Yes	Exceeds if 10% biosites?	Yes	No	Yes	No	Yes
# Monitors in park	0	Exceeds if 15% biosites?	Yes	No	Yes	No	Yes
Climate Region	Southwest	Exceeds if 20% biosites?	No	No	No	No	No
Park Name	Whiskeytown-Shasta-Trinity National Recreation Area	W126 (ppm-hrs)	16.27	9.22	13.00	8.66	11.65
		PZ (7-mo avg)	0.41	-1.07	-1.98	0.14	0.59
		Exceeds in Base?	Yes	No	Yes	No	Yes
Park Code	WHIS	Exceeds if 5%, injury=5?	Yes	No	No	No	No
Primary State	CA	Exceeds if 5% biosites?	Yes	Yes	Yes	Yes	Yes
Species in park?	Yes	Exceeds if 10% biosites?	Yes	Yes	No	Yes	Yes
# Monitors in park	0	Exceeds if 15% biosites?	Yes	Yes	No	Yes	Yes
Climate Region	West	Exceeds if 20% biosites?	No	No	No	No	No
Park Name	White Sands National Mounument	W126 (ppm-hrs)	15.30	9.45	9.64	7.61	9.36
		PZ (7-mo avg)	1.31	0.72	0.56	-0.60	0.80
		Exceeds in Base?	Yes	No	No	No	No
Park Code	WHSA	Exceeds if 5%, injury=5?	Yes	No	No	No	No
Primary State	NM	Exceeds if 5% biosites?	Yes	Yes	Yes	Yes	Yes
Species in park?	Yes	Exceeds if 10% biosites?	Yes	Yes	Yes	Yes	Yes

Park info		Criteria	2006	2007	2008	2009	2010
# Monitors in park	0	Exceeds if 15% biosites?	Yes	Yes	Yes	No	Yes
Climate Region	Southwest	Exceeds if 20% biosites?	Yes	No	No	No	No
Park Name	Wilson's Creek National Battlefield	W126 (ppm-hrs)	15.89	10.47	7.11	5.97	7.96
		PZ (7-mo avg)	-0.49	0.24	2.81	1.00	0.23
		Exceeds in Base?	Yes	Yes	No	No	No
Park Code	WICR	Exceeds if 5%, injury=5?	Yes	No	Yes	No	No
Primary State	MO	Exceeds if 5% biosites?	Yes	Yes	Yes	Yes	Yes
Species in park?	Yes	Exceeds if 10% biosites?	Yes	Yes	Yes	Yes	Yes
# Monitors in park	0	Exceeds if 15% biosites?	Yes	Yes	Yes	No	No
Climate Region	Central	Exceeds if 20% biosites?	No	No	Yes	No	No
Park Name	Wind Cave National Park	W126 (ppm-hrs)	18.06	10.88	5.85	5.36	5.46
		PZ (7-mo avg)	-0.28	-0.94	1.50	1.59	1.57
		Exceeds in Base?	Yes	Yes	No	No	No
Park Code	WICA	Exceeds if 5%, injury=5?	Yes	No	No	No	No
Primary State	SD	Exceeds if 5% biosites?	Yes	Yes	Yes	Yes	Yes
Species in park?	Yes	Exceeds if 10% biosites?	Yes	Yes	Yes	Yes	Yes
# Monitors in park	1	Exceeds if 15% biosites?	Yes	Yes	Yes	Yes	Yes
Climate Region	West North Central	Exceeds if 20% biosites?	No	No	Yes	No	No
Park Name	Wright Brothers National Memorial	W126 (ppm-hrs)	12.57	12.38	13.19	5.11	10.53
		PZ (7-mo avg)	1.15	-1.60	-0.77	-0.02	-0.43
		Exceeds in Base?	Yes	Yes	Yes	No	Yes
Park Code	WRBR	Exceeds if 5%, injury=5?	Yes	No	Yes	No	No
Primary State	NC	Exceeds if 5% biosites?	Yes	Yes	Yes	Yes	Yes
Species in park?	0	Exceeds if 10% biosites?	Yes	No	Yes	No	Yes
# Monitors in park	0	Exceeds if 15% biosites?	Yes	No	Yes	No	Yes
Climate Region	Southeast	Exceeds if 20% biosites?	Yes	No	No	No	No
Park Name	Wupatki National Monument	W126 (ppm-hrs)	21.67	17.82	19.57	10.22	13.14
		PZ (7-mo avg)	-0.54	-1.40	0.03	-1.48	0.39
		Exceeds in Base?	Yes	Yes	Yes	No	Yes
Park Code	WUPA	Exceeds if 5%, injury=5?	Yes	No	Yes	No	Yes
Primary State	AZ	Exceeds if 5% biosites?	Yes	Yes	Yes	Yes	Yes
Species in park?	Yes	Exceeds if 10% biosites?	Yes	No	Yes	No	Yes
# Monitors in park	0	Exceeds if 15% biosites?	Yes	No	Yes	No	Yes
Climate Region	Southwest	Exceeds if 20% biosites?	No	No	No	No	No
Park Name	Yellowstone National Park	W126 (ppm-hrs)	11.63	9.75	9.43	7.11	9.25
		PZ (7-mo avg)	-1.43	-1.50	-0.24	0.57	0.19
		Exceeds in Base?	Yes	No	No	No	No
Park Code	YELL	Exceeds if 5%, injury=5?	No	No	No	No	No
Primary State	WY	Exceeds if 5% biosites?	Yes	Yes	Yes	Yes	Yes
Species in park?	Yes	Exceeds if 10% biosites?	No	No	Yes	Yes	Yes
# Monitors in park	1	Exceeds if 15% biosites?	No	No	Yes	No	Yes
Climate Region	West North Central	Exceeds if 20% biosites?	No	No	No	No	No
Park Name	Yosemite National Park	W126 (ppm-hrs)	29.22	26.24	33.88	21.64	20.68
		PZ (7-mo avg)	0.71	-1.38	-1.47	-0.21	0.35
		Exceeds in Base?	Yes	Yes	Yes	Yes	Yes
Park Code	YOSE	Exceeds if 5%, injury=5?	Yes	No	No	Yes	Yes

Park info		Criteria	2006	2007	2008	2009	2010
Primary State	CA	Exceeds if 5% biosites?	Yes	Yes	Yes	Yes	Yes
Species in park?	Yes	Exceeds if 10% biosites?	Yes	Yes	Yes	Yes	Yes
# Monitors in park	5	Exceeds if 15% biosites?	Yes	No	No	Yes	Yes
Climate Region	West	Exceeds if 20% biosites?	No	No	No	No	No
		W126 (ppm-hrs)	21.68	17.57	17.66	12.78	16.92
Park Name	Zion National Park	PZ (7-mo avg)	-0.37	-1.40	-0.70	-1.43	0.16
		Exceeds in Base?	Yes	Yes	Yes	Yes	Yes
Park Code	ZION	Exceeds if 5%, injury=5?	Yes	No	Yes	No	Yes
Primary State	UT	Exceeds if 5% biosites?	Yes	Yes	Yes	Yes	Yes
Species in park?	Yes	Exceeds if 10% biosites?	Yes	No	Yes	No	Yes
# Monitors in park	0	Exceeds if 15% biosites?	Yes	No	Yes	No	Yes
Climate Region	Southwest	Exceeds if 20% biosites?	No	No	No	No	No

Table 7A-5: O₃ Estimates at 214 Parks for Recent Conditions and Just Meeting Alternative W126 Standard Levels

Park Name	2006-2008	Just meet 75ppb	15 ppm-hrs	11 ppm-hrs	7 ppm-hrs
Acadia National Park	6.01	0.30	0.30	0.30	0.30
Agate Fossil Beds National Monument	15.24	5.65	5.00	4.21	3.36
Alibates Flint Quarries National Monument	12.66	2.70	2.22	1.62	1.46
Allegheny Portage Railroad National Historic Site	10.88	0.41	0.41	0.36	0.32
Amistad National Recreation Area	8.35	1.66	1.59	1.51	1.49
Andersonville National Historic Site	13.03	2.27	2.27	2.10	1.38
Antietam National Battlefield	13.80	1.71	1.71	1.32	0.82
Apostle Islands National Lakeshore	4.53	0.81	0.81	0.81	0.79
Appomattox Court House National Historical Park	10.70	1.71	1.71	1.59	1.07
Arches National Park	16.69	4.51	3.65	2.52	2.19
Arkansas Post National Memorial	12.33	1.78	1.78	1.78	1.78
Assateague Island National Seashore	16.23	1.07	1.07	1.02	0.80
Aztec Ruins National Monument	12.06	4.17	3.42	2.37	1.95
Badlands National Park	10.14	4.77	4.62	4.44	3.58
Bandelier National Monument	12.22	3.23	2.55	1.67	1.44
Bandelier Wilderness	12.18	3.15	2.49	1.64	1.41
Bent's Old Fort National Historic Site	15.59	3.91	3.01	1.90	1.56
Big Bend National Park	10.30	1.64	1.54	1.41	1.38
Big Cypress National Preserve	6.48	2.51	2.51	2.42	1.98
Big Hole National Battlefield	9.82	2.41	2.20	1.94	1.78
Big South Fork National River and Recreation Area	12.57	3.37	3.37	2.32	1.36
Big Thicket National Preserve	9.22	2.17	2.17	2.17	2.17
Bighorn Canyon National Recreation Area	9.59	2.14	2.14	2.14	1.79
Biscayne National Park	6.50	3.20	3.20	3.10	2.63
Black Canyon of the Gunnison National Park	15.27	3.59	2.91	2.02	1.76
Blue Ridge Parkway	12.07	2.06	2.06	1.87	1.23
Bluestone National Scenic River	11.33	2.05	2.05	1.55	0.98
Booker T. Washington National Monument	12.58	1.74	1.74	1.61	1.08
Bryce Canyon National Park	18.13	4.72	3.90	2.79	2.43
Buffalo National River	9.70	2.30	2.30	2.09	1.88
Canaveral National Seashore	9.41	2.84	2.84	2.71	2.08
Canyon de Chelly National Monument	15.18	5.00	3.96	2.57	2.11
Canyonlands National Park	17.05	4.68	3.80	2.63	2.27
Cape Cod National Seashore	11.25	0.37	0.37	0.37	0.36
Cape Lookout National Seashore	10.46	2.10	2.10	1.99	1.45
Capitol Reef National Park	17.83	4.88	4.02	2.85	2.48
Capulin Volcano National Monument	14.10	3.34	2.61	1.70	1.47
Carlsbad Caverns National Park	13.41	4.54	3.72	2.67	2.42

Park Name	2006-2008	Just meet 75ppb	15 ppm-hrs	11 ppm-hrs	7 ppm-hrs
Casa Grande Ruins National Monument	15.01	5.22	3.73	2.06	1.67
Catoctin Mountain Park	14.49	0.32	0.32	0.32	0.32
Cedar Breaks National Monument	18.46	4.14	3.45	2.53	2.25
Chaco Culture National Historical Park	13.66	3.92	3.16	2.13	1.74
Channel Islands National Park	5.58	1.13	1.06	0.99	0.91
Chattahoochee River National Recreation Area	20.74	4.97	4.97	4.63	3.09
Chesapeake and Ohio Canal National Historical Park	13.40	2.01	2.01	1.71	1.10
Chickamauga and Chattanooga National Military Park	17.29	5.08	5.08	3.74	2.20
Chickasaw National Recreation Area	12.24	1.81	1.81	1.81	1.81
Chiricahua National Monument Wilderness	14.85	5.64	4.72	3.48	3.16
City of Rocks National Reserve	17.60	3.64	2.91	1.97	1.71
Colonial National Historical Park	15.34	2.55	2.55	2.38	1.63
Colorado National Monument	16.06	4.27	3.46	2.39	2.07
Congaree National Park	11.64	1.52	1.52	1.41	0.93
Cowpens National Battlefield	13.25	1.95	1.95	1.81	1.20
Crater Lake National Park	6.97	2.06	2.00	1.92	1.88
Craters of the Moon National Monument	13.87	3.33	2.82	2.22	1.96
Cumberland Gap National Historical Park	11.76	2.95	2.95	2.02	1.20
Cumberland Island National Seashore	9.22	0.82	0.82	0.78	0.59
Curecanti National Recreation Area	15.41	3.55	2.81	1.86	1.62
Cuyahoga Valley National Park	11.13	3.40	3.40	2.47	1.53
Death Valley National Park	29.76	2.72	2.53	2.32	2.10
Delaware Water Gap National Recreation Area	11.75	0.26	0.26	0.26	0.26
Devil's Tower National Monument	11.94	3.40	3.40	3.40	2.78
Dinosaur National Monument	16.24	4.51	3.78	2.79	2.34
Ebey's Landing National Historical Reserve	1.54	1.06	1.06	1.06	1.06
Effigy Mounds National Monument	6.68	0.97	0.97	0.97	0.97
Eisenhower National Historic Site	13.47	0.38	0.38	0.38	0.38
El Malpais National Monument	13.47	3.69	2.93	1.92	1.60
El Morro National Monument	13.61	3.87	3.08	2.02	1.68
Eleanor Roosevelt National Historic Site	8.57	0.03	0.03	0.03	0.03
Eleven Point National Wild and Scenic River	14.09	3.53	3.53	2.68	1.97
Everglades National Park	6.27	2.77	2.77	2.68	2.23
Fire Island National Seashore	14.98	0.47	0.47	0.47	0.47
Florissant Fossil Beds National Monument	19.52	5.65	4.01	2.22	1.73
Fort Bowie National Historic Site	16.76	6.26	5.13	3.64	3.25
Fort Davis National Historic Site	12.08	3.26	2.80	2.22	2.08
Fort Donelson National Battlefield	14.76	2.49	2.49	1.65	0.93
Fort Laramie National Historic Site	16.41	5.69	4.91	3.98	3.17
Fort Larned National Historic Site	10.16	1.78	1.75	1.70	1.69

Park Name	2006-2008	Just meet 75ppb	15 ppm-hrs	11 ppm-hrs	7 ppm-hrs
Fort Necessity National Battlefield	10.70	1.46	1.46	1.08	0.72
Fort Pulaski National Monument	6.86	2.06	2.06	1.98	1.59
Fort Union National Monument	10.80	3.35	2.87	2.26	1.92
Fort Washington Park	19.34	5.78	5.78	5.37	3.43
Fossil Butte National Monument	16.79	4.48	3.74	2.72	2.27
Fredericksburg and Spotsylvania Co. Battlefields Memorial National Military Park	11.79	2.22	2.22	2.05	1.35
Friendship Hill National Historic Site	11.47	1.98	1.98	1.44	0.93
Gateway National Recreation Area	12.34	0.45	0.45	0.45	0.45
Gauley River National Recreation Area	10.22	2.54	2.54	1.69	0.87
George Washington Memorial Parkway	17.34	4.18	4.18	3.86	2.43
Gettysburg National Military Park	13.75	0.39	0.39	0.39	0.39
Gila Cliff Dwellings National Monument	12.74	3.92	3.16	2.16	1.88
Glacier National Park	3.34	0.68	0.68	0.68	0.61
Glen Canyon National Recreation Area	17.59	5.05	4.11	2.84	2.43
Golden Gate National Recreation Area	1.61	0.59	0.56	0.53	0.49
Golden Spike National Historic Site	19.07	4.22	3.29	2.10	1.78
Grand Canyon National Park	17.88	4.96	4.09	2.92	2.56
Grand Teton National Park	11.67	2.68	2.50	2.27	1.95
Grant-Kohrs Ranch National Historic Site	7.14	2.46	2.24	1.97	1.72
Great Basin National Park	16.44	3.58	3.40	3.16	3.00
Great Sand Dunes National Park	14.31	3.20	2.51	1.65	1.43
Great Smoky Mountains National Park	14.65	3.35	3.35	2.43	1.48
Green Springs National Historic Landmark District	11.40	1.96	1.96	1.82	1.20
Greenbelt Park	19.98	5.84	5.84	5.44	3.54
Guadalupe Mountains National Park	12.69	4.10	3.53	2.79	2.61
Guilford Courthouse National Military Park	19.02	1.39	1.39	1.26	0.73
Gulf Islands National Seashore	15.23	5.01	5.01	4.66	3.09
Hagerman Fossil Beds National Monument	15.56	3.28	2.74	2.07	1.84
Harpers Ferry National Historical Park	13.95	1.91	1.91	1.57	1.03
Hohokam Pima National Monument	13.45	4.88	3.51	1.91	1.52
Home of F. D. Roosevelt National Historic Site	8.66	0.03	0.03	0.03	0.03
Hopewell Culture National Historical Park	13.87	3.43	3.43	2.50	1.58
Hopewell Furnace National Historic Park	16.42	0.30	0.30	0.30	0.30
Horseshoe Bend National Military Park	13.81	3.50	3.50	3.24	2.09
Hot Springs National Park	8.16	1.59	1.59	1.59	1.59
Hovenweep National Monument	16.40	5.23	4.22	2.85	2.37
Indiana Dunes National Lakeshore	8.35	3.87	3.87	3.05	2.17
Isle Royale National Park	5.45	0.91	0.91	0.91	0.89
Jean Lafitte National Historical Park and Preserve	12.87	3.36	3.36	3.34	3.28
Jewel Cave National Monument	11.99	6.33	6.33	6.33	5.05

Park Name	2006-2008	Just meet 75ppb	15 ppm-hrs	11 ppm-hrs	7 ppm-hrs
John D. Rockefeller Jr. Memorial Parkway	11.17	2.70	2.55	2.36	2.03
John Day Fossil Beds National Monument	5.58	2.28	2.21	2.13	2.09
John Muir National Historic Site	4.55	1.19	1.07	0.94	0.81
Joshua Tree National Park	29.01	2.93	2.58	2.16	1.86
Kennesaw Mountain National Battlefield Park	21.03	4.64	4.64	4.33	2.88
Kings Mountain National Military Park	14.46	1.20	1.20	1.12	0.74
Knife River Indian Villages National Historic Site	4.62	2.47	2.47	2.47	2.16
Lake Mead National Recreation Area	19.19	3.15	2.74	2.22	1.99
Lake Meredith National Recreation Area	12.67	2.70	2.22	1.62	1.46
Lake Roosevelt National Recreation Area	4.32	1.37	1.37	1.37	1.36
Lassen Volcanic National Park	17.83	1.80	1.70	1.57	1.45
Lava Beds National Monument	9.96	1.84	1.75	1.64	1.57
Little Bighorn Battlefield National Monument	9.69	2.24	2.24	2.24	1.87
Little River Canyon National Preserve	15.52	3.49	3.49	3.11	1.97
Lyndon B. Johnson National Historical Park	8.13	2.02	2.00	1.98	1.98
Mammoth Cave National Park	14.48	3.53	3.53	2.38	1.38
Manassas National Battlefield Park	12.99	3.20	3.20	2.98	2.03
Manzanar National Historic Site	42.52	1.95	1.78	1.58	1.39
Marsh-Billings-Rockefeller National Historical Park	5.38	0.38	0.38	0.38	0.38
Mesa Verde National Park	16.40	5.44	4.28	2.78	2.34
Minute Man National Historical Park	7.92	0.26	0.26	0.26	0.26
Mississippi National River And Recreation Area	6.56	0.86	0.86	0.86	0.86
Missouri National Recreational River	6.10	2.02	1.96	1.89	1.70
Mojave National Preserve	26.41	3.24	2.86	2.41	2.12
Monocacy National Battlefield	16.06	1.17	1.17	1.01	0.69
Montezuma Castle National Monument	21.85	9.96	7.20	3.75	2.85
Morristown National Historical Park	17.40	0.35	0.35	0.35	0.35
Mount Rainier Wilderness	4.34	1.76	1.76	1.76	1.76
Mount Rushmore National Memorial	11.52	6.01	6.01	6.01	4.79
Muir Woods National Monument	1.25	0.64	0.62	0.60	0.57
Natchez Trace Parkway	11.40	1.57	1.57	1.50	1.38
National Mall	19.47	8.26	8.26	7.66	4.80
Natural Bridges National Monument	17.09	5.06	4.12	2.84	2.40
Navajo National Monument	17.27	6.06	4.75	3.04	2.51
New River Gorge National River	10.84	2.21	2.21	1.53	0.85
Nez Perce National Historical Park	6.08	2.13	2.00	1.84	1.76
Ninety Six National Historic Site	14.17	1.31	1.31	1.21	0.80
North Cascades National Park	2.39	1.24	1.24	1.24	1.23
Ocmulgee National Monument	16.13	3.22	3.22	2.88	1.60
Olympic National Park	1.77	1.13	1.13	1.13	1.13

Park Name	2006-2008	Just meet 75ppb	15 ppm-hrs	11 ppm-hrs	7 ppm-hrs
Oregon Caves National Monument	6.35	1.44	1.42	1.40	1.37
Organ Pipe Cactus National Monument	11.69	4.10	3.29	2.28	1.98
Ozark National Scenic Riverways	13.82	3.51	3.51	2.59	1.86
Padre Island National Seashore	4.49	1.04	1.04	1.04	1.04
Palo Alto Battlefield National Historic Site	3.26	0.88	0.88	0.88	0.88
Pea Ridge National Military Park	10.33	1.63	1.63	1.46	1.28
Pecos National Historical Park	12.59	3.35	2.63	1.70	1.46
Petersburg National Battlefield	14.52	2.03	2.03	1.86	1.18
Petrified Forest National Park	16.92	5.43	4.17	2.55	2.07
Petroglyph National Monument	14.55	3.36	2.56	1.56	1.30
Pictured Rocks National Lakeshore	6.47	1.07	1.07	1.07	1.06
Pinnacles National Monument	14.03	0.86	0.79	0.72	0.65
Pipe Spring National Monument	17.49	4.01	3.34	2.44	2.17
Pipestone National Monument	4.94	1.57	1.57	1.57	1.44
Piscataway Park	18.55	5.66	5.66	5.26	3.36
Point Reyes National Seashore	1.50	0.42	0.41	0.40	0.39
Poverty Point National Monument	7.72	0.80	0.80	0.80	0.80
Prince William Forest Park	13.75	3.15	3.15	2.92	1.88
Rainbow Bridge National Monument	17.41	5.48	4.39	2.94	2.48
Redwood National Park	6.63	1.25	1.22	1.18	1.15
Richmond National Battlefield Park	17.04	2.83	2.83	2.55	1.39
Rock Creek Park	19.20	5.99	5.99	5.55	3.54
Rocky Mountain National Park	18.55	7.00	5.24	3.06	2.43
Saguaro National Park	14.34	6.68	5.15	3.25	2.75
Saint Croix National Scenic Riverway	5.48	0.79	0.79	0.79	0.78
Salinas Pueblo Missions National Monument	11.55	3.28	2.60	1.71	1.48
San Juan Island National Historical Park	1.32	0.89	0.89	0.89	0.89
Santa Monica Mountains National Recreation Area	15.19	3.22	2.63	2.02	1.49
Saratoga National Historical Park	7.08	0.05	0.05	0.05	0.05
Scotts Bluff National Monument	16.07	5.42	4.60	3.61	2.87
Sequoia-Kings Canyon National Park	43.35	1.98	1.74	1.48	1.24
Shenandoah National Park	11.00	1.77	1.77	1.62	1.09
Shiloh National Military Park	14.29	2.59	2.59	2.00	1.44
Sleeping Bear Dunes National Lakeshore	8.23	1.46	1.46	1.46	1.46
Stones River National Battlefield	15.27	3.89	3.89	2.69	1.58
Sunset Crater Volcano National Monument	19.34	7.42	5.65	3.40	2.77
Tallgrass Prairie National Preserve	9.22	1.70	1.70	1.70	1.70
Theodore Roosevelt National Park	6.09	3.45	3.45	3.45	2.86
Timpanogos Cave National Monument	19.03	5.90	4.70	3.01	2.47
Timucuan Ecological And Historic Preserve	10.40	0.30	0.30	0.29	0.21

Park Name	2006-2008	Just meet 75ppb	15 ppm-hrs	11 ppm-hrs	7 ppm-hrs
Tonto National Monument	22.44	8.54	5.93	2.91	2.25
Tumacacori National Historical Park	11.85	5.02	4.13	2.97	2.64
Tuzigoot National Monument	19.66	9.02	6.62	3.59	2.76
Valley Forge National Historical Park	17.76	0.29	0.29	0.29	0.29
Vanderbilt Mansion National Historic Site	8.75	0.04	0.04	0.04	0.04
Vicksburg National Military Park	8.84	0.96	0.96	0.96	0.96
Voyageurs National Park	4.54	0.96	0.96	0.96	0.90
Walnut Canyon National Monument	19.87	7.58	5.73	3.39	2.74
Whiskeytown-Shasta-Trinity National Recreation Area	12.67	1.03	0.95	0.87	0.78
White Sands National Monument	11.00	3.83	3.08	2.11	1.85
Wilson's Creek National Battlefield	10.66	2.90	2.90	2.18	1.61
Wind Cave National Park	12.47	8.13	8.11	8.08	6.44
Wupatki National Monument	18.61	7.14	5.49	3.37	2.78
Yellowstone National Park	9.99	2.64	2.56	2.45	2.10
Yosemite National Park	29.42	1.67	1.54	1.39	1.25
Zion National Park	18.70	3.79	3.13	2.26	1.99

Figures 7A-12 through 7-18 provide information regarding the geographic distribution of the 214 parks in this screening assessment. Figure 7A-12 provides a pie chart of the breakdown of all 214 parks into the 9 NOAA climate regions. Figures 7A-13 through 7A-18 provides the geographic distribution of the results of the 6 scenarios compared to the geographic breakdown of all 214 parks for all 5 years, at least 4 years, at least 3 years, at least 2 years, at least 1 year, and no years, respectively.

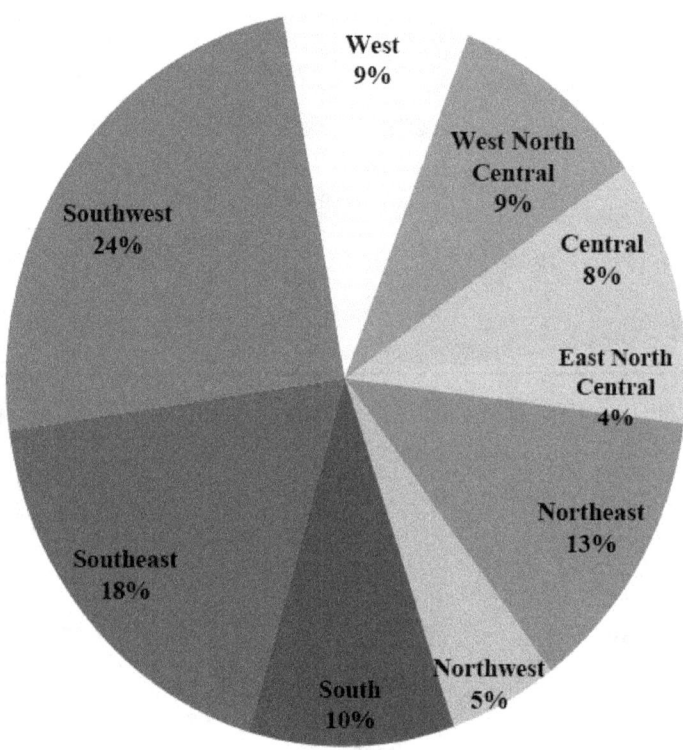

Figure 7A-12 Breakdown of 214 Parks by Climate Region

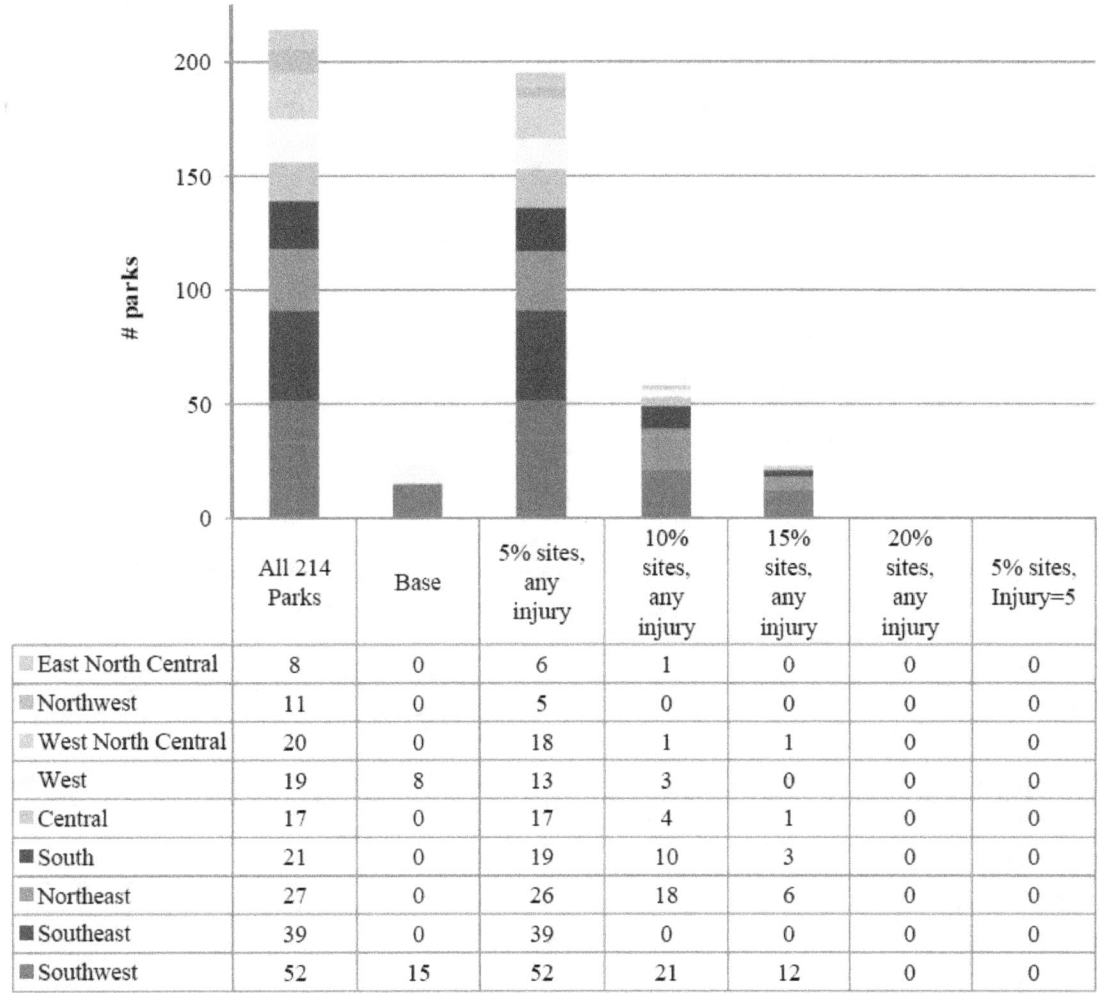

	All 214 Parks	Base	5% sites, any injury	10% sites, any injury	15% sites, any injury	20% sites, any injury	5% sites, Injury=5
East North Central	8	0	6	1	0	0	0
Northwest	11	0	5	0	0	0	0
West North Central	20	0	18	1	1	0	0
West	19	8	13	3	0	0	0
Central	17	0	17	4	1	0	0
South	21	0	19	10	3	0	0
Northeast	27	0	26	18	6	0	0
Southeast	39	0	39	0	0	0	0
Southwest	52	15	52	21	12	0	0

Scenario

Figure 7A-13 Parks Exceeding Benchmark Criteria for all 5 years (2006-2010) by Scenario and Climate Region

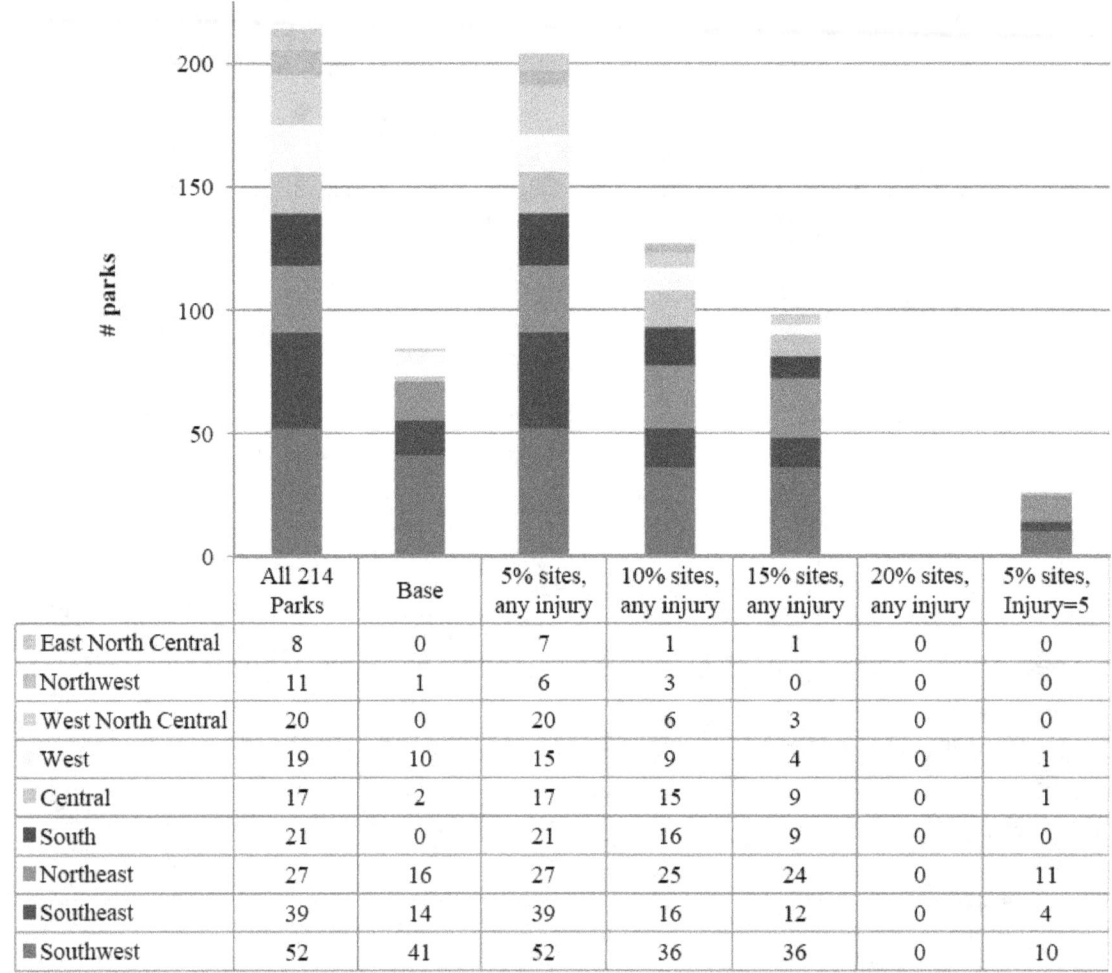

	All 214 Parks	Base	5% sites, any injury	10% sites, any injury	15% sites, any injury	20% sites, any injury	5% sites, Injury=5
East North Central	8	0	7	1	1	0	0
Northwest	11	1	6	3	0	0	0
West North Central	20	0	20	6	3	0	0
West	19	10	15	9	4	0	1
Central	17	2	17	15	9	0	1
South	21	0	21	16	9	0	0
Northeast	27	16	27	25	24	0	11
Southeast	39	14	39	16	12	0	4
Southwest	52	41	52	36	36	0	10

Scenario

Figure 7A-14 Parks Exceeding Benchmark Criteria for at least 4 years (2006-2010) by Scenario and Climate Region

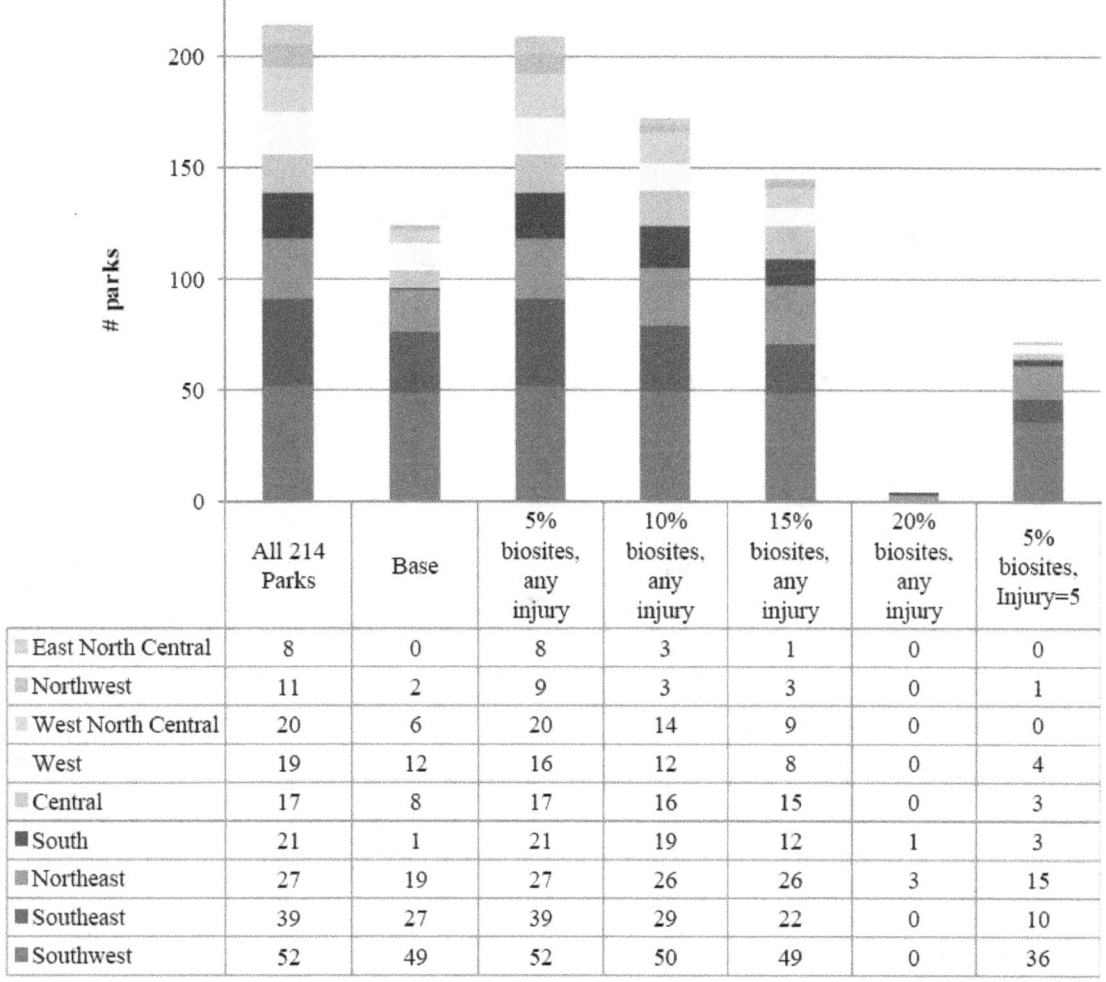

	All 214 Parks	Base	5% biosites, any injury	10% biosites, any injury	15% biosites, any injury	20% biosites, any injury	5% biosites, Injury=5
East North Central	8	0	8	3	1	0	0
Northwest	11	2	9	3	3	0	1
West North Central	20	6	20	14	9	0	0
West	19	12	16	12	8	0	4
Central	17	8	17	16	15	0	3
South	21	1	21	19	12	1	3
Northeast	27	19	27	26	26	3	15
Southeast	39	27	39	29	22	0	10
Southwest	52	49	52	50	49	0	36

Scenario

Figure 7A-15 Parks Exceeding Benchmark Criteria for at least 3 years (2006-2010) by Scenario and Climate Region

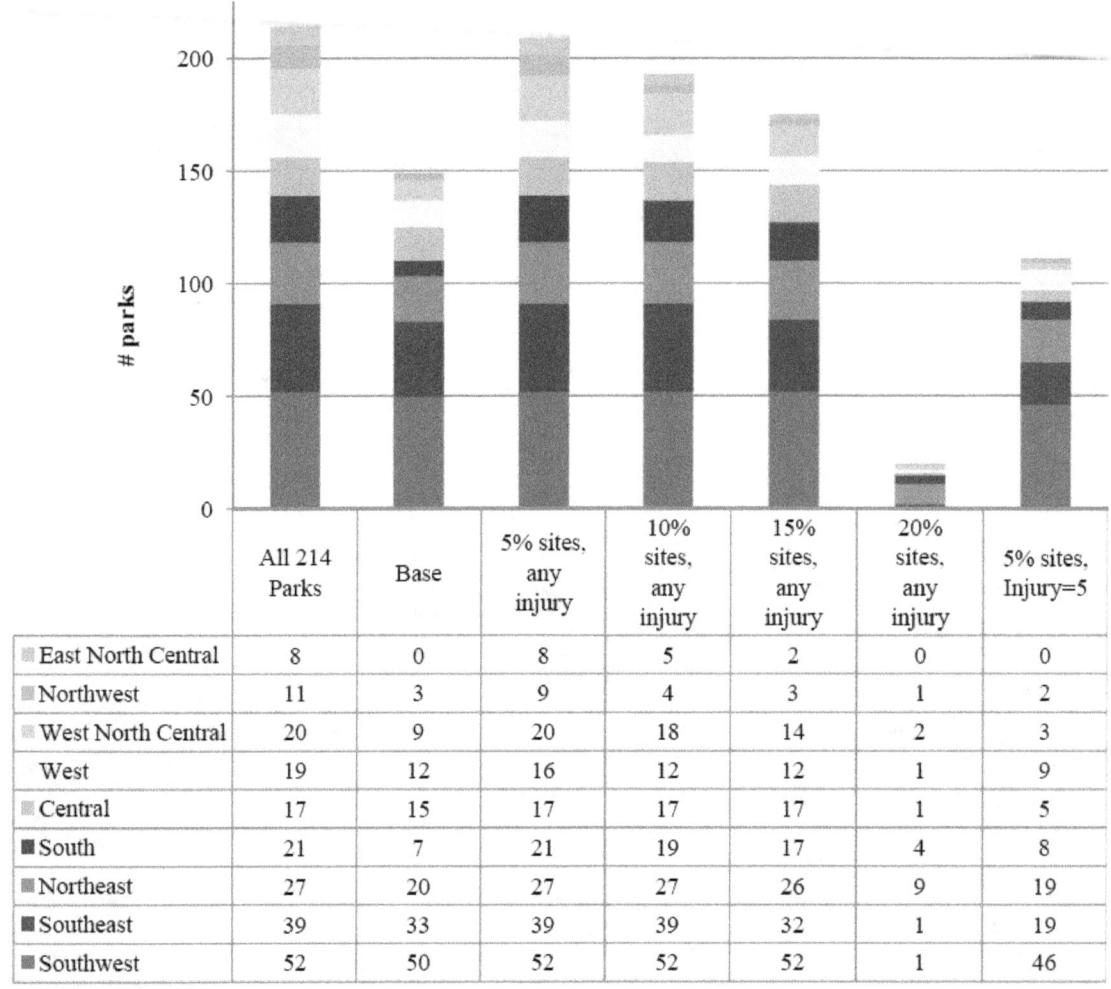

	All 214 Parks	Base	5% sites, any injury	10% sites, any injury	15% sites, any injury	20% sites, any injury	5% sites, Injury=5
East North Central	8	0	8	5	2	0	0
Northwest	11	3	9	4	3	1	2
West North Central	20	9	20	18	14	2	3
West	19	12	16	12	12	1	9
Central	17	15	17	17	17	1	5
South	21	7	21	19	17	4	8
Northeast	27	20	27	27	26	9	19
Southeast	39	33	39	39	32	1	19
Southwest	52	50	52	52	52	1	46

Scenario

Figure 7A-16 Parks Exceeding Benchmark Criteria for at least 2 years (2006-2010) by Scenario and Climate Region

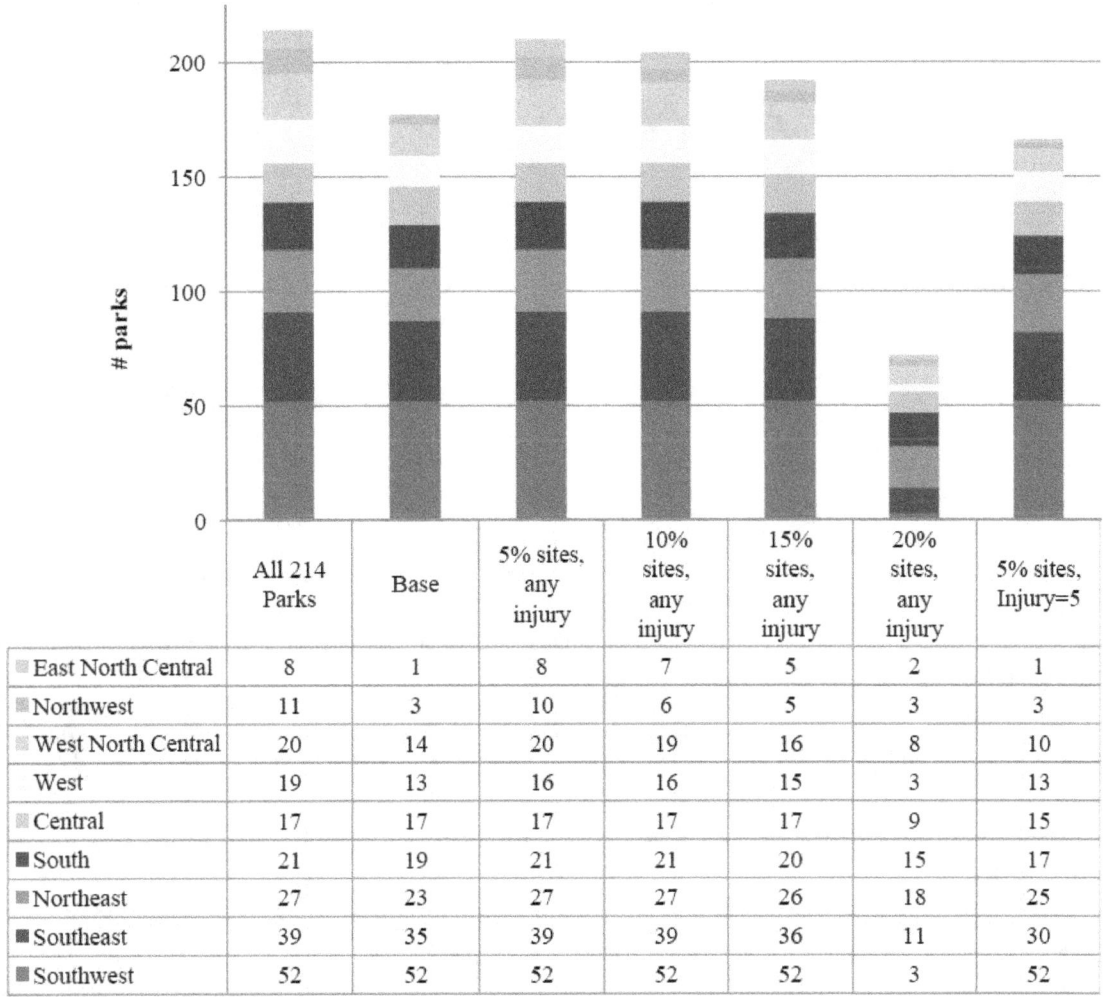

	All 214 Parks	Base	5% sites, any injury	10% sites, any injury	15% sites, any injury	20% sites, any injury	5% sites, Injury=5
East North Central	8	1	8	7	5	2	1
Northwest	11	3	10	6	5	3	3
West North Central	20	14	20	19	16	8	10
West	19	13	16	16	15	3	13
Central	17	17	17	17	17	9	15
South	21	19	21	21	20	15	17
Northeast	27	23	27	27	26	18	25
Southeast	39	35	39	39	36	11	30
Southwest	52	52	52	52	52	3	52

Scenario

Figure 7A-17 Parks Exceeding Benchmark Criteria for at least 1 year (2006-2010) by Scenario and Climate Region

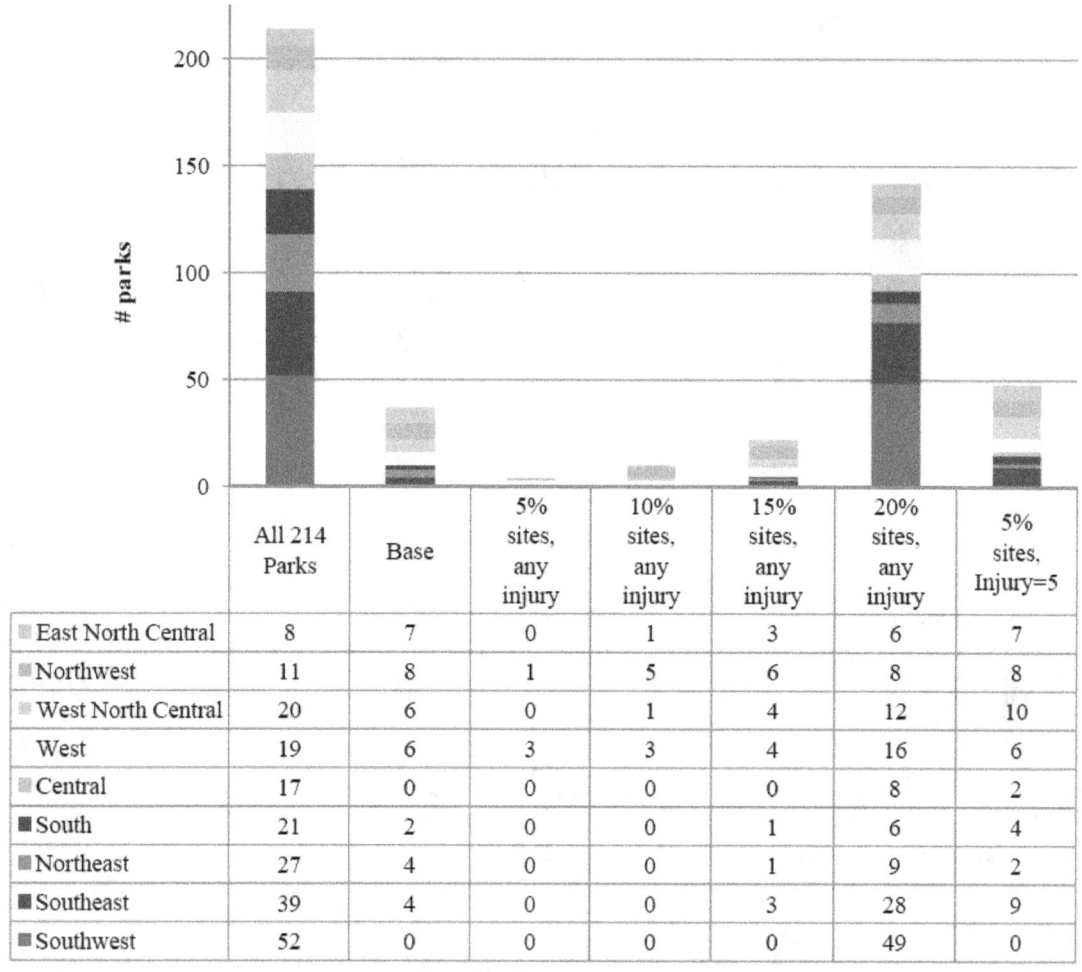

	All 214 Parks	Base	5% sites, any injury	10% sites, any injury	15% sites, any injury	20% sites, any injury	5% sites, Injury=5
East North Central	8	7	0	1	3	6	7
Northwest	11	8	1	5	6	8	8
West North Central	20	6	0	1	4	12	10
West	19	6	3	3	4	16	6
Central	17	0	0	0	0	8	2
South	21	2	0	0	1	6	4
Northeast	27	4	0	0	1	9	2
Southeast	39	4	0	0	3	28	9
Southwest	52	0	0	0	0	49	0

Scenario

Figure 7A-18 Parks Exceeding Benchmark Criteria for no years (2006-2010) by Scenario and Climate Region

APPENDIX 7B:
NATIONAL PARKS CASE STUDY LARGE SCALE MAPS

Canopy

No Data
0%
<20%
20%-40%
40%-60%
60%-80%
>80%

Figure 7B-1 GRSM Canopy Sensitive Species Cover

7B-2

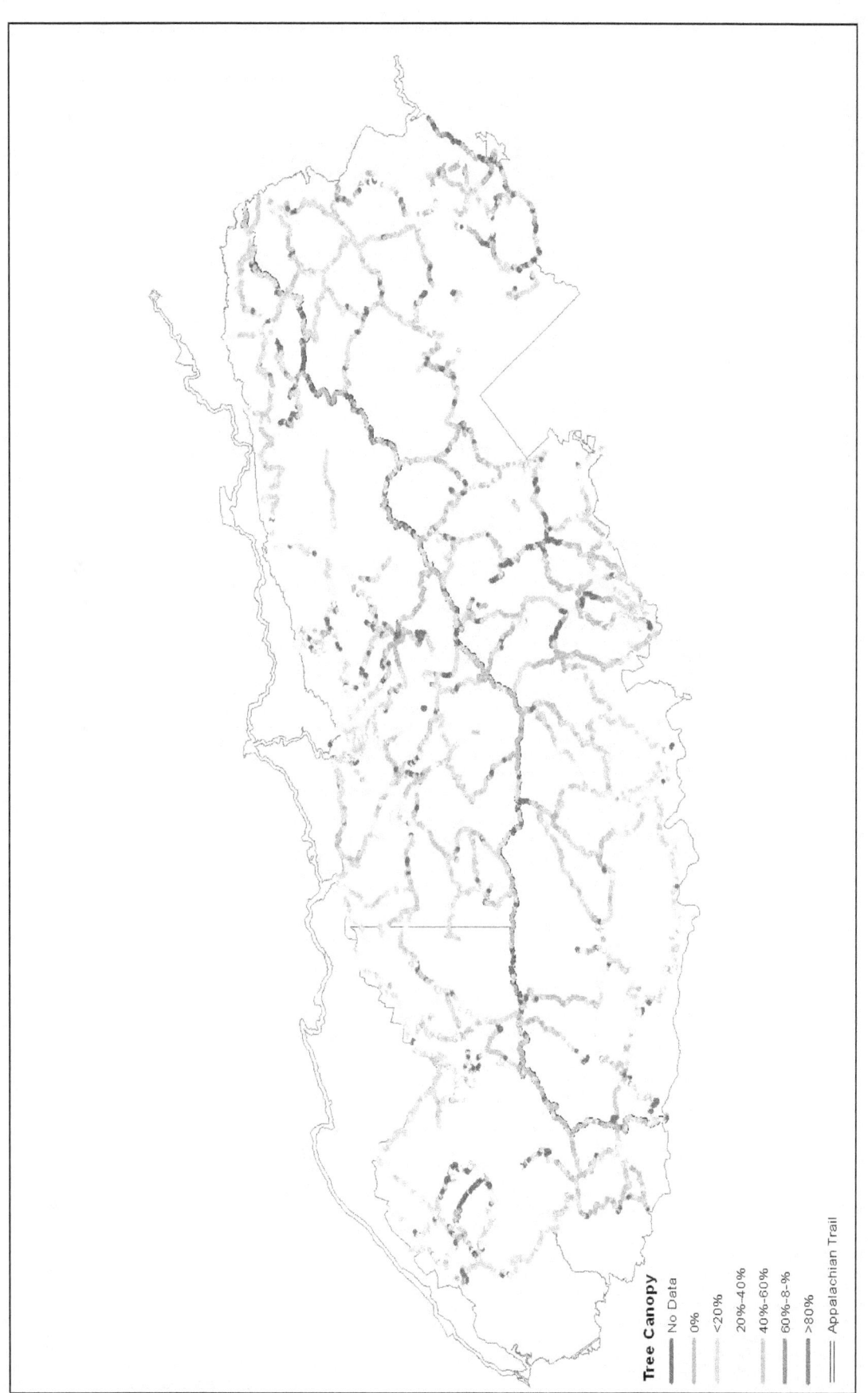

Tree Canopy

No Data
0%
<20%
20%-40%
40%-60%
60%-8-%
>80%
Appalachian Trail

Figure 7B-2 GRSM Canopy Sensitive Species Trail Cover

7B-3

No Data
0%
<20%
20%-40%
40%-60%

Figure 7B-3 GRSM Canopy Sensitive Species Overlooks (3km)

7B-4

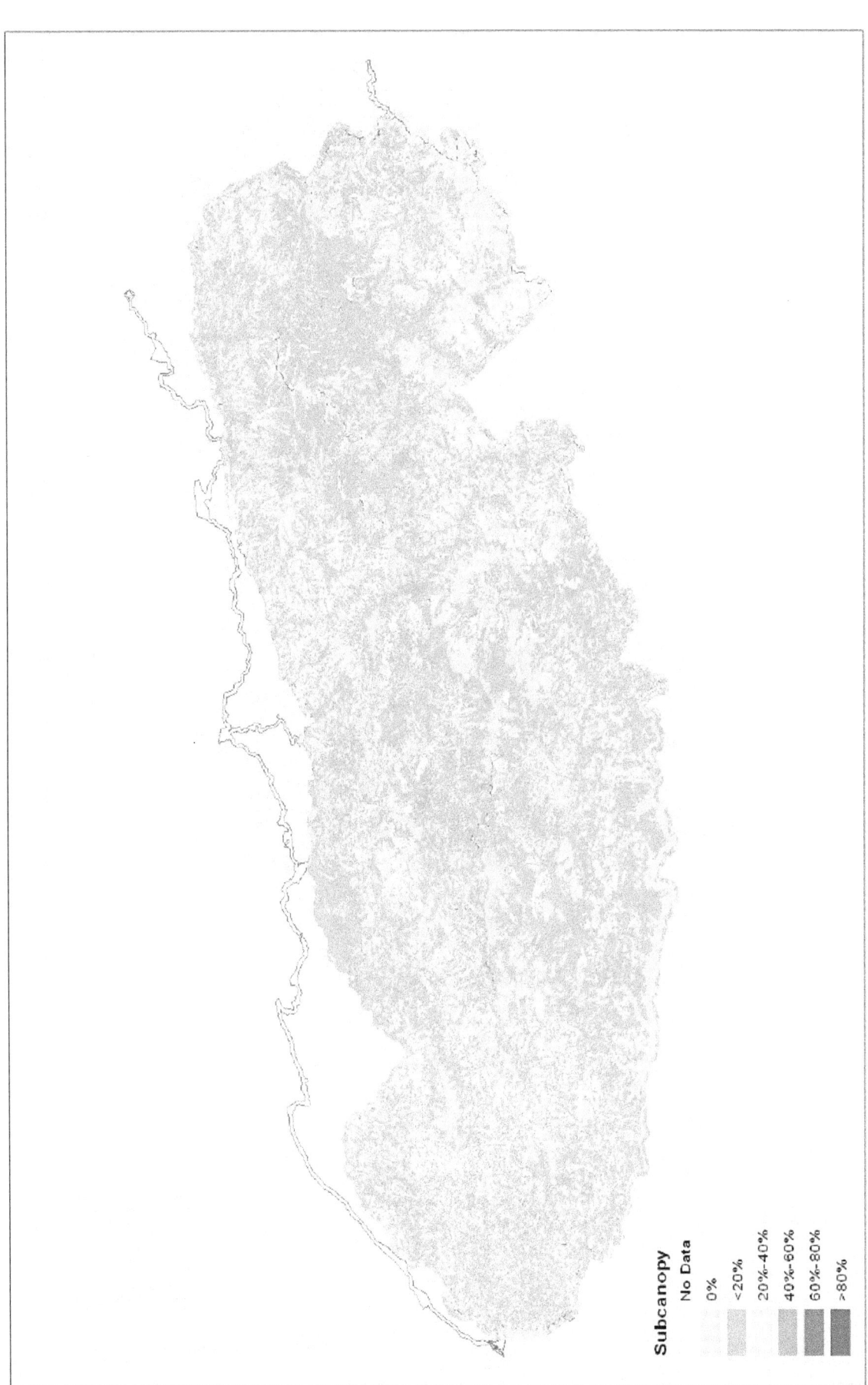

Subcanopy
No Data
0%
<20%
20%-40%
40%-60%
60%-80%
>80%

Figure 7B-4 GRSM Subcanopy Sensitive Species Cover

7B-5

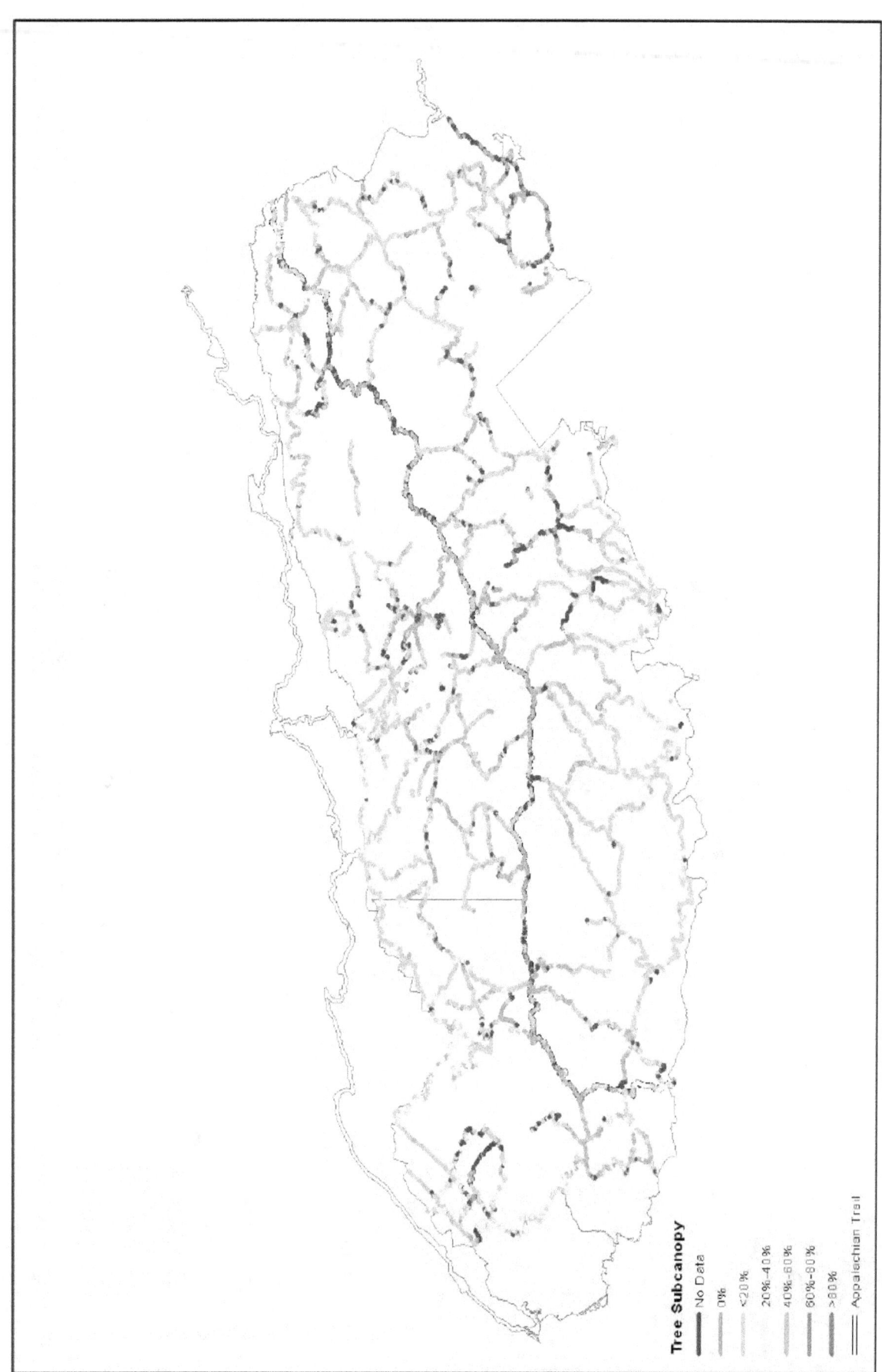

Tree Subcanopy
- No Data
- 0%
- <20%
- 20%-40%
- 40%-60%
- 60%-80%
- >80%
- Appalachian Trail

Figure 7B-5 GRSM Subcanopy Sensitive Species Trail Cover

7B-6

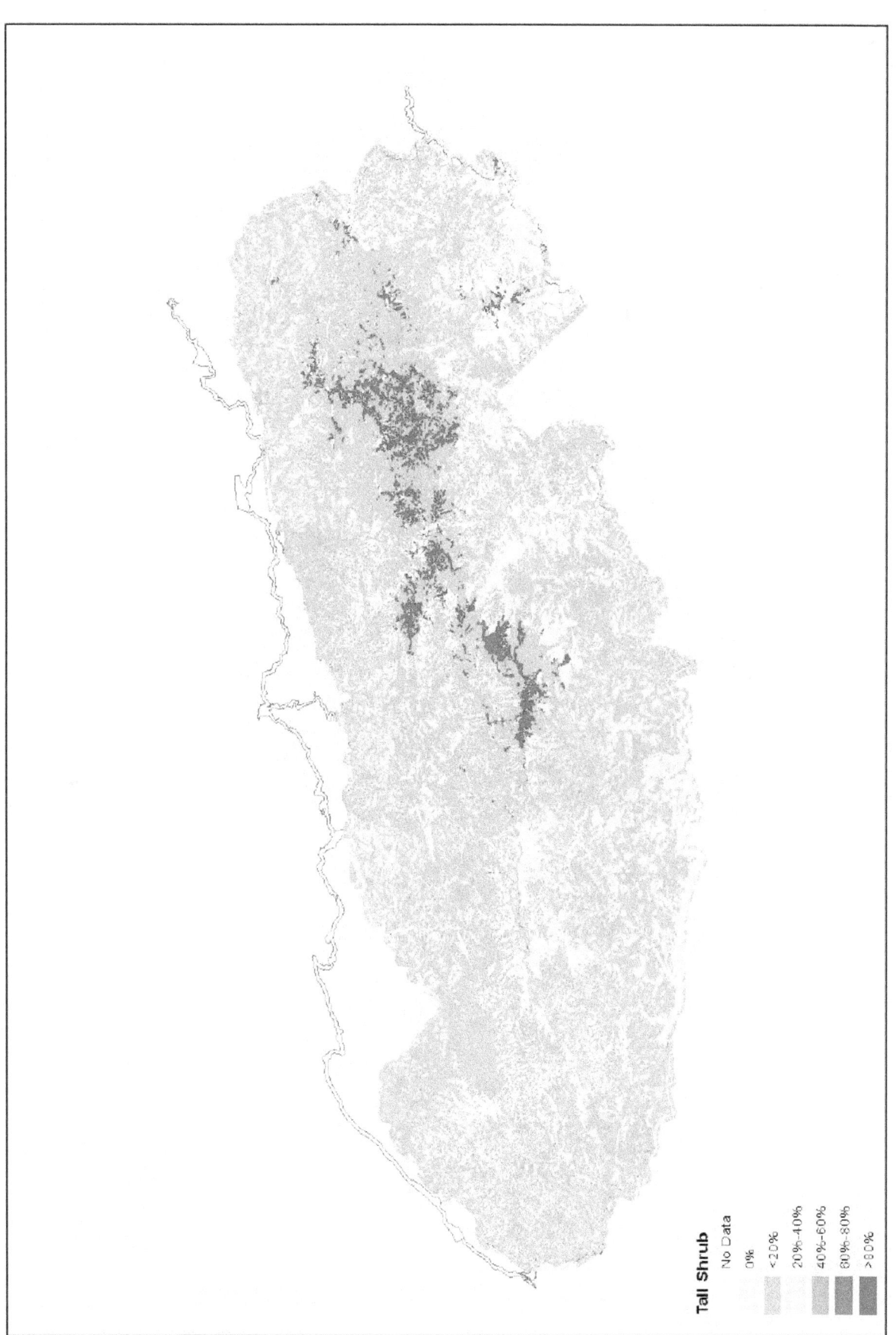

Tall Shrub

No Data
0%
<20%
20%-40%
40%-60%
80%-80%
>80%

Figure 7B-6 GRSM Tall Shrub Sensitive Species Cover

7B-7

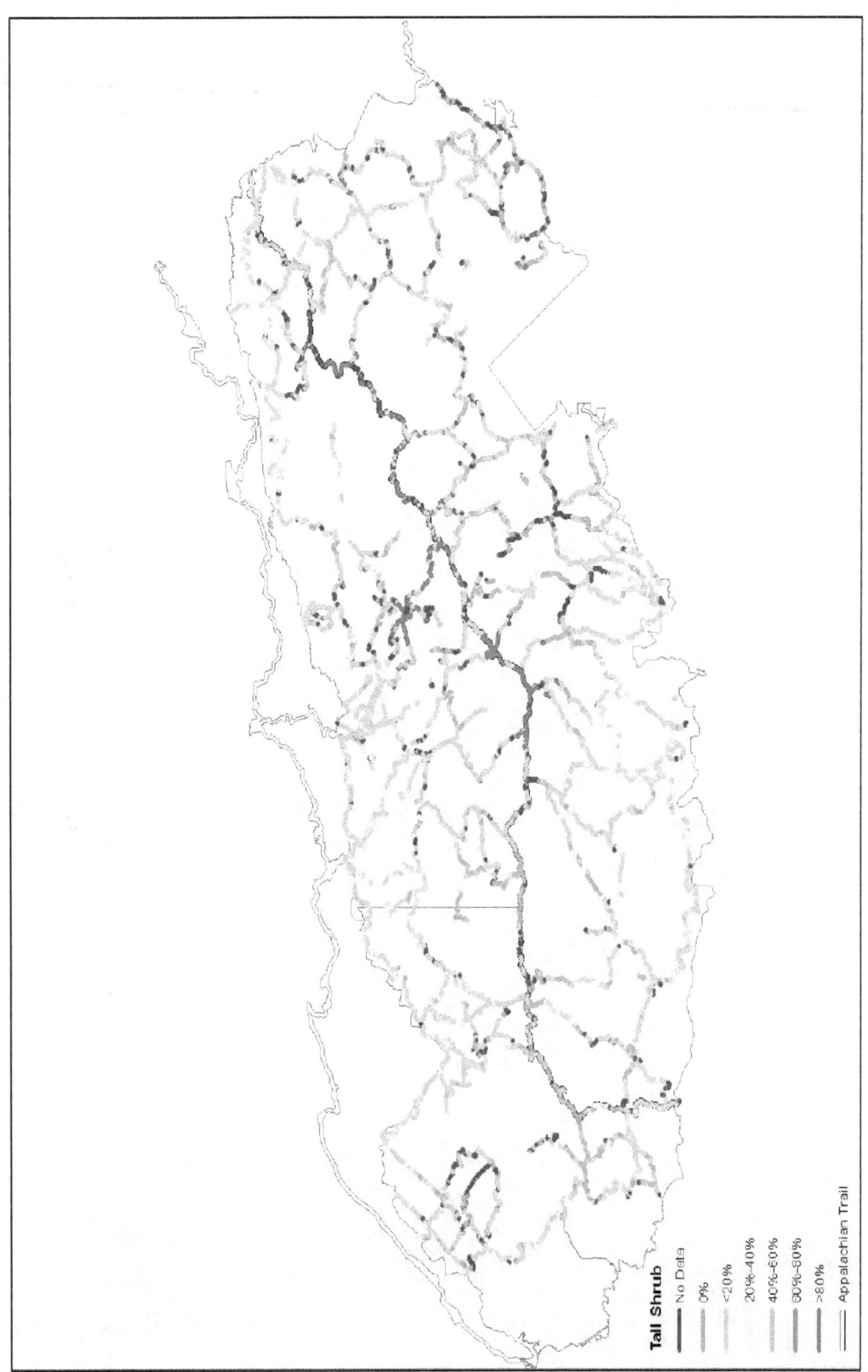

Tall Shrub

No Data
0%
<20%
20%-40%
40%-60%
60%-80%
>80%
Appalachian Trail

Figure 7B-7 GRSM Tall Shrub Sensitive Species Trail Cover

7B-8

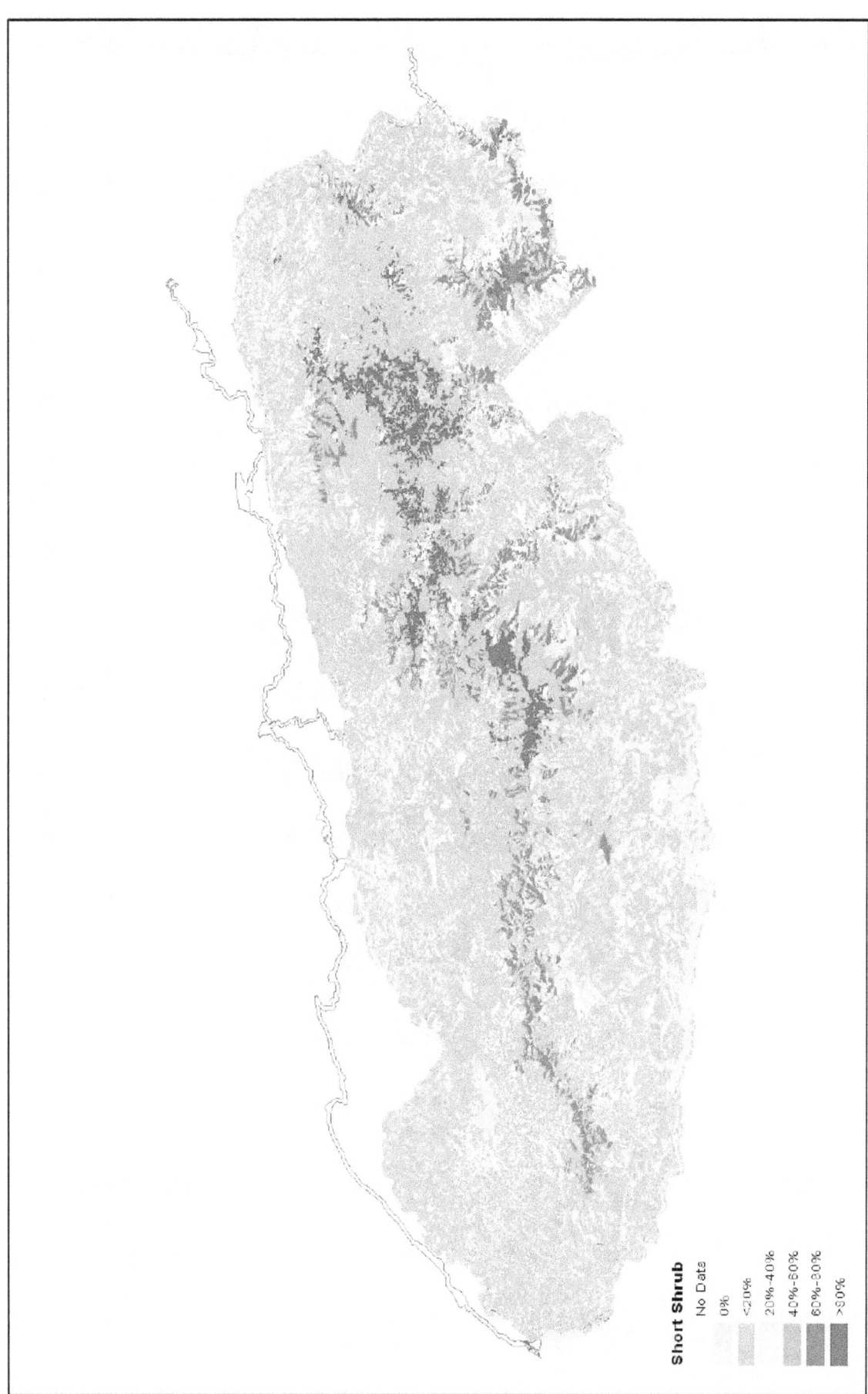

Short Shrub

No Data
0%
<20%
20%-40%
40%-60%
60%-80%
>80%

Figure 7B-8 GRSM Short Shrub Sensitive Species Cover

7B-9

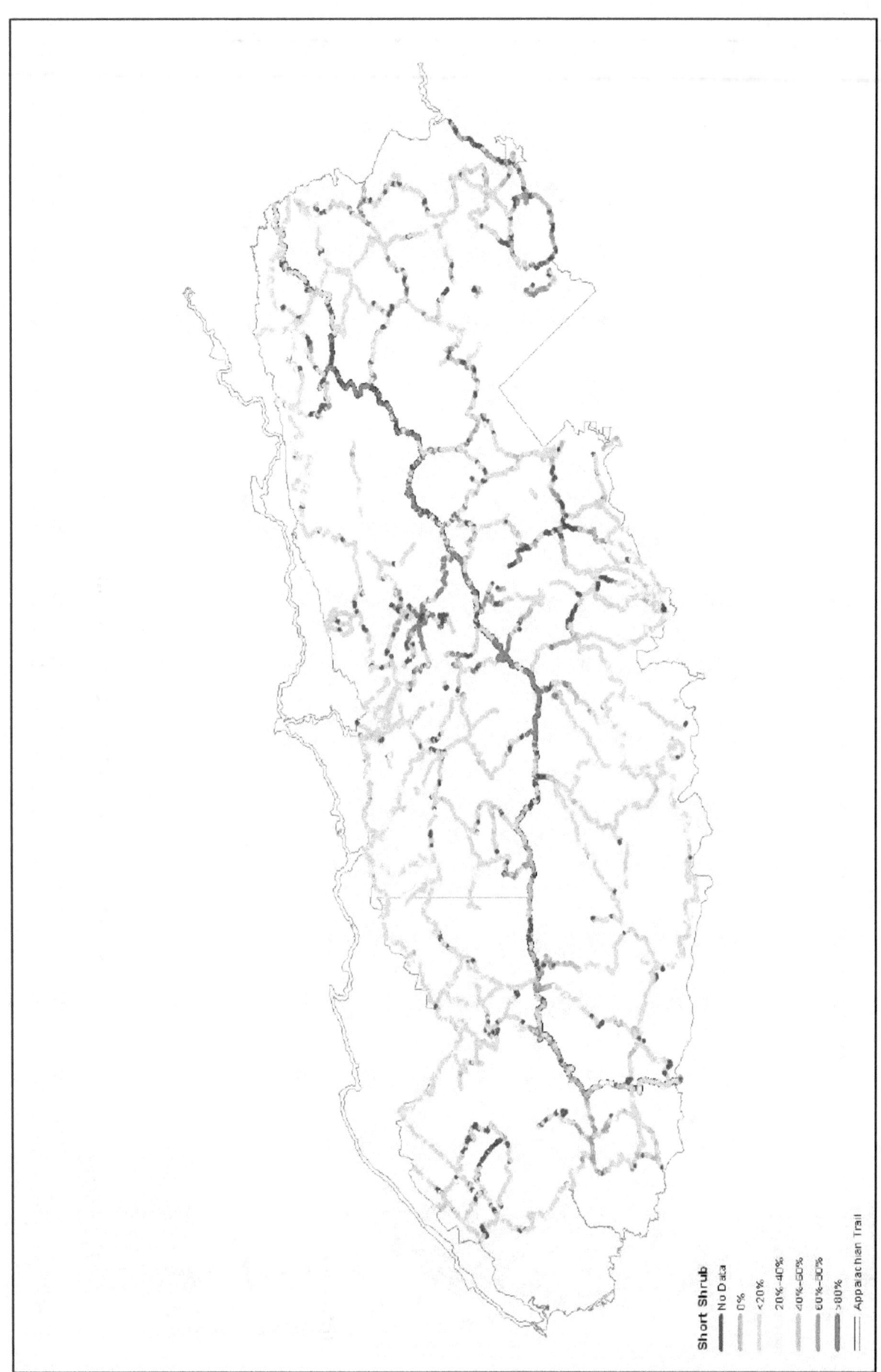

Short Shrub
— No Data
— 0%
— <20%
— 20%-40%
— 40%-60%
— 60%-80%
— >80%
— Appalachian Trail

Figure 7B-9 GRSM Short Shrub Sensitive Species Trail Cover

7B-10

Herbaceous

No Data
0%
<20%
20%-40%
40%-60%
60%-80%
>80%

Figure 7B-10 GRSM Herbaceous Sensitive Species Cover

7B-11

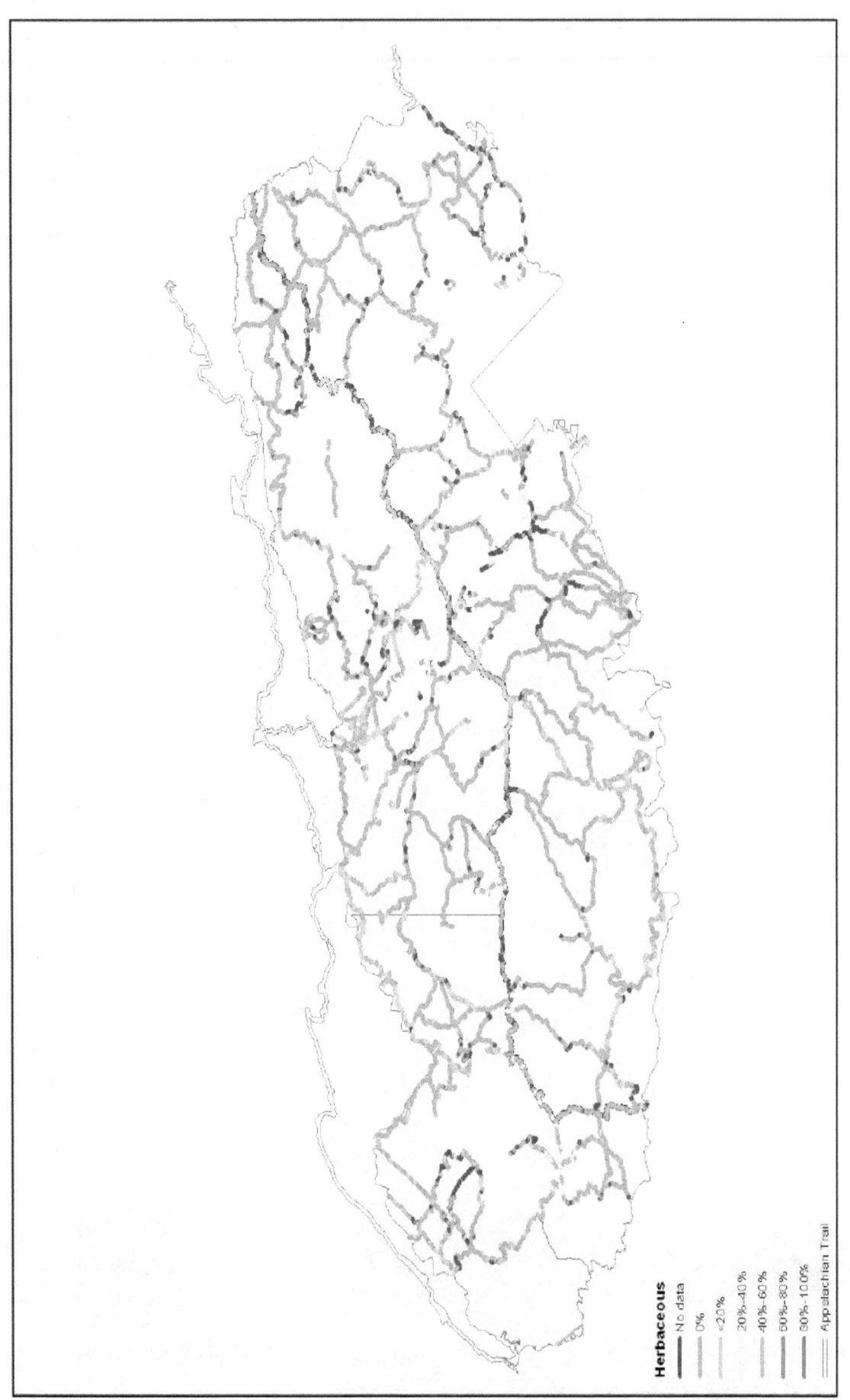

Herbaceous
— No data
— 0%
— <20%
— 20%-40%
— 40%-60%
— 60%-80%
— 80%-100%
═══ Appalachian Trail

Figure 7B-11 GRSM herbaceous Sensitive Species Trail Cover

7B-12

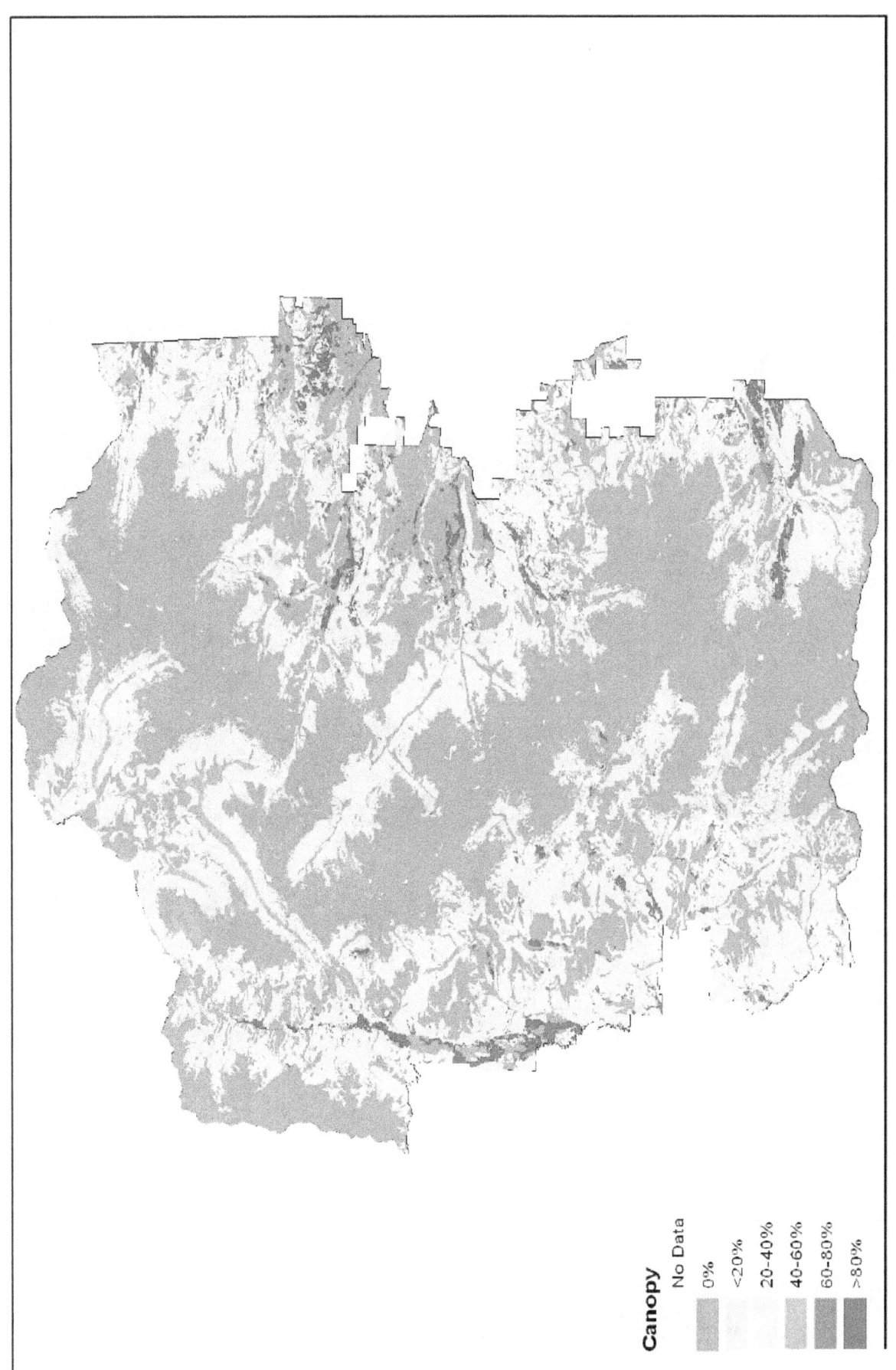

Canopy

No Data
0%
<20%
20-40%
40-60%
60-80%
>80%

Figure 7B-12 ROMO Canopy Sensitive Species Cover

7B-13

Canopy
— No data
— 0%
— <20%
— 20-40%
— 40-60%
— 60-80%
— >80%
— CDNST

Figure 7B-13 ROMO Canopy Sensitive Species Trail Cover

7B-14

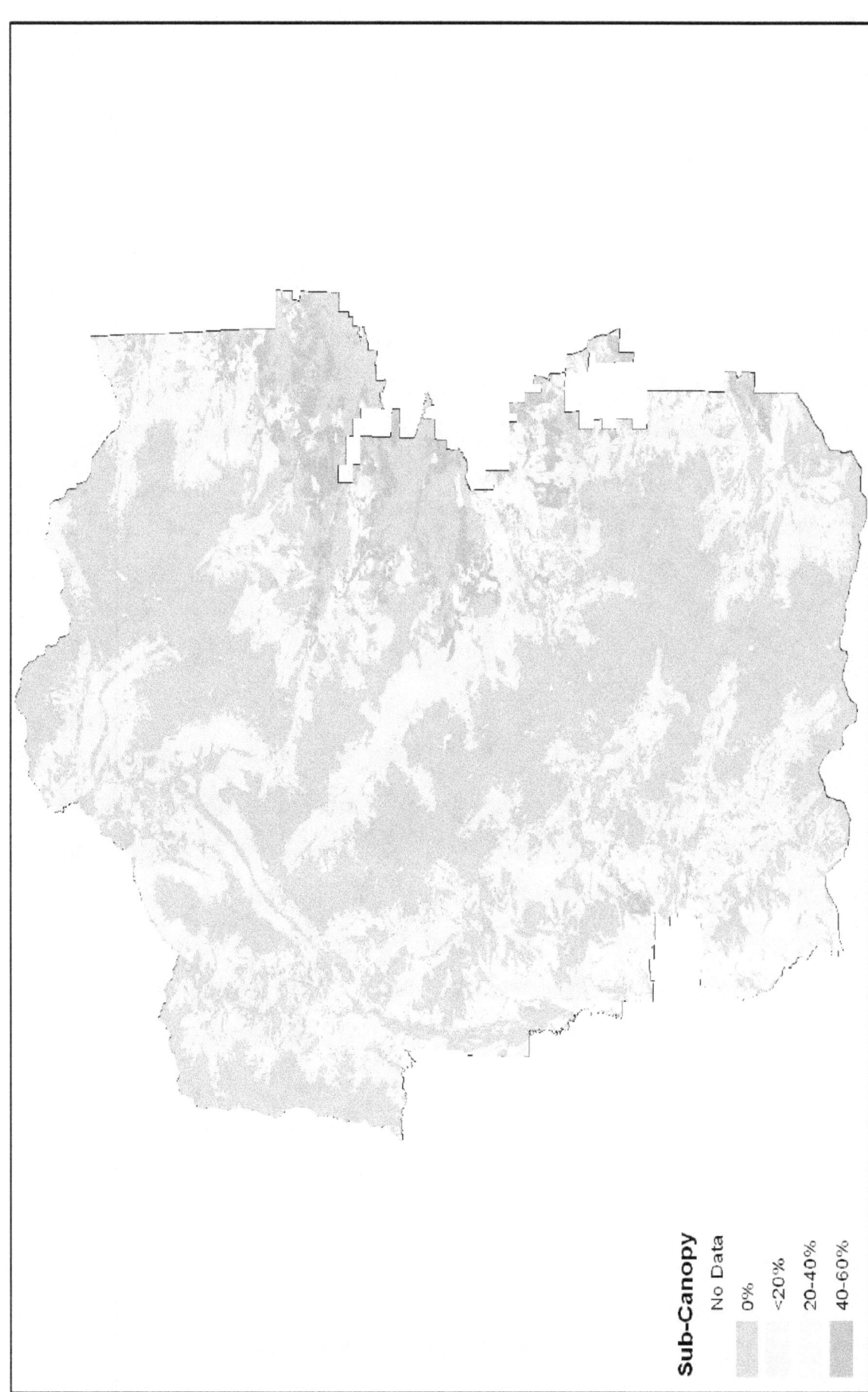

Sub-Canopy

No Data

0%

<20%

20-40%

40-60%

Figure 7B-14 ROMO Subcanopy Sensitive Species Cover

7B-15

Sub Canopy
— No Data
— 0%
— <20%
— 20%-40%
— 40%-60%
— >80%
— CDNST

Figure 7B-15 ROMO Subcanopy Sensitive Species Trail Cover

7B-16

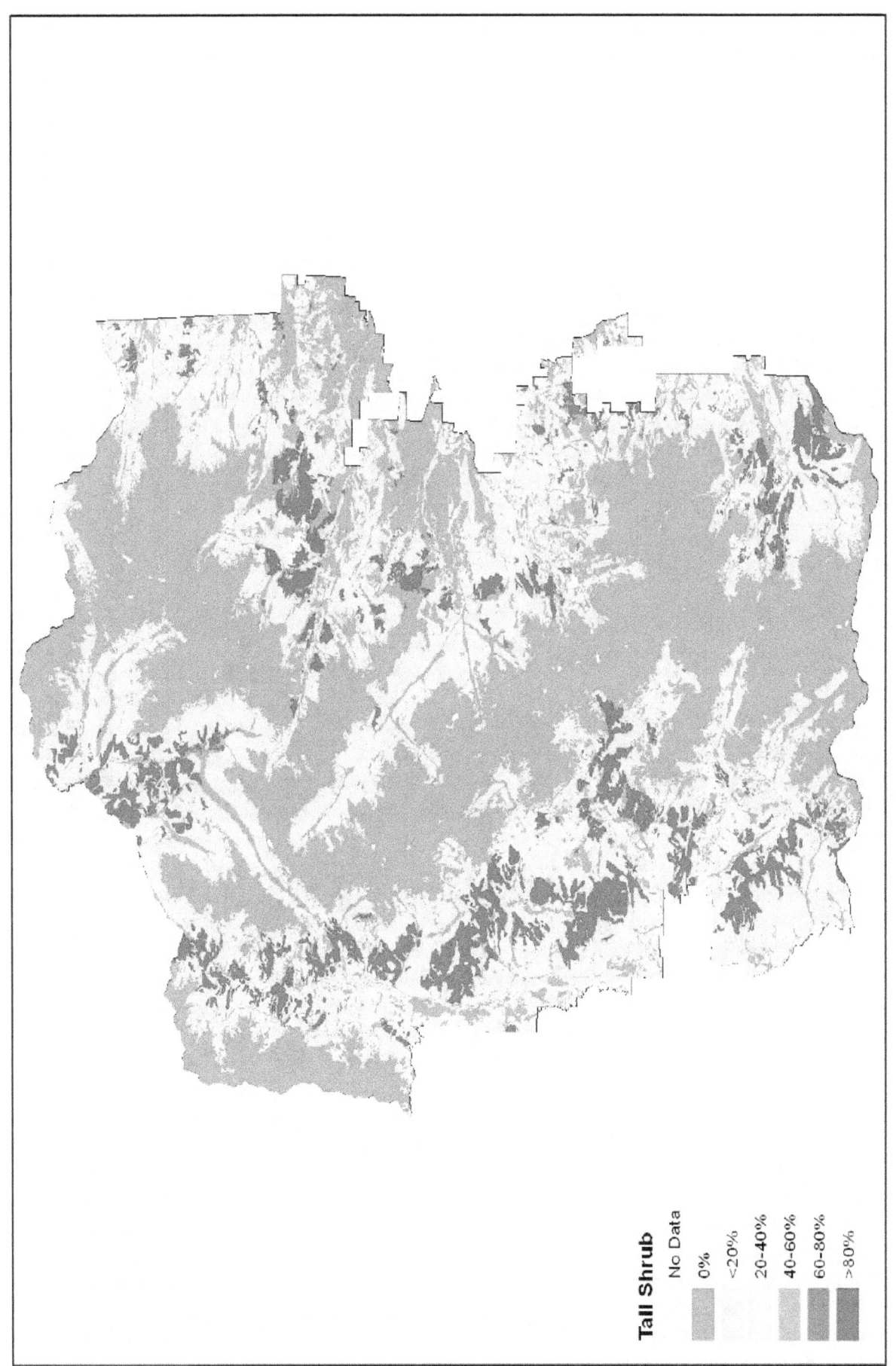

Tall Shrub

No Data
0%
<20%
20-40%
40-60%
60-80%
>80%

Figure 7B-16 ROMO Tall Shrub Sensitive Species Cover

7B-17

Tall Shrub
— No Data
— 0%
— <20%
— 20-40%
— 40-60%
— 60-80%
— >80%
═ CDNST

Figure 7B-17 ROMO Tall Shrub Sensitive Species Trail Cover

7B-18

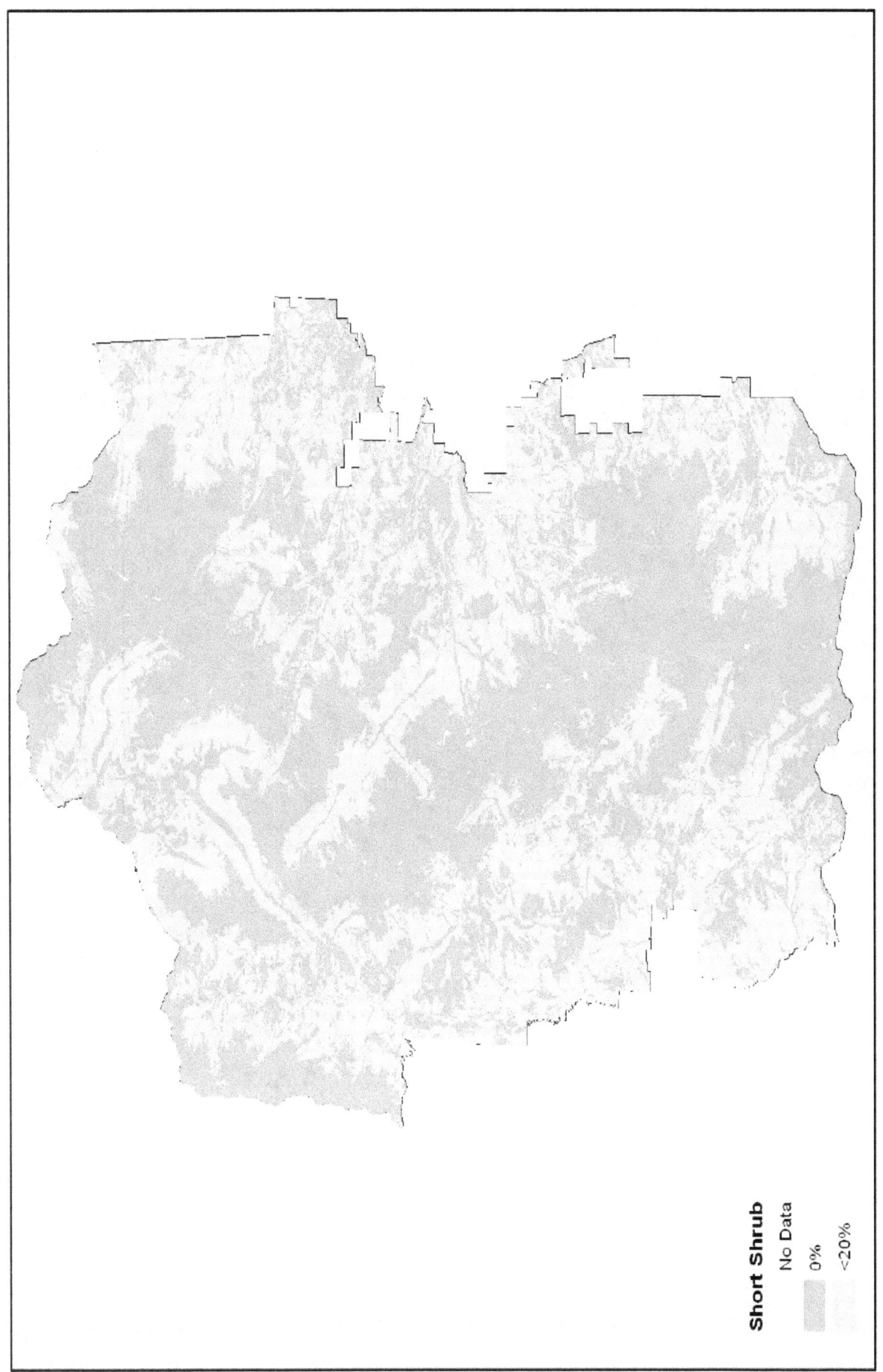

Short Shrub

No Data

0%

<20%

Figure 7B-18 ROMO Short Shrub Sensitive Species Cover

7B-19

Short Shrub

No Data
0%
<20%
20-40%
40-60%
60-80%
>80%
CDNST

Figure 7B-19 ROMO Short Shrub Sensitive Species Trail Cover

7B-20

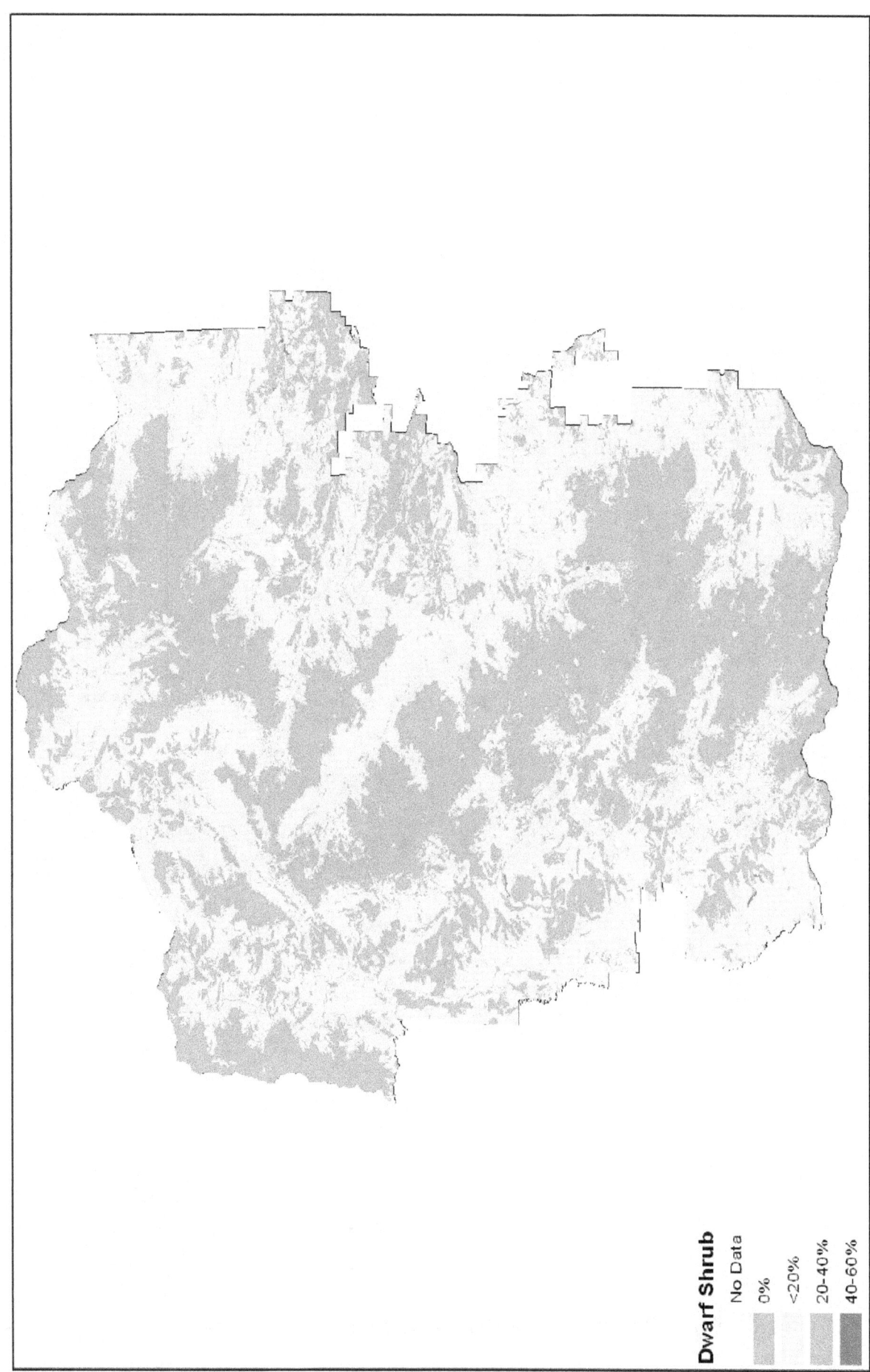

Dwarf Shrub

No Data

0%

<20%

20-40%

40-60%

Figure 7B-20 ROMO Dwarf Shrub Sensitive Species Cover

7B-21

Dwarf Shrub
No Data
05
.20%
20-40%
40-60%
60-80%
>80%
CDNST

Figure 7B-21 ROMO Dwarf Shrub Sensitive Species Trail Cover

7B-22

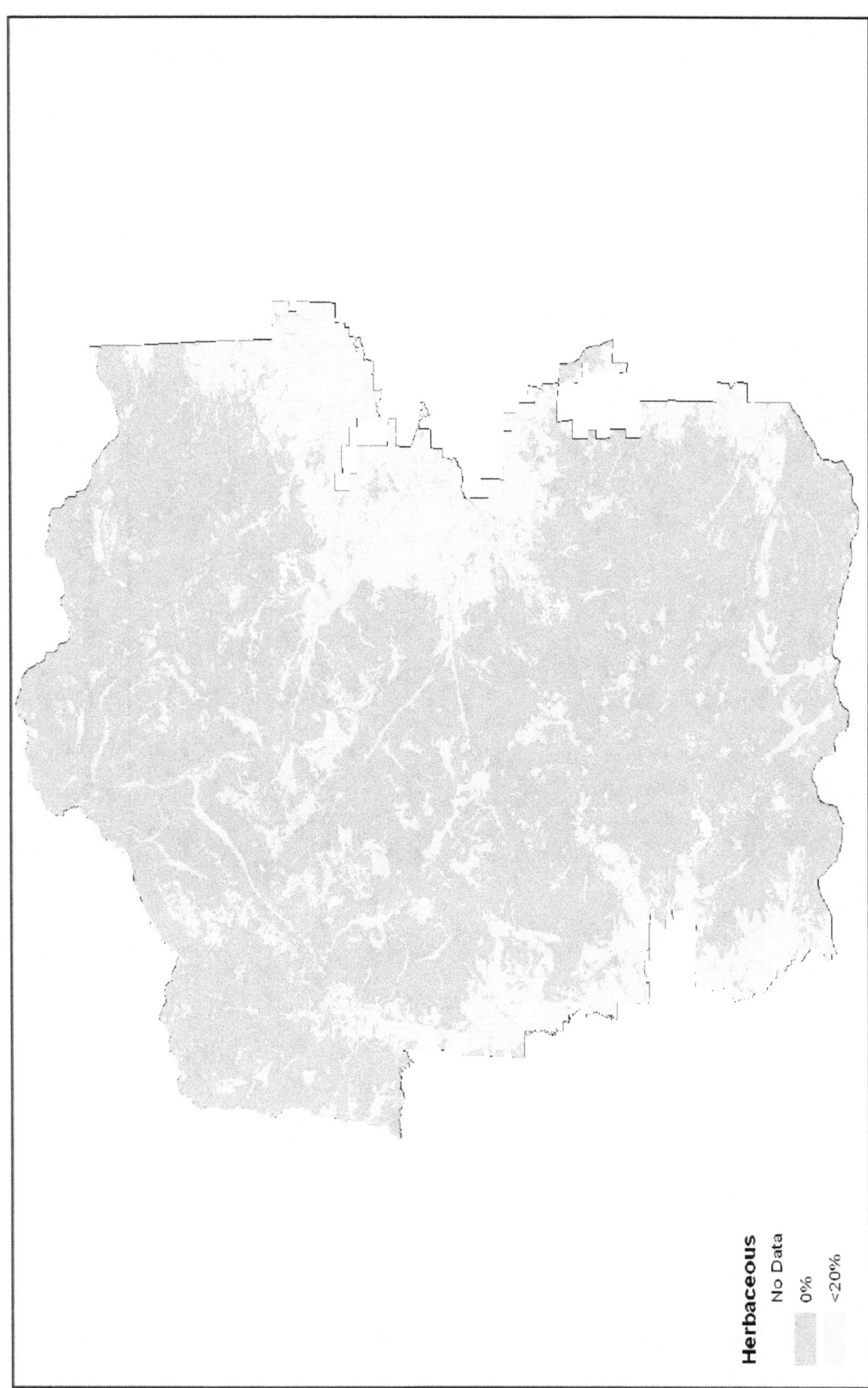

Herbaceous

No Data

0%

<20%

Figure 7B-22 ROMO Herbaceous Sensitive Species Cover

7B-23

Herbaceous
No data
0%
<20%
20-40%
40-60%
60-80%
>80%
CDNST

Figure 7B-23 ROMO Herbaceous Sensitive Species Trail Cover

7B-24

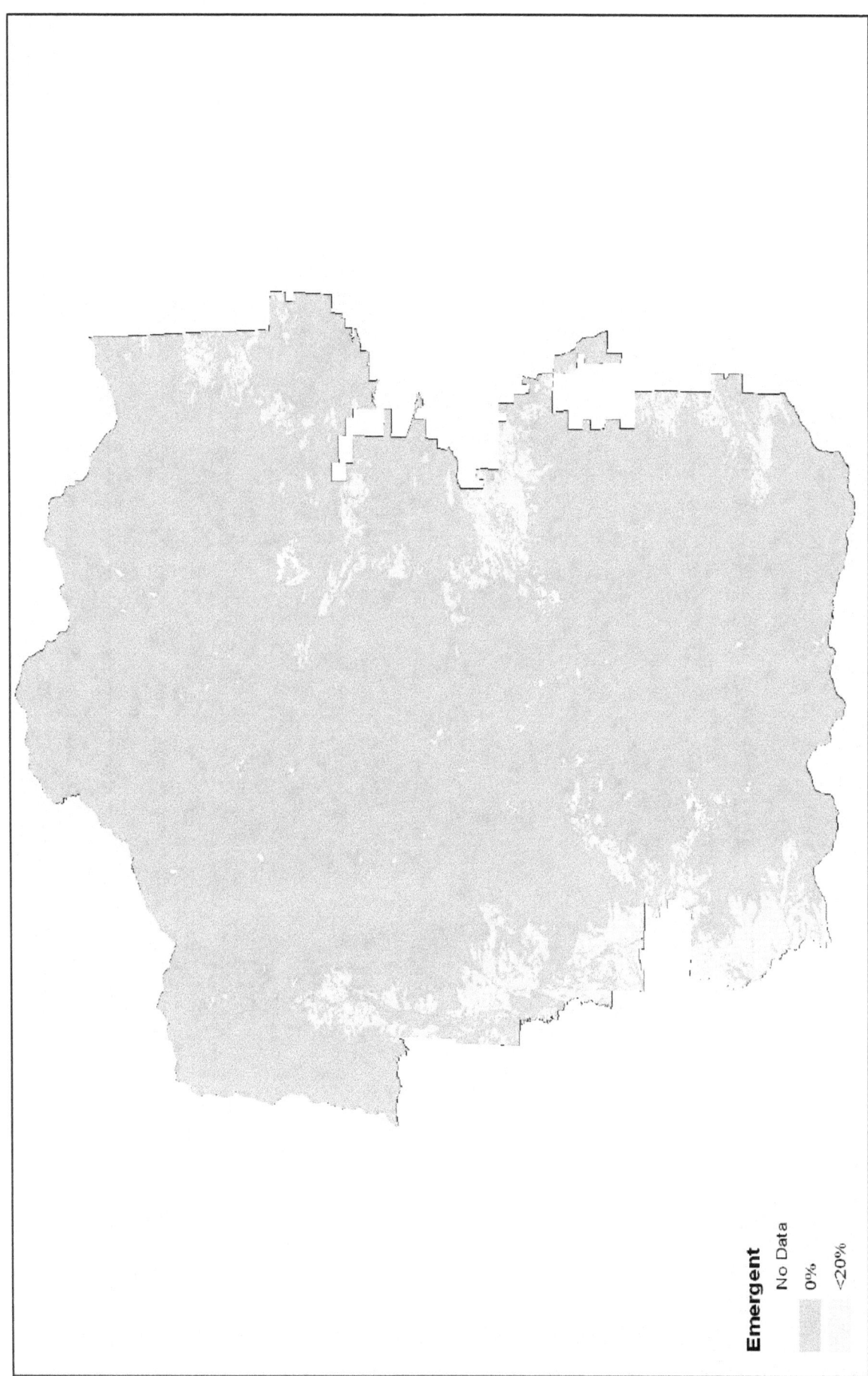

Figure 7B-24 ROMO Emergent Sensitive Species Cover

7B-25

Emergent

No Data

0%

<20%

Emergent Tree
— No Data
— 0%
— <20%
═ CDNST

Figure 7B-25 ROMO Emergent Sensitive Species Trail Cover

7B-26

Figure 7B-26 SEKI Canopy Sensitive Species Cover

7B-27

Canopy

No Data

0%

<20%

20-40%

40-60%

60-80%

Canopy

No Data
0%
<20%
20%-40%
40%-60%
60%-80%
>80%
John Muir Trail

Figure 7B-27 SEKI Canopy Sensitive Species Cover

7B-28

Tall Shrub

No Data

0%

<20%

20-40%

40-60%

Figure 7B-28 SEKI Tall Shrub Sensitive Species Cover

7B-29

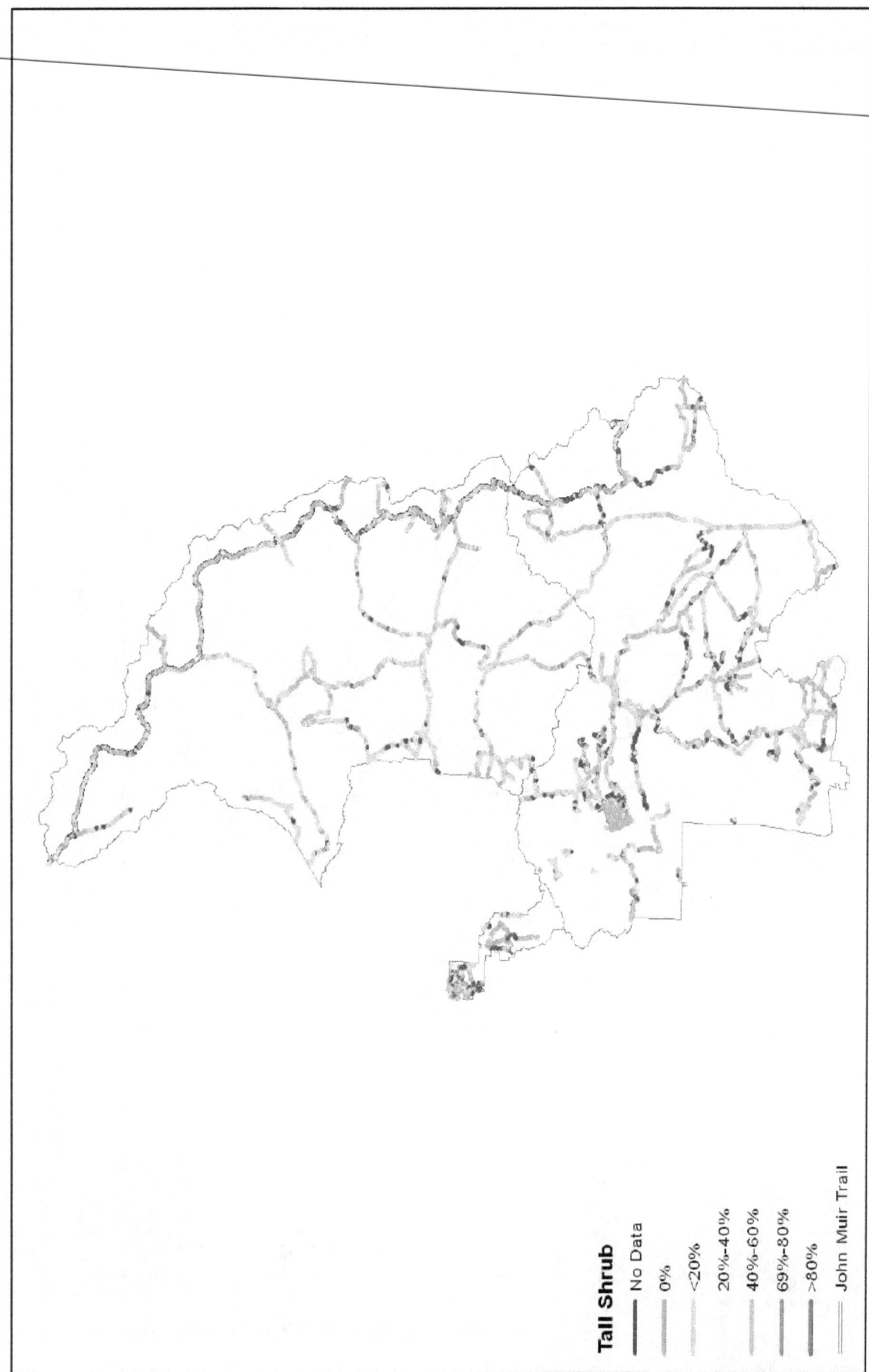

Tall Shrub
— No Data
0%
<20%
20%-40%
40%-60%
69%-80%
>80%
— John Muir Trail

Figure 7B-29 SEKI Tall Shrub Sensitive Species Trail Cover

7B-30

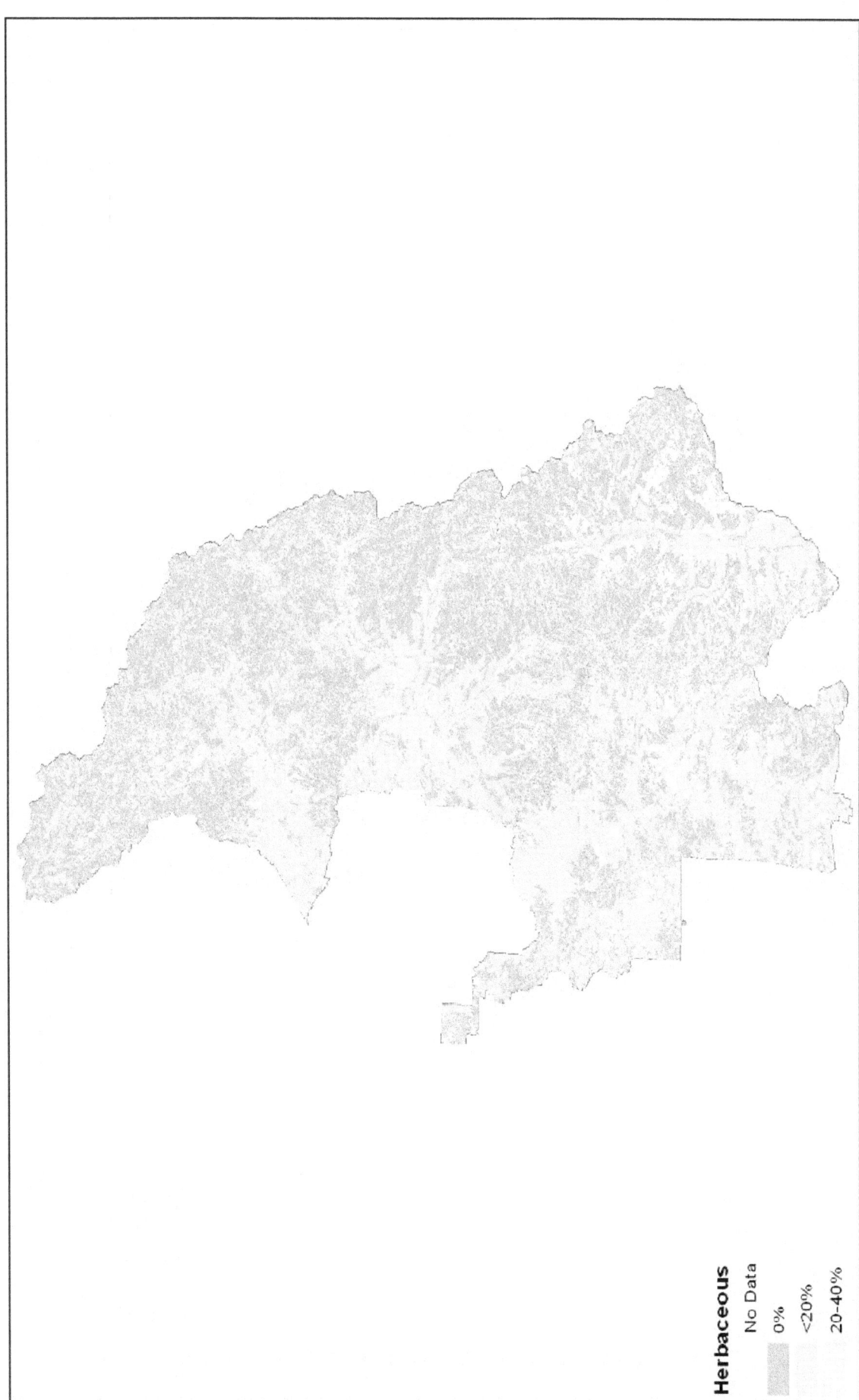

Herbaceous
No Data
0%
<20%
20-40%

Figure 7B-30 SEKI Herbaceous Sensitive Species Cover

7B-31

Herbaceous
— No Data
— 0%
— <20%
— 20%–40%
— 40%–60%
— 60%–80%
— >80%
══ John Muir Trail

Figure 7B-31 SEKI Herbaceous Sensitive Species Trail Cover

7B-32

United States
Environmental Protection
Agency

Office of Air Quality Planning and Standards
Health and Environmental Impacts Division
Research Triangle Park, NC

Publication No. EPA-452/P-14-003b
February 2014